INTEGRATED NETWORK MANAGEMENT, II

INTEGRATED NETWORK MANAGEMENT, II

Proceedings of the IFIP TC6/WG6.6 Second International Symposium on
Integrated Network Management
with participation of the IEEE Communications Society CNOM
and with support from the Institute for Educational Services
Crystal City, Washington, D.C., U.S.A., 1–5 April, 1991

Edited by

IYENGAR KRISHNAN
The MITRE Corporation
McLean, VA, U.S.A.

WOLFGANG ZIMMER
GMD Research Center for
Innovative Computer Systems & Technology at the
Technical University of Berlin
Berlin, F.R.G.

1991

NORTH-HOLLAND
AMSTERDAM · NEW YORK · OXFORD · TOKYO

ELSEVIER SCIENCE PUBLISHERS B.V.
Sara Burgerhartstraat 25
P.O. Box 211, 1000 AE Amsterdam, The Netherlands

Distributors for the United States and Canada:

ELSEVIER SCIENCE PUBLISHING COMPANY INC.
655 Avenue of the Americas
New York, N.Y. 10010, U.S.A.

ISBN: 0 444 89028 9

© IFIP, 1991

All rights reserved. No part of this publication may be reproduced, stored in a retrieval system or transmitted in any form or by any means, electronic, mechanical, photocopying, recording or otherwise, without the prior written permission of the publisher, Elsevier Science Publishers B.V. / Academic Publishing Division, P.O. Box 103, 1000 AC Amsterdam, The Netherlands.

Special regulations for readers in the U.S.A. – This publication has been registered with the Copyright Clearance Center Inc. (CCC), Salem, Massachusetts. Information can be obtained from the CCC about conditions under which photocopies of parts of this publication may be made in the U.S.A. All other copyright questions, including photocopying outside of the U.S.A., should be referred to the publisher, Elsevier Science Publishers B.V., unless otherwise specified.

No responsibility is assumed by the publisher or by IFIP for any injury and/or damage to persons or property as a matter of products liability, negligence or otherwise, or from any use or operation of any methods, products, instructions or ideas contained in the material herein.

Printed in The Netherlands.

PREFACE

Continuing the spirit of global cooperation established at our first landmark conference, the *Second International Symposium on Integrated Network Management* provides an international forum where vendors, system integrators, researchers, users, and standards developers are encouraged to exchange ideas, insights, and information. The program for the symposium is designed to address the many issues surrounding recent worldwide advances in the administration, operation, maintenance, and security of local and wide area communications networks, including data, voice, and video communications.

Keith Willetts, Chairman of the Board of the OSI Network Management Forum and Director of Network Management for British Telecom, will open the symposium with a keynote address on "Managing Global Information in the 90's." Throughout the symposium, industry leaders from participating vendor organizations will introduce leading edge technologies and corporate plans. And finally, panel discussions by distinguished experts such as Larry Bernstein (AT&T Bell Laboratories), Paul Brusil (MITRE Corporation), Vint Cerf (Corporation of National Research Initiatives), André Danthine (Université de Liège), Marshall Rose (PSI Inc.) and Keith Willetts will explore industry trends from a variety of perspectives.

The technical program covers a broad spectrum of topics in 9 tracks which are:

1. Standards and Architectures
2. Modeling and Design
3. Integrated Management
4. Implementation and Case Studies
5. Telecommunications
6. Artificial Intelligence
7. Distributed Systems
8. Usage Control
9. Management Information

Invited presentations by renowned experts will highlight each track. All of the papers have been carefully reviewed and selected to ensure unsurpassed standards of relevance and technical accuracy, while focusing on integrated network management concerns. The papers were selected from 110 submissions received from Africa, Asia, Australia, Europe, North America, and South America. Representing the diverse interests and concerns of a truly international audience, 24 of 50 papers accepted and two of the nine invited papers originate from outside the Americas. Each submission is the product of a stringent review process, evaluated by three globally dispersed referees from a pool of 120 referees selected from 158 nominations. Only those papers which withstood the scrutiny of this review process were considered by the 47-member Program Committee for presentation at the symposium and inclusion in the hardbound Proceedings. The Proceedings comprises all the technical papers including the invited papers and abstracts of panels sessions. The order of papers reflects the order of the presentations to be given.

The symposium also offers two days of tutorial sessions presented by recognized leaders from the industry and the academic community. Each session provides a theoretical as well as practical foundation of the subject area, further enhancing participation in the symposium's technical sessions. A unique feature of the symposium is the participation of Patrons who will provide demonstrations and presentations of their latest network management strategies, products and services. These presentations will be made available during times that do not conflict with other technical sessions.

Many people have helped in organizing the technical program of the symposium. We thank the authors of the papers, without whom this symposium would not have been possible and the members of the Program Committee for their help with paper solicitation and review. Our special thanks to Lorenzo Aguilar, Janet Cohen, Roberta Cohen, Will Collins, Heinz-Gerd Hegering, Serge Hurtubise, Kimberly Kappel, Mark Klerer, Aurel Lazar, Kenneth Lutz, Keith McCloghrie, Branislav Meandzija, Prodip Sen, Morris Sloman, Chris Sluman, Daniel Stokesberry, Liba Svobodova, Ole Krog Thomson, Jeremy Tucker, Yechiam Yemini, and Douglas Zuckerman who were in Berlin at the Program Committee meeting, in October 1990, to select the technical papers and finalize the technical program. Thanks also to GMD FIRST, Berlin, for hosting the Program Committee meeting. We also thank Carol McKinney, Dirk Claussen, Dietmar Koop, and Carsten Schulze, for their help with the handling of paper submissions, database maintenance, and many other tasks. And last but not least, we thank our employers, MITRE Corporation and GMD FIRST, for their generous support. ACTION MOTIVATION was instrumental in contributing to this highly successful event and we thank Michele Nessier and Rob Budinger for their efforts in organizing and managing many aspects of the symposium, particularly their help with correspondence to all of the authors and Program Committee members.

We hope that all of you will benefit from the technical program, and that you will capture the spirit of the complete *Network Management Week*.

Iyengar Krishnan and Wolfgang Zimmer
Program Co-Chairs
January 15, 1991

SYMPOSIUM COMMITTEES

EXECUTIVE COMMITTEE MEMBERS

Kimberly Kappel	Georgia Institute of Technology, USA, General Chair
Branislav Meandzija	MetaAccess Inc., USA, General Co-Chair
Paul Brusil	MITRE Corporation, USA, Past General Chair and Co-Chair for Advance Planning
Iyengar Krishnan	MITRE Corporation, USA, Program Co-Chair
Wolfgang Zimmer	GMD FIRST, Germany, Program Co-Chair
Lightsey Wallace	Hekimian Labs, USA, Vendor Program Chair
Keith McCloghrie	Hughes LAN Systems, USA, Internet Coordinator
Morris Sloman	Imperial College, United Kingdom, Program Coordinator
Douglas Zuckerman	AT&T Bell Laboratories, USA, IEEE/CNOM Coordinator
Melvyn Galin	MITRE Corporation, USA, Advisor

PROGRAM COMMITTEE MEMBERS

Lorenzo Aguilar	Hewlett-Packard, USA
William Bernstein	IBM Corporation, USA
Stephen Brady	IBM Corporation, USA
Janet Cohen	Butler Cox PLC, United Kingdom
Roberta Cohen	AT&T Bell Laboratories, USA
Will Collins	Codex Corporation, USA
André Danthine	Université de Liège, Belgium
Deborah Estrin	University of Southern California, USA
Rodney Goodman	California Institute of Technology, USA
Heinz-Gerd Hegering	Technical University of Munich, Germany
Chris Horn	Trinity College, Ireland
Christian Huitema	INRIA Unite de Recherche de Sophia-Antipolis, France
Serge Hurtubise	Bell Northern Research, Canada
Dipak Khakar	Lund University, Sweden
Peter Kirstein	University College of London, United Kingdom
Mark Klerer	AT&T Bell Laboratories, USA
Yoshi Kobayashi	IBM Japan Ltd., Japan
Alvyn Langsford	Harwell Laboratory, United Kingdom
Aurel Lazar	Columbia University, USA
Gesualdo Le Moli	Politico Milano, CREI, Italy
Kenneth Lutz	Bellcore, USA, IEEE/CNOM Liaison
Daniel Lynch	Interop Inc., USA

PROGRAM COMMITTEE MEMBERS (Continued)

Dave Milham	British Telecom, United Kingdom
Kazuyoshi Morino	NTT, Japan
Louis Pouzin	THESEUS, France
Marshall Rose	PSI Inc., USA
Prodip Sen	NYNEX, USA
Chris Sluman	SEMA Group PLC, United Kingdom
Daniel Stokesberry	NIST, USA, Advisor
Colin Strutt	Digital Equipment Corporation, USA
Carl Sunshine	The Aerospace Corporation, USA
Liba Svobodova	IBM Zurich Research Laboratory, Switzerland
Liane Tarouco	Institute of Informatics - UFRGS, Brazil
Ole Krog Thomson	Jutland Telephone, Denmark
Jeremy Tucker	Logica Communications Ltd., United Kingdom
Jil Westcott	BBN STC, USA, Advisor
Yechiam Yemini	Columbia University, USA

LIST OF REVIEWERS

L. Aguilar
S. Aidarous
B. Ahlgren
D. Bailey
S. Baker
J. Bannister
W. Bernstein
K. Batho
M. Bosch
S. Brady
P. Brusil
A. Chandna
J. Cohen
R. Cohen
W. Collins
H. Dastvar
D. Dimitrijevic
A. Dittrich
D. Estrin
R. Fellows
P. Francois
I. Frisch
N. Fujii
K. Garbe
R. Goodman
D. Grebovic
J. Hall
N. Hayward
W. Hiddink
H. Hegering
L. Higgs
C. Horn
S. Hurtubise
R. Hutchins
C. Huitema
P. Janson
C. Jard
T. Jeffree

T. Jeron
G. Juanole
P. Kapadia
K. Kappel
G. Kar
C. Kaufman
H. Kawai
D. Khakar
M. Klerer
Y. Kobayashi
M. Kosarchyn
Y. Koseki
L. LaBarre
H. Lam
A. Lazar
J. Le Boudec
G. Le Moli
K. Lutz
D. Lynch
J. Malcolm
K. McCloghrie
E. McCoy
B. Meandzija
D. Milham
R. Molva
K. Morino
G. Mouradian
A. Mouttham
K. Muralidhar
A. Murphy
K. Naemura
I. Neumeier-Mackert
B. Paillassa
R. Patton
B. Pehrson
A. Phillips
J. Pietras
E. Pinnes

S. Podar
P. Prozeller
A. Quirt
L. Raman
M. Rose
S. Sakata
R. Sandhu
A. Schwartz
A. Sciacca
P. Sen
D. Shurtleff
M. Sloman
C. Sluman
D. Stokesberry
C. Strutt
C. Sunshine
S. Susuki
C. Sundaramurthy
L. Svobodova
M. Sylor
N. Takahashi
H. Tanaka
D. Taylor
C. Teshigawara
J. Tucker
T. Usländer
R. Van den Heever
C. Vissers
L. Wallace
H. Wedde
J. Westcott
H. Wettstein
M. Willett
J. Winterbotham
H. Yahata
J. Yamahira
Y. Yemini
D. Zuckerman

TABLE OF CONTENTS

Preface v
Symposium Committees vii
List of Reviewers ix

INTRODUCTION

Integrated Network Management and the International Symposia   3
P. Brusil, The MITRE Corporation, USA
K. Kappel, Georgia Institute of Technology, USA
B. Meandzija, Meta Access Inc., USA

I. STANDARDS AND ARCHITECTURE

Invited Paper
Network Management is Simple:
You Just Need the "Right" Framework   9
M. Rose, Performance Systems International, Inc., USA

A. APPLYING STANDARDS

Interoperable Network Management:
OSI/Network Management Forum Architecture and Concepts   29
J. Embry, Opening Technologies, USA
P. Manson, Bell Northern Research Ltd., CANADA
D. Milham, British Telecommunications, UK

An Application of OSI Systems Management to
Intelligent Network Services   45
A. Tanaka, N. Matsumoto and K. Morino, NTT, JAPAN

DECnet/OSI Phase V Network Management   57
M. Sylor, Digital Equipment Corporation, USA

B. PANEL

Coalescing Management Standards   73
Moderator: D. Stokesberry
National Institute of Standards and Technology, USA

II. MODELING AND DESIGN

Invited Paper

Formal Description of Network Management Issues   77
G. Bochmann and P. Mondain-Monval, Université de Montréal
L. Lecomte, Computer Research Institute of Montréal, CANADA

A. REQUIREMENTS AND VERIFICATION

Network Management by Delegation   95
Y. Yemini and G. Goldszmidt, Columbia University, USA
S. Yemini, IBM T.J. Watson Research Center, USA

Network Management: An Alternative View   109
A. Pras, University of Twente, THE NETHERLANDS

Verification of Network Management System Configurations   119
D. Cohrs and B. Miller, University of Wisconsin-Madison, USA

B. MODELING FRAMEWORKS AND REQUIREMENTS

OSFA: An Object-Oriented Distributed Approach to Network Management   135
G. Raeder, MPR Teltech, Ltd., CANADA

The LOTOS Framework for OSI Systems Management   147
A. Kouyzer, Netherland Railways, THE NETHERLANDS
A. van den Boogaart, PTT Research, THE NETHERLANDS

A Development Environment for OSI Systems Management   157
S. Nakai, Y. Kiriha, Y. Ihara and S. Hasegawa
NEC Corporation, JAPAN

III. INTEGRATED MANAGEMENT

Invited Paper

Global Commonality in User Requirements   171
E. Adams, OSI Network Management Forum, USA

A. PANEL

Evolving User Requirements for Network Management — 185
Moderator: J. Cohen, Butler Cox PLC., UK

B. WAYS OF ACHIEVING INTEGRATION

Development of Integrated Network Management System NETM Based on OSI Standards — 189
M. Suzuki, R. Sasaki, R. Nagai, Systems Development Center, Hitachi, Japan and K. Mizuguchi, M. Saito and H. Kobayashi, Software Works, Hitachi, JAPAN

SNMP for Non-TCP/IP Sub-networks: An Implementation — 201
E. Duato, European Space Operation Center, GERMANY
B. Lemercier, BIM, BELGIUM

Evaluating Network Management Systems: Criteria and Observations — 213
E. Carter and J. Dia, The MITRE Corporation, USA

IV. IMPLEMENTATION AND CASE STUDIES

Invited Paper
Management by Exception: OSI Event Generation, Reporting, and Logging — 227
L. LaBarre, The MITRE Corporation, USA

A. IMPLEMENTING NETWORK MANAGEMENT

Communications Network For Manufacturing Applications (CNMA) — 245
W. Kiesel and K. Deiretsbacher,
Siemens AG Automation Group, GERMANY

Experience of Implementing OSI Management Facilities — 259
G. Knight, G. Pavlou and S. Walton
University College London, UK

The Architecture of LANCE: A Simple Network Management System — 271
M. Erlinger, Micro Technology, USA

B. CASE STUDIES

How Well Do SNMP and CMOT Meet IP Router Management Needs? — 285
S. Sanghi, AT&T Bell Laboratories and Columbia University, USA
S. Sengupta, Columbia University, USA
A. Chandna and G. Wetzel, AT&T Bell Laboratories, USA

Experience in Network Management: The Merit Network Operations Center — 301
K. Meyer, University of Southern California, USA
D. Johnson, Merit Computer Network, USA

Providing CMIS Services in DECmcc — 313
M. Densmore, Digital Equipment Corporation, USA

Network Management in the Space Transportation System Mission Control Center: Lessons Learned — 327
R. Durst, The MITRE Corporation, USA

C. PANEL

Polling vs. Event Reporting: What Are the Trade-offs? — 339
Moderator: K. McCloghrie, Hughes LAN Systems, USA

V. TELECOMMUNICATIONS NETWORK MANAGEMENT

Invited Paper
Management of Telecommunications Services Provided by Multiple Carriers — 343
W. Buga, AT&T Bell Laboratories, USA

A. ARCHITECTURE AND DESIGN

An Architecture for the Implementation of an Integrated Management System — 359
P. Senior, Bell Northern Research, UK
S. Harris, Roke Manor Research Ltd., UK
D. O'Sullivan, Broadcom, UK
Y. Trodullies, British Telecommunications, UK

*A Network Management Architecture For
SONET-Based Multi-Service Networks*   371
S. Kheradpir, W. Stinson and G. Sundstrom
GTE Laboratories Inc., USA

*A Study on an End Customer Controlled Circuit
Reconfiguration System for Leased Line Networks*   383
T. Yamamura, T. Yasushi and N. Fujii
NTT Transmission Systems Laboratories, JAPAN

B. PANEL

Can We Really Control Large Distributed Systems?   397
Moderators: K. Lutz, Bellcore, USA
D. Zuckerman, AT&T Bell Laboratories, USA

C. PLANNING AND OPERATIONS

A Nodal Operations Manager for SONET OAM&P   403
G. Berkowitz, P. Fuhrer, B. Gray, A. Johnston
and G. McElvany, AT&T Bell Laboratories, USA

*A Model And Tool For Integrated Network Planning
And Management*   413
A. Zolfaghari, Pacific Bell and Stanford University, USA
T. Ikuenobe, Stanford University, USA
S. Chum, Pacific Bell, USA

Achievement of Private Virtual Networks in a DQDB MAN   425
A. La Corte, A. Lombardo, S. Palazzo and D. Panno
Instituto di Informatica e Telecommunicazioni, ITALY

VI. REASONING AND KNOWLEDGE IN NETWORK MANAGEMENT

Invited Paper

Knowledge Technologies for Evolving Networks   439
S. Goyal, GTE Laboratories, USA

A. PANEL

Reasoning Paradigms for Integrated Network Management   465
Moderator: P. Sen, NYNEX, USA

B. KNOWLEDGE REPRESENTATION

A Multi-Agent System for Network Management   469
I. Rahali and D. Gaiti, Université P. et M. Curie, FRANCE

*Incorporating Non-Deterministic Reasoning in Managing
Heterogeneous Network Faults*   481
P. Hong and P. Sen, NYNEX Science and Technology, USA

A Modular Knowledge Base for Local Area Network Diagnosis   493
J. Schröder and W. Schödl, University of Stuttgart, GERMANY

*Combining Knowledge-Based Techniques and Simulation
with Applications to Communications Network Management*   505
P. Smyth, J. Statman and G. Oliver
Jet Propulsion Laboratory, USA
R. Goodman, California Institute of Technology, USA

C. APPLICATIONS OF KNOWLEDGE-BASED SYSTEMS

*A Knowledge-Based System for Fault Localisation in Wide
Area Networks*   519
M. Frontini, J. Griffin and S. Towers
Hewlett-Packard Laboratories, UK

*RAC & ROLE - A Knowledge-Based System for Network
Trouble Administration Defined by Rapid-Prototyping*   531
M. St. Jacques, D. Stevens, J. Sipos and L. Lau
GTE Data Services, USA

*Automated Knowledge Acquisition from Network Management
Databases*   541
R. Goodman, California Institute of Technology, USA
H. Latin, Pacific Bell Systems Technology, USA

VII. DISTRIBUTED SYSTEMS MANAGEMENT

Invited Paper

Open Distributed Processing and Open Management   553
P. Linington, University of Kent at Canterbury, UK

A. CONCEPTUAL ASPECTS

Using RPC for Distributed Systems Management   565
L. Aguilar, Hewlett-Packard, USA

Dealing With Scale in an Enterprise Management Director   577
C. Strutt, Digital Equipment Corporation, USA

Delegation of Authority   595
J. Moffett and M. Sloman
Imperial College of Science, Technology and Medicine, UK

B. DISTINGUISHED EXPERTS PANEL

*Trends in Integrated Network Management:
Convergence or Divergence?*   609
Moderator: V. Cerf,
Corporation of National Research Initiatives, USA

C. CASE STUDIES

Management in a Heterogeneous Broadband Environment   613
J. Hall and M. Tschichholz, GMD FOKUS, GERMANY

*Gemini: An Environment for Distributed Network Management
Applications*   625
J. Read, Hewlett-Packard Information Architecture Group, USA
J. Balfour, P. Brun, P. Hyland, P. Mellor and P. Toft
Hewlett-Packard Laboratories, UK

Design of the Netmate Network Management System   639
A. Dupuy, S. Sengupta, O. Wolfson and Y. Yemini
Columbia University, USA

VIII. USAGE CONTROL

Invited Paper

Objective-Driven Monitoring — 653
S. Mazumdar, IBM T.J. Watson Research Center, USA
A. Lazar, Columbia University, USA

A. SYSTEMS MANAGEMENT FUNCTIONS

Performance Management in an EMA Director — 679
M. Anwaruddin, Digital Equipment Corporation, USA

Diagnosis of Connectivity Problems in the Internet — 691
M. Feridun, BBN Systems Technology Division, USA

Secure Management of SNMP Networks — 703
J. Galvin, Trusted Information Systems, Inc., USA
K. McCloghrie, Hughes LAN Systems Inc., USA
J. Davin, MIT Laboratory for Computer Science, USA

B. DISTINGUISHED EXPERTS PANEL

Same as VII - B — 715

C. ACCOUNTING MANAGEMENT

Design Considerations for Usage Accounting and Feedback in Internetworks — 719
D. Estrin, University of Southern California, USA
L. Zhang, Palo Alto Research Center, Xerox Corporation, USA

A Hierarchical Domain Concept as a Main Part of an OSI Accounting Model — 735
E. Bötsch, Leibniz-Rechenzentrum, GERMANY

Design of an Open Network Billing Application — 747
B. Ambrose, Broadcom, IRELAND
D. Mahony, Trinity College, IRELAND

IX. MANAGEMENT INFORMATION

Invited Paper

Distribution of Managed Object Fragments and Managed Object Replication: The Data Distribution View of Management Information — 763
S.M. Klerer and R. Cohen, AT&T Bell Laboratories, USA

A. MODELING ASPECTS

Design Concepts for a Global Network Management Database — 777
R. Valta, Leibniz-Rechenzentrum, GERMANY

Composite Managed Objects — 789
A. Dittrich, GMD FOKUS, GERMANY

A Model for Object Relationship Management — 801
K. Klemba and M. Kosarchyn, Hewlett-Packard, USA

B. DISTINGUISHED EXPERTS PANEL

Same as VII - B — 813

C. IMPLEMENTATION ASPECTS

OSI Management Information Base Implementation — 817
S. Bapat, Racal-Milgo, USA

The Concept of the Network Management Information Base in CNM, the TRANSDATA Network Management Scheme — 833
C. Rauh, Siemens Nixdorf Informationssysteme, GERMANY

MINT: an OSI Management Information Support Tool — 845
T. Nakakawaji, K. Katsuyama, N. Miyauchi and T. Mizuno
Mitsubishi Electric Corporation, JAPAN

AUTHOR INDEX — 857

INTRODUCTION

From the Chairs...
INTEGRATED NETWORK MANAGEMENT AND THE INTERNATIONAL SYMPOSIA

Branislav N. MEANDZIJA*, Kimberly W. KAPPEL**, and Paul J. BRUSIL***

*MetaAccess Inc., USA
**Georgia Institute of Technology, USA
***The MITRE Corporation, USA

1 MANAGEMENT PERSPECTIVES

In the eighties we witnessed the proliferation and explosive growth of networking on a global, world-wide scale, enabled through and coupled to the massive quantitative advances and qualitative break-throughs in telecommunications and computer industries. Network management was key to providing reliable communications service to the end-users in an efficient and cost-effective manner. However, management solutions were only specific to each vendor's networking product environment.

When we held the first IFIP-sponsored "International Symposium on Integrated Network Management" in Boston in 1989, standards for enabling integrated and interoperable network management across multiple vendor networking resources were in the heat of the development in regional and international arenas. While some thought that developing these standards was the most difficult part on the road to integrated management, many now realize that today's standards will enable only our first infantile steps down this road.

Now, as we introduce the Second International Symposium on Integrated Network Management, the need for enterprise-oriented management across data and telecommunications applications and distributed system resources continues to show even greater exponential growth. We face yet another key time in the history of network management: incorporating the standards into products aimed at providing coherent, integrated network management across future, standards-based, multi vendor components as well as existing proprietary components. Multi-vendor demonstrations in Japan,

Europe and North America have shown that the time is upon us when users can competitively procure network management products in any of several countries and be confident that they can interoperate with comparable products in other world regions.

Accordingly, our theme for this second symposium, "Worldwide Advances in Integrated Network Management," was chosen to reflect increasing user needs, the rapid advances in development of both standards and products, and the trend for harmony among the key contributing groups. While convergence toward common approaches is gaining acceptance among several corners of the broad management community, including international standards developers, regional and vendor consortia-based implementation agreements bodies, government procurers and commercial users, more awareness for convergence focusing activities needs to occur in other parts of the management community, such as those addressing the management of heterogeneous, information processing data and applications.

2 NETWORK MANAGEMENT IN THE NINETIES

The understanding of management today is tightly coupled to the issues centered around the modeling, structure, and representation of management information in general, and more specifically related to issues concerning the design and implementation of management information bases. A variety of problems and solutions to these issues can be found in these proceedings. While these issues are close to being completely sorted out, their direct application to defining the actual management information to be used in management systems is only slowly beginning. The lack of a rich library of publically available management information is a painful reminder of how few steps we've taken down the road to integrated network management.

As we look forward, it is our opinion that the nineties will witness further exponential growth of networking in general, coupled to the growth of a new phenomenon, the phenomenon of multi-vendor, heterogeneous distributed processing and distributed computing. That growth will create chaotic situations in the existing and newly emerging world-wide computer networks and distributed systems, especially if they are not consistently controlled via standard, ubiquitous, integrated total system management. Such total system management should prove to require only modest extensions to the network management tools now being standardized. However, the lack of standard management information for the plethora of new manageable resources will loom even greater as the most significant near term impediment to integrated management of distributed, heterogeneous, multi-vendor, shared resources in the variety of fast expanding, complementary and cooperating information technology domains, including system, application and network domains.

Later in the nineties, as we get a better grip on management information for all resources in the total environment and as convergence toward common management tools for this environment becomes wide spread, the community's focus will shift more towards the operations and control aspects of total systems management. Currently, most systems apply only to management of networking resources. Furthermore, these systems are passive and offer little more than pretty interfaces to raw, or at best only partly aggregated and/or correlated, data in management information bases.

Distributed system managers will need the ability to actively control hardware and software modules of all sorts. That ability would enable them to react quickly to the anticipated massive changes in the total distributed system environment which will be taking place in the nineties. Active control should give managers the opportunity to participate proactively in the growth of their networks and their interconnected computing and application environments, as well as in the management process itself, in ways best suited for their particular management task. That would provide for a natural way of taking advantage of the most valuable resources of management, the human resources and their accumulating experience base. Systems offering active and easy, automated, integrated control of computer networks and distributed systems, however, will not exist for some time.

There are two ways in introducing active management. One approach is to build systems capable of dealing with most foreseeable management problems. Currently, this approach is prohibitively expensive in terms of performance and development time and cost. Given the overall lack of management experts who could train such systems to provide general automation of management for general environments, it would appear that at best we could get automation for the heuristics associated with managing but a handful of specific environments. Some believe that fuzzy logic control technology, which has been having success in dealing with systems that are difficult to characterize analytically, could be a boon.

The second approach for active management is to give managers, as much as possible, the ready ability to alter the functioning of the network and management system. That can be accomplished through capabilities to easily redesign, reimplement, reinstall, and reintegrate major software modules at run time. Computer Aided Software Engineering (CASE) tools have been created for most issues related to the design and implementation of management information bases. The nineties will witness application of CASE principles to the areas of operations and control of networks and distributed systems.

Hegel's Dialectical Principles,
- everything is in a process of change,
- everything is becoming more complex,
- everything is increasing in numbers which leads to jumps in quality,
- everything has contradictions as its very essence,

are today more true then ever in the world of networking and distributed processing. Management systems, distributed systems, and networks which best accommodate these principles will be dominating the development and growth in the nineties.

The book you hold in your hands contains the essence of the technical program of the Second International Symposium on Integrated Network Management. It is an accurate representation of the state of the art and constitutes the second milestone in our quest for mastering networks, in particular, and distributed systems, in general.

3 FUTURE EVENTS

As the management world evolves in these and other hitherto unexpected ways, the ongoing series of biennial international symposia will continue to foster and to promote non-partisan forums among individuals of diverse and complementary backgrounds to encourage international information exchange about all aspects of network and distributed systems integrated management.

To broaden the scope of these symposia, the International Federation for Information Processing (IFIP) Working Group (WG) 6.6, the main organizer of this international symposia series, has entered into a close working relationship with the IEEE Committee for Network Operations and Management (CNOM) and the Institute for Educational Services (IES). In a reciprocal agreement, IFIP WG 6.6 will participate in organizing the IEEE Network Management and Operations Symposium , NOMS '92, just as IEEE CNOM is participating in this symposium (for more information on NOMS'92, contact Alan R. Johnston, AT&T Bell Laboratories, Room 2B-071, 480 Red Hill Road, Middletown, NJ 07748-3052).

In order to achieve our eventual focus on total systems management, we hope to expand the scope of our reciprocal agreements in the future to include other major organizations involved with other aspects of distributed system management.

The next international symposium on integrated management will be held in the spring of 1993 (for more information write to Ms. Kimberly Kappel, College of Computing, Georgia Institute of Technology, Atlanta, GA 30332-0280). In 1991, IFIP WG 6.6 together with IEEE CNOM will also be organizing an International Workshop on Distributed Systems Operations and Management. This workshop takes place annually in late October. The first Workshop was held in October 1990 in Berlin. The second will be held in October of 1991 in Santa Barbara (for more information write to Robert Hastings, MetaAccess Inc.,P.O.Box 21956, Santa Barbara, CA 93121-1956).

4 ACKNOWLEDGEMENTS

Many people and organizations have made this symposium possible. We would like to thank the Executive Committee which has created this Second International Symposium on Integrated Network Management, the Program Committee which has created the outstanding program, the Program Co-Chairs Wolfgang Zimmer and Kris Krishnan for leading the Program Committee, the Institute for Educational Services for their support when it was needed most, and the IEEE CNOM and their officers Doug Zuckerman and Ken Lutz for their tireless work.

Two factors have been especially crucial in the success of organizing this symposium: the quality of the technical papers and the inclusion of vendor technical presentations. Our thanks go to all the authors who submitted papers for consideration. The papers published herein, which form the base of our technical program, are truly outstanding.

The prominence and diversity of the vendor patrons who agreed to present their network management plans and products constitute a unique and memorable feature of this symposium. We greatly appreciate the support of these vendor organizations. Thanks go to these Corporate Patrons, as well as to the Corporate Friends, for their contributions and help in completing symposium preparations.

Finally, we acknowledge support and the participation of all the people who have come to the symposium. We welcome the opportunity to continue working with all the staff, and organizations who have made this, our first symposium, a success. We shall see you again in 1993.

I
STANDARDS
AND ARCHITECTURE

Network Management *is* Simple:
you just need the "right" framework!

(Invited Paper)

Marshall T. Rose

Performance Systems International, Inc.

US

Although the International standards community has been diligently working on network management technology for several years, empirical evidence may suggest that its approach is fundamentally "wrong": an object-oriented approach might be fine for fourth-generation-languages, but it is a poor basis when trying to manage a real network. Another approach, the *Internet-standard Network Management Framework*, which emphasizes simplicity in design and scalability in deployment, has proven quite successful in achieving multi-vendor, interoperable network management solutions. This paper is purposefully written to be provocative: although consensus has been reached in the International standards community as to the framework for network management, practical experience, involving management of large production networks, appears to dictate philosophies and techniques which are fundamentally opposed to the OSI approach.

1 Introduction

Readers of these proceedings should be well-versed in the need for solutions that provide vendor-independent, interoperable network management. Indeed, as open systems (embodied by either the Internet or OSI suite of protocols) have risen to the forefront of networking, so has the need to manage networks which use these technologies. In the interest of brevity, this introduction omits a discussion of the history of networking, open systems, or the protocol suites which compose open systems. Instead, we merely note that just as today's enterprises need networking technology which is vendor-independent and fully-interoperable across product and vendor lines, so do today's enterprises need network management capabilities which exhibit the same properties.

Readers of these proceedings are likely well-versed in the politics of open systems and in particular of network management within open systems. Indeed, some pundits have gone so far as to publish amusing treatises outlining the deficiencies of the standardization process. This paper is also critical of the process, but in a serious tone, as the author feels that the standardization process for OSI network management has made a tragic mis-step.

In this paper, we present an alternate perspective on network management, one which has proven successful in the market, namely the *Internet-standard Network Management Framework*. Throughout the presentation a comparison is made to the OSI approach to network management, and several myths about network management are hopefully "debunked". In the concluding remarks, a plea is made for a drastic change in the course being taken by OSI network management.

2 Models and Architectures

Models and architectures are like "motherhood and apple pie", but not nearly as respectable. Every system has them, and, in a highly-political and market-crazed industry, there are few objective criteria which can be used to distinguish good design from poor design. Nonetheless, let us ask ourselves what singular property should be promoted by a network management technology and proceed from there.

Clearly, network management must be *ubiquitous* if it is to be truly useful. That is, each and every network node must be capable of being a *managed* node. However, there are two mediating concerns:

- network nodes exhibit great differences: some are super-computers or workstations or terminal servers, others are routers, and still others are media devices, such as a bridge or modem; and,

- users purchase network nodes, not for the purposes of management, but for other purposes (e.g., a super-computer is purchased for "number crunching", not because it can run a network management protocol).

These concerns provide a tremendous focus, which can be summarized as the *Fundamental Axiom*:

The impact of adding network management to managed nodes must be minimal, reflecting a lowest common denominator.

Thus, if we are to achieve ubiquitous management capabilities, then we must design our management technology so that it can be realized on the least "capable" devices in the network. Consistent with this, when the technology is realized on a network node, it should not adversely affect the performance of that node. In brief, the Fundamental Axiom argues for a technology which can be implemented and executed efficiently on a managed node.

Further, since there will be many more managed nodes in the network than management stations, economy of scale argues for placing the computational and network burden on the stations rather than the managed nodes. Thus, for each design decision in which functionality must be split between the management stations and the managed nodes, the decision will be made in favor of reducing the functionality required of the managed nodes.

Of course, the easiest way to achieve efficiency is through *simplicity*: if we can develop a simple design, capable of meeting the basic needs of network management, then efficiency is a likely outcome. Further, by emphasizing simplicity, we also hope to achieve the scalability in deployment which is our chief goal.

2.1 The Internet-standard Network Management Framework

Although the time-frame in which OSI will achieve dominance is widely debated, it is clear that the *Internet Suite of Protocols*, commonly referred to as TCP/IP, has achieved the promise of open systems as today's *de facto* standard for open networking.

It has only been since mid-1987 that work began towards standardized management capabilities for the Internet suite. Further, despite both knowledge of, and participation in, the International standardization process, the community working on the Internet-standard Network Management Framework produced a system quite different from the OSI framework. The differences in the two network management approaches can be viewed as a micro-example of the distinctions between the Internet and OSI protocol suites: the Internet approach is tightly focused, the problem area is well-defined and the technical approach is geared towards that problem.

The key advantage of such an approach is that it produces *tractable technology*, which is the fundamental building block for a mature market. Consider:

- an important characteristic of a mature market is the competition of robust products;

- in order to develop robust products, it is necessary to deploy the technology in a large number of operational environments and subsequently "harden" it; and,

- in order to deploy technology, it is necessary to be able to implement that technology within a reasonable time-frame.

Although anything can be viewed as "a simple matter of coding", experience has shown that simpler technologies are more tractable, and in being so exhibit a greater likelihood of being implemented in a timely fashion. Clearly, the technology comprising the Internet suite of protocols is much more tractable than the correspondent OSI technology; and, just as clearly, the Internet suite has been widely-implemented, deployed in numerous operational environments, and has a tremendous number of mature products in market competition.

Thus, whilst the OSI approach to management has been to develop an all-encompassing model and respective service in an effort to architect a "complete" solution, the Internet-standard network management framework has been specified, agreed upon, widely-implemented, and is already delivering useful service in the market.

The current framework is based on three documents:

- the Internet-standard *Structure of Management Information* (SMI) [1];
- the Internet-standard *Management Information Base* (MIB) [2]; and,
- the *Simple Network Management Protocol* (SNMP) [3].

Readers of these proceedings should be well-versed in the terms SMI and MIB as used in the OSI framework. However, the use and scope of these terms in the Internet-standard framework is somewhat different.

2.2 Structure of Management Information

The OSI framework views the network management problem as solved with an *object-oriented* paradigm. Objects defined in this framework have *attributes*, generate *events*, and perform *actions*. Further, objects are scoped by numerous hierarchies, e.g., for the purposes of inheritance or containment. Towards this end, ASN.1 is used as an object-oriented specification language to define these characteristics, along with a set of *Guidelines for the Definition of Managed Objects* (the ISO GDMO).

In contrast, the Internet-standard framework views managed objects as little more than simple (scalar) variables residing in a virtual store. The SMI is used as the schema for this database, and provides a naming relationship between managed objects (to provide for efficient traversal of the MIB), along with the syntax for the object, and a default access level. Towards this end, a subset of ASN.1 is used, as a notation to provide concise information to implementors.

In examining this difference in perspective, it should be observed that the Internet-standard approach is simply a formalized restating of the *remote-debugging* paradigm commonly used in earlier management technology. Although the panacea of object-oriented programming is quite attractive to some, others argue that both past and present experience show little to recommend it. If the basic needs of network management can be met without this seemingly large investment, then there is little reason to require all managed nodes to implement it. In brief, "neat" architectures are, by themselves, unimpressive; in contrast, useful architectures are needed to solve real problems in real networks. Argumentative readers should consult the Appendix for further exposition.

2.3 Management Information Base

As might be expected, definitions of managed objects often differ greatly between the two approaches. The most marked difference however is the minimalist perspective taken by the Internet-standard approach. For example, the Internet-standard MIB defines a scant 111 objects for the core aspects of the Internet suite of protocols, divided into 8 groups.

To provide for future growth, the Internet-standard SMI provides three extensibility mechanisms:

1. addition of new standard objects through definitions of new versions of the Internet-standard MIB;

2. addition of widely-available, but non-standard, objects through an "experimental" subtree; and,

3. addition of private objects through an "enterprises" subtree.

All three of these mechanisms have received use:

1. the successor to the Internet-standard MIB, MIB-II [4], was produced some two years later and contains 170 objects;

2. as of this writing there are some 20 experimental MIB modules under development (for management of everything from the CLNS, to FDDI, to SMDS); and,

3. numerous vendors have developed their own extensive private MIB modules which are specific to their product line. This is inevitable, and in fact, quite necessary: until such time as network devices are sold generically — in white boxes with blue lettering (e.g., "Router") — there will always be vendor-specific objects that are unique to each product.

In 1990, an interesting development has been the use of the experimental forum for "de-osifying" MIB modules defined by the ANSI and the IEEE. That is, the essence of a given MIB module is restated using the language of the Internet-standard SMI. In comparing MIB modules written using the two frameworks, several key differences can be found.

First, a MIB module conforming to the Internet-standard SMI has *no* optional objects. Instead, related objects are grouped, and these groups form the basic unit of conformance. For example, in the Internet-standard MIB, there is a TCP group: implementation of *all* the objects in that group is mandatory for *all* managed nodes which implement TCP.[1]

The rationale for this is straight-forward: if there are a large number of optional managed objects, then there is likely to be a decrease in the overlap between the objects supported by managed node implementations and manager implementations, and interoperability suffers.

Second, experience with the Internet-standard network management framework has shown that it is better to define a core MIB module first, containing only essential objects; later, if experiences demands, other objects can be added.

---

[1] However, as a consequence of the Fundamental Axiom, a management station must be prepared to interact with managed nodes that implement only a subset of the desired managed objects.

The rationale for this is also straight-forward: it helps implementors to focus on core issues. By stressing the basics first, a relatively large number of implementations can be produced from different sources. As management of real networks takes place, practical experience, and not the mercurial whims of a committee, is the best judge as to what additional objects are needed.

Finally, traps (roughly equivalent to events in the OSI framework) are used sparingly. Discussion of this is postponed until the next section, when an overall philosophy for using traps is presented for the Internet-standard network management framework.

3 Protocols and Mechanisms

Having now taken the top-level view of the Internet-standard Network Management Framework, it is time to see how the Fundamental Axiom affects the design of the underlying technology.

3.1 Transport Mappings

SNMP assumes only a basic CL-mode transport service, whilst the OSI Common Management Information Protocol (CMIP) uses a connection-oriented model as a part of an application layer entity.

The OSI approach is general, in that network management uses the same framework as all the other OSI applications. However, network management is unlike "normal" applications and requires special mechanisms. For example, with either the CO-mode transport or network services, sophisticated algorithms are used to ensure reliability of the transmission. Unfortunately, these mechanisms, which are designed to hide problems in the network, are intended for applications like file transfer and electronic mail. When the network is collapsing, this behavior is most likely inappropriate when trying to manage the network! CL-mode access means that there are no hand-shaking packets required before management can occur, and that each management application controls the level of retransmission. This allows for different retransmission schemes based on the requirement of each management application.

Further, implementation of a full OSI stack in addition to management instrumentation is simply *problematic* on smaller networks nodes. The likelihood, or even desirability, of a full OSI implementation on a MAC-level bridge is quite small: the additional resource cost (processor, memory, and programming) poses a large burden in a market segment which is highly competitive. Users want network management capabilities, but they certainly aren't going to be happy if it substantively increases price or decreases performance.

One response might be to suggest either an abbreviated stack (e.g., layering CMIP directly on top of the LLC), or use of a proxy. However, neither approach is satisfying: the former approach lacks generality in that network management cannot occur beyond each LAN segment; and, the latter approach may lose information if the proxy protocol is sufficiently different from the OSI framework.

In contrast, observe that by combining a simple protocol with minimal requirements on the underlying stack (a CL-mode transport service), a complete management stack can be easily implemented in virtually any network node. For example, experience has shown that the burden of adding a CL-mode network and transport protocol along with the SNMP to a MAC-level bridge is quite small.

Finally, from a theoretical perspective, considering that the basic unit of commerce in a network is the datagram, it should be easy to grasp that the management protocol must have a direct mapping onto an underlying datagram service.

3.2 Operations

SNMP performs operations on scalar variables, whilst CMIP performs operations on collections of objects. In terms of functionality it should be noted that the both SNMP and CMIP use request-reply interactions, and that the object-oriented operations of CMIP can be modeled using the few operations of SNMP (e.g., the CMIP `action`, `create`, and `delete` operations can be performed by the SNMP `set` operation).

However, by imposing a "scalar view" on the managed nodes, instrumentation for, and implementation of, SNMP is much simpler. As such, management stations can layer whatever "high-powered" approach they want for management, but managed nodes need only implement the simplest paradigm, that of remote-debugging.

In practice, use of a CL-mode transport service does not impose undue complexity upon well-design management stations: even use of a CO-mode service is not a complete guarantee of reliability. The well-known (but often little understood) *end-to-end argument* [5] shows that ultimate responsibility for reliability resides with each application. As such, mapping operations onto a CO-mode service achieves little benefit in a robust management station.

Finally, although CMIP offers linked replies (for so-called *incremental reading* of large results), this still places a significant implementational burden on the managed node. The SNMP approach, in which management stations can dynamically adapt to the abilities of the managed node (through a well-defined error response), is less invasive.

3.3 Data Representation

Only a subset of both ASN.1 and the BER are used by SNMP, in contrast to the full generality allowed by CMIP.

In particular, a scalar object ultimately resolves to either an INTEGER, OCTET STRING, or OBJECT IDENTIFIER. Further, the rows of a conceptual table contain only scalars. Experience has shown that this limitation is not a serious restriction as the basic data structures of computing are numbers, byte strings, and structures. Hence, one can easily emulate the semantics of the remaining ASN.1 data types. For example, a BOOLEAN can be represented by an INTEGER taking the value zero (for false) and non-zero (for true). Similarly, a BIT STRING can be represented by an OCTET STRING in which some of the bits in the last octet might be marked as "reserved" if the string of bits is not a multiple of eight in length.

By omitting support for arbitrarily complex data types, substantives savings can be achieved — in terms of program development, code size, and execution time.

Finally, SNMP mandates the use of the definite-length form encoding with the BER, and, whenever possible, primitive encodings are used rather than constructed encodings (e.g., for OCTET STRINGs). Again, the restrictions here are intended to simplify the option space to allow for faster systems with greater interoperability.

3.4 Identifying Management Information

SNMP employs a novel mechanism for identifying management information. At the coarsest level, managed objects are of two kinds: *columnar* objects, which exist in conceptual tabular structures; and, non-columnar objects. Each conceptual table contains zero or more rows, and each row may contain one or more scalar objects — termed columnar objects. It must be emphasized that, at the protocol level, relationships among columnar objects in the same row are matter of convention, not of protocol: instances of columnar objects which appear as operands in SNMP are treated separately.

All object instances are identified by an ASN.1 OBJECT IDENTIFIER. An instance of a non-columnar object is identified by appending a zero-valued sub-identifier to the object name. For example, the one and only instance of the sysDescr object (as defined in the Internet-standard MIB) which resides in a managed node is identified by:

sysDescr.0

or, more concretely:

1.3.6.1.2.1.1.1.0

For each columnar object, there is an associated definition which describes how the instance-identifier is formed. This collection of one or more sub-identifiers is appended to the name of the columnar object. For example, in the Internet-standard MIB, instances of the columns of the interfaces table are identified by using the value of the ifIndex column. So, the instance of ifDescr associated with the first interface is identified by:

ifDescr.1

or, more concretely:

1.3.6.1.2.1.2.2.1.2.1

There is an important observation on naming in SNMP: by naming instances using OBJECT IDENTIFIERs, a *lexicographic ordering* is enforced between all object instances: for instance names a and b, one of three conditions uniformly holds: either $a < b$, $a = b$, or $a > b$. This is the *only* retrieval methodology which a managed node must support and is in stark contrast to the CMIP approach, which introduces complex scoping and filtering mechanisms to identify object instances.

As a first comparison, recall that since SNMP treats operands separately, it is straight-forward for a single management operation to reference two objects in different parts of the information tree. With CMIP, additional management traffic is often required as a single scope may contain too many irrelevant objects.

However, in order to appreciate the true power of the SNMP approach to naming, we must examine how traversal is performed. SNMP provides a powerful get-next operator, which takes one or more operands, and for each returns the name and value of the object instance immediately following. Since each operand is simply an OBJECT IDENTIFIER, and needn't correspond to an existing instance, management stations may exploit the naming scheme to the fullest.

There are several traversal strategies which can be realized using this single retrieval methodology:

- To see if a particular non-columnar object exists, a management station uses the get-next operator on the object name (not the instance name), e.g.,

    get-next (sysDescr)

    If the instance for the desired object is returned, the management station has the desired information; otherwise, if some other instance is returned, then the management station has determined that that the object is not supported by the managed node. This is particularly useful when retrieving a collection of non-columnar objects in a single operation.

- Since the result returned by the get-next operator is an instance name and correspondent value, a management station can use the result of one invocation as the operand to subsequent invocation of the get-next operator.

    In the simplest case, to examine all objects supported by the managed node, a managed station could simply start with

    get-next (0.0)

    and traverse the entire space of object instances until the last instance is returned. The get-next operator, when provided with an operand that is lexicographically greater than all instances in the managed node, will return a well-defined error response.

A more useful example of traversal is to consider how columnar objects are examined:

- At first cut, a tabular sweep can be started by invoking the `get-next` operator with the names of the columnar objects of interest, e.g.,

    `get-next (ipRouteDest, ipRouteIfIndex, ipRouteNextHop)`

    which will return the instances of these objects that correspond to the "first" row in the corresponding conceptual table. To find the next row in the table, these instance names can be used as operands for the next invocation, e.g.,

    ```
    get-next (ipRouteDest.0.0.0.0,
              ipRouteIfIndex.0.0.0.0,
              ipRouteNextHop.0.0.0.0)
    ```

    The sweep might continue until an instance of a different object type is returned, e.g.,

    `get-next (ipRouteDest.192.33.4.0) -> ipRouteIfIndex.0.0.0.0`

    An important side-effect of this strategy is that the instance-identifiers are, by and large, opaque handles — the management station needn't interpret them as they are simply strings passed back and forth through SNMP. (Historically, such a strategy has proven particularly successful for "stateless" applications which utilize a CL-mode transport service.)

- Of course, given an appropriate definition of instance-identification, a sophisticated manager might sweep only a portion of a table, e.g.,

    ```
    get-next (ipRouteDest.192,
              ipRouteIfIndex.192,
              ipRouteNextHop.192)
    ```

    can be used to start a sweep in the IP routing table looking for those destinations with IP addresses that start with 192. When a result is returned which does not have this prefix, the sweep terminates.

Again, it must be emphasized that managed nodes need implement only a single retrieval methodology based on lexicographic ordering. The management station can layer whatever strategy it wishes on top of this. In contrast, a node being managed with CMIP must be prepared to accept arbitrary scoping and filtering directives. As such, it is not possible to bound the complexity involved in processing a single CMIP operation, whilst each SNMP operation is of fixed (and almost certainly smaller) complexity.

Of course, there are occasional arguments about "efficiency" when conceptual tables are retrieved in bulk using SNMP. Unfortunately, these arguments are often put forward without much exploration of possible implementation strategies. For example, the author and his colleagues have developed a parallel algorithm for bulk retrieval using SNMP[6]. The algorithm works in two phases:

- In the first phase, all instances of a single columnar object are retrieved.

    This is achieved by issuing several invocations of the `get-next` operator, each with several operands. The naming space is divided up by generating sub-identifier suffixes. For example, the first request might appear as:

    ```
    get-next (ipRouteNextHop,
              ipRouteNextHop.63,
              ipRouteNextHop.127,
              ipRouteNextHop.159,
              ipRouteNextHop.192)
    ```

    The algorithm keeps track of the invocations that are outstanding, along with the `OBJECT IDENTIFIER` bounds that are active in each. For each result returned in an invocation: if the result is outside its assigned bounds, that result is removed; otherwise, the name and value of that instance is stored, and possibly the bounds is split into two (depending on the observed performance of the managed node and the network).

    When all of the bounds have been removed, all instances of the single columnar object have been retrieved and the second phase begins.

- In the second phase, instances of any other desired columnar objects (in the same conceptual table) are retrieved.

    This is achieved by use of the SNMP `get` operation. The instance-identifiers are taken from the results determined in the first phase.

The algorithm uses an adaptive retransmission strategy in order to maximize throughput whilst avoiding overrunning the agent. It should be noted that the transmission strategy is one in which ordering of invocations is, by and large, unimportant. Thus, intermittent network lossage results in retransmission only of lost traffic, traffic corresponding to other parts of the table is neither retransmitted nor delayed.

Initial implementation of the algorithm in a prototype environment shows a speedup of roughly an order of magnitude over the serial retrieval approach.

3.5 Traps

SNMP uses *trap-directed* polling to report extraordinary events, and as noted earlier, a CL-mode transport service is used for transfer. The management station is then responsible for initiating further interactions with the managed node in order to determine the nature and extent of the problem. Of course, since traps are sent unreliably, low-frequency polling is performed by the management station. Many feel that this choice has proven to be an effective compromise between trap-based and polling-based approaches.

Trap-based approaches suffer from requiring substantive resources on the part of managed nodes, management stations, and the network:

- It requires resources to generate the trap. If the trap must contain a lot of information, the managed node may be spending too much time on the trap and not enough time on it's primary task (e.g., routing packets for the user).

- If the trap requires some sort of acknowledgement from the manager, this places further requirements on the managed node's resources. With CMIP, in which a CO-mode transport service is used, either a connection is maintained between the management station and the managed node (which has serious implications as to how many managed nodes a station can manage), or a connection must be established immediately prior to sending the trap.

- Of course, use of a CO-mode transport service is no guarantee as to the reliability of traps — network failures can *always* prevent traps from getting through.

- Further, if several extraordinary events occur, a lot of network bandwidth is tied up containing traps, which is hardly desirable if the report is about network congestion. A managed node might use *thresholds*: traps are generated only when the occurrence of an event exceeds some threshold.

    Unfortunately, this means that the managed node must usually spend substantial time determining if an event should generate a trap, and these cycles are usually spent during the *critical path* (e.g., during the forwarding loop for a router). In contrast, the management station could make this decision in deciding whether a trap should generate an alert for the operator. That is, the same functionality is present in the management system, except that the management station, and not the managed node, has the burden of implementation.

In any event, the managed node has only a very limited view of the network, so it is arguable as to whether it can provide "the big picture" on the problem by using traps. For all these reasons, traps are used sparingly in the Internet-standard network management framework.

With trap-directed polling, the impact on managed nodes remains small; the impact on network bandwidth is minimized; problems can be dealt with in a timely fashion; and, the management station is in control as it determines what "the big picture" really is.

4 Conclusions

Given these criticisms of the OSI approach to network management, is it any wonder that some network management wags often use the catch-phrase:

> *If the answer is OSI network management, then the question must have been mis-stated!*

As the discussion has shown, from both a theoretical and practical perspective, it would appear that the OSI approach may be tragically detached from reality.

Why is this so? Perhaps the fundamental reason can be seen from the seemingly harmless introductory statement:

> ... just as today's enterprises need networking technology which is vendor-independent and fully-interoperable across product and vendor lines, so do today's enterprises need network management capabilities which exhibit the same properties.

The problem is that although open network management is needed, the model, architecture and design of such technology needn't be identical to that of the other applications in the protocol suite. Network management is unlike other applications:

- network management is often time-critical;
- network management must continue to work during catastrophic network failure, even when all other applications fail; and,
- network management should be deployed on each and every network node, if it is to be truly useful.

These characteristics all argue for a paradigm different than that of the other applications which provide file service, message handling, directory service, and the like.

4.1 The Problems of the Real-World

It must be emphasized that

> *The problems of the real-world are remarkably resilient towards administrative fiat.*

Thus, citing "successful" implementation agreements for OSI management, well-intentioned government mandates for OSI, and other bits of "excellent progress", are largely unimpressive: users probably want a robust market with competing mature products that solve their problems. Users may be willing to put up with the standardization process, but *only* if it is a means towards the end of a robust market. Standards, per se, are probably of no interest to users without robust products which implement those standards.

Over the last four years, there has been some experience with the OSI framework in the Internet, involving considerable support from both vendors and the U.S. Federal government. However, products have been sadly lacking. In contrast, the Internet-standard network management framework has been solving problems quite handily and has developed an enthusiastic following amongst network managers throughout the International community using the Internet suite. With the introduction of SNMP over OSI-based transports[7], some are beginning to suspect that the OSI networks of tomorrow will be managed with SNMP and not CMIP.

This has lead to the so-called "soft-pillow" defense for the OSI approach: namely that SNMP will manage network nodes, and perhaps CMIP will be used for communication between management stations. However, the market may be unwilling to accept a dual protocol approach: it may be more attractive to use the SNMP for all management tasks — simply to avoid the loss-of-leverage inherent with the dual approach.

4.2 A Plea for Sanity

Perhaps it is time to re-think the OSI approach to network management and take a more pragmatic, workable solution? Much can be learned from successful systems, and good things often come from copying these successes. Perhaps the International standards community can profit by adopting the Fundamental Axiom and using the Internet-standard network management framework as the basis for OSI network management.

Acknowledgements

In presenting this paper, the author is acting as an interpretive historian: the Fundamental Axiom of the Internet-standard network management framework, and the philosophy which embodies it, is the work of many others, most notably Case, Davin, Fedor, and Schoffstall, who present an elegant argument for the Fundamental Axiom in [8]. The author's presentation has been to focus these issues and present a concise, albeit painful, criticism of the OSI approach to network management.

Keith McCloghrie of Hughes LAN Systems provided substantive comments, and his contributions are greatly appreciated. Ole J. Jacobsen of Interop, Inc. provided a thorough proof-reading of a manuscript, and his help is greatly appreciated. Finally, the author apologizes for shamelessly using a colloquialism in the first sentence of Section 2.

References

[1] Marshall T. Rose and Keith McCloghrie. *Structure and Identification of Management Information for TCP/IP based internets*. Request for Comments 1155, DDN Network Information Center, SRI International, May, 1990.

[2] Keith McCloghrie and Marshall T. Rose. *Management Information Base Network Management of TCP/IP based internets*. Request for Comments 1156, DDN Network Information Center, SRI International, May, 1990.

[3] Jeffrey D. Case, Mark S. Fedor, Martin L. Schoffstall, and James R. Davin. *A Simple Network Management Protocol*. Request for Comments 1157, DDN Network Information Center, SRI International, May, 1990.

[4] Marshall T. Rose (editor). *Management Information Base Network Management of TCP/IP based internets: MIB-II*. Request for Comments 1158, DDN Network Information Center, SRI International, May, 1990.

[5] J.H. Saltzer, D.P. Reed, and D.D. Clark. End-to-End Arguments in System Design. *Transactions on Computer Systems*, 2(4):277–288, November, 1984.

[6] Marshall T. Rose, Keith McCloghrie, and James R. Davin. *Bulk Table Retrieval with the SNMP*. Request for Comments 1187, DDN Network Information Center, SRI International, October, 1990.

[7] Marshall T. Rose (editor). *SNMP over OSI*. Request for Comments 1161, DDN Network Information Center, SRI International, June, 1990.

[8] Jeffrey D. Case, James R. Davin, Mark S. Fedor, and Martin L. Schoffstall. Network Management and the Design of SNMP. *ConneXions—The Interoperability Report*, 3(3):22–26, March, 1989. ISSN 0894-5926.

Appendix: A Cautionary Tale

The story which follows is unattributed, as reported by Manny Farber to the author. Any individual who can prove authorship of the story is entitled to a particularly *fine dinner* at the expense of the author!

Once upon a time, in a kingdom not far from here, a king summoned two of his advisors for a test. He showed them both a shiny metal box with two slots in the top, a control knob, and a lever. "What do you think this is?"

One advisor, an engineer, answered first. "It is a toaster," he said. The king asked, "How would you design an embedded computer for it?" The engineer replied, "Using a four-bit microcontroller, I would write a simple program that reads the darkness knob and quantizes its position to one of 16 shades of darkness, from snow white to coal black. The program would use that darkness level as the index to a 16-element table of initial timer values. Then it would turn on the heating elements and start the timer with the initial value selected from the table. At the end of the time delay, it would turn off the heat and pop up the toast. Come back next week, and I'll show you a working prototype."

The second advisor, a computer scientist, immediately recognized the danger of such short-sighted thinking. He said, "Toasters don't just turn bread into toast, they are also used to warm frozen waffles. What you see before you is really a breakfast food cooker. As the subjects of your kingdom become more sophisticated, they will demand more capabilities. They will need a breakfast food cooker that can also cook sausage, fry bacon, and make scrambled eggs. A toaster that only makes toast will soon be obsolete. If we don't look to the future, we will have to completely redesign the toaster in just a few years."

"With this in mind, we can formulate a more intelligent solution to the problem. First, create a class of breakfast foods. Specialize this class into subclasses: grains, pork, and poultry. The specialization process should be repeated with grains divided into toast, muffins, pancakes, and waffles; pork divided into sausage, links, and bacon; and poultry divided into scrambled eggs, hard-boiled eggs, poached eggs, fried eggs, and various omelet classes."

"The ham and cheese omelet class is worth special attention because it must inherit characteristics from the pork, dairy, and poultry classes. Thus, we see that the problem cannot be properly solved without multiple inheritance. At run time, the program must create the proper object and send a message to the object that says, 'Cook yourself.' The semantics of this message depend, of course, on the kind of object, so they have a different meaning to a piece of toast than to scrambled eggs."

"Reviewing the process so far, we see that the analysis phase has revealed that the primary requirement is to cook any kind of breakfast food. In the design phase, we have discovered some derived requirements. Specifically, we need an object-oriented language with multiple inheritance. Of course, users don't want the eggs to get cold while the bacon is frying, so concurrent processing is required, too."

"We must not forget the user interface. The lever that lowers the food lacks versatility, and the darkness knob is confusing. Users won't buy the product unless it has a user-friendly, graphical interface. When the breakfast cooker is plugged in, users should see a cowboy boot on the screen. Users click on it, and the message 'Booting UNIX[2] v. 8.3' appears on the screen. (UNIX 8.3 should be out by the time the product gets to the market.) Users can pull down a menu and click on the foods they want to cook."

"Having made the wise decision of specifying the software first in the design phase, all that remains is to pick an adequate hardware platform for the implementation phase. An Intel 80386 with 8MB of memory, a 30MB hard disk, and a VGA monitor should be sufficient. If you select a multitasking, object oriented language that supports multiple inheritance and has a built-in GUI, writing the program will be a snap. (Imagine the difficulty we would have had if we had foolishly allowed a hardware-first design strategy to lock us into a four-bit microcontroller!)."

The king had the computer scientist thrown in the moat, and they all lived happily ever after.

Although it is believed that this story was originally written to parody "modern" software engineering practices, this cautionary tale can also be seen as being quite germane to the OSI approach to network management.

---

[2]UNIX is a trademark of AT&T.

I
STANDARDS AND ARCHITECTURE

A
Applying Standards

INTEROPERABLE NETWORK MANAGEMENT: OSI/NM FORUM ARCHITECTURE AND CONCEPTS

Jock Embry, Opening Technologies and OSI/NM Forum Architecture Team[1]
Peter Manson, Telecom Canada and OSI/NM Forum Architecture Team[2]
Dave Milham, British Telecom and OSI/NM Forum Architecture Team

This paper summarizes the architecture and key concepts that have been adopted by the OSI/Network Management Forum for interoperable network management. A number of organizations are incorporating this architecture and specifications into their network management products and systems. It is expected that, as this architecture develops, it will be the basis for the multi-vendor exchange of network management data for the next 5 to 10 years.

The first set of specifications that the Forum has published is identified as Release 1. The purpose of the Release 1 package is to provide a complete set of specifications defining basic interoperability for alarm management, and basic configuration management.

The Forum was formed in 1988 as an open, non-profit corporation. Its objective is to promote the rapid development, acceptance, and implementation of standards for interoperable network management. The Forum's work is intended to align with the international standards set forth by the International Organization for Standardization (ISO) and recommendations issued by the International Telegraph and Telephone Consultative Committee (CCITT). If ISO and CCITT have not yet stated a position on a specific issue, the Forum will propose one and use it on an interim basis. The interim proposal will be submitted to the international standard committees for consideration. Once a position has been approved by ISO and CCITT, the Forum is committed to conform to it.

This paper covers the key concepts necessary to enable the interoperation of several network management systems using an open interface. This covers the full range of networks that must be managed, including computer and communication networks, local area networks (LANs), wide area networks (WANs), international networks, voice (public and private) networks, and packet data networks. There is some difference here between the Forum and the work by the standards organizations. Both ISO and CCITT are defining management standards with a focus on managing particular kinds of networks: ISO is concentrating on how to manage Open System Interconnection (OSI) networks, and CCITT is concerned with how to manage telecommunications networks. The Forum is trying to apply those management standards to manage any network.

---

[1] Mr. Embry is a principal in Opening Technologies, and represents Unisys in the OSI/NM Forum.

[2] Mr. Manson is employed by Bell-Northern Research Ltd., and represents Telecom Canada in the OSI/NM Forum.

The concepts covered in this paper include the conformant management entity (CME): any management system that supports the interoperable interface defined by the Forum; managed objects: the abstract representations of the real resources to be managed; and the interoperable interface: a defined set of protocols (using the OSI seven layer protocol stack), over which particular messages about managed objects are exchanged.

This paper is based on information from published and working Forum documents and from material in published ISO and CCITT documents.

The OSI Network Management Forum Release 1 package is composed of a set of interrelated documents, described below. Each document describes an aspect of Forum interoperable network management, and some documents provide a framework on which other documents are based. *[Note: These descriptions are taken directly from the Forum documents.]*

1. Glossary [1] - provides short definitions of key terms and provides references to the documents where those terms are completely defined and placed in context.

2. Architecture [2] - presents the problem of interoperable network management in terms of a general model which is examined from a number of different perspectives. Other Forum documentation builds upon the concepts described in the Forum Architecture, hence new readers of Forum documentation may find it helpful to read this specification first, referencing the Forum Glossary [1] where necessary.

3. Protocol Specification [3] - specifies the elements of the OSI/NM Forum Interoperable Interface Protocols. The protocol suites are designed to facilitate communication between equipment of different vendors, suppliers and networks. The Forum protocol suite is based on International Standards (all seven layers, including CMIS [11] and CMIP [12] among others) and agreements reached in regional profiling organizations. As such, it encompasses agreements reached in a broad community designed to ensure interoperability.

   Addendum to the Protocol [10] - provides revisions and errata to reference the IS version of CMIP, and correct several problems. The major addition is the Forum Protocol Profiles Appendix.

4. Application Services [4] - defines a number of generic models, as well as protocol and procedures to enable CMEs to transmit network management functions (i.e. exchange messages to perform management functions). Specification of both common (e.g., Event Reporting and Event Logging) and specific functions (e.g., Alarm Reporting) are provided. The models and management services defined in this document make use of the protocol elements specified in the Protocol Specification [3].

5. Object Specification Framework [5] - provides guidelines and a notation for defining managed object classes, attributes, name bindings, notifications and operations. The material presented in this document is intended for use by designers in developing object specifications for the Library [6].

6. Library [6] - serves as the definitive source for the definition of managed object classes, name bindings, and attributes. These definitions are based on the guidelines specified in the Object Specification Framework [5] document.

7. Managed Object Naming and Addressing [7] - provides requirements for the naming and addressing of managed object instances. This documentation extends and supercedes the naming sections in the Object Specification Framework [5], the Architecture [2], and is reflected in the Library [6].

8. Shared Management Knowledge [8] - defines mechanisms such that conformant management entities can achieve a common understanding of the management protocols, procedures, and capabilities that are required for the exchange of management information.

9. Conformance Requirements [9] - this document provides a summary of Network Management product conformance related requirements.

The Forum architecture has been developed to meet four objectives. The first is to provide the framework to support interoperability of systems that manage communications and computer networks. The reason for developing standards is to allow systems to be designed independently, yet to interact productively. For network management systems, it is necessary to identify what data the management systems need to exchange and then to determine the best ways of defining that exchange. A successful exchange of data demands that all parties agree to do some things the same way.

The second objective of the Forum architecture is to allow freedom for different types of system implementation. Different management systems are designed with varying objectives in terms of function, performance, cost, and environment, for example. It is both necessary and healthy to encourage system distinction in some areas.

Third, the Forum architecture must be flexible and extensible to the management of a full spectrum of networks, regardless of size or kind. A wide variety of today's networks need to be managed -- ranging from a LAN that has only a handful of users to public voice networks with millions of users. Network management standards must recognize and support all the networks that require management.

The fourth objective of the Forum architecture is to align with ISO and CCITT. One of the tenets upon which the OSI/NM Forum was founded is that it will support established international standards and recommendations as set forth by ISO and CCITT. In many cases, the Forum may be advancing work beyond what has been accepted, but it is generally in the same direction that is expected from ISO and CCITT. Adjustments will be necessary over time as recommendations become adopted as formal standards.

Except for a few cases, networks are composed of parts from a number of vendors and providers. Operators of those heterogeneous networks, however, want to manage them with a consistent set of policies, practices, and tools. Interoperable network management should enable network owners to operate their networks more efficiently and consistently.

CONFORMANT MANAGEMENT ENTITY

Conformant management entity, or CME, is the term used by the Forum to refer to a network management system that supports the Forum's standards for exchanging management data. Figure 1 shows a CME in context with other key elements of the architecture. The Forum is not trying to define how management systems should be built, but how they should interoperate. The focus is on the interoperable interface -- that point where two different management systems meet to exchange management data. This interface, which uses a particular set of communications protocols to exchange messages, is referred to as the P+M interface, as described in more detail later in this paper.

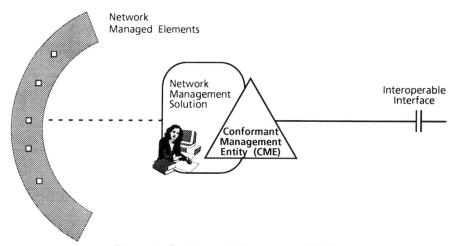

Figure 1. Conformant Management Entity

The network management solution itself may contain all of the elements that normally are considered part of a management system, such as computer hardware; operating system and data base management software; communications software (e.g., protocol engines, layers, implementations); application software; expert systems (e.g., inference engines, rule bases, knowledge bases); data bases and files; internal communications networks; and human interfaces such as screens and displays, keyboards, mice, and track balls (hardware) and screen generators, graphic generators, and window managers (software). All of these elements may not be present in any one system, depending on the particular requirements and design of that system. The management solution also includes network management users, since they contribute to solving the problems of network management. Beyond the interoperable interface, a network management system is treated as a black box to allow as much implementation freedom as possible.

There is also an underlying network to be managed that is composed of such elements as computers, switches, communication processors, multiplexers, and circuits. These network elements make up a set of resources that must be managed in certain ways to meet network objectives.

Several key aspects of network management system design are not specified by the Forum. It is in these areas that systems are expected to differ widely to support different kinds of networks, requirements, technology, and markets, and to provide product differentiation. One of these aspects is the technology on which the system is based. This includes the computer hardware, its operating system, and the system design approach (object, message, data base). Another aspect is the level of distribution or centralization. Management systems can be implemented as single, centralized systems on one machine; they can be fully distributed across hundreds or thousands of processors; or they can fall somewhere in between.

Another unspecified aspect of network management system design is the human interface. What information is presented to the user and how it is presented represent areas for product differentiation. Although other standards efforts are addressing the human interface (e.g., X-Windows and IBM's Common User Interface), the Forum is not pursuing this issue now. Nor is the Forum addressing how the management system is to actually connect to the underlying network it manages, although the Forum architecture document describes how the interoperable interface concept can be used recursively.

A single CME by itself is not very useful -- somewhat like one hand trying to clap. In general, common management is desired over two or more networks when the networks are interconnected, overlap, or have some other relationship. The the parts of the outer ring in Figure 2 depict communication and computer networks that are being managed by a set of interoperable CMEs. As the figure shows, the managed networks may or may not be interconnected to each other.

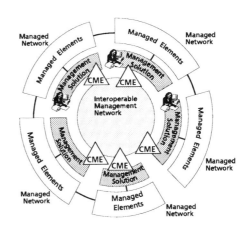

Figure 2. General Architectural Model

The interoperation of the CMEs is shown through an interoperable management network. This would actually consist of the interoperable interface between each pair of CMEs that are interacting. Although the figure shows humans at various places around the management solution, it is likely that the human interface from only one or a few CMEs would be used by the network operator. Other CMEs would provide data to and from that CME for presentation. To the extent that interoperability between management systems can be achieved, the number of different, inconsistent interfaces that a network operator must use can be reduced.

MANAGED OBJECTS

The Forum has adopted an object-oriented approach to representing management data, following the approach of ISO and CCITT. This approach uses many of the concepts from object-oriented programming and object-oriented design. It does not, however, borrow all of the elements of these concepts; and, in some cases, the terms and concepts that are used have been modified.

A managed object is an abstraction or representation of a resource that is to be managed across the interoperable interface. Rather than directly manipulating network elements and resources, the approach is to manipulate an abstraction of the resource. Some other (unspecified) mechanism maintains the relationship between the managed object and the actual resource. Figure 3 shows that the managed objects are considered to be in the CME, and the resources to be managed are outside the CME, generally in the managed network. A resource can be anything that it is useful for management purposes, such as individual cards in a piece of equipment, cables between buildings, software modules, data bases, subnetworks, and host computer complexes. Managed objects can also represent resources in the management solution and the management network as well. A managed object could also be an abstraction of one or many other managed objects.

Because a managed object is the abstraction of some resource, the skill in defining a managed object is to obtain the level of abstraction that includes all of the details and features necessary for management, but not so much detail that the key aspects get lost.

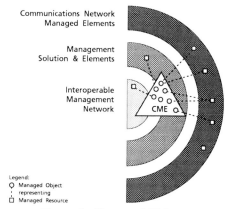

Figure 3. Managed Objects

A single managed object may represent a single network resource or many resources. The same network resource may be represented by a single managed object or by several different managed objects, each of which may provide different views of the resource. It may be that a particular resource is not represented by any managed object; that does not mean that the resource does not exist, only that it is not available for interoperable management. Some managed objects may be available solely for management purposes and not represent an actual network resource. Examples of this situation include event logs and sieves.

Managed Objects and CMEs

Managed objects are "made visible" by one CME to another across the interoperable interface, as shown in Figure 4. This means that a CME can accept messages addressed to the managed object or issue messages from that managed object, or do both.

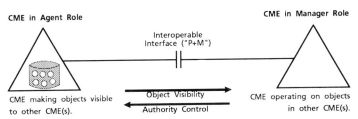

Figure 4. CME Authority Relationships

The CME that makes objects visible is considered to be acting in an agent role, because it is an agent for the resources that the managed object represents. The CME that operates on managed objects in another CME is said to be acting in a managing role because it is managing or exercising control over the resources that the managed object represents. Any particular CME implementation can act in the agent role, the manager role, or both.

All of the managed objects made visible by a CME constitute what is called the management information base (MIB). Managed objects are the only things seen across the interoperable interface; thus, the MIB represents all of the management information a CME acting in the agent role has or will allow another CME to see.

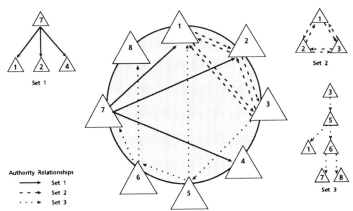

Figure 5. CME Authority Relationship Sets

Because CMEs can act in both the agent and manager roles, they can be in complex arrangements, as shown in Figure 5. The arrows show an authority relationship between CMEs, where the head of the arrow indicates a CME acting in the agent role, and the tail shows a CME acting in the manager role. CMEs can participate in many authority relationships. Authority relationships may be organized into sets to reflect organizational, functional, geographical, technological, or other groupings. Set 2 in the figure shows three CMEs that are acting in a peer-to-peer fashion, each as both agent and manager for the other two CMEs. Set 3 shows a hierarchy of CMEs.

Definitions of Managed Objects

Managed object classes are a way of grouping things that have similar properties; a managed object instance is a particular thing that can be named and talked about. For example, the IEEE Computer Society is an instance of the general class "Professional Society." Classes can be viewed as a taxonomy, as biologists would think of it. Instances are then the particular individuals of their class. Managed object classes are defined independent of any particular instance. A managed object class is intended to capture some set of properties considered significant for interoperable management. The four properties that make up the definition of a managed object class are described below.

Attributes can be thought of as data elements and values. Each attribute has a particular definition (semantics) and format (syntax). There is provision for a wide variety of attributes, ranging from simple bit strings or integers to complex structures with variable length strings, subdivided elements, and so forth. The full power of Abstract Syntax Notation One (ASN.1) is available for defining attributes; however complex attributes can be more difficult to manipulate than simple ones.

Attributes can be single or set-valued. A set-valued attribute is a set of values without any ordering. Typical attributes are the data rate for a circuit or the amount of main memory on a computer. An example of a set-value attribute is "user labels", which allows one or more arbitrary text strings to be associated with an object.

Management operations are operations that can to be applied to a managed object instance. General operations apply to all classes; they include create and delete instances, and get and set attribute value; unique operations apply to individual classes, for example, concurrently setting several particular attributes to certain values.

Behaviour exhibited by a managed object instance includes responses to management operations or behaviour based on the resource that the object represents. For example, if a computer goes "down," then the operational state attribute of its managed object instance would have the value "disabled." The behaviour definition of a managed object would specify that whenever the underlying resource is unusable, the operational state attribute must have the "disabled" value.

Notifications are those messages emitted by a managed object instance to a managing CME. Similar to operations, there are both general notifications (e.g. report attribute change) and notifications that are specific to an object class, such as "transmission alarm."

Managed object classes are defined in a structured fashion. The Forum Object Specification Framework document [5] defines how managed object classes are to be defined. The Forum Library [6] contains actual definitions.

Note: Since the Forum completed work on Release 1, ISO has advanced the corresponding document, Guidelines for the Definition of Managed Objects [16] to Draft International Standard (DIS) status. The Forum is currently studying how to migrate the Release 1 Library to GDMO, and the implications for interoperability.

The Forum, ISO, and CCITT use hierarchies (or "trees") for defining objects. An **inheritance tree**, shown in Figure 6, depicts relationships between managed object classes and introduces the notions of superclasses and subclasses. Each class has one superclass. A class inherits all of the properties (attributes, behaviour, operations, and notifications) defined for its superclass, and then additional properties are added. This hierarchy applies strict inheritance, which means that the subclass is assigned all properties, and deletion of properties is not allowed. The ultimate superclass is called TOP, from which everything is derived.

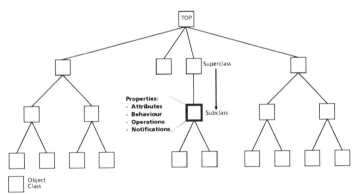

Figure 6. Managed Object Class Tree

The idea behind the inheritance tree is that general classes are defined close to TOP, such as "equipment." Then other classes, such as "multiplexers," are defined as a subclass of "equipment." There has been much discussion about the proper use of classes and inheritance. The use of multiple inheritance (more than one superclass) and other inheritance issues have been studied in the Forum, with a set of modeling guidelines documented in the "J-Team" report [15].

Figure 7 shows a **naming tree**, which depicts relationships between managed object instances for the purposes of naming. A name is the way of uniquely identifying each managed object instance. Instance names are based on a hierarchy of containment relationships. Each instance is considered to be "contained" within some other instance, called its superior. The contained objects are called subordinates. The end of the tree is a node called "ROOT," which is the container of all instances.

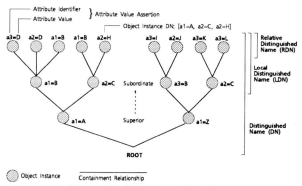

Figure 7. Object Naming Tree

Within a superior instance, all subordinates are uniquely identified by a relative distinguished name (RDN) ("relative" as in relative to its superior). An RDN is formed by the identifier of an attribute, (sometimes called the distinguishing or naming attribute), and some value. The combination of attribute and value must be unique for each managed object instance having the same superior. For example, the managed object in the upper left of the figure uses the attribute identified as "a3" for naming; its RDN then is "a3=D".

A complete name for a managed object instance, called a distinguished name (DN), consists of a sequence of RDNs, starting with the ROOT, and working to the instance itself. Thus, all DNs are unique, and each managed object instance has a unique name.

Managed object instance names are created whenever an instance is created. These names do not have to be registered or even be made public. They do have to be exchanged between two interoperating management systems.

The naming tree is also called the management information tree (MIT), or sometimes the containment tree. Some caution is needed when talking about a containment tree: All names are based on containment, but not all forms of containment are necessarily used for naming.

For each managed object class, one or more rules must be written to identify a superior class and the distinguishing attribute. These rules are called name bindings.

A general naming architecture is described in detail in the Forum Managed Object Naming and Addressing document [7]. A number of issues having to do with naming and addressing managed objects -- such has how to identify the CME that makes a particular managed object instance visible -- still are under study.

The *registration tree*, shown in Figure 8, depicts the general mechanism for producing globally unique identifiers for things to which people and systems around the world will need to refer. There is one registration tree in the world; branches are assigned to different organizations to administer and further refine, as needed.

Registering something in this tree means having it assigned a unique sequence of numbers, starting from ROOT. (The registration tree ROOT is not the same as the naming tree ROOT.) This sequence is represented in ASN.1 as a special data type, called an object identifier (not to be confused with the name of a managed object instance). Once an object definition is complete, each new managed object class, attribute, notification, name binding, etc. is assigned a unique ASN.1 object identifier, placing them in the registration tree.

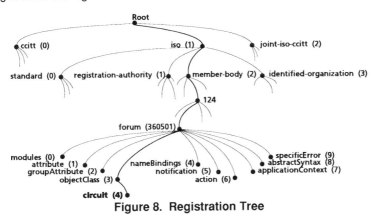

Figure 8. Registration Tree

After ASN.1 object identifiers are assigned and published, the definitions they refer to are never changed. When the inevitable need for correction or refinement occurs, it will be necessary to register a thing as new, with the assignment of a new ASN.1 object identifier. The question of how to deal with obsolete registered things, or at least how to mark them as being replaced with newer registrations, is now being considered.

INTEROPERABLE INTERFACE

Protocols

The Forum has selected a protocol stack, based on accepted standards and the OSI seven layer model. These protocols are defined in detail in [3,9] and in the applicable standards documents. Two network/data link/physical substacks have been designated to date: X.25 (WAN) and the Institute of Electrical and Electronics Engineers (IEEE) 802.3 (LAN). Other network layers are being considered, with the expectation that they will be able to interoperate at the application layer, given there is a common transport layer.

Application Layer Services and Protocols

The structure of the application layer depicted in Figure 11 includes general service elements defined by ISO and CCITT. The Association Control Service Element (ACSE) is used to set up and release associations between CMEs. The Remote Operations Service Element (ROSE) is used to remotely invoke operations and receive correlated responses. The Common Management Information Service Element (CMISE) is the basis for exchanging management data between CMEs. It provides the mechanism for creating and deleting objects, manipulating attributes, and processing notifications.

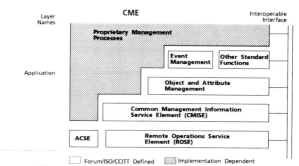

Figure 11. Application Layer Protocols

The Forum Application Services document [4] defines generic object and attribute management functions based on CMIS. Above those is a set of generic event management functions. Proprietary management processes in each CME will then use the event management, object management, and attribute management functions plus other standardized management functions to provide the full interoperable interface.

Configuration Management

Using CMIS, the Forum has defined functions that support basic configuration management. These functions are dependent on the role being played by each CME.

When operating in the manager role, a CME can request that managed objects be created and deleted, Relationships between managed objects can be established or modified by operating on various relationship attributes. For example, the circuit managed object class has two attributes, aEndpointName, and zEndpointName, to identify the two ends of a circuit. A managing CME can set those attributes through the interoperable interface to reconfigure the one or both ends of a circuit. Other attributes carry configuration information, which can be inspected and manipulated across the interoperable interface.

When operating in the agent role, a CME is expected to be able accept and process requests from the managing CME, such as create or delete a managed object. In addition, events may be reported that have configuration significance, such as the automatic or manual creation of a managed object. Changes to attributes, including those that carry relationship or configuration information, may be reported, even if they are caused by a local operator, or some other stimulus.

Fault Management

The Forum has also defined functions that support fault management. These functions are dependent on the role being played by each CME.

When operating in the manager role, a CME can request that events be reported immediately through the interoperable interface, or captured on a log, for later retrieval. It is also expected to be able to accept unsolicited event messages, such as various alarms.

When operating in the agent role, a CME reports events, such as transmission alarms, according to the criteria set up by the managing CME. It will also report changes to attributes, such as the operational state, that have significance for alarm management.

Generic Event Management

The Forum Application Services document [4] defines generic event management services as shown in Figure 12. This definition introduces two kinds of managed object classes, or "support objects" -- event reporting sieves and event logs.

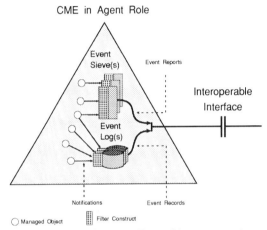

Figure 12. Generic Event Management

An event reporting sieve acts as a notification filter for each object and allows a remote CME to control exactly which notifications from which objects it wishes to receive. Filtering can be done by managed object class, object instances, notification type, and so forth. Services provide for starting, stopping, suspending, resuming, modifying, and reporting on sieves.

An event log is a repository on an agent CME that a managing CME can use to store events. As with a sieve, filtering of events to log can be by managed object class, instance, notification type, and so forth. Services provide for starting, stopping, suspending, resuming, modifying, and reporting on a log process, as well as retrieving entries (event records) from an event log.

Both event reporting sieves and event logs have a filtering mechanism. Through the interoperable interface, a remote CME can define a filter so that only certain notifications are accepted. For example, a filter can specify that only enrol and attribute change notifications from managed object classes x, y, and z, are to be accepted.

The figure shows a subtle but important terminology distinction. Managed object instances emit *notifications.* Notifications are always automatically available to go into event reporting sieves and/or event logs. Those *notifications* that pass through an event reporting sieve are sent across the interoperable interface as **event reports;** those that go into an event log become **event record** managed objects. The same notification can go into many filters and event logs.

As seen by the interoperable interface, the control of event reporting sieves and event logs is the same as for any other managed object. They are, for example, created with CMIS M-CREATE requests and modified by using CMIS M-SET requests.

Example

Figure 13 shows an example of two CMEs exchanging information about a circuit. The circuit managed object class is defined with administrative state and operational state attributes. When the administrative state is "unlocked", use of the circuit is allowed. Similarly, its operational state is "disabled" when it cannot be used because of some failure; and "busy" when it is in use.

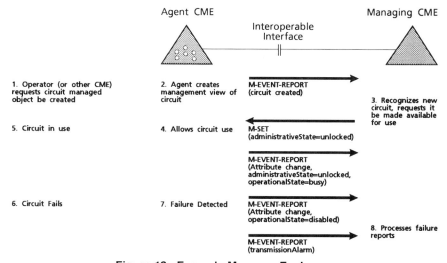

Figure 13. Example Message Exchange

1. Some process (operator command, etc.) requests the creation of a circuit managed object instance.

2. The CME acting in the agent role creates a new instance of a circuit managed object. An M-EVENT-REPORT (ENROL) message is sent to the managing CME, indicating that a circuit managed object was created.

3. In order to allow the circuit to be used, the managing CME issues an M-SET request, specifying that the administrative state attribute is to be changed to "unlocked" (assuming the circuit was created in the "locked" state).

4. The agent CME allows the circuit to be used. Since the value of an attribute changed, it issues an M-EVENT-REPORT request (ATTRIBUTE CHANGE), specifying that the administrative state attribute changed to "unlocked". It is assumed that circuit immediately goes into use, so the same message is used to indicate that the operational state has changed to "busy".

5. The circuit is in normal use.

6. Some time later, the circuit fails.

7. The agent CME issues an M-EVENT-REPORT, specifying a transmission alarm, with details of the failure. Since the circuit is considered unusable, the agent CME issues another M-EVENT-REPORT request (ATTRIBUTE CHANGE), specifying that the operational state attribute changed to "disabled".

8. The managing CME receives the M-EVENT-REPORTs, and processes them as necessary for corrective action or presentation to a network operator.

CURRENT AND FUTURE FORUM WORK

Work is underway to define new Forum releases, by extending specifications, in the following areas:

- Conformance testing - tools to verify conformance to Forum specifications.
- Interworking - allow interworking between various different protocol stacks.
- Testing management - initiate, control, and analyze diagnostic tests.
- Performance management - measure and manage performance.
- Trouble ticketing - exchange trouble reports (tickets).
- Usage records - collect and exchange usage data.
- Security of management - provide security for the interoperable interface.
- Path tracing - trace the source of problems.
- Standards alignment - provide migration strategies to international standards to allow coexistence and backwards compatibility.

SUMMARY

The Forum network management concept is based on the seven layer OSI protocol stack; an object-oriented paradigm; conformant management entities; the exchange of standard messages about managed object instances using a standard protocol; and open global registration. A formal language, is used to define the managed object classes, attributes, messages, and so forth. The Forum has defined protocols and generic messages and an initial set of managed object classes. Forum Release 1 provides a basic infrastructure for interoperability, with a focus on alarm management, and initial configuration management.

REFERENCES

1. OSI/Network Management Forum: Forum Glossary, Issue 1, FORUM 005, January 1990.
2. OSI/Network Management Forum: Forum Architecture, Issue 1, FORUM 004, January 1990.
3. OSI/Network Management Forum: Protocol Specification, Issue 1, FORUM 001, January 1989.
4. OSI/Network Management Forum: Application Services, Issue 1.1, FORUM 002, June 1990. [Replaces Issue 1]
5. OSI/Network Management Forum: Object Specification Framework, Issue 1, FORUM 003, September 1989.
6. OSI/Network Management Forum: Forum Library of Managed Object Classes, Name Bindings and Attributes, Issue 1.1, FORUM 006, June 1990. [Replaces Issue 1]
7. OSI/Network Management Forum: Managed Object Naming and Addressing, Issue 1, FORUM 007, June 1990.
8. OSI/Network Management Forum: Shared Management Knowledge, Issue 1, FORUM 009, June 1990.
9. OSI/Network Management Forum: Forum Release 1 Conformance Requirements, Issue 1, FORUM 008, June 1990.
10. OSI/Network Management Forum: Protocol Specification - Addendum 1 to Issue 1, FORUM 001, June 1990.
11. ISO/IEC 9595 : 1990, *Information Processing Systems - Open Systems Interconnection - Common Management Information Service Definition, including Addendum 1 and Addendum 2.* ("CMIS")
12. ISO/IEC 9596 : 1990, *Information Processing Systems - Open Systems Interconnection - Common Management Information Protocol Specification, including Addendum 1 and Addendum 2.* ("CMIP")
13. ISO 8824 : 1987 with DAD1, *Information Processing Systems - Open Systems Interconnection - Specification of Abstract Syntax Notation One (ASN.1)*
14. ISO 8825: 1987 with DAD1, *Information Processing Systems - Open Systems Interconnection - Specification of Basic Encoding Rules for Abstract Syntax Notation One (ASN.1)*
15. OSI/Network Management Forum: J-Team Technical Report on Modelling Principles, Issue 1.0, Draft 6 [May 1990]
16. ISO DIS 10165-4 : June 1990, *Information Processing Systems - Open Systems Interconnection - Structure of Management Information, Part 4: Guidelines for the Definition of Managed Objects.* ("GDMO")

Jock Embry is a founder of Opening Technologies, a company established to support the implementation of open systems. Mr. Embry serves as a technical representative to the OSI/NM Forum for Unisys and is currently chair of the Forum Architecture team. Prior to that, he was a senior staff member at MCI, developing architectures for network management. He has been a senior manager at MCI and Satellite Business Systems, developing network management systems for large world wide networks.

Peter Manson is a member of scientific staff at Bell-Northern Research Ltd., currently on an assignment with Bell Canada, a member of Telecom Canada. He represents Telecom Canada in the OSI/NM Forum, and acted as editor of the Forum Architecture document. Mr. Manson's responsibilities include the development and analysis of emerging network management standards in ISO, the OSI/NM Forum, and elsewhere, and analysis of how these standards could be applied to the needs of Telecom Canada member companies. Prior to his position at BNR, he developed networking software for Xios Systems Corporation. In 1986, Mr. Manson earned a Bachelor of Mathematics degree in Computer Science from the University of Waterloo.

Dave Milham is a Section Head in British Telecom, responsible for the development of OSI Management Standards and Architecture. He headed the OSI/NM Forum Architecture Team from its inception in November 1988 through early 1990. He has had considerable experience within British Telecom, in the development of switching systems, signalling systems, packet switching systems and office automation products.

AN APPLICATION OF OSI SYSTEMS MANAGEMENT TO INTELLIGENT NETWORK SERVICES

A. Tanaka, N. Matsumoto and K. Morino

NTT Network Systems Development Center
1-2-1, Uchisaiwai-cho, Chiyoda-ku, Tokyo 100, Japan

Abstract
   This paper discusses the applicability of OSI Systems Management to intelligent network services, and presents a service management model, based on the concepts of OSI Systems Management, and demonstrates how service management can be integrated with network management.

1. INTRODUCTION

   Network management systems have evolved over the years from individual systems designed for a particular type of network element, e.g., telephone switches and transmission equipment, or for a particular management function, e.g., alarm surveillance and traffic management, toward network management systems capable of integrating a set of distributed systems. The main objective of the integration is to provide timely accurate operation information for customers. One of the key steps to realizing the integration of management information from distributed management systems is through the standardization of management information. Both OSI Systems Management [1-4] and Telecommunications Management Network (TMN) [5] provide a basis upon which integrated management applications are implemented.
   At the same time, public networks have been providing new services, e.g., Free Dial Service and Dial Q2 Service, to fulfill the requirements from both public and private business enterprises. In these supplementary telecommunication services, the customers, for example, define the members of a group, modify a schedule of physical destination telephone numbers, and get traffic statistics. In other words, the customers can design their own logical networks on a physical network. Some of the service management information is based on network management information. The concepts of OSI Systems Management can, therefore, be applied to service management.
   This paper discusses the applicability of OSI Systems Management to intelligent network services, presents a service management model, and demonstrates how it can be integrated with network management.

2. CURRENT STATUS OF SUPPLEMENTARY SERVICES IN JAPAN

The major supplementary telecommunication services now in service and under development are summarized below. These are provided by NTT based on the architecture of the Intelligent Network.

2.1 Free Dial Service

This service [6] is similar to 800 Service in the United States and the number of contracts is 150,000 as of November 1990. The major contractors are department stores and large retail stores. A contractor is given logical telephone numbers, Advanced Network Numbers (AN), in addition to his physical telephone numbers. When a caller dials 0120 identifying that the call requests Free Dial Service, followed by an AN, the call is connected to the contractor.

Also provided are optional services such as Call Forwarding Service where the physical destination number is changed according to the schedule defined by the contractor, Call Allocation Service where physical destination numbers are selected according to the ratio defined by the contractor, and Line Control Service where the number of active physical destination lines is administratively modified.

2.2 Dial Q2 Service

This service provides a mechanism to collect the billing data for information providers, who provide, for example, news service, weather forecasting service, voice mail service, telephone consulting service and teleconference service. The number of contracts is 2,000 as of November 1990.

2.3 Virtual Private Network Service

This service, under authorization process, defines virtual private networks on a physical network. A member of a contractor makes a telephone call with only an extension number, and the contractor may change the extension numbers. Also provided are optional services such as Call Forwarding, Screening of Originating Calls and Screening of Terminating Calls.

2.4 ISDN Supplementary Services

Two of the ISDN Supplementary Services defined by CCITT Recommendations, Closed User Group (CUG) and Private Numbering Plan (PNP) services are under contemplation.

The CUG service enables a contractor to define groups and the members of each group to restrict access from and to other groups. A member may belong to up to and including 8 groups. A member may or may not originate calls beyond the group, and may or may not receive calls from other groups depending upon the conditions defined by the contractor. Members of a group may be restricted to communicate with each other within the group.

The PNP service enables a member of a group to originate a call to another member of the same group with only an extension number defined by the contractor.

3. FUNCTIONAL MODEL FOR PROVIDING SUPPLEMENTARY SERVICES

These supplementary services are provided based on the following model of Intelligent Network.

3.1 Major Functions of Intelligent Network
The major functions of the Intelligent Network are summarized below.

(a) Identification of service calls

On receiving a call request, the network distinguishes a call requesting a supplementary service from a basic call. If the call requests a supplementary service, then, the network identifies the type of the requested supplementary service. When the subscriber's lines are dedicated to a specific supplementary service, all the calls from these lines are regarded as calls requesting the specific service. When the subscriber's lines are not dedicated to a specific service, identification numbers are to be dialed, for example, 0120 for Free Dial Service, and 0990 for Dial Q2 Service. This function is executed by local switches.

(b) Translation of numbers

In many supplementary services, members are given logical numbers. A caller dials the logical numbers instead of the physical numbers. In the case of Free Dial Service, the logical number is the Advanced Network Number. The contractor may change the schedule of the physical destination numbers through the control terminal in the customer's premises. The second function of the network is the translation of the logical numbers to the physical numbers, which is executed by the higher layer of the Intelligent Network.

(c) Collection of additional information

In addition to the service identification numbers and the logical numbers, the originators' physical numbers, and, in some cases, the passwords are necessary. The collection of the above information and billing data is executed by local switches.

(d) Application of connecting conditions

From the view point of connection establishment, the supplementary services define the conditions of the establishment. The fourth and biggest function is to store the conditions in the network, to modify the conditions through the control terminals, to answer the requests from the control terminals, and to process traffic data. This function is executed by the higher layer of the Intelligent Network. For example, Call Forwarding Service specifies the conditions of the physical destination numbers which are decided according to a given schedule. Screening of Originating and Terminating Calls specifies the conditions of the rejection.

3.2 Architecture of Intelligent Network

The Intelligent Network consists of the transport layer, which provides Service Access Points (SAP) with the function of service call identification, in addition to the functions of traditional public networks, and the intelligent layer, which provides the functions of service management and control [7]. Figure 1 shows the configuration of the Intelligent Network, which is applied to the supplementary services described in Section 2.

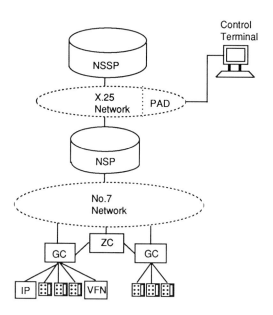

Figure 1. Intelligent network configuration

The traditional public network consists of local switches called Group-unit Center (GC), which provides the SAP functions, transit switches called Zone Center (ZC), transit trunks, subscriber's lines, and other devices. The devices are telephone sets, and Vender Feature Nodes (VFN) such as voice storage equipment, audio conference equipment, and database systems. These equipment or systems controlled by the Network Service Control Point (NSP) are called Intelligent Peripherals (IP).

The intelligent layer consists of NSP and Network Service Support Point (NSSP). The NSP receives the formulated conditions from the NSSP, translates numbers, makes traffic data to be sent to the NSSP. The NSSP receives the requests from the control terminals, formulates the related conditions, processes the traffic data, and answers the requests.

The control terminals are personal computers or work stations, located in the customer's premises or in the telephone offices. They have communication functions with an NSSP and with humans.

4. ANALYSIS OF OSI SYSTEMS MANAGEMENT FROM SERVICE MANAGEMENT PERSPECTIVE

OSI Systems Management standards, designed for network management, provide mechanisms for the monitoring, control and coordination of resources [2-4]. The OSI protocol standards provide mechanisms for communicating information pertinent to those resources, which are viewed as managed objects with defined properties [1].

OSI Systems Management can be applied not only to network management but also to service management. The key OSI Systems Management concepts are analyzed from the service management viewpoint.

4.1 Managers and Agents

The interactions which take place between management application entities are abstracted in terms of management operations and notifications issued by one entity to the other, a manager and an agent, as shown in Figure 2. An agent performs management operations on managed objects as a consequence of management operations communicated from a manager. An agent may also forward notifications emitted by managed objects to a manager.

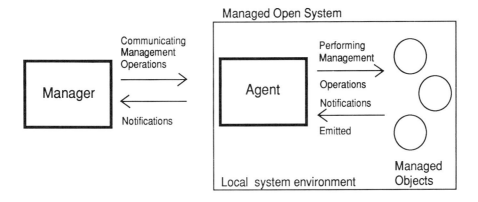

Figure 2. Manager, agent and object interactions

In the case of service management, a control terminal will play the role of a manager and an NSSP in an Intelligent Network will play the role of an agent. The management information maintained in the NSSP includes information about telephone sets. It, however, excludes information about switching systems because the customers are not concerned with the switching systems. In turn, the NSSP will play the role of a manager and an NSP will play the role of an agent. The management information maintained in the NSP includes information about the switching systems.

4.2 Managed Objects and Attributes

A managed object is the management view of a resource that is subject to management, such as a layer entity, a connection or an item of physical communication equipment.

A named set of managed objects sharing the same set of attributes, notifications and management operations is a managed object class. A managed object class that is a specialization of another class is known as a subclass of that class (its superclass). Managed object classes are arranged hierarchically where the hierarchy is organized on the basis of the class specialization.

A managed object of one class can contain other managed objects of the same or different classes. This relationship is called containment. Contained managed objects are known as subordinate managed objects. The containing managed object is called the superior managed object of its subordinate managed objects. The containment relationship is used for naming managed objects. A naming tree is a hierarchical arrangement of managed objects where the hierarchy is organized on the basis of the containment relationship.

Attributes are properties of managed objects. An attribute has an associated value, which may have a simple or complex structure.

In the case of service management, some of the managed object classes will be the same as those for network management, for example, log, traffic record and physical telephone set. Other managed object classes, e.g., logical telephone set, are specific to service management. The same discussion is applied also to attributes. For example, physical telephone number and user's name of the telephone set are common to both network management and service management. Logical telephone number is specific to service management.

4.3 Operations and Notifications

OSI Systems Management provides two kinds of management operations; those which can be sent to a managed object to be applied to its attributes, and those which apply to the managed object as a whole. A managed system may be requested to perform an operation across several managed objects.

The following operations can be sent to a managed object to be applied to its attributes:
  Get attribute value,
  Replace attribute value, Set-to-default value,
  Add member, Remove member.

The following management operations apply to managed objects as a whole:
  Create, Delete, Action.

Managed objects emit notifications when some internal or external event occurs. Whether or not notifications are transmitted externally in protocol, or logged, depends on the management configuration of the open system.

Service management can be modeled through operations on managed objects and attributes, and therefore the management operations described above are sufficient for and can be applied to service management.

5. SERVICE MANAGEMENT DESIGN BASED ON OSI CONCEPTS

5.1 Network Configuration for Service Control

The network configuration model of a TMN defined by a CCITT Recommendation [5] and that for service management will be combined into one network configuration model as illustrated in Figure 3.

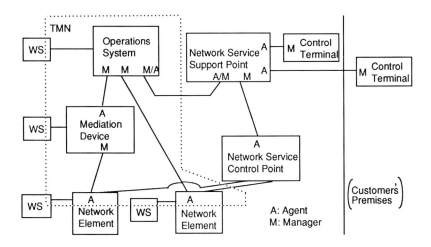

Figure 3. Integrated network configuration model for network management and service management

Network management information, for example, can be retrieved by the NSSP through the Operations System.

The information aspect of service management is discussed and the model of the service management interface between the Network Service Support Point (NSSP) and the control terminal is outlined in the following subsections.

5.2 Functional Requirements to Service Management

A set of new values of control information for a new customer, which is generated at a control terminal in a telephone office, is created and stored in an NSSP. The set of stored control information for a canceling customer will be deleted from the NSSP, the request of which is issued by the control terminal. These requirements are met by the Create and Delete operations. A customer may retrieve the values of some control information, modify the values of control information, e.g., a schedule of physical destination numbers, and add/remove the values of control information, e.g., the members of a group. These requirements are met by the management operations applied to the attributes.

There are other requirements to service management. A customer may request traffic statistics, e.g., the mean duration time of calls and the total number of calls over a specified period of time, which are calculated from past traffic records. The activation of a set of control information may be requested in advance, where the delayed operations will be performed at specified time in future. These requirements are met by defining managed objects, which are described below.

5.3 Managed Objects and Operations/Notifications

First of all, there is a need to distinguish between two kinds of managed objects in order to secure the extensibility of relevant software products. One of them is common to several services and/or represents physical resources. The other is specific to each service and does not represent physical resources.

For the supplementary services, the common managed object classes are Physical network, Analog telephone set, Digital telephone set, and Control terminal. The specific managed object classes are Logical network and Logical telephone set for one of the supplementary services.

The managed object instances of Logical network, Logical telephone set and other contained managed objects for a specific service are created in the NSSP when a contract is made. These object instances are deleted when the contract is canceled.

Secondly, each control function may require other types of managed object classes depending on the necessary functionality.

The Traffic summarization object observes the attribute values of traffic records and calculates the mean duration time of calls and so on. The conditions of the observation and calculation are specified by the attributes of the summarization object or by the parameters of an operation communicated from the manager in a control terminal. The outputs of the Traffic summarization object are emitted as notifications to the manager, according to the schedule or upon the completion of the calculation, and stored in the Log object as Traffic summary record objects. The attribute values of the Traffic raw record object, the information source of which is a switching system, and the Traffic summary record object are observed by the Traffic summarization object.

The managed object class of Delayed performer is introduce to fulfill the requirements of the delayed operation. It creates a delayed request object, e.g., Delayed create request object and Delayed delete request object, when it receives a request communicated from the manager in a control terminal. The attributes of the delayed request objects include Performing time. The Delayed operation performer periodically compares Performing time with current time, and will perform the specified operation when they are equal. It deletes the delayed request object upon the completion of the performance.

Some other managed object classes including Call forwarding scheduler object and security related objects are also defined.

Figure 4 illustrates the major part of inheritance hierarchy for the supplementary services. Figure 5 illustrates the naming tree for the managed objects in Figure 4.

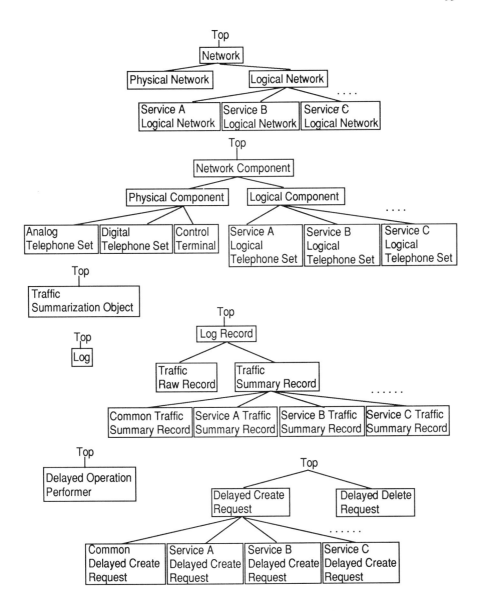

Figure 4. Inheritance hierarchy for supplementary services

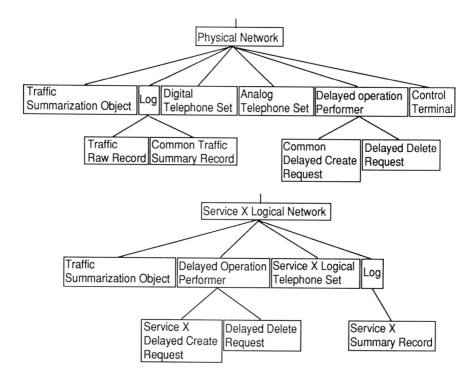

Figure 5. Naming tree for supplementary services

These network management/service management object classes and attributes are planned to be implemented in future software product releases.

6. CONCLUSIONS

We have confidence that OSI Systems Management can be applied not only to network management but also to service management with the additional definitions of managed object classes, attributes types, notification types and actions types. Applying OSI Systems Management to both network management and service management will accelerate their integration, which will make network operations more efficient, enhance the utility value of networks, and create new services.

There are still issues to be studied further. One of them is the timely extension of the definition of the managed object classes because new service control requirements will be received from customers and operators in the future. Another issue is how to encourage network management/service management designers to understand and use the concepts of OSI Systems Management because not all the designers are familiar with them.

ACKNOWLEDGEMENT

The authors wish to thank Mr. H. Iizuka and Mr. M. Koyama for giving this opportunity, and Mr. H. Itoh and Mr. S. Kayaguchi for technical discussions.

REFERENCES

[1] ISO/IEC 9595 and 9596, Open Systems Interconnection - Common Management Information Service Definition and Protocol Specification
[2] ISO/IEC DIS 10040, Open Systems Interconnection - Systems Management Overview
[3] ISO/IEC DIS 10165, Open Systems Interconnection - Structure of Management Information
[4] ISO/IEC DIS 10164, Open Systems Interconnection - Systems Management
[5] CCITT Recommendation M.30, Principles for a Telecommunications Management Network
[6] S. Amiya and H. Uesaka: "Making Start in Intelligent Network Architecture," NTT Review, Vol. 2, No.2, 1990
[7] H. Itoh and J.Kuwabayashi:"Intelligent Network Management for Customer Controlled Services," ICC '89, 1989

DECnet/OSI Phase V Network Management

Mark Sylor
Digital Equipment Corperation, 550 King St. LKG2-2/W10, Littleton, MA 01460-1289

DECnet/OSI Phase V is the latest phase of the Digital Networking Architecture. Phase V incorporates a new Network Management Architecture based on DEC's Enterprise Management Architecture (EMA). This paper gives an overview of some of the key features and functions of Phase V Network Management, placing special emphasis on the lessons learned during the development of the architecture. Because Phase V Network Management is closely based on OSI System Management, these lessons, and the architected solutions to problems developed in Phase V will be of general interest

1 Introduction

DECnet/OSI Phase V is the latest phase of the Digital Networking Architecture (DNA). It expands DECnet to include OSI protocols while remaining backward compatible with earlier phases of DECnet.

DNA has had network management capabilities ever since Phase II (1976) [1]. As we began working on Phase V, it was clear that changes in the Network Management Architecture were needed. Primarily Network Management had to be **extensible**. We expected new subsystems would be added, such as FDDI LANs and TCP/IP support, moreover, we wanted to manage things making up the network other than DECnet nodes (for example bridges and terminal servers). Simultaneously we wished to manage these components in a **consistent** fashion.

While Phase IV management could have been extended to solve these problems, we felt it was time to revisit the basic Network Management architecture and see if there was a better approach.

The result of that re-examination was an overall structure that divided the network management system into two fundamental kinds of components, *entities*, the basic components of the network that had to be managed, and *directors*, the software systems used by managers to manage the entities. Directors and entities would communicate with each other using a *management protocol*, so the entities could be remotely managed. Functions common to a group of entities are provided by an *agent*. This division, along with a description of how the management interface to an entity would be defined was specified in the Entity Model[2]. A summary of the Entity Model can be found in [3].

Essentially, the Entity Model defines an object oriented view of the entities where an entity is defined to have:
- A hierarchically structured name

- A collection of operations that can be performed on that entity, including a set of attributes that can be read or modified, a set of actions that can be performed on the entity, and a set of events the entity might report to interested directors.
- A specification of the behavior of the entity in relationship to the service the entity provides to the network. This is usually specified as some abstract state machine or through pseudo-code.

Readers familiar with OSI System Management will find the Entity Model very similar to OSI's Structure of Management Information, SMI[4,5]. This is not a coincidence.

2 Entities

The entity model described *how* to specify the management of an architected subsystem, but left the actual work to other specifications. Of the various Phase V specifications, The Network Management Specification[6] is central to Phase V Management.

2.1 The Node

The obvious entity to begin with is a single computer system, called a Node, in the DECnet/OSI network. A node is a computer system, where the bounds of that system are kept suitably vague. A single processor system, a multiprocessor system, a system with or without disks, even a VAXcluster^{TM} can be considered to be a single node.

The node entity has only a few functions in management. First and foremost, it is a global entity which serves as the parent entity for a large number of modules (and their children). Second, a node has an identity, a name and an address that allow it to be managed remotely. Third, a node plays a major role in system initialization and start-up.

A node has an identifier attribute called its Address. This address is actually the application layer address(es) of the node's agent. DECnet/OSI supports multiple protocols at any of the seven layers, and it allows a protocol in one layer to operate over more than one protocol at the next layer down. For example, DNA Session Control can operate over either the Phase IV compatible NSP or the OSI Transport TP4 protocol at the transport layer. Each protocol has its own addressing conventions. Thus a node address is actually a set of protocol towers, where each tower defines a sequence of protocols, each with its associated addressing information. A protocol tower provides all the information needed by a director to connect to the node's agent, and issue management directives to the node or any of its children.

Users and network managers generally don't refer to nodes by their address. Not only are they hard to remember, moving the node from one place in the network to another generally changes the address. Thus each node has a user friendly identifying attribute called Name, a DECdns fullname. DECdns is DECnet/OSI's distributed name server [see below and 7]. The node knows its name and address. Each node's name is stored as a DNS entry, and one of the entry's DNS attributes holds the node's address. Thus, any director can look up the node's name in the DNS and look up the address associated with it, and use any one of the towers to connect to the node's agent.

For backward compatibility with DECnet Phase IV, a node also has an attribute called its Synonym, which is a six character, Phase IV style node name. If a node has a synonym name, that name is entered in a special directory in the DNS name space as a soft link to the node's Phase V name. A *soft link* is a form of alias or indirect pointer, from one name to another that allows an entry to be reached by more than one name.

Each network layer address of the node (a node can have more than one) is encoded in a standard way as a soft link to the node's name. This allows a manager (or director) to *back-translate* a node address into the equivalent node name, making many diagnostic problems much simpler.

DECnet/OSI includes many features that allow most nodes to autoconfigure their address. Network layer addresses consist of an area address and a 48-bit ID. This ID can be obtained from a ID ROM chip on many devices (for example each Digital 802.3 LAN device has one). Area addresses can be learned by end nodes listening to the messages sent by the routers adjacent to the end node. Higher level addresses used by management are architecturally defined constants.

Names of nodes are chosen by the managers and users of the network. The manager tells the node its name using the Rename action. The argument of Rename is the new name of the node. Rename is a good example of a case where an action is a more appropriate operation to express a function than a Set operation. Renaming a node is a fairly complicated operation. Not only is the Name attribute changed, but the appropriate information must be stored in the right place in the DNS name space. Along the way, there are many places where the operation can fail. Actions allow those errors to be reported back to the manager together with enough detail on what went wrong so that the manager can decide what corrective action to take. This is difficult to achieve with a simple Set operation.

2.1.1 Identity

One of the more difficult configuration problems to track down is when two nodes in a network have either the same name or the same address. DECnet/OSI has a number of management features to prevent this from occurring, or to detect it when it does occur. Each node has a unique 48-bit ID. This ID is presented as an attribute of the node. If a node should have more than one ID ROM, for example a router with 2 Ethernet interfaces, then a simple algorithm chooses one of them to be the node's ID. The ID is *spatially unique*. Special manufacturing procedures ensure that no two ID ROMs have the same ID in them. However, an ID is not tied to the same node uniquely over time. Hardware devices can be removed from one machine, and inserted in another. Indeed, this is a common diagnostic procedure.

A space and time unique value that can be associated with a node is provided by a UID (Unique identifier) service. UIDs combine a spatially unique ID with a timestamp in such a way that no two generated UIDs will ever have the same value [8]. The UID is stored in non-volatile storage (if the node has some) so the UID remains constant across system reboots. Nodes without non-volatile storage will generate a new UID on every reboot.

When any of the four attributes (Name, Address, ID or UID) is changed, an event is reported by the Node indicating what has changed. For example, a common way in which a node is duplicated on the network is to copy the disk where a node stores its system image, name, address and UID, and then boot the copy on some other machine. When the disk is booted on the second machine, it would see that the node's ID had changed, and an event would be generated. While this would not *prevent* the duplicate node from booting, it would allow the manager to *detect* that a duplicate node may be on the network.

I am often asked, "Which of these four attributes *really* identifies the node?" The answer is "None of them, silly!" Under various scenarios, any (or even all of them) could be changed, and yet the node would remain "the same". For this reason we have said, "A node is a named set of user expectations."

2.1.2 Start-up

A node is responsible (with the hardware) for system start-up and booting. We have defined four node states: Dead, where the node is down, Booting, when the node is in the initial stages of software start-up, Off, where the node is initializing itself and its internal configuration, and On, when the node has "completed" enough initialization so it can at least be managed remotely.

Booting is a transient state, and one that is highly system specific. It is initiated by a hardware, software, or power failure, or by a console request. The main task of booting is to load a *system image*, start it running, and get it to a known state. If the system has a disk, or equivalent storage, it can be booted from there. If the system lacks those, it can be booted across the network using the MOP Down Line Load protocol[9]. MOP is layered directly over the Data Link protocols, and thus a node can only be booted from an adjacent node. Yet it is possible to boot an entire network through a *rolling boot* by booting successive systems outward from the management system. MOP is generally implemented as part of the hardware or firmware in microcode, and thus does not require a working operating system.

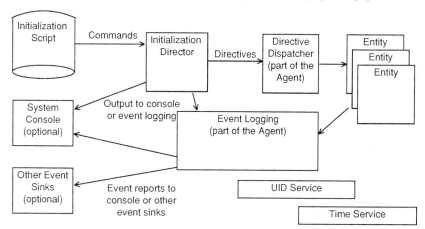

Figure 1 The Node at the Big Bang

When Booting completes, the node changes to the Off state. This transition is called the *big bang*. In the first instant after the big bang the node has at least the following things available, as shown in figure 1:

- a working clock and time service
- a UID generator
- the node entity (and possibly some of the node's child entities), together with its agent (which includes both the directive dispatcher and event logging).
- an *initialization script*, a series of management commands to configure the system. This can be in the form of a text NCL command file (see below), or it can be a *compiled*

script, one that has been encoded as a series of CMIP requests. MOP can be used to Down Line Load an initialization script.
- an *initialization director*, which reads the script and invokes the directives in the order given. Errors and other output may be displayed on a console (if the system has one) and/or reported as events.

Somewhere in the initialization script (probably near the end) the node is Enabled, which changes its state to On, i.e. it can be managed remotely.

2.2 Modules

While the node models a computer system as a whole, a node also has many subsystems that may or may not be configured within it. The DECnet/OSI subsystems are organized into the seven OSI layers. More than one subsystem is allowed within a layer. For example, in the data link layer both the CSMA/CD subsystem (for 802.3 LANs) and the HDLC subsystem (for serial data links) are supported. In Phase V management, a subsystem, or closely related group of subsystems, is called a module. A node never has more than one instance of a module contained within it. The manager can create and delete modules. This allows the manager a great deal of flexibility in configuring the node to serve a particular purpose.

In DECnet/OSI Phase V, the specification of the management of each module is an integral part of the architecture of the subsystem. Responsibility for defining the module's management interface was moved from the network management architect to the subsystem architects.

Placing responsibility for the management of a subsystem in the architecture of that subsystem has made the specifications clearer and more complete. In Phase IV, the implementors had to read two specifications, the specification of the subsystem they were implementing, and the network management specification to understand how management and the subsystem interacted. For example, both specs had to be read to understand exactly what a counter counted, or exactly what a characteristic attribute (such as Buffer Size) really meant. This led to many problems as the two specifications could be inconsistent or incomplete. Merging the two, and placing responsibility in one person's hands made writing an internally consistent subsystem much easier.

The sheer size of DECnet/OSI Phase V management would have made it impossible for a single person to design the management of the whole system. To give you some idea of the magnitude of the effort, there were at a recent count 24 different modules that could be in a node, and this number is increasing as new subsystems are designed. Those modules defined 91 different classes of entities with 1250 attributes, 172 events and 231 actions. The size of Phase V management has both good and bad points. On the one hand, the manager has a great deal of detailed control and monitoring capabilities. But on the other hand, the sheer size of those capabilities will be initially daunting. Fortunately for the manager, most of the characteristic attributes (the primary means of controlling a node) have default values which were chosen to work for most nodes in most networks. Moreover, each kind of node is shipped with a default initialization script that initializes the node into a typical configuration. Only a few values, mainly concerned with naming and addressing that cannot be autoconfigured, must be chosen by a manager.

3 Folklore and Guidelines

As the entities for DECnet/OSI Phase V were defined, a collection of *folklore* grew on how typical design issues could or should be solved. As with any folklore, these guidelines were passed from one architect to another, either verbally, or as selected portions of the management specifications were copied from one subsystem to another. This folklore is continually changing, as new, better solutions are found. Much of the folklore has already been described [10]. Some other guidelines are described below.

When an entity is initialized (in particular when a node is booted) it typically goes through three stages, entity creation, entity configuration and entity enabling.

Creating an entity is kept simple, the only ordering constraint being parent (superior) entities must be created before their children. Usually, the only arguments on the create action select the variant *type* of the entity. For example, when Routing is created, its major type is whether it is a router or an end system. The type of an entity cannot be changed except by deleting the entity and creating a new one.

Once an entity is created, it is configured by modifying its characteristic attributes. Since most characteristics have defaults that will "work", this step is only complex if the manager wishes to configure an entity to meet specialized needs.

When all the entities are properly configured, each entity is enabled. Enabling an entity makes it available to its users. Note the entity is always assumed to be manageable. It also causes the entity to attach to the services it uses. In general, entities should not require the entities be enabled in any particular order. This is usually implemented by having an entity open any services it uses, and then wait for the lower level services to come On. Entities should verify that the lower level services do exist, and have been configured correctly when the entity is enabled.

We model the usage of a layer's services by a higher layer by a *Port* entity. Ports are children of the lower layer module. Ports are not created by the manager, but are created when a user of the service issues an *OpenPort* call, indicating it wishes to use the service. Ports are a generalization of SAPs and CEPs described in the OSI Reference Model. Because a layer can use many lower layer services, and a service can be used by many higher layer services, a port is an example of a relationship entity representing a many-many relationship[10]. It is also an entity in its own right, because a Port typically includes attributes describing how the service is being used e.g., usage counters).

When a client opens a port, there are often options to be selected. Examples of these options include sets of features, standard profiles, quality of service selections, etc. While the OpenPort call allows the client to pass these as arguments, passing all these options binds the client to the service. Changes to the service (new options) require changes to all clients. Worse, these arguments must come from somewhere, either the client implementor must hard code values (obviously a bad idea) or the ultimate end user must supply them. The end user is unlikely to understand all the options, nor even have a complete list. Defaults are one way to solve this, but we felt that a single default was inadequate to the needs. The solution adopted was to have the network manager create a *Template* entity to represent a set of related option values. OSI System Management has a similar concept with its *Initial Value Managed Object* (IVMO). The template is identified by a simple name (like UK-GOSIP-Profile) at each layer. One of the characteristics of a template is the name of the template to use when opening a port on a lower service. The client of a service would pass a template name when it opens a port, and any parameters it wishes to override. The service creates the Port, and uses the options in the template entity to set up the port's initial state. Thus an end user can specify a complete set of options for all the layers with a single, simple parameter, yet the network manager retains control.

4 Supporting Mechanisms

Network Management in DECnet/OSI is supported by a number of services provided as part of DECnet/OSI. Some are obvious, such as Session Control's communications services supporting DNA CMIP. Some, while interesting in their own right, are omitted due to lack of space. Three supporting modules are described below:
- Event Logging
- DNS Name Server
- Application Loopback

4.1 Event Logging

When an event occurs in an entity that a manager may wish to be informed of, the entity emits an event report. The event logging module provides a service that will transmit event reports from the reporting entities to one or more *sink applications* which in EMA are considered to be a certain kind of director. Event logging in Phase V is based on the similar concepts to that provided by Phase IV. Because the main use of event logging is for reporting faults, event logging does not guarantee delivery of event reports to the sink application. Figure 2 shows the event logging architecture[11].

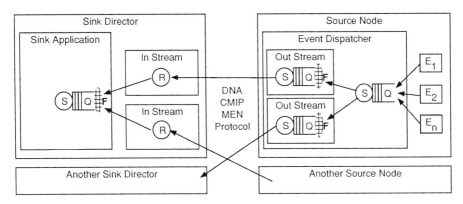

Figure 2 Event Logging

When an event occurs within an entity (E_i) in a *source node* the entity invokes the *PostEvent* service provided by the *event dispatcher* (a part of the node's agent). When posting an event, the entity supplies its name, the type of the event, all the arguments related to the event, a timestamp of when the event occurred, and a UID assigned to the event. UIDs assure that each event can be uniquely identified, so if a sink application should receive more than one copy of an event report, it can detect the duplication. Timestamps allow the event reports to be ordered in time (an important step in determining causality). A time service (DTSS) is used to synchronize clocks across the network. It provides a consistent view of time for correlating observations. An important feature for management is the inclusion of an inaccuracy bound on the timestamp.

The PostEvent service formats an event report and places it in an *event queue* (Q). Event queues are limited in the amount of memory they use, and thus limit the number of events that can be held in the queue. Because it is possible that events can be placed in the queue at a rate faster than the *queue server* (S) can process them, the queue can fill, and any new events placed in the queue will be lost. The Events Lost event is recorded as a *pseudo-event* in the queue (it appears as an event report with the name of the entity holding the queue). An argument of the Events Lost event is the number of events lost, and it is never lost (the count is incremented instead).

The queue server for the event dispatcher compares each event report against a *filter* (F) associated with an *outbound stream*. The filter lists a collection of entities and events that are *passed* through the filter, or which are *blocked* by the filter. Event reports passing the filter are placed in an event queue within the outbound stream. Each outbound stream's queue server sends events to a corresponding *inbound stream* in the sink. Multiple outbound streams can be set up by the manager, and thus events can be sent to many sinks. Outbound streams are modeled as entities in their own right, and standard management operations (Create, Get, Set) are used to configure them.

Each inbound stream in a sink has an *event receiver* (R). Inbound streams are generally created when a connection request is received from an outbound stream. Events received by the receiver are compared against a sink filter and queued to the sink application. Thus the events from multiple inbound streams are merged.

The protocol between the outbound stream and the inbound stream is the CMIP Management Event Notification (MEN) protocol. The protocol operates over a reliable connection (using either NSP or OSI Transport). The connection can be made based either on the DNS name of the sink, or the address of the sink. Unconfirmed event reports are used, as the protocol does not guarantee delivery of events, and the use of a reliable connection means no messages will be lost in the network. Events may still be lost due to failures of the sink or of the source node.

4.2 Name Server

In Phase IV, each node in the network contained a remote node table which mapped node names to node addresses. While a simple concept, in a N node network, each node would have to have N entries in its tables, and thus the network manager had to set up N^2 table entries in the entire network. In large networks this became a significant chore. To ease this the Digital Naming Service (DECdns) was designed[12]. Each node has a DECdns Clerk which talks to a collection of DECdns Servers. The servers store the name to address table. The complete set of names stored in all the name servers is called a *namespace*. For a large network, a single server could not hold the entire namespace (as that would consume far too many resources). The namespace is divided into *directories*, each name server holding only some of them. So that a Server is not a single point of failure, a directory may be *replicated* in multiple servers. The copies are kept loosely consistent by a process called *skulking* the directories.

While DECdns eliminates the need to set up N^2 entries in the network, it introduces a new problem, the management of DECdns. Most nodes are DECdns Clerks. Clerks are quite simple to manage, and it includes many autoconfiguration features. The management of servers is more complex. They must be sited near the Clerks and hold the right directories so look-ups are efficient, yet the number of replicas of a directory should not be excessive so that the skulking operation completes in a timely manner. Thus we have traded a large, time consuming, but relatively simple chore (setting up N^2 table entries) for a smaller but more technically sophisticated task (configuring name servers).

I expect this to be a common situation. New features designed to *simplify* management will actually transform a simple, time consuming, people intensive task into one that does not require large numbers of people but which is more complex (at least in the event of failure). Sophisticated network management features can reduce the number of people needed to support a network, but they may *increase* their required technical sophistication. And they will certainly increase the complexity and sophistication of both the system and the management software.

4.3 Application Loopback

The Application Loopback module provides a simple test of two node's ability to communicate, figure 3. A manager or director on any node (M) in the network, directs a source node (S), to set up a connection to a third destination node (D) and to exchange messages with it over that connection. The Application Loopback Protocol describes the connection establishment rules, and the format of all messages exchanged. The Loop action directive has request arguments that select the destination and control the number and data pattern of the messages exchanged. The response (if the test succeeds) include the start and end time of the test. Any errors are reported as exceptions, and include the reason for the error, the number of messages exchanged, and the start and end times of the test. The application loopback protocol is backward compatible with DECnet Phase III and IV.

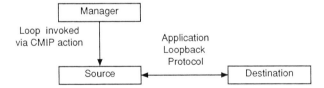

Figure 3 Application Loopback Protocol

5 Directors

In the EMA model, the software used by network and system managers is called a director. Two general kinds of directors were envisioned as being necessary for DECnet/OSI Phase V.

5.1 DECmcc/Director

The first kind of director needed was a high function director that was configurable and extensible. This concept merged and extended many of Digital's previous network management products and is called the DECmcc/Director. The DECmcc/Director has been described elsewhere [12,14].

5.2 NCL, the Director

The second kind of director envisioned was a family of simple, specialized directors, each tailored to do a single job. Among these were various kinds of event sinks, programs that would receive events from the network, and do simple things with them, log them to files, display them on terminals, etc. The other basic director envisioned was a simple command line based director. It would be universally available and would allow a manager to invoke the basic directives (Show, Set, various actions, etc.) on any entity that conformed to the overall EMA DECnet/OSI Management Architecture. This director replaced the NCP program in Phase IV and was called NCL.

5.3 NCL, the Language

NCL (Network Command Language) is used with two quite distinct meanings. It refers to both the command language and the basic director[12]. NCL, the language, was designed to be the command language syntax for both the NCL director and the DECmcc director. That way, users would be able to easily transfer any knowledge they had learned from one director to another. The DECmcc Director includes an iconic, windows oriented user interface and this was expected to be the main interface to it. But it also includes a text terminal oriented user interface following NCL as a back-up so the manager would not be tied to a workstation.

The syntax of NCL is straightforward. A command consists of a verb (which identifies the operation to be performed), the name of an entity (or entities if the name is wildcarded), a list (possibly empty) of arguments, and a list of qualifiers to the command (possibly empty). For example:

NCL> Show Node Capn OSI Transport All Counters, -
    By User Smith, Password secret

where
- Show is the verb, and identifies that the Get operation is to be performed.
- Node Capn OSI Transport is the entity name.
- All Counters is the only argument to this command. For a Show command, the arguments are a list of attributes or attribute groups to be gotten and displayed. In this case, the Counters attribute group has been requested.
- By User Smith, Password secret are the qualifiers to the command. These particular qualifiers provide access control information in the form of a user name and password. Qualifiers are used infrequently, have the same meaning when applied to any command, and are distinguishable from arguments by beginning with a preposition.

The language is English-like, full words are used, cryptic abbreviations and cybercrud are avoided, and the use of punctuation is kept to a minimum. This leads most NCL commands to be verbose. We have eased this somewhat by providing the user with features such as: abbreviation of keywords to their initial unique string, defaults for the entity and qualifiers, and symbol substitution. The goal for NCL was consistency and extensibility to new entities as they were designed, esthetic considerations were secondary (we say in jest, NCL is equally ugly to all users),.

The most important goal of NCL was that it be extensible to new entities as they were defined. Thus the keywords of NCL (the verbs, the entity class names, the argument and attribute names, etc.) are defined by the architect of the entity, not by the NCL implementor. NCL's command parsing and display functions are driven by the contents of a dictionary which contains the keywords, their meaning, i.e., how to map them into CMIP, and information on how the values in the command are encoded. The dictionary is in turn loaded from information provided in a management specification language (MSL) which is a registered description of the entity defined in the DNA architecture. Thus as new EMA or DECnet/OSI entities are defined, or older ones extended, new versions of NCL or its dictionary can be provided that allow them to be managed with NCL (the program) or DECmcc without having to reimplement either director.

6 DNA CMIP

The management protocol for DECnet/OSI Phase V is DNA CMIP. CMIP is an evolution of the Phase IV Management Protocol (called NICE). CMIP includes the Set, Show (a.k.a. Get), and event report operations defined in NICE. The main differences between the two are:
- Treatment of other operations. In NICE each operation required a new kind of message, in CMIP, a general extension mechanism, the Action, is provided.
- Naming. NICE supported a limited number of entities (8) and provided a rudimentary naming hierarchy based on the notion of "qualifying attributes". CMIP supports hierarchical entity names, and is essentially unlimited in the number of entities it can deal with. Similarly CMIP is much more extensible in naming attributes, attribute groups and event reports.
- Encoding. CMIP uses ASN.1 (an ISO standard TLV encoding) while NICE used fixed fields and a private TLV encoding of attributes and arguments.

DNA CMIP is not quite the same as the IS version of OSI CMIP, although it was based on the 2nd DP draft of the CMIP standard. There are two reasons for this:
- First and foremost was one of timing. At the time that we had to freeze the functions expressed in the protocol (and hence the protocol itself), there were a few "bugs" in the ISO draft, and many unanswered questions. For example, a standard way of defining an extensible collection of arguments to an event report has only recently been drafted in OSI's SMI. Rather than wait for all of these questions to be answered, we chose to answer them as best we could. We plan to "migrate" to the standard CMIP and SMI at a later date. Since the fundamental concepts of OSI CMIP and DNA CMIP are closely aligned, this migration is expected to be straightforward (and transparent to any user).
- Secondly, DNA CMIP operates over a DNA protocol stack, not over a pure ISO stack. This allows directors on Phase IV systems to manage Phase V systems.

DNA CMIP can be viewed as two separate protocols. One is used by a Director to invoke a directive (Get, Set, Action, etc.) on an entity (or entities). This sub-protocol is called

Management Information Control Exchange (MICE). The other is used by an entity (or entities) to report events to a director. This Protocol is called Management Event Notification (MEN). The two protocols operate over separate connections. There are a number of reasons for this:
- The times at which the associations are connected differ. A MEN association is brought up when an entity wishes to report an event, and is thus controlled by the agent, whereas a MICE association is brought up when a director (or manager) wishes to invoke an operation on an entity, and thus is controlled by the director. Attempting to share control of association establishment was not worth the complexity.
- Whenever an association is shared by two different users, the problem of fairly allocating resources to the two users must be addressed. Since transport connections deal with this questions between connections, the addition of a multiplexing protocol at the application level (with an attendant flow control mechanism) was again felt to be too complex. After all, transport connections are not (or should not be) expensive.

7 Conclusions

DECnet/OSI Phase V Network Management offers many lessons for architects and developers. Because it is so close to OSI System Management most of these lessons can be directly applied to OSI.

The basic decisions on *how* Phase V management was developed have worked well. Defining the Entity Model first provided a framework of consistency among all of the architectures. Developing a management protocol (CMIP) expressing the basic concepts in the Entity Model in conjunction with the model puts the protocol in its proper place. I.e., the needs of the model have driven the protocol definition, not the other way around. Giving responsibility for defining the management of a subsystem to the architects of that subsystem has made each subsystem more complete and coherent. As problems were found in the model based on lessons learned during the specification of entities, any needed changes to the Entity Model were applied as fixes to those problems.

We were all surprised by the number of entities, attributes and operations in Phase V. Phase V has many more features and functions that Phase IV and that, together with giving the manager more complete control over those features, explains why the size has increased. Use of intelligent defaults, autoconfiguration and most importantly self-management features reduces most management to dealing with specialized requirements. Still, I believe that providing a management system that deals effectively with this scope remains a fruitful area of study.

As we developed EMA-Phase V management we found that it could be applied to a much wider scope of problems than traditional "network management". EMA, the Entity Model, the protocols, and the mechanisms built for Phase V management could be applied equally well to many problems in system and application management [16]. This came to be expressed in a simple slogan.

EMA - It's not just for networks anymore!

8 References

1. N. LaPelle, M. Seger, M. Sylor, "The Evolution of Network Management Products", *Digital Technical Journal*, vol. 1, no 3, Sep 86, pp. 117-28.

2. Digital Equipment Corporation, *EMA Entity Model*, 1991.

3. M. Sylor, "Managing DECnet Phase V: The Entity Model, *IEEE Networks*, Mar 88, p 30ff.

4. ISO/IEC DIS 10165-1, *OSI Management Information Services - Structure of Management Information - Part 1: Management Information Model*.

5. ISO/IEC DIS 10165-4, *OSI Management Information Services - Structure of Management Information - Part 4: Guidelines for the Definition of Managed Objects*.

6. Digital Equipment Corporation, *DNA Network Management Functional Specification*, V5.0.0.

7. Digital Equipment Corporation, *DNA Naming Service Functional Specification*, V2.0.0.

8. Digital Equipment Corporation, *DNA Unique Identifier Functional Specification*, V1.0.0.

9. Digital Equipment Corporation, *DNA Maintenance Operations Protocol Functional Specification*, V4.0.0.

10. M. Sylor, "Guidelines for Structuring Manageable Entities", *Integrated Network Management I*, B. Meandzija and J. Westcott (ed.), Elsevier Science Publishers, 1989, pp. 169-183.

11. Digital Equipment Corporation, *DNA Event Logging Functional Specification*, V1.0.0.

12. S. Martin, J. McCann, D. Oran, "Development of the VAX Distributed Name Service", *Digital Technical Journal*, vol. 1, no 9, Jun 89, pp. 9-15.

13. C. Strutt & D. Shurtleff, "Architecture for an Integrated, Extensible Enterprise Management System", *Integrated Network Management I*, B. Meandzija and J. Westcott (ed.), Elsevier Science Publishers, 1989, pp. 61-72.

14. Digital Equipment Corporation, *DECmcc System Reference Manual*, Order No. EK-DMCC1-RM.

15. Digital Equipment Corporation, *DNA Network Command Language Functional Specification*, V1.0.0.

16. L. Fehskens, "An Architectural Strategy for Enterprise Network Management", *Integrated Network Management I*, B. Meandzija and J. Westcott (ed.), Elsevier Science Publishers, 1989, pp 41-60.

I
STANDARDS
AND ARCHITECTURE

B
Panel

COALESCING MANAGEMENT STANDARDS

Moderator: Dan STOKESBERRY, National Institute of Standards and Technology, USA

ABSTRACT

OSI Management Standards are within the scope of work of many groups within many organizations. The work includes the development of:

(1) International base standards within ISO and CCITT,
(2) Implementation agreements within the OIW, EWOS, & AOW,
(3) Implementation agreements by the NM Forum, and
(4) Conformance tests by COS et al.

There is great interest in the availability of network management products today, but a full set of base standards, implementation agreements, and conformance tests will not be available for several years. Therefore, the standards communities have undertaken a phased approach to the development process. Each group has defined its short-term goals and a work plan to achieve its goals.

There are, however, interdependencies between the work of the various groups and overall success is contingent upon a common, clear understanding of the role and the scope of activity of each group as well as sufficient synchronization of effort and harmonization of work among the groups.

The panel, composed of leaders of these groups, will explore the status of the specific activities of each group and the present degree of cooperation among the groups. They will offer predictions on significant future achievements and/or road blocks to success.

II
MODELING AND DESIGN

Integrated Network Management, II
I. Krishnan & W. Zimmer (Editors)
Elsevier Science Publishers B.V. (North-Holland)
© IFIP, 1991

Formal Description of Network Management Issues[*]

Gregor v. Bochmann[a], Pierre Mondain-Monval[b] and Louis Lecomte[c]

[a,b]Département I.R.O, Université de Montréal, CP 6128, Succursale A, Montréal, Québec, Canada, H3C 3J7

[c]Computer Research Institute of Montreal (CRIM), 3744, rue Jean-Brillant, Bureau 500, Montréal, Québec, Canada, H3T 1P1

Abstract
   This paper presents an object-oriented design methodology and a supporting object-oriented language for the specification of applications in the field of network management. A general object-oriented software design methodology is presented. An example of application of this methodology for the specification of fault management functions in the case of a real, industrial, transmission system is then presented. Last, we present the desired features of an object-oriented language, MONDEL, that we are currently defining for the specification of distributed systems.

1. Introduction

   Appearances of very large scope networks, and the possibility of interconnecting multiple and various networks (telephone networks, private and public packet switching networks, Integrated Services Digital Networks, and LocalArea Networks) allows a great number of applications and users to communicate using a profusion of services. Apart the operational aspects of the network components, current efforts focus on the management aspects of the network: (1) network surveillance; e.g., how to know what is happening in the network, which users are using the services, (2) network control; e.g., how to offer the services in the most efficient way, how to recover from failures.
   So far, the efforts of the various telecommunication and information processing standardization groups have defined (1) some abstract

---

[*] This work is supported by a joint research contract between Bell Northern Research and Computer Research Institute of Montreal.

representation of a network (components, users, applications, events) to be included in a global, distributed, data base, the Management Information Base (MIB), (2) the different functions to be provided (configuration management, fault management, accounting, and many others), and (3) some basic services to support the previously cited functions (e.g., ISO Common Management Information Service [ISO89]). Considering the great versatility of networks, these standards must achieve a high level of abstraction, genericity, flexibility, and extensibility, as well as large applicability and acceptance among the networking community.

The aim of this paper is to show how these high-level issues can be formally described for real systems while staying in line with the concepts defined by the standardization groups. In particular, we use an object-oriented approach, and represent the standards concepts by means of objects. This paper is organized as follows: in Section 2 we present an object-oriented design methodology; in Section 3 we show how this methodology was applied to a real, industrial, transmission system, and Section 4 deals with the language issues involved in such an exercise. Section 5 provides our conclusions both on the methodological and language aspects.

2. Methodology for specification development

This section presents an object-oriented software design methodology for the specification of applications in the field of network management and other areas. The goal is to provide some simple, efficient, means to structure the process leading from some informal application specification to a formal object-oriented design.

This methodology is based on the concept of objects; i.e., that software applications can be designed based on the principle of aggregating data items together with operations performed on them. The application can be seen as a composition of such objects. Advantages of this design approach are numerous, though it still stays a very informal, intuitive, process based on human expertise. This methodology was defined [Mond90] with a specific object-oriented target language, MONDEL, whose main features are described in Section 4. However, we believe it is general enough to be applied to any object-oriented language.

Second, this methodology acknowledges the need to relate object-oriented concepts to the more classical, function oriented, design practices [Ward89, Bois89]. Thus our methodology uses some Entity-Relationship concepts already widely applied to software design.

A last key point is to specialize this methodology for the specific case of applications related to network management. Considering the starting point of the design process, we take in account the type (and amount!) of information usually available to designers of such applications; especially, it seems highly desirable to stay as close as possible to the practices of telecommunication and information processing standardization groups [T1M189c ,ODP, ISOnm1, ISOnm2, ISOnm3 ,ISOnm4].

Current object-oriented design methods usually comprise a certain number of design steps. Though not all practitioners agree on the number and the denomination of these steps, all come up with approximately the same philosophy [Booc86, Meye87, Bail88, Bail89, Jalo89, Bail90].We identified four major steps which are explained in more detail below. They can be iterated until a satisfactory design is obtained:

1. A preliminary step first consists of the identification of the general problem area, of the specific aspects we want to handle, and also on the aim of the expected design; this step is an informal one, but it should lead the designer to separate the different aspects included in the application, and also to precisely define the intended purposes of the model to be elaborated.

2. A second step consists of the identification of the key components of the so-called application domain; this step takes as input any information describing the functionalities to be provided, and should produce as a result a set of so-called entities, together with relationships among them, that can be easily mapped onto object-oriented languages constructs.

3. The third step is concerned with the allocation of the functions to be provided as operations offered by objects identified in the previous step; some functions will also uncover some new objects which must be integrated in the application domain.

4. The last step is concerned with the definition of the behaviours of the objects; these behaviours consider the possible sequences of operations calls as well as the necessary processing associated with each operation; complex objects can be specified as smaller "applications"; i.e., the entire design process can be applied to each individual object as it is applied to the global application.

2.1. Step 1: Problem definition

This preliminary step is necessary to help the designer to consider some important points.

Network Operation, Administration, Maintenance, and Provisioning (OAMP) is a very complex field due to the intertwinning of very different functionalities: data transmission, configuration management, error recovery, service optimization, security control, accounting, and many others. This step should help to state the intended functionalities of the application to be designed and to focus on the relevant aspects.

Complexity is also due to the large variety and number of components: various pieces of equipment, multiple services, numerous users and applications. Also, each component must present some of the different functionalities previously stated. With respect to the intended functionalities, not all components need to be specified. Also, the level of details to be considered for a given component may vary.

Due to the geographical distribution of networks, as well as evolution over time, application specifications must achieve a high-level of abstraction and genericity: the specifications must stay valid for a wide variety of equipment from different manufacturers, as well as for various configurations and services which evolve over time.

The application to be specified may concern an existing domain. For instance, network management applications are usually defined for existing networks. Therefore, the application domain already exist and the new functions must cope, at the appropriate abstraction level, with existing components. Generally, the design of an application starts with a (possibly empty) given domain, and new components are added. Such an approach promotes re-use of specifications since it does not isolate a given application form existing and future ones.

The last point to consider is the intended purposes of the specifications. We use the term system model to denote the specification of a domain together with its functionalities. Such a model may be used for various purposes:

- formal verification, to check the consistency of the design, and/or the correctness of algorithms;

- documentation for some hardware and/or software architecture;

- simulation, for user training, for future system development analysis, or as a prototype before building a larger scale system;

- performance analysis;

- (automated) software production.

Different formalisms may be required to achieve these different goals. Even with the same formalism, different specification styles may be adopted. For instance, "intentional" or "extensional" styles might be preferred. Also, the level of details to be included in the model greatly depends on its intended use. The results of this preliminary step are more of a set of guidelines for the following steps than formal results. They should help the designer to focus on relevant purposes and to determine the level of abstraction required for each component. Though this is presented as a preliminary step, our experience showed that this should be constantly considered throughout the entire design process.

2.2. Step 2: Domain definition

The domain definition step is intended to capture the relevant elements of the existing or foreseen domain, together with their essential characteristics. The result should be a description of the domain as a set of entities together with the various relationships among them that are relevant to the functionalities and purpose of the model. This step can be refined in the following sub-steps:

1. The first one is to identify entities of interest, together with their specific characteristics. Various entities exist, and are usually characterized with specific properties covering different aspects such as state, value, and structure. These entities can be represented as objects and attributes whatever the target language is.

2. A second sub-step is to identify the various relationships existing among these entities. They usually cover different aspects of the application to specify:

    - structuring aspect, represented by the aggregation relationship, often stated as "is-part-of" or "is-made-of" relationships; this relationship helps to define appropriate abstraction levels; i.e., whether a given entity must be handled as a whole or as a set of components; aggregation can be static or dynamic;

    - typing aspect, represented by the inheritance relationship, often stated as "is-a" relationship; different entities in the domain may share some common characteristics which can be specified in a generic template (or type); templates can be specialized for various purposes; e.g., to specialize inherited features or to add some new ones;

    - functional aspects: many identified relationships stem from the functionalities the designer intends to specify; also, some of these relationships are static while some are dynamic.

    Object-oriented languages allow a designer to formally specify these relationships: aggregation and functional relationships are represented by means of attributes, while subtyping is well captured by the inheritance mechanism.

3. The next sub-step is a step of consistency checking. Consistency checking applies to both entities and relationships:

    - entities having common attributes might be specialized instances of more generic types; common characteristics may be grouped within common templates;

    - an entity might include orthogonal aspects; therefore, a given entity may be considered as the combination of two or more generic ones separately specified;

    - entities must have attributes representing their relationships with others; specific entities may also be defined to represent these relationships;

    - relationships may lead to define entities functionalities;

- relationships must be considered with respect to the inheritance lattice; a relationship stated for general templates might not hold as such when coming to more specialized entities, and thus should be more precisely defined.

The principle here is to provide some basic means (1) to structure the domain (inheritance and aggregation relationships), and (2) to examine the domain according to the identified functionalities (specific relationships).

4. A last sub-step is to (re)write a design documentation where the identified components appear clearly. Since a formal model may not be suitable for human purposes, a textual, informal, documentation should closely match the system model.

2.3. Step 3: Functions definition

The function allocation step intends to distribute the required functionalities among the identified entities. There are four important aspects to this process:

1. Since entities are represented as objects, the functions they have to perform are defined as operations they must offer. An operation is formally defined as a procedure or a function; i.e., with some input and output parameters, which must be objects too. These parameters must be defined if possible.

2. Having identified the operations, the designer must allocate them to some objects. For each operation, the designer must consider which entity seems the "most natural" one to offer the operation. Several points are to be considered:

   - when an operation can be offered by several entities, it might be better to allocate it to a common ancestor in the inheritance lattice;

   - when an operation does not seem to fit a particular entity, or seems to naturally belong to very different entities, or seems to correspond to some cooperative processing by several entities, it might be convenient to allocate it to a new "support" entity introduced for the purpose of allocating the operation;

   - operations and parameters types should be specified with respect to the entities offering them; also, when operations are inherited, operations names and parameters types may have to be refined.

3. The next point is to consider the support entities introduced during the allocation process and to integrate them within the application domain.

The same process as in Step 2 should be reapplied; this favors the re-use of software specifications since the resulting structured domain can be re-used when defining future applications in the same field.

4. Since this step leads to define new entities and operations they offer, the textual specification can be rewritten so all identified entities appear clearly and their interactions are expressed in terms of operations.

2.4. Step 4: Behaviours definition

This last step consists of the definition of the behaviours of the various entities for the allocated operations. This process mainly depends on the knowledge and expertise of the designer in the specific field. However, some general principles can be applied:

1. For each object, it is first necessary to define the accepted sequences of operations; a general skeleton can be defined for that purpose.

2. For each operation offered by a given object, there are two possibilities:

    - the behaviour for that operation is simple enough so it can be easily specified with the language statements; the necessary processing is described in terms of state changes, attributes modifications, and interactions with other objects;

    - the behaviour is complex, and a refinement process can be applied to it; the technique is to consider the processing to be performed as a restricted application which can be specified by applying the design process from Step 1 to 4, until all components are fully specified.

3. An example: Network Fault management

We present in this section an application of the previously defined methodology to a real, industrial, transmission system.

3.1. Step 1: Problem definition

This example aims at specifying some standard network management functions [T1M189b] in the case of a real, industrial, transmission system [FD86, Leco90]. Also, conformance with a general network model defined by some telecommunication standards group [T1M189a] is required (Fig. 1).

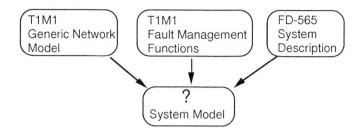

Figure 1: Problem definition

Though this example applies to a specific system, it has a significant scope since it includes all the components designers of network management applications have to deal with:

- an abstract, standardized, description of the domain [T1M189a],

- an abstract, standardized, description of the functions [T1M189b], and

- a real, industrial, system description [FD 86].

The general problem of network management is defined in [T1M189a]. The set of nodes (called *Network Elements* or NE's for short) providing the transmission service is connected to a set of surveillance systems (called *Operation and Surveillance Systems* or OS's for short) through a *Network Management Network*. The real system to be specified is an instance of a *Network Element*. Only aspects related to fault management, as defined between a *Network Element* and an *Operation and Surveillance System* (Fig. 2), by the relevant standards [T1M189b], are to be defined. Fault management functions in [T1M189b] are defined as a set of messages exchanged between the NE and the OS. They include standard parameters, and the functions to be implemented are denoted by the expected sequences of sent and received messages, and the behaviours in reaction to them.

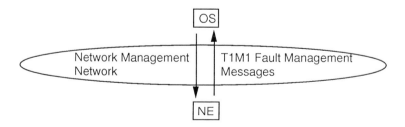

Figure 2: Fault management functions

Therefore, we did not consider in our specification the "normal" behaviour of the transmission system (such as connection establishment, data transfer, ...) but only (1) the physical structure of the system, as a specialized instance of a more general, abstract, network element, and (2) the "faulty" behaviour.

3.2. Step 2: Domain definition

The result of the second step is the description of the physical structure of the system, as an instance of the generic network element, in terms of its sub-components which also are instances of the generic model (Fig. 3). In particular, we included as the leaves of the aggregation tree the so-called *AlarmPoints* of the real system, which actually are part of the basic components, the so-called *Units*. But since we are concerned only with fault management, we can "project" these specific parts onto the domain we are interested in. Therefore, other aspects (e.g., data transmission) are not modelled. On this figure, aggregation is shown by a plain arrow while inheritance is shown by a dashed arrow.

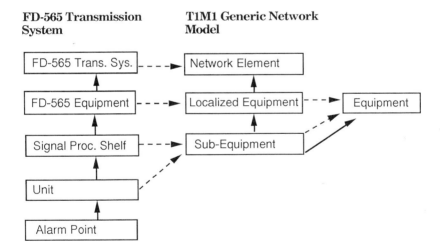

Figure 3: Domain definition

3.3. Step 3: Functions definition

The results of this third step is the definition of support objects accepting operations corresponding to the messages they might receive. The operations parameters strictly correspond to those of the messages. For instance, when considering the alarm reporting functionalities, we defined two support objects, the *NEAlarmReportControl* and *OSAlarmReportControl* objects. One of them represents the interface of the network element and the other one represents the interface of the surveillance and operation system.

Since we described the physical structure as an aggregation tree, we also defined the *NEAlarmReportControl* object as a hierarchy of components matching the physical structure (Fig. 4). We foresaw that messages exchanged between the NE and OS would also have to be exchanged among the components of the system hierarchy. The resulting support objects are then integrated within the application domain as suggested on Fig. 4. Each level of the *NEAlarmReportControl* has a specific, functional, relationship with the corresponding level of the system physical structure.

In our object-oriented language, there are several ways to represent such a functional relationship:

1. by use of attributes, which allows each object to call operations offered by others,

2. through inheritance, by defining that a *RealAlarmPoint* object will inherit attributes, operations, and behaviours of the *AlarmPoint* and *AlarmPointReportControl* objects, or

3. by use of a specific *control* construct which allows one object to define more constraints on acceptance of operations and more processing in response to operations calls of another object declared as a *controlled* attribute [Boch90a].

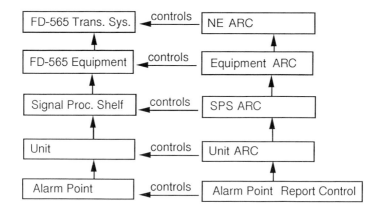

Figure 4: Alarm Report Control (ARC) support objects

Considering the functionalities described in the standards, we also allocated two operations, *Enable* and *Disable*, both supported by the *AlarmPointReportControl* object, in order to enable and disable the alarm reporting capability of the *AlarmPoint* object.

3.4 Step 4: Behaviours definitions

The last step is the most difficult one since (1) the system description does not include any behaviour specification related to T1M1 fault management functions, and (2) the T1M1 standards do not describe any detailed behaviour of an object such as an alarm control point. We include in that paper the behaviour we defined for the $AlarmPoint$ and $AlarmPointReportControl$ objects. Basically, the $AlarmPoint$ and *AlarmPointReportControl* objects behave as described on Fig. 5.

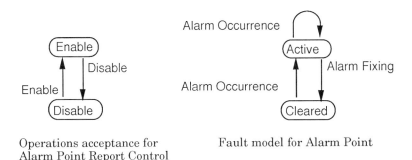

Operations acceptance for   Fault model for Alarm Point
Alarm Point Report Control

Figure 5: Alarm Point and Alarm Point Report Control behaviour

Their behaviours are represented by two finite state machines running in parallel: the first one relates to the occurrence of an exception within an *AlarmPoint* object (the two states are *active* and *cleared*) while the second one relates to the alarm reporting capability of the *AlarmPointReportControl* object (the two states are *enabled* and *disabled*). The combined behaviour is quite simple: an alarm must be reported when the *AlarmPoint* object is in the *Active* state the AlarmPointReportControl object is in the Enable state. The mapping of such finite state machines onto object-oriented languages constructs is shown in [Boch90a].

4. Issues concerning specification languages and development tools

We have developed an object-oriented specification language called MONDEL [Boch89] which supports the development methodology described above. In many respects, MONDEL ressembles other existing object-oriented languages. It has, however, certain properties which make it particularly suitable for describing real-time control systems, such as the example of Section 3. One of these properties is the fact that object instances and classes can be naturally represented in the language. Another aspect is its formal definition [Barb90a] which provides the basis for the construction of development tools, while certain other aspects are discussed below.

4.1. Overview of MONDEL

In MONDEL, everything is an object. Therefore the entities, their attributes and the relationships identified during Step 2 of the methodology are represented in MONDEL as objects. Relationships may be represented in different ways, as discussed in [Boch90a]. The multiple inheritance scheme of MONDEL supports in a direct manner the "is-a" relations identified in the design.

In contrast to many other object-oriented languages, MONDEL distinguishes between persistent and non-persistent objects. Persistent objects are like entries in a data base; they remain present until they are explicitly deleted, and they can be interrogated by database-oriented statements which identify the object instances in the database which belong to a given class and have specified properties. Entities and relationships identified during Step 2 are usually represented as persistent objects.

Concerning the design information related to Step 3 of the methodology, MONDEL has the well-known concept of operations (sometimes called "methods") which are defined for a given object class and provided by each instance of that class to be called by other object instances. In order to call an operation, the calling object has to know the identity of the called object. The type checking feature of MONDEL is able to detect many design errors related to the parameters of the called operations and the returned results.

In contrast to many object-oriented languages, communication between objects is synchronous, that is, a calling object is blocked until the called object executes a RETURN statement, which may include the delivery of a result. Synchronous communication is better suited for specifications of higher level of abstraction, since cross-over of messages at interfaces can be largely avoided [Boch90b].

Concerning Step 4 of the methodology, MONDEL provides a number of statements to express the order in which the operations can be accepted by an object, or for specifying the actions to be performed when an operation call is accepted by the object. The concept of an ATOMIC operation is introduced which represents a "transaction" in the sense of databases, that is, it represents a set of actions which are either all performed without interference from other "transactions", or undone if an exception condition is encountered.

The language has exception handling similar to ADA. The actions of different objects are usually performed in parallel; it is also possible to define several parallel activities within a single object.

The statements of the language have the flavor of a high-level programming language, except for the database-oriented statements mentioned above. However, it is also possible to write assertional specifications by defining input and output assertions for operations, or by defining INVARIANT assertions which must be satisfied at the end each "transaction".

4.2. Relation with other specification formalisms

For applications, such as network management, where many standard documents must be considered during the system design, it is important to find a common description formalism which is suitable to express all relevant design issues, is close to the formalism used in the standard documents, and formal enough to allow computer-aided design tools. We have tried to make MONDEL compatible with several notations used in the context of OSI standardization, as explained in the following (for more details, see [Boch90a].

The ASN.1 notation [ISO87] is commonly used for the description of the application layer PDU's and their parameters. The ASN.1 constructs of SEQUENCE (or SET), SEQUENCE (or SET) OF, CHOICE as well as most predefined ASN.1 types have direct counterparts in MONDEL.

The notation for Remote Operations (ROSE) [ISO88] corresponds to the definition of operations that can be invoked on objects. Depending on the intention of the specifier, a remote operation including the return of a result can be presented in MONDEL as a single synchronous operation call, or as two operation calls corresponding to the two PDU's exchanged between the calling and called objects according to the ROSE protocol. The exception handling notation of MONDEL may be used to describe the error situations and error handling in the calling object.

For the description of OSI management, a notation for describing object classes has been developed in the standardization community [ISOnm4]. Most of its concepts can be directly related to the MONDEL notation, although certain concepts have no direct counterpart. It is important to note, however, that the OSI notation does not provide a formal framework for defining the behaviour of objects (i.e. Step 4 of the methodology).

Finite state diagrams, often found in communication protocol and service specification, can be translated into MONDEL using various schemes, one of which is shown in the example of Section 3.

4.3. Use of formal descriptions and related development tools

Formal, as well as informal, specifications are used in various ways. The specification of a system is the basis for the implementation of that system.

It is also the basis for the selection of test cases and the reference for the analysis of test results. But first of all, the specification itself must be validated, possibly against a more abstract specification and requirements document. In the case of protocol specifications, these issues are further discussed in [Boch90c].

In order to partially automate the above development activities, various support tools can be used. It is important to note, however, that specification languages without a formal definition (in particular, natural language) make the construction of automated support tools very difficult. A survey of tools developed for use with protocol specifications can be found in [Boch87]. Many of these tools are intended for specifications written in one of the so-called Formal Description Techniques (FDT's). For the MONDEL language, a formal language syntax and semantics has been developed and is the basis for several development tools which are shortly described below.

A MONDEL compiler verifies the syntax and static semantics of MONDEL specifications, including the type checking rules. It also translates the specification into an intermediate form which can be used for the execution by an interpreter written in Prolog [Will90]. This allows a user-guided or automatic execution of MONDEL specifications in a simulated environment which is very useful for interactively validating specifications.

While simulated executions of specifications are helpful for finding errors, they are not able to show the absence of errors. A complementary approach of exhaustive validation may prove the absence of errors, but is usually much more difficult to realize. Work on exhaustive validation of MONDEL specifications is in progress [Barb90b]. It is restricted to a (useful) subset of the language and exploits the fact that this subset can be translated into Petri nets. The validation methods and algorithms available for Petri nets can therefore be used on the translated specifications, or be adapted to operate directly on the MONDEL specifications.

5. Conclusions

Our conclusions during that experiment are manifold; they deal with standards representation issues, with general software design issues, as well as with the required features for a formal language aiming at supporting such an area.

- On the standardization issues, it appeared that the generic models provided by the standardization groups are greatly improved by the use of a formal, common, representation. The object-oriented model achieves such a purpose. For instance, many extensions to ASN.1 have been defined to provide some formal, high-level, and precise definition of management issues. The point to be considered here is that the language we defined easily supports these ASN.1 extensions and can be used throughout the entire design process.

- On the design process itself, we believe that a common design model powerful enough to represent aspects such as Entity-Relationship concepts, Object, and Functions is absolutely necessary to provide formal models matching the various standard documents. The design methodology presented in Section 2 takes into account efforts of standardization groups.

- Beside its capability to represent standardization aspects, the model must be formal; i.e., its semantics must be formally defined if one wants to use it for software design, formal verification, simulation, and automated implementation. Such a formal model must also be supported by adequate tools.

Acknowledgments

The authors want to thank M. Shurtleff, J.M. Serre, A. Bean, and D. Wood from Bell Northern Research for their useful comments and advice.

6. References

[Bail88] S. Bailin, *An object-oriented specification method for ADA*, in ACM *Proceedings of the Fifth Washington Ada Symposium*, June 1988.

[Bail89] S. Bailin, *An object-oriented requirements specification method*, CACM, Vol. 32, N. 5, May 1989.

[Bail90] S. Bailin, *Remarks on object-oriented requirements specification*, 1990.

[Barb90a] M. Barbeau, G. v. Bochmann *Formal semantics of MONDEL*, Progress Report Document N. 11 for CRIM/BNR project, June 1990.

[Barb90b] M. Barbeau, G. v. Bochmann *Verification of high-level language specifications: a Petri net based approach*, Progress Report Document N. 12 for CRIM/BNR project, June 1990.

[Boch87] G. v. Bochmann, *Usage of protocols development tools: the results of a survey*, Invited Paper, 7th IFIP Symposium on Protocol Specification, Verification, and Testing, Zurich, May 1987.

[Boch89] G. v. Bochmann, M. Barbeau, A. Bean, M. Erradi,L. Lecomte, *CRIM/BNR Project---The specification Language MONDEL* Progress Report Document N. 2 for CRIM/BNR project, April 1989.

[Boch90a] G. v. Bochmann, S. Poirier, P. Mondain-Monval, M. Barbeau, *System specification with MONDEL and relation with other formalisms*, Progress Report Document N. 13 for CRIM/BNR project, June 1990.

[Boch90b] G. v. Bochmann, *Specifications of a simplified Transport protocol using different formal description techniques*, to be published in Computer Network and ISDN Systems.

[Boch90c] G. v. Bochmann, *Protocol specifications for OSI*, Computer Network and ISDN Systems, April 1990.

[Bois89] H. Bois, *Une methode de developpement de logiciels fondee sur le concept d'objet et exploitant le langage ADA*, Universite Paul Sabatier, Toulouse, France, October 1989.

[Booc86] G. Booch, *Object-oriented development*, IEEE Transactions on Software Engineering, February 1986.

[FD86] *FD-565 Optical Fiber Digital Transmission System: System description*, Northern Telecom Limited, 1986.

[ISOnm1] DP 10165-2 *Systems Management - Object Management Function*.

[ISOnm2] DP 10165-2 *Structure of Management Information -Definition of Support Objects*.

[ISOnm3] DP 10165-3 *Structure of Management Information -Definition of Management Attributes*.

[ISOnm4] DP 10165-4 *Structure of Management Information -Guidelines for Managed Object Definition*.

[ISO87] DIS 8824 *Specification of Abstract Syntax Notation One (ASN.1)*, 1987.

[ISO88] DIS 9072-1 *Remote Operations, Part 1: Model, Notation, and Service Definition*, 1988.

[ISO89] DIS 9595 *Common Management Information Service Definition*, 1989.

[Jalo89] P. Jalotte, *Functional refinement and nested objects for object-oriented design*, IEEE Transactions on Software Engineering, Vol. 15, N. 3, March 1989.

[Leco90] L. Lecomte, P. Mondain-Monval, G. v. Bochmann, *Un modèle orienté objet pour le système de transmission NT FD-565*, Progress Report Document N. 8 for CRIM/BNR project, June 1990.

[Meye87] B. Meyer, *Reusability: the case for object-oriented design*, IEEE Software, March 1987.

[Mond90] P. Mondain-Monval, G. v. Bochmann, *An object-oriented software design methodology*, Progress Report Document N. 7 for CRIM/BNR project, June 1990.

[ODP] *Working Document on Topic 6.1 - Modeling techniques and their use in ODP*, ISO/IEC JTC1/SC21 N 3196, December 1988.

[T1M189a] Committee T1-Telecommunications Standards Contribution, *Operation, Administration, Maintenance, and Provisioning (OAMP): A generic network model for interfaces between operations systems and network elements*, Doc. Number T1M1.5/89-010R2, July 1989.

[T1M189b] Committee T1-Telecommunications Standards Contribution, *Fault Management Messages*, Doc. Number T1M1.5/89-011R2, July 1989.

[T1M189c] Committee T1-Telecommunications Standards Contribution, *Modelling guidelines*, February 1989.

[Ward89] P.T. Ward, *How to integrate Object orientation with structured analysis and design*, IEEE Software, march 1989.

[Will90] N. Williams, G. v. Bochmann *Description technique d'un simulateur pour le langage MONDEL*, Progress Report Document N. 10 for CRIM/BNR project, June 1990.

II
MODELING AND DESIGN

A
Requirements and Verification

NETWORK MANAGEMENT BY DELEGATION

Yechiam Yemini[a]*, German Goldszmidt[a]*' and Shaula Yemini[b]

[a]Computer Science Department, Columbia University, New York, NY 10027

[b]IBM T.J. Watson Research Center, POBox 704, Yorktown Heights, NY 10598

Abstract
Emerging management frameworks tacitly centralize responsibilities into managers, with agents playing restrictive support roles. Managers can delegate to agents only primitive monitoring and control tasks. A non-trivial management task requires that managers micro-manage agents through primitive steps, resulting in ineffective and costly distribution of management responsibilities. Centralization leads to additional difficulties as managers become failure-prone management bottlenecks, whose loss can result in paralysis. It can lead to serious limitations on manageability of emerging complex and high-speed networks, where distribution of management function is mandatory. This paper proposes a novel model for effective distribution of management responsibilities. Managers delegate to agents execution of management programs prescribed in a management scripting language. Management programs include primitives to permit agents to monitor and control local managed objects effectively without unnecessarily involving managers. Agents can coordinate management activities to avoid hazardous interactions. The proposed manager-agent delegation (MAD) model greatly extends the OSI model permitting more flexible, powerful and effective distribution and coordination of management responsibilities among agents and managers.

[Keywords: network management, management model, OSI model, fault and performance management, delegation, remote-procedure-call, client-server architecture]

1. INTRODUCTION

Industrial enterprises are increasingly dependent upon large scale complex networked systems serving as their information backbones. The likelihood and costs associated with faults or performance inefficiencies of these systems increase with their scale and complexity. Effective manageability of such systems, to ensure fault-free, efficient and secure operations, is thus a vital need of industry. The management problem presents a broad range of non-trivial technical challenges (see [DSYBC89] for sample problems). The state of the art is one of significantly growing efforts by vendors, standard committees [ISO89, SNMP90] and the research community [IFIP89, KMW90, Rose91] to establish manageable systems. Unfortunately, typical networked systems have not been designed to be manageable. Nor are the fundamental technical problems involved in accomplishing manageability fully understood and their

---
Part of the work of this author was done while at IBM's T.J. Watson Research Center.
* Work performed under DARPA contract F29601-87-C-0074.

research is still in embryonic stages.

A typical network management system [Rose91] involves *agents*[*], responsible to monitor and control *managed-objects* of the network, and *managers,* responsible to collect dynamic status data from the agents, interpret the data and direct the agents (e.g., how to handle fault scenarios). Managers and agents use a *management protocol* to coordinate their activities. The management protocol utilizes a *management-scripting language* (MSL) to describe the monitoring and control operations to be applied to managed objects. The program written in the MSL is called the *management script.* The MSL includes support for structuring management information (SMI) and for accessing and operating on it. In what follows, we use the term "manageability" to describe the ability to monitor and control networked systems behaviors.

The key question that this paper aims to address is: how should management responsibilities be distributed among managers and agents to accomplish effective manageability. The OSI CMIS management model [CMIS89], for example, distributes management responsibilities through the following manager-agent interaction primitives: managed object instance creation (M-CREATE) and deletion (M-DELETE), retrieval (M-GET) and setting (M-SET) of managed objects attributes, reporting of events (M-EVENT-REPORT) and invocation of actions (M-ACTION). These are illustrated in figure 1, below. Thus, for example, a manager could use M-GET to retrieve certain managed-object attributes values. The manager could use these monitored values to decide on appropriate control/diagnostic measures which may be invoked by setting the values of certain managed-object attributes (M-SET) or activating certain agent actions via M-ACTION.

Figure 1: The CMIP protocol structure.

Managers and agents can establish an association which defines the specific subset of these capabilities that will be used in their protocol interactions [IEEE89]. Thus, an association of type <Event> permits only event reporting while association of type <Monitor/Control> supports M-GET, M-SET and M-ACTION but not M-EVENT-REPORTING.

Standardization efforts have focused primarily on the management protocol [SNMP90, CMIS89] and the SMI model to support it. The management model that they assume is only partially elaborated. The management needs and styles to be used are tacitly assumed. In particular, the manager/agent software architecture required to support the model is left implicit. Such questions as: how should agents and managers be designed and implemented to support effective manageability? How should management responsibilities be distributed among managers and agents? How should management activities be coordinated among managers and agents? How should agents monitor and control managed object behaviors? are not answered. As we shall see, these questions lead to non-trivial design problems that must be resolved for effective manageability.

This paper aims to elaborate these and other fundamental problems arising in the design of manageable systems. In section 2 we focus on the micro-management problems arising in current management protocols. In section 3 we outline the remote delegation model. In section

---
[*] This paper draws no distinction among agents native to managed objects vs. proxy agents (see [Rose 91]).

4 we consider the design of management scripting languages to support effective monitoring and control. In section 5 we focus on the problems of management coordination among agents and managers. Section 6 briefly describes some of the fundamental problems involved in agent design and in section 7 we present some conclusions.

2. THE PROBLEM OF MICRO-MANAGEMENT

2.1 Protocol Expressiveness Can Seriously Limit Manageability

Consider an agent charged with monitoring and control of a link (or a virtual circuit) object. A typical management scenario may consist of detection of certain unusual link conditions, execution of diagnostic testing to identify the cause of the conditions and procedures to handle the problem. This is illustrated by the following management script[*] :

Example 1 Management script for link failure

```
On (link.control_stat>normal_stat)&(link.q_length>normal_q);
                /*trigger on composite event*/
Begin                   /*management handling actions*/
link.handle_congestion; link.test;. /*quick-fix + diagnose*/
........
If link.failure then
     Begin
           recover(link.failure_type);....
           notify(Manager,link.failure_params)..
     End
End
```

Typical management scripts would similarly use the occurrence of an exceptional event, typically signaled via threshold conditions, to identify potential troubles and trigger diagnostic and corrective management actions. Both detection of the event and execution of the action involve data and procedures entirely local to the agent and managed link object. The manger must thus use the management protocol to transfer execution responsibilities to the agent to compute the script. What does it take to accomplish execution of this script using CMIP?

The manager must first establish an appropriate association with the agent. It needs to establish mechanisms to detect the occurrence of the triggering events. Suppose that the manager polls the respective variables link.control_stat and link.q_length using M-GET and evaluates the event condition. [This may be the case with agents whose associations do not include M-EVENT-REPORT capabilities]. Polling, however, provides poor event detection mechanism. Events may occur highly infrequently and hold for a relatively short random time (or require response within short time of occurrence). In order not to miss such occurrence, the manager would have to poll the agent at a frequency determined by the expected holding time (or permissible tolerance to detection latency). A high polling frequency can exhaust the manager communication bandwidth and cycles. Intermittent events, (e.g. failures followed by device rebooting [DSYBC89]) might be unobservable through polling. In summary, polling can result in highly ineffective division of event detection responsibilities among manager and agents.

These intrinsic limitations on polling lead the CMIP designers to provide a richer set of event detection capabilities than adopted in SNMP. Using M-EVENT-REPORT, a manager could be notified of the occurrence of such threshold events as in example 1. However, in the absence

---
[*] Scripts will be described using a self-explanatory pidgin-programming-language format to render examples reasonably realistic and formal.

of mechanisms to specify composite events, the manager would have to seek independent notifications of the two events "(link.control_stat > normal_stat)" and "(link.q_length> normal_q)" and decide on their conjunction. This division of event detection responsibilities leads, again, to inefficiencies and hazards. Event notifications may be randomly delayed and thus the conjunction event may go unnoticed even if its components have been adequately reported. Agents and manager must spend bandwidth and cycles to divide detection responsibilities.

Once an event is detected, the management script calls upon certain diagnostic handling to be executed. This require execution of certain diagnostic actions (e.g.,invoking link.test), checking certain managed variables (e.g., ..If link.failure..) or invoking a corrective action (e.g.,"recover(link.failure_type)"). Each of these computations may be accomplished via an appropriate CMIP primitive. Thus diagnostic-testing may be invoked using M-ACTION and the value of link.failure may be retrieved by the manager via M-GET [There may be, however, a problem in invoking a procedure that requires parameters as in recover(link.failure_type)]. Both SNMP and CMIP provide, however, limited or no capabilities to combine primitives to handle composite scripts. Thus, the manager must step the agent through the execution of the script. In the example the manager must step the agent through the execution of "link.handle_congestion; link.test;....", via two (or more) invocations of M-ACTION. This is followed with a retrieval of "link.failure" (using M-GET) to evaluate the conditional statement and with additional calls upon the agent to complete the rest of the script. In other words, the manager must micro-manage the execution of the script by stepping the agent through it.

Micro-management is due to the inability of management protocols to compose primitive management actions into scripts. Micro-management can lead to highly ineffective and costly management systems. Costly communication bandwidth and managers cycles may be required to accomplish even simple management tasks. These costs place severe restrictions on manageability, barring all but trivial management scripts from being executable. Micro-management means that all management functions must be centralized into the manager. Managers are rendered most vulnerable to network failures as even simple failures could load managers bandwidth and cycles bringing it down. Once the manager is down, local agents cannot accomplish recovery as they must wait for management instructions. Thus even a minor failure may potentially lead to an avalanche failure of the very management system.

Centralization of management is inconsistent with the emerging needs of complex, large-scale or high-speed networks. In such networks the complexity of management needs and the speed at which response may be required may render centralization ineffective or even impossible. Nor is management centralization consistent with the tradeoffs suggested by current and future technology trends. Agent processing cycles are abundantly available and increasingly so as one climbs in the protocol hierarchy towards managing high-layers entities. The tacit assumption in the design of current management protocols that agents lack any management smart and thus all smart should be left to manager cycles may reflect the tradeoffs represented by traditional voice and teleprocessing networks of the past rather than networks of the future.

The micro-management problem can be resolved via **delegation** of entire composite management scripts by managers to agents. Thus, in example 1, a manager would pass the entire script to an agent, responsible for the link object. The agent would then execute the composite script without manager intervention (except when the script calls upon one). Indeed, the agent must be ultimately responsible to execute all the primitives of the script from event detection to invocation of local action. Management involvement in execution is neither necessary nor useful and is only mandated by the limitations of current management protocols. This paper proposes a model for delegation of management scripts between managers and agents, and briefly describes some of its implications. In the following section we review existing models of process interactions and evaluate their suitability for resolving the micro-management problem.

2.2 Models for Manager/Agent Interaction

The remote procedure call (RPC) [BN84] interaction paradigm is a central model for structuring distributed systems. In this model, a server exports a number of fixed procedures that can be invoked by remote clients. Upon a remote call, the caller is suspended, the remote procedure is executed, and its results are returned to the caller, which then resumes its execution.

It is natural to ask whether the RPC mechanism suffices to solve the micro-management problem. Let us examine this question via example 1. Suppose that the entire management program in example 1 is encoded, as part of the agent code, in terms of a particular action, say, link_handler. A manager will need to execute a remote procedure call (RPC) invoking execution of the script by the agent. Does this solution offer an adequate model for management?

There are two fundamental problems associated with the solution above. The first problem is rooted in the synchronous invocation semantics of RPC mechanisms. A management script will typically involve actions to be triggered by independent event occurrence s in the agent, asynchronously with the manager's invocation of the RPC. Clearly, the manager's own execution should not be tied with the execution of the management script. An RPC mechanism would block the manager until the completion of the manager script.

A second limitation of the RPC model for supporting manager agent interactions is the static nature of the exported procedures. Thus, there is an underlying assumption that programming management scripts is done by the *agent/managed-object designer* at agent-design time (just as server calls are part of server design) rather than at management design time as part of incorporating the agent/managed-objects in a network. Designers of the agent (server) must thus predict and code as agent procedures the entire range of composite management scripts in which this agent may be usefully involved. It is typically not possible to predict at agent-design time all the possible management scenarios in which it may be involved, nor is it desirable as it limits the manageability afforded by the product. Furthermore, coding the scripts into the agent will significantly increase its complexity and costs.

R*emote evaluation* [SG90] (REV) is another interaction paradigm for processes in distributed systems. It allows transferring a procedure (written in a LISP dialect) from a client to a server where an interpreter will execute it. It thus overcomes the problem of fixing at design time of the server, all the server procedures that can be invoked remotely. However, the execution of the remotely evaluated procedure is still synchronous with the execution of the caller (manager), i.e., the REV procedure is executed upon its receipt by the server (agent).

In summary, the RPC model fails to provide an adequate solution of the micro-management problem as agent designers must predict and code all management scripts required as part of the agent. The remote evaluation model goes a step further permitting delegation of programs to remote execution. Both models involve synchronous procedure-call interactions whereby the agent executes the management program upon invocation while the caller is blocked until completion. This is not adequate for management.

3. REMOTE DELEGATION

The delegation model supports composite management programs to be transferred by managers to agents that execute them. The management programs are designed and coded as part of a specific management platform, rather than as part of individual agent design (although agent designers and vendors can provide useful libraries of such programs). Management program can be dynamically delegated by managers to agents, permitting managers to flexibly adapt management responsibilities assigned to agents.

3.1 The Delegated Program

A delegated program is a management script, written in a *Management Scripting Language,* (MSL -- see next section) which is transferred from a manager to an agent. MSL includes primitives for monitoring and control of managed objects and for composition of these primitives into programs. Additionally, the language includes primitives to support interaction between management programs and agent services. These services include, for example, notification of certain event occurrences, used to invoke handling actions by the management program. Also, these services may enable coordination of actions with other delegated management programs. The delegated program can use those constructs in order to provide management capabilities to its creator.

For instance, example 1 of the previous section, can be handled as follows. The management script is coded as a management program in an appropriate MSL, e.g., one that provides the appropriate primitives for this case. The program is passed to the agent and instantiated for execution. The delegated program first requests an agent triggering service for the event E = (link.control_stat >normal_stat) & (link.q_length > normal_q). The agent will monitor the event occurrence and when it detects it, would notify the delegated program. The delegated program will be waiting for the trigger (the event notification) and then proceed with the remainder actions: "link.handle_congestion ...".

Delegation enables a manager to transfer the responsibility for performing the functions in the delegated management script to a remote agent, and to perform these functions independently of the manager's execution. This capability enables a manager to augment, during execution time, the functionality of the agent.

Several important issues must be addressed regarding the remote delegation capability. For example, what is the language(s) in which the delegated programs are written? What novel language structures can usefully support delegation? How is execution life-time of a management program controlled (e.g., when is it invoked, terminated)? How can a manager control execution of programs that it delegated? How does a management program obtain access to managed objects? How can an agent control concurrent accesses to management information? How can a management program coordinate its execution with other management programs and with managed objects? We briefly examine some of these questions in the following. A more detailed exposition is included in [GYY91].

3.2 The Lifetime of a Delegated Program

Managers must be able to continue their execution independently of the management programs that they delegate. Thus, a delegated program must have an independent lifetime. The manager must retain control over a delegated program to allow it to terminate, suspend or resume it. Thus, for example, the manager may delegate a program that monitors and reports periodically certain managed variables. At certain times the manager may wish to suspend reporting to focus resources on other critical problems. Once these are handled, reporting should be resumed. Management-programs lifetime control constructs should support such functionality.

Management programs life-time can be best conceptualized as a *thread*[*] of execution in the agent's environment. A thread or light-weight process is a unit of execution over which the

---

[*] Note that thread is used for explanation only. The model does not require a thread implementation.

agent's environment can exercise control: it can be instantiated, suspended, resumed and terminated. Delegation may be considered as creating a remote thread in the agent environment, associated with a management script. The life-time of these *delegated management threads* (DMT) can be controlled by the manager. From a protocol's perspective, life-time control may be accomplished via generalizations of CMIS constructs such as M-CREATE and M-DELETE. These constructs are typically used to control the life-time of instances of inert managed objects. From this analogy, we may consider a DMT as an "active management object". From an implementation perspective, a delegated thread should execute concurrently with the agent, and requires run time support. The life-time of a DMT should be synchronized with the life-time of managed objects that it requires. This means that managed objects accessed by a DMT (whether active or suspended) must be allocated and maintained in the environment of the agent to ascertain that accesses by the DMT to these objects are properly supported. This support may be provided explicitly by the agent, or implicitly by the underlying supporting execution environment. In section 6 we discuss additional considerations related to the agent's structure and run time support requirements.

3.3 Delegation and Access Rights

A DMT program must have some means to obtain information and to interact with the agent and its environment. The DMT should obtain *controlled-access* to the managed objects of the agent. Indiscriminate access to the address space of the agent by a delegated program, should be prevented to avoid concurrency conflicts. Thus, the behavior of the agent, as far as its own internal state is concerned, is not disturbed by the delegated threads. The delegated thread should only have access to the attributes of the managed objects which have been explicitly declared as exported.

The delegated program should be able to invoke locally, the same (or a subset of the) actions that its creator could invoke remotely. Thus, the delegated-thread should inherit access capabilities from its creator. Additionally, the DMT should have the capability of sending messages to its creator, e.g. event reports, and even invoking actions in its creator. Furthermore, a DMT program should be able to inherit the role of receiving event reports which were previously been routed to its creator.

4. MANAGEMENT SCRIPTING LANGUAGES

A management scripting language is used to specify management programs to be delegated from managers to agents. The language must support delegation and coordination of management programs, and provide constructs for event specifications as Boolean expressions and for event handling mechanisms. In [GYY91] we describe the Monitoring, Coordination and Control (MC^2) language, designed to support these functions. In this section we focus on novel language design requirements associated with management languages.

4.1 Grading of Management Languages to Control Complexity

Delegated management programs must be compatible with the computational capabilities supported by respective agents and the processors in which they execute. These computational capabilities may vary greatly among networked systems. Agents controlling modems or multiplexors may offer very limited management capabilities, while those controlling switches or computing systems may support extensive manageability. A management system must thus provide for a spectrum of possible agent requirements and capabilities. This has been recognized in working implementation agreements for the OSI management protocols [IEEE90] by using the concept of *association-type*. An association-type defines a restricted sub-language of the CMIP protocol. For example, an association of type <events> restricts agents to event reporting only over that association. We characterize sub-languages of a management scripting language as *grades*.

Grading establishes a lattice order over restricted management sub-languages. This is illustrated in figure 2. The shaded rounded rectangles indicate language grades. At the bottom of the grading lattice one finds the null language (no manageability). The four association types [IEEE89] are organized above it, each defining a different language of management primitives. Additional primitives and composition operators used to support delegation generalize and extend these capabilities creating higher grades. At the top of the lattice, the management language is a full-fledge, general purpose programming language which has been augmented to support management constructs.

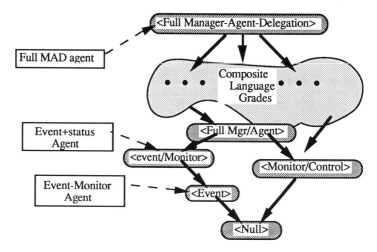

Figure 2: A Lattice of Management Languages Grades

The cost of deploying manageability capabilities offered by different grades is in agent complexity (memory and cycles requirements). Lower grades are adequate for simple agents as illustrated in the rectangular boxes on the left. An agent responsible for a simple device (e.g., modem, multiplexer) and responsible only for event monitoring would support the language offered by the grade (association type) <Event>. Agents responsible for more complex systems (e.g., a switch or a workstation) involve substantial management responsibilities and require richer management scripting languages. Micro-management arises when manager and agents responsible for performing complex management functions are restricted to the use of language grades not possessing sufficient expressive power to capture these responsibilities.

The MAD model can be conceptualized as extending the OSI association types to offer higher management language grades to support broader needs of management. These extensions provide extended language primitives (to monitor, coordinate and control) as well as composition operators to create management scripts and delegate them to agents. This extension is depicted in the figure above.

5. THE PROBLEM OF COORDINATION

Distribution of management functions among agents and managers requires effective mechanisms to coordinate management actions. A few problems can arise as a result of inadequate coordination of management activities.

5.1 Non-determinism of management actions

Consider an agent to whom a number of management scripts have been delegated (by one or more managers) sharing the same triggering event E. When E occurs, the respective management actions must be executed by the agent. Different execution orders may result in substantially different results. Consider for instance the following

Example 2: Non-determinism of management actions

An X.25 controller within a switch includes an agent responsible for managing the controller operations (mostly virtual circuits (VC) operations). The agent has been delegated three management scripts by different managers, to be triggered upon the event E = "buffers-full" (i.e., all VC buffers are full):

A1 = run local diagnostics to check possible link or controller failures, reboot the controller upon fault; A1 is delegated by a vendor-provided switch manager.

A2 = evaluate and report performance parameters, abort some VCs and set limits on local resources usage; A2 is a congestion handler delegated by the network control center.

A3 = dump buffers contents and reset flow-control on VCs; A3 is a flow-control action pursued by network flow-control protocol (acting in a manager role).

This example illustrates typical forms in which management actions distributed through networked systems can lead to unpredictable management actions. First, any order of execution pursued in applying the three actions will lead to very different behaviors. Thus, if A1 is executed first, the controller may be rebooted and then when A2 evaluates the performance parameters it would sample and report to the global manager different values then those reported by executing A2 first and then A1. As a result the controller may pursue unobservable intermittent failures as it reboots and fails again. Global management seeking reporting through A2 could never know about the problem.

Second, the roles of managers and agents is effectively established through the very design of network devices and protocols. A global network manager is not the only entity acting in a role of a manager. Typically, devices would be equipped with local vendor-provided software functioning in explicit management roles. Thus, local automated recovery systems, monitoring and diagnostics software also function in management roles. Management roles too are often implicitly built-into the very specifications of certain protocol entities. Thus, flow control procedures of the protocol entity may be triggered by conditions monitored by the protocol entity. In designing an overall management system it is critical that all management actions are considered and that possible interactions among such actions and those pursued by management of a network control center are carefully coordinated.

Third, the distribution of management functions among different management entities is typically established through ad-hoc evolution. Such distribution of function may contribute to very difficult coordination problems, or even render certain devices and protocol entities unmanageable.

Non-determinism can (and should) be avoided by constraining interactions among concurrent management programs. Two management programs are *concurrent* if they may be simultaneously executed; they are said to be *associative* if they lead to the same results independent of their execution order. Obviously, the design of a management system must aim to assure that concurrent management programs are associative. The execution orders of non-associative concurrent management programs must be carefully controlled to ensure that the results are deterministic. This bears some similarity to the problem of serializability [BHG87] of transactions in database systems. However, unlike the problem of serializability, where arbitrary serializable execution orders are acceptable, deterministic management requires control over the specific ordering of concurrent actions. The specific ordering of non-associative management program executions can reflect management hierarchy (e.g., prioritize programs of higher-level managers) or critical priorities of the program (e.g., execute first higher priority handlers). The role of the management scripting language is to help minimize the occurrence of, and control the execution orders of non-associative scripts.

5.2 Atomic Management of Distributed Objects

Managed objects are often distributed. At the physical layer of a network, links are monitored and controlled by distributed entities. At higher layers virtual circuits may be distributed throughout a collection of network switches and two end-points. Monitoring and controlling distributed managed objects lead to a class of coordination problems.

Example 3: An X.25 virtual-circuit manager.
Consider an X.25 network [Tane89] where a manager would like to respond to loss of capacity, or heavy congestion, by reducing the thruput classes assigned to virtual circuits (VCs) along the following management script:
On heavy_load
   For all active VCs,
   If (VC.thruput > 2.4Kbs)**then** VC.thruput:= 2.4Kbs

(This may be accomplished by sending M-SET to the nodes maintaining the VCs).

The first problem is that of *atomic-monitoring* of events associated with the states of distributed managed objects. The event "heavy_load", for example, may be definable in terms of the states of a collection of virtual circuits. These distributed variables used in event detection are sampled at different times. Proper detection may require that the temporal window over which monitoring occurred is limited. The actual temporal tolerance associated with distributed detection may depend on the event specifics. Thus, changes in certain local status variables describing a VC condition may occur over fractions of a second. A given event associated with these variables may or may not be tolerant of temporal difference in the readings of the variables within seconds apart. Similarly, certain high-level distributed events may occur over periods of days. Their detection may be tolerant on readings of managed variables taken within significant time apart. The problem of atomic monitoring in general is that of assuring that distributed objects are monitored within the temporal tolerance permitted by the respective management actions.

A dual problem is that of *atomic-change* in distributed managed objects. In the example, the execution of "VC.thruput:= 2.4Kbs" means that all distributed parts of the VC must be set simultaneously (i.e., within certain temporal tolerance). The problem bears similarity to problems of atomicity [BHG87] in distributed transaction systems and may utilize classical solution techniques. It presents, however, issues distinct from those arising in the context of traditional transaction systems. First, it is often possible (even desirable) to tolerate some degree of inconsistency, rather than pay the price of consistency (in complexity and time). Second, there may be tight temporal bounds on accomplishing atomic actions.

5.3 Coherent Management, Controlled Interference

Management actions may interfere with each other, leading to potential hazards. A management system is coherent when interferences are properly controlled. Interference is typically caused when evaluation of an event or an action causes changes in managed object attributes triggering other events/actions. Coherence may be accomplished by minimizing and controlling interference. For example, interference among event detection may be avoided by preventing side-effects of detection.

It is, however, undesirable to prevent interference among actions and events in general. In the example, the event "`heavy_load`" triggers changes of the thruput class (set to 2.4Kbs) for some VCs. A user of a VC (e.g., TP4 protocol [Tane89]) may require notification of such changes to ensure proper handling. In the absence of TP4 handling of these thruput class changes, the network may instantly crash as a result of the changes. Interference is thus essential to assure communication of changes among managers. Interference is also important in supporting other forms of cooperative management. For example, in a hierarchical management organization, a high-level manager may capture and prevent attempts of lower-level managers to cause certain actions.

However, interference may lead to hazardous behaviors. For example, management actions may form a loop, each triggering evaluation of the next one in the loop. This would lead agents to a management livelock. It is thus necessary for managers to control possible interference among management programs and ascertain management coherence. A management system should support coordination primitives for interference control.

6. AGENT ARCHITECTURE

Agents execute management programs, maintain delegated management threads and coordinate these functions with the behaviors of managed objects. In this section we consider briefly some of the fundamental problems involved in agent's design.

6.1 Managed Objects Access

Agents must control the access of management programs to monitor and control managed objects. Agent functions may be integrated with the design of the managed objects or separated via vendor-provided object interfaces. In either case, agents must support managed-objects interfaces to export management capabilities. These interfaces may be implemented using module-interconnection constructs such as private components in a C++ class [Lipp90] or an Ada package [Ada82].

Control actions are typically accomplished via changes of managed objects or invocations of procedures that they export. Again, standard module interfacing techniques may be used to support these type of access. However, some control functions may require that attempts to change an object or invoke certain procedures be aborted. To support these constraints we must assure that all attempts to pursue such actions are pursued via the appropriate interfaces and that the required mechanisms are supported [GYY91].

6.2 Events handling

Management events can be defined in terms of Boolean operators over attributes of managed-objects. DMTs may require triggering services upon event occurrence. This means that the agent must support capabilities to evaluate event conditions and trigger DMTs upon event occurrence. The agent must also support coordination of these triggering by DMTs, as discussed in sections 3 and 5.

The evaluation of events by the agent will be delayed relative to their occurrence at the real object. Delays may be introduced via the detection mechanisms used by the agent. For example, an agent may be periodically polling the real-object state to obtain values traces and then evaluate event conditions over them. Polling may cause significant delays between event occurrence and detection. Additional delays may be caused by the mechanism used by the agent to communicate events to DMTs. Clearly, event detection offered by an agent can only provide an approximation of the actual event occurrence. This may lead to potential difficulties in handling events as response time to certain events may be critical. Efficient event handling mechanisms as well as capabilities to tune detection and handling to meet real-time constraints are thus important components of agent design.

6.3 Agent Structure

Figure 3 depicts the relationships between a manager, agent, delegated program and management objects of the network. A manager uses the delegation protocol to transfer and control execution of delegated management programs. Management programs execute in the environment of the agent, and obtain access to managed objects and additional control services from the agent. In particular, the agent supports triggering and coordination services. The agent's private data space should be protected from interference by delegated programs via access control mechanisms.

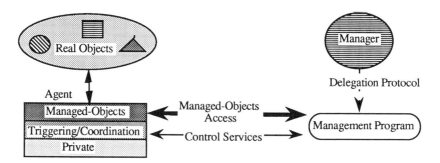

Figure 3: Manager-Agent Interactions via Delegated Management Program

7. CONCLUSIONS

Current approaches to manageability, as pursued by standardization efforts, reflect a restrictive view of the subject. Current management models centralize all management functions within managers with agents playing minor roles. Managers must micro-manage agents through management scripts. Micro-management results in inefficient and costly management, seriously limiting manageability. Centralized management results in intrinsically unreliable network management (managers are turned into sensitive communications and processing bottlenecks whose loss can result in paralysis). Centralized management is neither commensurable with the management needs of emerging networks (e.g., managing high-speed increasingly complex networks), nor is it reflective of emerging cost/performance tradeoffs as agents may require and support as powerful processing as managers.

Management models must thus support flexible and effective distribution of management functions among managers and agents. The manager-agent delegation (MAD) model proposed in this paper offers an extension of the OSI model to support such flexible and efficient distribution. Management programs can be composed by the designers of managers, using a management scripting language, and delegated to the agents dynamically. These management programs can then execute locally at the agent without involving the manager unnecessarily.

Management programs access managed objects via controlled agent interfaces. The agent also provides event notification and management coordination services for the delegated management programs. The capabilities and limitations of agents may be reflected in the management language grade that they use. The MAD model extends the manageability constructs developed by standards efforts, by adding higher, more powerful grades.

REFERENCES

[Ada82] Reference Manual for the Ada Programming Language, D.O.D, Ada, Joint Program Office, July 1982.

[BGLSYY91] Bacon D., Goldberg A., Lowry A., Strom R., Yellin D., and Yemini S., "Hermes A Language For Distributed Computing", Prentice Hall, 1991.

[BHG87] Berenstein P., Hadzilacos V., and Goodman N., "Concurrency Control and Recovery in Database Systems.", Addison Wesley, 1987.

[BN84] Birrell A. and Nelson P., "Implementing Remote Procedure Calls", ACM TOCS, 2-1, Feb. 84.

[CMIS89] ISO, "Information Processing systems - Open System Interconnection - Management Information Service Definition Part 2: Common Management Information Service Element", December 1989.

[DSYBC89] Dupuy A., Schwartz J., Yemini Y., Barzilai G. and Cahana A., "Network Fault Management: A User's View", in [IFIP89].

[GKY89] Goldszmidt G., Katz S. and Yemini S., "Distributed System Debugging", IBM Research Technical Report, January 1989.

[GYY91] Goldszmidt G., Yemini Y. and Yemini S., "Management by Delegation", Working Draft.

[Hold89] Holden, D., "Predictive Languages for Management", in [IFIP89].

[IEEE90] "Working Implementation Agreements for Open Systems Interconnection Protocols", Boland, F.E., Editor, IEEE Computer Society Press,Vol. 2, Number 2., 1990.

[IFIP89] "Proceedings of the IFIP TC6/WG 6.6 Symposium on Integrated Network Management", North-Holland, B. Meandzija & J. Westcott (eds.), May 1989.

[ISO89] Information Processing Systems - Open Systems Interconnection - Systems Management Overview, December 1989

[KMW90] "Network Management and Control", 1990 Plenum Press, Kershenbaum A., Malek M. and Wall M., editors, proceedings of the Network Management and Control Workshop, September 1989.

[Lipp90] Lippman S.B., "C++ Primer", Adison Wesley, 1990.

[ROSE91] Rose M.T., "The Simple Book, An introduction to Management of TCP/IP-based Internets", Prentice Hall, 1991.

[SCHW87] Schwartz M., "Telecommunications Networks, Protocols, Modeling and Analysis", Addison Wesley, 1987.

[SG90] Stamos J.W. and Gifford D.K., "Implementing Remote Evaluation", IEEE Transactions on Software Engineering, Vol 16, No 7, pp 710-722, July 1990.

[SNMP90] Case J.D. et al., "Simple Network Management Protocol (SNMP)", RFC 1157, Network Information Center, SRI International, CA, May 1990.

[Tane89] Tanenbaum A., "Computer Networks", Second Edition, Prentice Hall, 1989.

Network management: An alternative view

Aiko Pras

Tele-Informatics and Open Systems group, University of Twente, Enschede
The Netherlands, email: pras@cs.utwente.nl

Abstract
This paper argues that network management should not be restricted to the operational phase of networks, but should also include the design phase. Only than it is possible to relieve the network manager. The paper discusses the nature of the design process, and addresses the relation between designing 'normal data' and 'management' protocols. The example of the OSI network layer is used to illustrate this relation.

Keywords
Design process, cyclic approach, CMIS, network layer, routing

1. PROBLEM DESCRIPTION

In the last decade there has been an impressive growth in the number of data networks. It is expected that this growth will continue during the next couple of years. As a result of this growth, many people have been connected to one or more data networks. In many cases these networks had impact on the work of these people; remote printers and fileservers were introduced and electronic mail replaced part of the ordinary mail system.

Although this evolution has many positive aspects, the worrying fact is that people become more and more dependent upon the correct operation of their networks; small failures within these network will have severe consequences and may even halt the work. The task of the network manager is therefore becoming more and more important. Not only should he / she cope with the continuous growth of the network, but also with the request for an increased reliability. No wonder that network management symposia are well attended!

Fortunately a lot of work is being performed to support the network manager: managed objects have been defined, protocols are underway that allow remote access of these objects (CMIP and SNMP), human interfaces are under development and even usage of expert systems is being considered. As a result of this work, the network manager will be able to monitor and modify all objects within the network. In case of failures, the network manager will be equipped to detect these failures and take appropriate actions.

The problem that remains however, is what actions are 'appropriate', and what actions make things even worse. What events require immediate operator's intervention, and what events may be neglected.

The work on network management that is mentioned above, hardly addresses this question. For example, the relation between a time-out in one part of the network and a required configuration change in another part of the network, is not specified. As a consequence, the network manager himself should answer this question; he / she must have a detailed knowledge of the network protocols plus an understanding of their underlying ideas.

An alternative of having omniscient network managers, is to pay more attention to management in the design phase of a network. As a result, the network should perform a number of management task automatically. For example: a self-managing network should detect and isolate malfunctioning systems (fault management), react upon topology changes and initialize new systems (configuration management). The task of the traditional network manager will than be restricted to strategic (long-term) management, attachment of new systems plus the replacement of equipment that could not be repaired automatically.

The purpose of this paper is twofold. First, it advocates that more attention be paid to management issues in the design phase of a network. Second, it tries to give an idea in what part of the design process management aspects should be addressed.

2. DESIGN OF THE NETWORK SERVICE PROVIDER

The purpose of this section is to show how management relates to the design process. The scope of this section is the example of a general purpose network service provider. The ideas expressed in this section could as well be applied to other levels of the OSI Reference Model.

A design usually starts from the specification of the user requirements. At this level of the design, the provider's internal structure is not considered [1]. The top-level specification is therefore a *service specification*, describing the provider as a black-box. The design process must gradually transform this black-box into a 'transparent' box. As a result, the functions that must be performed within the provider will become clear. Also the mapping of these functions upon physical components must be considered.

Section 2.1 discusses the technique of decomposing an N-service provider into an N-protocol plus an (N-1)-service provider. Section 2.2 shows how an N-protocol must be refined before a viable implementation can be obtained. Section 2.3 argues that, for the inclusion of management, it may be necessary to perform the techniques of Sections 2.1 and 2.2 in a *cyclic fashion*.

2.1. Vertical refinement

According to the OSI Reference Model, the network service provider can be decomposed into a number of data link service providers, interconnected by a network layer protocol (Figure 1).

 The purpose of this decomposition, is to describe the complex network service provider in terms of its less complex functional ingredients. It is expected that the implementation of these ingredients is simpler than the implementation of the monolithic network service provider. An advantage of this decomposition, is that the dependencies between the functional ingredients are well defined. The data link service providers for example, transfer data on behalf of their users in a transparent way; they do not interpret this data nor do they know the purpose of the data exchange. On the other hand, the network layer entities are not aware of the internal structure of the data link service providers. These facts make this decomposition technique so important.

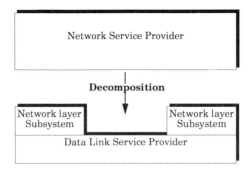

Figure 1: Decomposition of the Network Service Provider

The decomposition technique is usually applied in a recursive fashion. For example, the complexity of the data link service provider still prohibits an immediate implementation. The data link service provider must therefore also be decomposed. *The design is thus refined in a vertical fashion* (Figure 2).

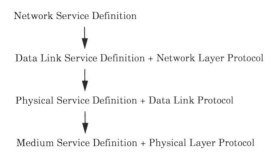

Figure 2: Vertical Refinement

2.2. Horizontal refinement

Not only the data link service provider, but also the network layer protocol requires an additional number of refinement steps. This is, because the top-level protocol specification abstracts from a number of 'details'. Examples of these 'details' are:
- PDU coding: Coding can be considered as the protocol's syntax, it can be addressed in later design stages than the protocol's semantics (rules). Example: first the decision is made that each PDU carries the source address. Than a decision is needed whether the source addresses have fixed or variable lengths, whether address length fields are needed etc.
- Parameter settings: Accurate parameter settings can only be given at later design stages, where more knowledge is available concerning acceptable protocol implementations. Example: the setting of the 'life-time control field' depends

(amongst others) upon the speed of the Intermediate System's forwarding process.
- Routing: the highest level protocol definition may specify *that* routing is needed, but refrain from *how* routing should be performed.

'Details', such as mentioned above, must be elaborated in a number of subsequent design steps. To distinguish these steps from the previously mentioned vertical refinement steps, we will call these steps *horizontal refinement* steps.

While some of the protocol 'details' (e.g. coding) must be resolved before viable implementations can be build, other 'details' (e.g. parameter settings) can be resolved at installation time. In this case, the equipment should support *local management* functions and the (human) network manager 'walks around to resolve the details'. In fact, the network manager *implements those parts of the protocol that have earlier not been refined*.

2.3. Cyclic approach

Vertical and horizontal refinement are both exponents of a top-down design process[1]. In order that such a process be complete, the top-level specification should be complete. It is thus necessary that the top-level specification addresses *all* requirements. Due to the conflicting nature of some of the requirements (e.g. low-cost versus high reliability), in real designs it may be impossible to have all requirements completely specified at the design start. As an alternative, the design could be performed in a *cyclic fashion*.

In a cyclic design, the first cycle is restricted to the most important requirements. In subsequent cycles, new requirements are added and existing requirements refined. This process continues until, in the last cycle, all requirements have been addressed[2] (Figure 3).

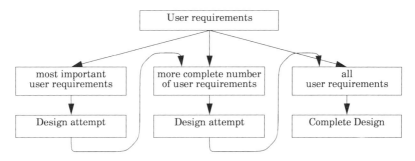

Figure 3: Cyclic approach

---

1. We use the term 'top-down' to indicate that the architecture is transformed (via a number of steps) into a realisation.
2. In real designs, the requirements may change in time. If this is the case, the design never completes and the last cycle does not exist. Note that, in case the first cycles address only a limited number of requirements, it may be better to withhold from the realisation until a more complete picture is obtained.

The advantage of this approach, is that the dependencies between requirements gradually become clearer. Subsequent cycles may therefore start from better (in quantity and quality) requirements. Networks should also be designed in cycles. User data aspects should be handled in the first cycles. *Management aspects should be added in subsequent cycles.* The effect of these subsequent cycles is twofold:
1: the layer protocols may be refined, and / or
2: the layer structure may be refined.

The first aspect is the cyclic variant of horizontal refinement. This time however, there is no need to know all user requirements in advance. In the remaining text, horizontal refinement is used in the sense of this variant. Section 3 shows an example.

The second aspect leads to a modification of the structure that resulted from the previously discussed 'vertical refinement' (Figure 2). The result of this modification, is a decomposition of the network service provider into a network layer protocol, a Data Link Service Provider (DLSP) *plus a new service provider for the transfer of network layer management information* (Figure 4). Compared to our previous structure, the DLSP is still the same. The network layer protocol however, additionally performs network management functions. Examples of such management functions are fault and configuration management. The data for these management functions is exchanged over the 'Management Transfer Service Provider' (MTSP).

From a functional point of view, the MTSP and the DLSP are independent. As opposed to the DLSP, the MTSP may support the reliable transfer of arbitrary sized user data (e.g. routing tables). The MTSP may also support code conversion and cover a wider area than the DLSP. In fact, the service provided by the MTSP can be the same as the *Common Management Information Service* (CMIS) [2].

From an implementational point of view, the MTSP and DLSP may remain independent. However, it is also possible to map the MTSP and DLSP upon the *same* physical components. Such a shared use of components introduces a dependency that was absent at the (higher) functional level. ISO's Management Framework [3] follows this approach.

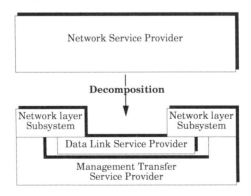

Figure 4: Decomposition into separate underlying services

3. NETWORK LAYER MANAGEMENT

This section applies the theory of *cyclic protocol refinement* to a real-life example. The example that will be taken, is the example of the connectionless network layer. It will be shown that the first design cycle addresses aspects related to the exchange of ordinary user data, and that subsequent cycles address management aspects (Figure 5).

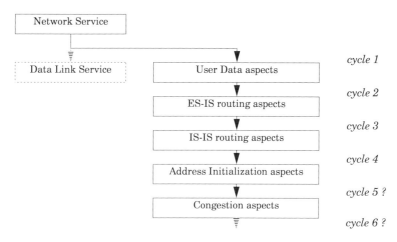

Figure 5: Cyclic development of the network layer

3.1. Connectionless network protocol

The most important aspect of the network layer, is its ability to transfer user data on an end-to-end basis. The designers of the network layer addressed this aspect first. The result of this was the ConnectionLess Network Protocol (CLNP) standard [4]. Besides of the coding rules, the standard defines *what* functions are needed to transfer user data. The standard does not define *how* these functions should look like (apart from some exceptions). Nor does it specify how the knowledge is obtained to control these functions. Obtaining such knowledge is considered to be a management action that should not be addressed in the first cycle of the design.

The statement that CLNP refrains from specifying *how* the network layer functions look like, requires some explanation:
- CLNP defines that routing should be performed. How routing should be performed is not defined, except for some direct consequences for the structure of the normal data units. These consequences are that all PDUs should carry a destination address, and that fields may be included to carry a (partial or complete) source route. No decision on a specific routing scheme is made however.
- CLNP recognizes the fact that data units may need to be segmented. A description of the control fields that are needed to perform segmentation, is included. However, CLNP does not define how systems obtain the knowledge whether segmentation is needed or not.
- CLNP recognizes that intermediate systems may need to discard data units be-

cause of congestion. How intermediate systems detect congestion, is not specified however.
- CLNP recognizes that intermediate systems may need to discard data units because the destination is not reachable. How intermediate systems detect reachability, is not specified however.
- The service, as well as the protocol standard requires that addresses and / or entity titles be associated with each network layer subsystem. How these addresses and titles should be assigned, is not specified however.

3.2. ES-IS routing protocol

After the user data aspects of the network layer were defined, a subsequent cycle was started to address the management aspects. As a result of this, the network layer protocol was refined[1].

The second cycle was based upon the observation that the number of End Systems (ESs) connected to a LAN, is much higher than the number of Intermediate Systems (ISs) connected to the same LAN. It was therefore decided to keep the ESs as simple as possible by choosing a routing scheme that would not require storage of routing tables by ESs. The major part of the routing problem was delegated to the ISs, even if this would make ISs more complex [6].

The following routing strategy was chosen. Each Intermediate System advocates its routing capabilities by regularly multicasting an IS-Hello PDU to all ESs. This Hello PDU contains configuration information of the IS (data link address plus network entity title). Upon receipt, an ES stores the information of at least one IS. This information may be used to reach that IS. All data units that should be transmitted by the ES, will be transmitted to that IS first. It is now the responsibility of the IS to make an appropriate routing choice.

Whenever an IS detects that a better route exists (e.g. the destination is on the same LAN as the source), the IS informs the source of this better route. The ES may decide to use this information for future data units, or ignore it.

The above routing strategy is standardized in the ES-IS routing protocol [5]. This protocol can also be used to transfer configuration management information from ESs to ISs. This information serves two purposes:
- to tell the ISs how a particular ES can be reached.
- to inform the ISs of the existence of particular ESs. This information can be used by ISs to decide whether a data-unit can be forwarded, or should be dropped[2].

3.3. IS-IS routing protocol

After the ES-IS routing aspects were addressed, a third cycle was performed. This cycle was based upon the observation that a world-wide OSI network may consist of a large, but fluctuating number of IS. This fluctuation is a result of the fact that new ISs may be added, old ISs be removed and that existing ISs may fail or come back into service. Since these configuration changes may have impact on the routing decisions, a number of standards were developed that addressed the problem of routing outside the local environment (= between Intermediate Systems).

A first standard defines the routing framework [6]; this standard gives the re-

---

1. The result of this was not a new version of the ISO 8473 standard, but additional standards to be used in conjunction with ISO 8473.
2. Note that CLNP requires this information to be available

quirements for the subsequent design step. Among the aspects that are defined in this framework are the notions of administrative and routing domains. As a result of this cycle, two other standards will emerge. The first one [7] defines how routing must be performed within a single routing domain, the second one [8] defines how routing must be performed between domains.

The resulting protocols can be considered as important configuration management protocols. The availability of these protocols will ease the task of the (human) network manager.

3.4. Future protocols

In this subsection we will mention some areas in which the development of additional network layer standards can further ease the task of the network manager. Since some of these standards will rely upon CMIS (or another service provider for the transfer of management information), refinement of the layer structure may be needed. Some of these areas have already been identified during the design of CLNP.

3.4.1. Assignment of network layer addresses

Each ES supports one or more Network Service Access Points (N-SAPs). Each N-SAP should have a unique address. These addresses are usually assigned by the network manager and need to be stored somewhere in the system. Common locations to store this knowledge are the system's hard-disk or an EEROM. In case the knowledge gets lost or in case the knowledge needs to be modified, the network manager needs to intervene.

Availability of CMIP or SNMP may not change this situation. This can be understood by realizing that these protocols are application layer protocols, and thus require some kind of underlying service to be available. If no N-SAP addresses are available, provision of the network service is impossible and higher layer protocols (e.g. CMIP) can not operate.

To allow for the automatic assignment and maintenance of N-SAP addresses, a special protocol is required. Development of such a protocol is proposed by ISO-IEC / JTC 1/ SC6 / WG2 [9].

3.4.2. Congestion control protocols

Each network, no matter what size, may become overloaded. To recover from overload situations, congestion control procedures are needed. Such procedures usually specify that data units must be discarded by ISs.

The CLNP mentions two aspects that relate to congestion. The first aspect is that ISs may discard data units due to local congestion. The second aspect is that data units must be discarded in the case of expiration of their lifetime. This aspect is related to global congestion (correct operation of higher layer protocols may also depend upon maximum lifetimes).

CLNP does not specify however how the initiating ES determines which lifetime should be used for the transmission of a data unit. Obtaining such knowledge, may be the task of a future congestion control standard.

3.4.3. Accounting protocols

In the past decade, the costs of using networks have not played a decisive role. In the local area environment, the network owner (usually a company or university) had often central budgets to pay all network costs. In the wide are environment, until now users could often rely upon research networks that were sponsored by governments and big companies (e.g. Arpanet, Earn). Now that both the size and usage of the networks have increased, passing on the costs to the individual users has become more important.

It is desirable that accounting protocols will be designed in the near future. These protocols should not only collect accounting information on a retrospective base, but should also take action on a prospective base. An example of the latter is a command to a gateway to drop all PDUs that come from a specific source.

4. CONCLUSIONS

This paper proposes to relieve the network manager by paying more attention to management aspects in the design phase of networks. The writer believes that by doing so, networks will get better characteristics concerning fault-tolerance, security, the capability of self-configuration, etc. Since the distinction between 'normal data aspects' and 'management aspects' is hard to make, there should be a close collaboration between the designers of the 'normal layer protocols' and the 'management protocols'. The result of this collaboration should be the production of new standards and RFCs in subsequent design cycles.

5. REFERENCES

1. Bogaards K., "A Methodology for the Architectural Design of Open Distributed Systems", PHD thesis, University of Twente, Enschede, The Netherlands, 1990
2. ISO-IEC 9595 - Common Management Information Protocol
3. ISO-IEC 7498/4 - Basic Reference Model - Part: 4 Management Framework
4. ISO-IEC 8473 - Protocol for providing the Connectionless-mode Network Service
5. ISO-IEC 9542 - End System to Intermediate System routing exchange protocol for use in conjunction with the protocol for the provision of the connectionless-mode network service
6. ISO-IEC 9575 - OSI Routing framework
7. ISO-IEC 10589 - Intermediate System to Intermediate System Intra-Domain routing exchange protocol for use in conjunction with the Protocol for providing the Connectionless-mode Network Service
8. ISO-IEC / JTC1 / SC6 N6120 - Intermediate System to Intermediate System Inter-Domain Routeing Exchange Protocol.
9. ISO-IEC / JTC1 / SC6 N5849 - Proposal for a New Work Item: Dynamic discovery of OSI NSAP addresses by End Systems

Verification of Network Management System Configurations

David L. Cohrs and Barton P. Miller

Computer Sciences Department, University of Wisconsin – Madison, 1210 W. Dayton Street
Madison, Wisconsin, USA 53706

Abstract
The size and complexity of current computer internets are increasing the need for automated network management. Internets have become too large, and individual administrative domains too autonomous for a centralized system. Distributing the network management system decentralizes control over the configuration of the parts of the system. We address this problem through the use of a high level, formal specification language, NMSL. NMSL allows network administrators to describe their network environment and its relationship to other environments. The NMSL system verifies network management specifications, and configures network management systems given a verified specification. Configuration can be performed using standardized protocols, such as SNMP.

This paper presents a model for network management systems, and a method for verifying specifications of these systems. We divide the verification problem into three parts: capacity, protection, and configuration verification. We also provide a way to distribute the verification process, and a way to summarize information for propagation across domain boundaries. We discuss the performance of our implementation of this system, and describe our future research directions.

1. INTRODUCTION

Computer internets are growing in size and complexity, and the need for automated network management systems is growing as well. If the network is small, the task of network management can be performed by a human administrator, using simple, ad hoc tools. In a large network, managing the network management system becomes a significant problem in itself. The NMSL (Network Management Specification Language) system addresses the complexity of managing larger internets by providing a language to specify the configuration of network management systems, and a way to configure network management systems from a specification.

A large network management system has problems similar to those of other networking applications. It instantiates processes throughout the network. The processes communicate via network management protocols, which must be configured correctly. These processes must be configured to perform the correct internal operations to make queries at the correct

---

The authors' e-mail addresses are dave@cs.wisc.edu and bart@cs.wisc.edu.
Research supported in part by an AT&T Ph.D Scholarship, NFS grant CCR-8815928, ONR grant N00014-89-J-1222, and a Digital Equipment Corporation External Research Grant.

time, to answer queries correctly, or to reject invalid queries. Additionally, these processes are divided along the lines of *administrative domains*. Domains reduce the amount of sharing and coordination possible in configuring the network management system. Each administrative domain must remain autonomous – administrators and the owners of a domain are generally not willing to give up control over the management of their networks. However, because management processes need to communicate between these domains, a mechanism is necessary to help coordinate this communication.

The NMSL system addresses the problems of distributed network management configuration. The goal of NMSL is to reduce the errors present in the network management system due to incorrect configuration. We employ a high level specification language to achieve this goal. A specification of a network management system can be verified, and a verified specification can be used to directly configure the processes and data bases used by the network management system. The verification process is distributed along domain boundaries, allowing administrative domains to hide any information about themselves that do not pertain to the relationships between domains. An overview of the NMSL system can be found below, with additional information in [5, 6].

1.1. NMSL Overview

The NMSL system provides network administrators with a language, NMSL, for describing the configuration of their network management system and the networks that they manage. NMSL specifications depend on a model of network management systems structure, shown in Figure 1.1. In this model, a network management system consists of management *processes* that interact via a management protocol. The processes maintain the state of the network management system in a set of *objects*. Objects are the data or management information maintained by the network management system. The processes manage the hardware attached to the network, which we call *systems*. Systems include all types of network-attached hardware, from mainframe computers and workstations to bridges. Systems are grouped into *domains*. Domains define the administrative boundaries in network management. In existing network management systems, administrative domains are set up in a hierarchical manner, so we support this structuring of domains in our model. Domains are also allowed to overlap, allowing systems (and the processes that manage them) to be members of more than one domain. Overlaps are also permitted in existing network management systems, like SNMP. An example specification is shown in Figure 3.1.

NMSL allows the specification of *objects* and *processes* in terms of definitions and instantiations. Object specifications define the management information, including structural information (abstract data types), access permissions, naming, and containment. Objects are specified using standard ASN.1 notation[7]. Process specifications define a process's operation, including the types and frequencies of queries, the objects the process must be able to access, and the objects the process allows other processes to access.

NMSL uses *system* and *domain* specifications to model the physical layout of the network management system. System specifications describe the individual properties of the hardware, e.g. a computer, router or bridge, attached to computer network, and the software that runs on that hardware. They include the instantiations of processes and configuration information, such as the number and types of networks to which the system is connected.

Domain specifications describe the administrative grouping of systems and processes. They define a boundary between administrative organizations. Domain specifications list the systems and domains that are members of the domain, and processes that operate on behalf of the domain. Domains can partially overlap as long as the overlap does cause a self-reference.

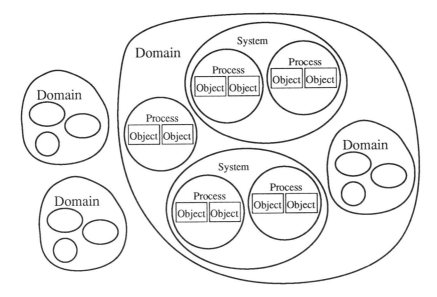

Figure 1.1. The NMSL Network Management System Model

There are two roles in which NMSL can be used, for describing the network management configuration of an internet, and for configuring the management processes described in the specification. We call these two roles the *descriptive* and *prescriptive* roles of NMSL.

The NMSL compiler is central to both roles of the NMSL system. In its descriptive role, the compiler takes as input the full specification of an administrative domain, and verifies its validity. Validity is defined in Section 2. Some parts of the specification may include references to the specifications of other administrative domains. In this case, the verifier creates a new specification for this domain, describing only its interface with other domains. The *external* specification is propagated to external NMSL verifiers. Information about the internal structure of a domain is not propagated across domain boundaries.

Given a valid specification, the NMSL compiler generates prescriptive output in the form of commands to configure the network management processes within the domain. These commands can take many forms, so the NMSL compiler is has the ability to generate many types of output. The systems and processes of the domain are configured using implementation dependent modules, that speak the protocols and have the permission necessary to reconfigure the network management system. Ideally, standard management protocols, such as SNMP[4] or CMIS/CMIP[8,9] will be used to reconfigure the network management system. Therefore, the NMSL system provides for both the description of the network management system through a specification language, and a method for enforcing that description through automated configuration.

2. THE CONSISTENCY MODEL

The consistency model states the conditions that must be met for a given specification to be correct. It divides these conditions into three categories: *capacity*, *protection*, and *configuration*. These categories correspond to the three types of relationships specified in a NMSL specification. Any specification, for example, a specification of the error messages a router sends and the conditions under which they occur, are defined in terms of these three relationships. The capacity condition states that a service provider must have enough capacity to handle the requests of all of its clients. The protection condition states that a client must be given permission to perform the requests that it makes. The configuration condition states that individual configuration statements in a specification must meet global constraints. If these three conditions are met in a specification, the specification is consistent.

The existence of administrative domains complicates the problem of determining the consistency of a network management specification. The autonomy and privacy of administrative domains does not allow all the information of a domain's specification to be sent to a central location for the consistency check. Copying all of the specifications to a central location is bad for performance as well. To solve these problems, we divide a domain's specification into two logical components, its internal specification and its external specification. The internal specification defines how the parts of the domain, including subdomains, interact with each other. The external specification defines the how this domain interacts with other domains. Dividing the consistency check along domain boundaries allows information hiding, reduces the search space as compared with a single, centralized check, and distributes the work involved.

In sections 2.1 through 2.3, we describe each of the consistency categories, capacity, protection, and configuration. Section 2.4 describes the effect that domains of administration have on this basic model, and how the internal and external specifications are derived.

2.1. Capacity

The goal of the capacity model is to determine, as quickly as possible, if each service provider has the capacity to provide the services needed by its clients. This is a form of the classic capacity planning problem[3].

Capacity planning provides a systematic approach to modeling and predicting the capacity of a system, in our case, a network management system. To form a capacity planning model, one must determine the parameters that characterize the system workload, and the parameters that are required to predict the system's future performance. Designing a capacity planning model requires the creation of an initial model for the system workload, validation, and modifying the model if it does not adequately model the capacity of the system.

In our capacity planning model, we wish to obtain a reasonable answer to the capacity question by use of a simple, easy to understand model. This lead us to employ a system of closed-form equations to solve the capacity planning problem. Closed-form equations have two characteristics we find important. First, since the users of this system will not be performance experts, they need a simple, easily understandable model with simple parameters. Second, this model is important because our capacity problem is part of a larger automated consistency proof, which requires a yes or no answer, and also must execute quickly.

After serious consideration, we determined that standard modeling tools such as queuing networks were not appropriate for our task. Given our desire for fast execution in the face of a large data set, a queuing model would not be appropriate. Furthermore, standard modeling technology was not necessary with our simple model, and enhancing our model to include forms of feedback made the problem too difficult for current tools.

The capacity model assumes a *client/server, request/response* based system. Each network management process is considered a client or a server (or both). Network management standards[4, 10] refer to these processes as *managers* and *agents*, respectively. These processes are instantiated on systems (hardware) throughout the network being managed. The systems of the network are divided into administrative domains, implying that the processes are also divided into administrative domains. Administrative domains are allowed to nest or overlap, but a domain may not contain itself.

The capacity model employs an independent, discrete distribution of the frequency of interactions, requests and responses, between network management processes. Interactions are measured in queries per second (qps). We assume the messages involved are sent reliably. We also assume that response time is not a factor in determining the frequency of requests.

Given a group of clients, we can determine the aggregate load that the group places on a server. This aggregate load and its distribution are important for determining the probability of overloading a server and for propagating information between domains. Given the frequency distribution of the clients, we can determine the average and peak load and can calculate the discrete aggregate distribution of the group. The method used for determining consistency depends on the constraints of the client processes. Other uses are described in Sections 2.1.1 and 2.4.

To keep the model simple and the problem tractable, a few concepts are not included in our model. We do not model the fine-grained operation of the message delivery protocols. Behavior is considered on an instantaneous basis. The model also excludes response time. Because we are not modeling the behavior of the entire network, but just the end-to-end behavior of the processes, we do not have enough information to determine response time. Thus, because we do not model these interactions, the results obtained by this model should be used as only one step in determining the capacity of a system.

2.1.1. Calculating Capacity

We are interested in determining the capacity, in queries per second, of servers, and in determining the load clients will place on each server, *i.e.* whether each server's capacity will be exceeded. The load can be measured in several ways. We are interested in the *average utilization*, the average number of queries per second clients place a server, and the *peak utilization*, the maximum number of queries a server can receive per second. We are also interested in the probability that a server's capacity will be exceeded. The utilizations depend on the distribution of requests from the clients, which in turn depend on the number of *modes* of operation of each client. If only one client is being considered, we call its distribution a *simple distribution*. A group of clients has an *aggregate distribution*.

We divide the calculations into two cases: a single client querying a single server, and a group of clients querying a single server. The case of a single client simultaneously querying a group of servers (multicast), while an interesting from a reliability point of view[2], is not used in current network management systems.

The single client/single server case is easy to calculate. For example, consider the interactive client and server shown in Figure 2.1; the average utilization is 4 *qps*. The peak utilization is 20 *qps*, and the aggregate distribution is the same as the simple distribution. In general, for a single client and a single server, the average rate is just the sum of the rates for each mode. Determining whether this example is consistent is a matter of determining if the server can withstand the load of its client. In the example in Figure 2.1, the server, which can receive 20 *qps*, has the capacity for both the average and the peak rate of requests. Therefore, the probability that the server's capacity will be exceeded is zero.

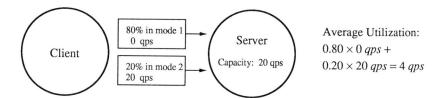

Figure 2.1. A Single Client/Single Server Configuration

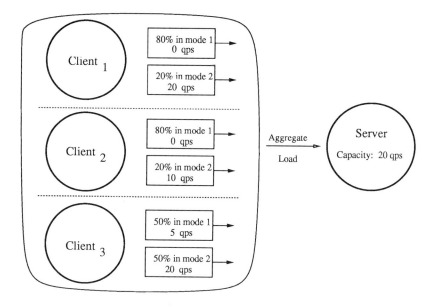

Figure 2.2. A Multi-Client/Single Server Configuration

In the multiple client/single server case, we simply add the loads of the individual clients to get their average and peak request frequency. For example, Figure 2.2 shows three clients $Client_1$, $Client_2$, and $Client_3$. Their average request rate is 18.5 *qps*. In this case, the server, which can handle 20 *qps*, will not be overloaded in the average case. The peak request rate is the sum of the peak rates for each of the clients, 50 *qps*. This is greater than the capacity of the server. The distribution of the aggregate load of these three clients has 8 modes, and is determined using simple probabilities. Basically, we take all combinations of the modes of the three clients. For example, to calculate aggregate Mode 1, we take the combination of Mode 1 of $Client_1$, Mode 1 of $Client_2$ and Mode 1 of $Client_3$. The probability of operating in this aggregate mode is 0.32, and the number of queries per second is 5 *qps*. The other modes are calculated similarly. The resulting distribution is

Mode 1:	0.32	5 *qps*		Mode 5:	0.08	15 *qps*
Mode 2:	0.32	20 *qps*		Mode 6:	0.08	30 *qps*
Mode 3:	0.08	25 *qps*		Mode 7:	0.02	35 *qps*
Mode 4:	0.08	40 *qps*		Mode 8:	0.02	50 *qps*

From this aggregate distribution, we can see that the probability that the server will be overloaded at any given instant is 0.28 (the sum of the modes with query rates greater than the server's capacity of 20 *qps*).

In general, for n clients, $C_1, C_2, \ldots, C_n$, each with m operating modes, or request rates, $F_i(m)$, the probabilities of being in each of the modes, $P_i(m)$, and one server, S, the average aggregate load is

$$\sum_{i=1}^{n} \sum_{j=1}^{m} P_i(j) F_i(j).$$

The maximum load is

$$\sum_{i=1}^{n} \max_{j=1}^{m} (F_i(j)).$$

The aggregate distribution is calculated by taking all combinations of the n clients and m modes. For each mode, we determine the probability of operating in that mode by multiplying the probabilities of the combination of modes of the n clients. We determine the query rate by summing the query rates of the clients for the given combination of their modes. Because there are n clients, each of which has m modes of operation, the time to calculate the aggregate distribution using a naive algorithm is $O(m^n)$.

We can reduce this to polynomial time by using an approximation, and restrict aggregates to have exactly N modes. If we restrict the number of modes allowed in an aggregate to a constant number, and if we calculate the aggregate iteratively, the time to calculate the approximate probability is $O(n \times N^2)$. Details are shown in [6]. In a related algorithm[12], it was shown that by increasing N this approximation can be made arbitrarily close to the real probability of exceeding the server's capacity.

2.2. Protection

The purpose of the protection model is to verify that all operations performed by a client on a server are permitted by that server. For example, if a client specification states that it requests routing tables from a server, the specification for that server must state that this client has permission to perform that operation. The protection model includes the property that each request performs a single operation on a single object.

To verify the protection condition, we must show that each client has permission to perform each of its queries. To perform this verification step, we developed a theory for protection conditions, described in [6]. To apply this theory, we convert the access specifications in a NMSL specification into a set of logical relations, and prove that there are no inconsistencies in this set. That is, we prove that for every request, there exists a corresponding permission. The problem is simplified by the finite problem space – we only need to prove this condition for a given specification, and not for all possible specifications. This allows an iterative proof method. We use a Prolog dialect, CLP(**R**)[1], to automate this iterative proof.

To hide the information specific to a given domain from other domains, we need to generate a new specification and new logical relationships from the set of requests that should be satisfied by permissions from other domains. The algorithm for producing this external specification is described in [6].

2.3. Configuration

The configuration model is used to verify parts of NMSL specifications relating to the configuration of network managers. We call each of these a *configuration condition*. This classification includes all parts of a specification not modeled by the capacity or protection models. The configuration conditions have the common characteristic that they are general requirements placed on the entire specification by a single element of the specification. Some examples of configuration conditions are verifying that: the protocol used to communicate between two management processes is the same, the maximum number of routers allowed in a network is less that some maximum, and all requests sent to a process are smaller than the maximum request size accepted by that management process.

Two examples of specifications defining configuration conditions are shown in Figure 2.3. These examples are fragments of a NMSL specification. The first example shows the hardware configuration of a computer on our local network, defining several configuration conditions. The cpu type and operating system version define constraints on the types of programs that can be run on this computer. The interface definition sets conditions for maximum transmission rates and packet sizes.

The second example specifies the requirement that the number of computers that route packets between networks (i.e. those computers that run routing processes), be three or less. This type of configuration condition is important for networking products that have limits on the number of routers allowed on an individual network. They can also be used to enforce administrative requirements. Limiting the number routers on a network to one, for example, would enforce the requirement that that network cannot be split into two physical subnetwork without intervention from the central administrator.

These conditions can also be used at runtime to configure network management processes. For example, by listing the interfaces a computer has and the networks to which it is attached, a network management server can be configured to monitor these interfaces. Different instantiations of the management process can be configured in different ways, depending on these conditions. This use is part of the prescriptive role of NMSL, mentioned in Section 1.

To verify the configuration conditions, we must show that each of the conditions in a specification are met. The method used is similar to that use for verifying permission

```
system "dip" ::=                    domain "wisc-cs" ::=
  cpu mips;                           requires {
  opsys ultrix version 2.1;             count(process Router) >= 1 and
  interface se0 {                       count(process Router) <= 3;
    net wisc-twisted;                 }
    type ethernet-csmacd;           end domain "wisc-cs".
    speed 10 Mbps;
  }
end system "dip".
```

Figure 2.3. Examples Of Configuration Conditions

conditions. To verify conditions concerning amounts, or those concerning numbers of entities (*e.g.* objects or systems), we employ equations, in the manner similar to that used for verifying the capacity of the system.

2.4. Domains

Our model thus far has implicitly assumed that we will evaluate the specifications in a single, central location. This centralized method can cause two problems. First, the number of logical relations to evaluate would be quite large. Second, it would require that each organization provide detailed information about their internal computing environment. Both of these problems occur at the domain boundary, and share a common solution, information hiding. The goal of information hiding is to summarize the specification at domain boundaries and propagate a small subset across the boundary.

Information hiding is necessary to preserve the autonomy of each domain, and ensure that private management information is not visible to the outside. It can also prevent information explosion. We employ three methods for providing information hiding. First, we include only those parts of a domain's specification that reference another domain. Second, we remove private and common information. Third, we summarize the remaining information.

When determining which parts of a domain's specification to include in the external specification and which are private or common, we consider the low level, logical relations, not the high level NMSL specification. Those relations that reference only objects that are in the local domain and cannot be referenced by another domain, can be excluded from the external specification, because they are private. Examples of such relations are those that concern the frequency of requests a client in the domain makes to a server within the same domain, the permissions involved with those requests, and the configuration conditions of individual systems within the domain. Relations concerning process specifications can also be excluded; all that is important to the external specification is the behavior of the instances. We must include relations that have an indirect effect on other domains. An example of an indirect effect is if a client queries a server that serves both clients both inside and outside the domain. Common information includes relations describing the MIB; such information need not be propagated.

To summarize the capacity specifications of a domain, we determined the aggregate load that the domain's client processes place on other domains. To do this, we must determined which clients contribute to the domain's aggregate load, and then calculate the aggregate load. Such clients include those that refer to servers in another domain. Clients that refer to a server that serves both the local the other domains must also be included in the summary; these clients cannot be considered private. At this point, the only remaining logical relations for clients are those that make external references. Given these external references, we use the aggregation method described in Section 2.1 to calculate the aggregate load.

We cannot hide the specifications of individual servers in the way we described above, because any aggregate of the servers' capacities would lose information about the individual servers' capacities. This would allow invalid specifications to pass through the verification phase, despite inconsistencies. For example, if all of the clients query the same server in a group of servers, they could overload that server, but if the servers' capacities are aggregated, this inconsistency will not be recognized.

To summarize the protection specifications, we propagate relations listing the objects referenced, but specify the domain, not a process instance, as the initiator of the reference. Once again, the permission relations must be propagated unchanged. Summarization of configuration specifications follow the rules we described for the capacity and protection specifications.

3. APPLYING THE CONSISTENCY MODEL

In practice, network management systems perform high level operations, such as detecting error and faults, and noting exceptions. These operations often cause messages to be sent from a network management server to an application, such as a network operation center (NOC) tool, so that the network operator can take appropriate action. While this is a small part of what a network management system can do, it is the most common current application. This section shows an example of one way error reporting can be performed, the way to specify the example in NMSL, and the relationship between the specification and the three verification categories.

To see how errors propagation is handled in our Consistency Model, we use a simple example. Consider a router connecting two networks, and a management station (another computer) on one of these networks. The router, in addition to routing packets, runs a network management server or agent. The network management stations runs several network management applications – these all communicate with the agent via SNMP[4]. The agent supports trap management, as defined in the SNMP standard, and sends trap notices to a trap management application on the management station. A specification of this example is shown in Figure 3.1. The gateway is the system gw and the management station is called

```
system "gw" ::=
  cpu cisco;
  opsys cisco version 2;
  interface ie0 {
    net wisc-research;
    type ethernet-csmacd;
    speed 10 Mbps;
  }
  interface ie1 {
    net wisc-twisted;
    type ethernet-csmacd;
    speed 10 Mbps;
  }
  supports mgmt.mib;
  process snmpd("dip");
end system "gw".

system "dip" ::=
  cpu mips;
  opsys ultrix version 2.1;
  interface se0 {
    net wisc-twisted;
    type ethernet-csmacd;
    speed 10 Mbps;
  }
  process trap_hdlr("gw");
end system "dip".
```

```
process snmpd(HOST: string) :=
  supports mgmt.mib; -- full MIB
  exports mgmt.mib {
    to "wisc-cs"; access ReadOnly;
    rate { mode 1 1.00 10 qps; }
  }
  sends traps {
    to HOST;
    using protocol "snmp";
    port "snmp-trap";
    rate { mode 1 1.00 1 qph; }
    provides {
      "packetsize" <= 484 octets;
    };
  }
end process snmpd.

process trap_hdlr(HOST: string) :=
  receives traps {
    from HOST; port "snmp-trap";
    using protocol "snmp";
    rate { mode 1 1.00 1 qph; }
    requires {
      "packetsize" <= 1024 octets;
    };
  }
end process snmp_trap_handler.
```

Figure 3.1. Example Of A Trap Propagation Specification

`dip`. The agent and application processes are `snmpd` and `trap_hdlr` respectively. In this specification, the rate of interaction is at most one trap per hour (1 qph).

For traps to be sent correctly, several conditions must be met:

(1) The agent must specify the recipient of the trap messages (the application running on the network management station).
(2) The agent must give the recipient permission to view any trap related data objects.
(3) The recipient must be interested in receiving trap messages, *i.e.* it must permit the agent to send it the traps.
(4) The agent and the recipient must speak the same protocol and agree on a rendezvous point (e.g. an IP port number).
(5) Packet sizes must be within acceptable bounds.
(6) The rate of interaction (sending of trap messages) must be within the limits of the configured system.

These requirements were arrived at by examining the contents of a `TRAP-PDU` message in SNMP, which is sent from an agent to an application, as well as general requirements of network management interaction. A brief inspection of the requirements given above show that they fall into the three categories of capacity, protection and configuration.

Some of these requirements are difficult to determine, especially (6), but a reasonable value can be determined for interaction rates based on examining the mean time between failure characteristics of the the routing hardware. Obtaining such information for the agent itself is not a subject of this paper.

As this example shows, the NMSL specification written by a network administrator need not be divided into these three parts. However, for the purpose of verification, the compiler/verifier takes the input, high level specification, and generates constraints in these three forms.

4. IMPLEMENTATION AND PERFORMANCE

We have implemented a verifier for the Consistency Model described in Section 2. Verification is a two step process. First, a high level, NMSL specification is compiled into a low level set of logical relations, like the ones described in Section 2. Next, these logical relations are passed to a proof checker that uses the rules we described to find inconsistencies in the specification. After a brief description of the the compiler and verifier, we present the performance of the current implementation.

The compiler is written in C, and provides an interpreted extension language. The compiler's internal parser enforces only the syntactic structure of the specifications. An early version of the basic structure of NMSL specifications is described in [5]. This syntactic structure is a list of clauses, with the ability to group clauses into blocks. Examples of NMSL specifications are shown in Figures 2.5 and 3.1. The use of our extension language has allowed more rapid implementation of the specification language, and reduces the turn-around time when debugging the semantic routines.

In our implementation we represent logical relations as statements of CLP(**R**)[1]. CLP(**R**) is a Constraint Logic Programming language that provides a logic programming model similar to PROLOG, but with a more general proof mechanism than that used by PROLOG. The CLP(**R**) mechanism includes the ability to solve equations over the real numbers, which is useful in proving the capacity conditions. Given these logical relations, we use a set of deduction rules, also written in CLP(**R**), to determine if the capacity, protection, or configuration conditions are violated in the specification. Violations are reported to the administrator.

Several issues are important to the performance of the NMSL verifier. We need to quantify the sizes of the input specifications that the NMSL verifier is expected to process. Because we are concerned about information explosion at the domain boundaries, we need to determine how much information can be hidden within a domain, and how much must be propagated. The time needed to verify a specification is also of importance. To determine this, we must look at the execution time of the NMSL compiler and the time it takes CLP(**R**) to verify a specification.

At the time this paper was being written, the automated network management system in our department was in a early state. Because of this, we have written specifications for the for a representative management structure based on part of our current network configuration. The specifications were written for an SNMP environment.

The tests are straightforward. For each specification, we determined its size based on a simple line count. Each line corresponds, on average, to a clause in NMSL. We then compiled the specification and counted the number of logical relations it took to represent the NMSL specification. We also inspected the logical relations and determined the size of the internal and external specifications, in terms of their size in logical relations. Note that our implementation does not currently perform this separation automatically. Last, we used CLP(**R**) and the deduction rules we described in Section 2 to verify the specifications, and measured the time needed for verification.

The results of our tests are shown in Table 4.1. The first test is an empty file. This test shows the startup costs for the compiler to process the extension language statements. The second test is a specification of processes and systems within a domain, but describes no data objects. This test file shows the time it takes to compile and verify a specification with no data objects. The third is a full specification, including processes, systems, data objects, and domain groupings. By comparing the results of the second and third tests, we can determine the effect the number of data objects in a specification has on its verification time. The second and third tests included 7 system specifications, 3 process specifications, 2 server and one client, and 2 domains. The server processes are instantiated on each of the systems, the client at one central location. Of the domains, the first includes 6 of the systems, the second, the remaining system. The object specifications in the third test are those defined in the complete RFC1066 MIB[11]. The internal and external parts shown in Table 4.1 are divided using the criteria discussed in Section 2.4. These tests were run on a Sun 4/110, running SunOS 4.0.3 with 8 Megabytes of main memory. The file system buffer cache was primed before the tests were run, so disk activity was not a factor. The verification time for Test 3 is shown in minutes and seconds.

Several conclusions can be drawn from these simple tests. The sizes of specifications are intended to be large; the specification for a single domain will be well over 1000 lines long. The number of logical relations for a domain correspond roughly to the size of the input specification (Test 3 includes empty and comment lines). More importantly, the size of the external specification is kept small compared to the internal specification's size. Because the data object specifications in Test 3 came from a standard MIB, we were able to exclude them from the summary as well. Therefore, our summarization methods seem to be effective.

The compiler has a reasonable but noticeable startup overhead, due in part to the use of the interpreted extension language. However, after this time is factored out, it processes specifications quickly. A very important result is the effect the number of data objects in the specification has on the total verification time. In Test 2, the verification step executed

Table 4.1
Performance of the NMSL Verifier

	Test Number		
	1	2	3
NMSL Specification size (lines)	0	210	1215
Total Number of Logical Relations	0	191	569
Internal Part	0	174	1198
External Part	0	17	17
Compile Time (sec)	1.22	2.06	3.52
Verification Time (sec)	0.00	0.98	3:50.30

quickly, while in Test 3, many data objects needed to be checked for the protection conditions. Because there are so many data objects in the SNMP MIB, this becomes the main factor in the verification time. The implementation in CLP(R) does not prove the equivalent of a lemma, and reuse such results throughout the rest of the proof. This results in re-proving partial results each time they are needed. Some enhancements were made to the implementation to reduce this effect, but additional work is needed in this area.

5. SUMMARY

In this paper, we discussed the problems caused by the complexity and autonomy in modern network management systems with administrative domains. To solve these problems, we apply a formal specification language. The specification language, NMSL, addresses these problems by providing a way to formally specify the configuration of a network management system, and a mechanism to automatically verify a specification. We use a model for network management systems, based on four concepts: administrative domains, systems, management processes, and managed objects. Network management systems fit into this model, and NMSL uses this model to verify specifications. This method divides the verification problem into capacity, protection, and configuration conditions. Separate methods are used to verify each of these conditions.

We also distribute the verification problem along administrative domain boundaries. We use summarization to reduce the size of external specifications that are propagated across these boundaries. The distribution and summarization provided by the NMSL verification model are very important to preserving the autonomy of individual domain, and reducing the overhead of copying specifications across the network.

The current results of our work is encouraging. Our models for capacity and protection verification are well defined. We use closed-form equations are used for determining the capacity of servers. We also use a simple logic to verify protection conditions. Similar methods are used for proving the capacity conditions.

We have used the NMSL verifier to verify some representative specifications. Our tests show that specifications for small domains can be processed by the NMSL verifier in reasonable time. Our initial performance tests make several important points. The mechanisms we proposed for summarization cause a great reduction in the size of external specifications. The verification step also executes in a reasonable time for simple specifications. However, the performance of the verifier is affected greatly by the number of data objects present in the specification. This was caused by re-proving partial results in the verification proof.

Several issues remain to be addressed. The performance of the verifier for large numbers of data specifications is not as good as we had hoped. The cause of the problem has already been determined, but we need to investigate ways to avoid the problem. Next, we are turning to the other role of the NMSL system, which uses a verified specification to configure the network management system. Work is underway to set up an environment for testing the configuration role of NMSL.

NMSL and the techniques described are also appropriate for specifying and managing distributed systems other than network management. We plan to investiate applying NMSL to these other systems and services.

6. REFERENCES

[1] N. Heintze, et al, *The CLP(R) Programmer's Manual,* Dept. of Computer Science, Monash University, Clayton, Victoria, Australia (1987).

[2] K. P. Birman, "Replication and Fault-Tolerance in the ISIS System," *Proceedings of the Tenth ACM Symposium on Operating Systems Principles,* pp. 79-86 Orcas Island, Washington, (December 1985).

[3] L. Bronner, "Overview of the Capacity Planning Process for Production Data Processing," *IBM Systems Journal* 19(1) pp. 4-27 (1980).

[4] J. Case, M. Fedor, M. Schoffstall, and J. Davin, "A Simple Network Management Protocol," RFC 1067, IETF Network Working Group (August 1988).

[5] D. L. Cohrs and B. P. Miller, "Specification and Verification of Network Managers for Large Internets," *ACM SIGCOMM 89,* pp. 33-44, Austin, TX, (September 1989).

[6] D. L. Cohrs, "PhD Thesis," Computer Sciences Technical Report, University of Wisconsin – Madison (June 1991).

[7] Information Processing Systems – Open Systems Interconnection, "Specification of Abstract Syntax Notation One (ASN.1)," ISO 8824, International Organization for Standardization (December 1987).

[8] Information Processing Systems – Open Systems Interconnection, "Management Information Service Definition," ISO DIS 9595/2, International Organization for Standardization (1988).

[9] Information Processing Systems – Open Systems Interconnection, "Management Information Protocol Definition," ISO DIS 9596/2, International Organization for Standardization (1988).

[10] Information Processing Systems – Open Systems Interconnection, "Basic Reference Model Part 4 – OSI Management Framework," ISO DIS 7498/4, International Organization for Standardization (1989).

[11] K. McCloghrie and M. Rose, "Management Information Base for Network Management of TCP/IP-based Internets," RFC 1066, IETF Network Working Group (August 1988).

[12] D. C. Verma, Private correspondence.

II
MODELING AND DESIGN

B
Modeling Frameworks and Requirements

OSFA: An Object-Oriented Distributed Approach to Network Management

G. Raeder

MPR Teltech, Ltd., 8999 Nelson Way, Burnaby, B.C., Canada V5A 4B5

Abstract
Current developments in communications technology are having a profound impact on the operation and management of networks. A new type of management system architecture is needed to cope with this challenge. This paper shows how the properties of distributed object-oriented systems match the needs of integrated network management and how this technology is being applied within MPR Teltech to a new network management architecture, the Open Systems Framework Architecture (OSFA). Dynamic features such as object replacement, migration and replication, and distributed mechanisms for object storage, synchronization, and name resolution are shown to be essential in the context of network management. While corresponding base technology support is not yet well developed, current events indicate that such support will soon become available.

Keywords: Distributed object-oriented systems, network management systems, open architecture, framework architecture, mediation technology.

1. INTRODUCTION

Current developments in communications technology are having a profound impact on the operation and management of networks. A new type of management system architecture is needed to cope with this challenge.

The technological push comes from the introduction of *digital technology*, such as ISDN, which adds considerable complexity compared with analog systems. Associated with this advance is a diversity of products, creating a need for effective management of *heterogeneous, multivendor* networks. The situation is exacerbated by the convergence of telecommunications and computer networks, with a need for products that can *bridge technologies*. The network components embed varying degrees of *intelligence*, requiring the management system to be open to different levels of control. To be able to support these developments, a flexible and open management framework must be established that can withstand diversity and technological change.

Organizationally, these developments imply a *distributed, end-to-end* approach to management. Intelligent, *self-managed* components allow a high degree of autonomy for the distributed elements, but an *integrated* view of the total network must still be supported by the management system. Most networks do not exist in isolation, but are part of a *global* collection of networks, and management information and operations must be global in their applicability. Continuous change and growth of the network is the normal situation, and thus the management system must provide an *extensible* framework. *Customer control* of parts of the network is another requirement that stresses the need for flexible, open solutions.

Resolving all these issues, and doing so at a reasonable cost, is the challenge of integrated network management. Several international standards bodies are currently putting much effort into the development of standards in the field. Some areas have already been standardized, whereas others will take several more years to complete. The standards define an organizational view of network management and the information structures and interchanges involved, but it is up to each management system to implement an architecture that can support this view.

The main features of the next generation network management architecture can be characterized by a few keywords, describing what the architecture must offer:

- *integration* of heterogeneous systems, services, and networks for interoperation and unified presentation;
- *distribution* of systems, applications, and control;
- *dynamics*, both in terms of network change and growth and in terms of responding to customers' functional needs;
- *openness* to communicate with other systems and to incorporate external components;
- *standards* compliance for achieving openness and integration.

The picture that emerges is one of a *framework* architecture for management systems rather than just a new, powerful system. A framework can be filled in with a variety of components that support the diversity of equipment and technologies within a distributed environment. Furthermore, the components can change over time while the framework remains the same. The framework approach offers an organic, resilient solution that will provide the flexibility and openness required for effective management, and at the same time enable incremental development, controlling development cost.

In this paper, we first investigate (Section 2) in some detail the emerging technology of object-oriented distributed systems and what it offers for integrated network management. Next (Section 3), we present the Open Systems Framework Architecture (OSFA), MPR Teltech's application of this technology for network management, an example of its use (Section 4), and some initial experiences (Section 5). Section 6 concludes the paper.

2. OBJECT-ORIENTED DISTRIBUTED SYSTEMS AND NETWORK MANAGEMENT

During the past decade, a number of research projects have established a fairly coherent view of a class of systems called object-oriented distributed systems (OODS). These are software architectures that span computing nodes (single- or multi-processor) in a network, and where the software is organized according to the object-oriented perspective on software construction. The literature abounds with descriptions of object-oriented principles and technology, and how features such as encapsulation and inheritance help promote system reliability and maintainability and software reuse. However, until now most of the discussion and tools have focused on non-distributed software development only, as witnessed by the lack of distributed features in recent object-oriented languages such as Eiffel and C++. Judging from current developments in applications and such base technology as operating systems and databases, the next major theme in software technology will be object-oriented *distributed* systems. Let us examine the distinguishing features of these systems and how they apply to integrated network management.

Fine-grained Decomposition. Objects do in general provide a *finer-grain* granularity of system decomposition than the process concept available in operating systems. Several objects may map onto one process, but this depends on the underlying support architecture. There may be different levels ("sizes") of objects—smaller "light-weight" objects within larger "process" objects. A flexible, multi-level object model supports OSI-based network management, which is based on object-oriented modelling.

Concurrency. Objects execute in parallel, allowing considerable *concurrency* to be exploited in applications. Even objects within a process can achieve concurrency via *threads* (context switching within a process). Concurrency implies that some concept of *transactions*, or atomic operations across objects, normally must be supported.

High-level Messages. Objects are tied together at a very *high level*, by means of a message protocol rather than language-based interfaces or shared memory. This provides a means for *encapsulation*, in that individual components in a system can be exchanged and modified without changes propagating to other parts of the system. As a result, quite diverse technologies, both software and hardware, can be packaged to fit the framework, and *heterogeneous* collections of components can therefore be supported more easily than with traditional approaches. This *integration* capability is invaluable when bridging technologies in network management systems.

Dynamic Reconfiguration. High-level message-based interconnections also enable dynamic system changes to occur. Objects can be *replaced* with new versions, often while the system is running. Objects can be made *mobile*, i.e., allowed to migrate from one computing node to another. Because the message mechanism is a separate component controlling all interchanges, messages can be redirected to the new location without the sender needing to know about the change. Other features, such as *multicasts* and *replicated* objects can also be localized in the message communication mechanism. This affords *scalability*. As the system grows, new objects can be integrated and existing objects can be moved between nodes to offset new loads, but without having to change themselves. No architectural modifications are necessary as the system expands. The result is that it is possible to implement management systems that at the same time meet the demands for maintainability, for robustness and availability, and for flexibility toward the customers' changing needs.

Naming and Location Transparency. A key mechanism supporting dynamic reconfiguration is *name resolution*. Name resolution refers to the process of translating a symbolic *name* into a globally valid object *address*. Often designated *directory* objects are used to perform name resolution for other objects. Thus, the name used for communicating with an object does not tie the receiver to a specific computing node. The addressing mechanism can itself find the whereabouts of an object. In other words, addressing of objects is *location transparent*. This feature allows object migration and replacement to take place without changes propagating to the whole system.

Failure Recovery. When objects are distributed among different computing nodes, the inter-object message protocol must handle communication *failures*. Failure recovery must also be visible at the application level, since all information interchange is via messages, and node break-downs are likely to be reflected in altered application behaviour. The high-level, yet separate, message mechanism allows different approaches to failures to be installed in different groups of applications. For network management systems, a powerful approach to failure handling is central for achieving high reliability and availability.

Security. Open, distributed systems are especially prone to security attacks. The issue of security must therefore be dealt with and included as a basic feature of the architecture. One common mechanism is to require that, in addition to the receiver's address, senders must pass a *capability* along with a message, containing access rights that the receiver can check. For higher levels of security, encryption of messages may be needed. Security in distributed systems is currently being addressed in the context of OSI network management.

Decentralization of Services. In addition to providing a basis for *distribution* of applications, there is also a push towards *decentralization* of services. For example, many features that traditionally were located in some kind of kernel, like scheduling policies and parts of the addressing mechanism, can now be moved out to server objects. Objects need only basic support for operation execution and message exchange. This further contributes to the flexibility and expandability of the system.

Persistence. Distributed objects, such as representations of OSI managed objects, are often expected to have long lifetimes and to survive break-downs in the underlying system. To achieve this, they must be *persistent*, i.e., backed up on permanent storage. Conversely, objects that are of ephemeral use may be *transient*. Object persistence is the only way to store permanent information, replacing traditional databases.

Active Objects. Some objects are *passive*, in that they only perform actions on behalf of other objects. *Active* objects in addition have their own sequence of control, allowing them to initiate requests to other objects. Applications normally contain both kinds of objects, corresponding, respectively, to entities that are more naturally modelled as data repositories, and processing entities such as managers in the OSI management model.

Mediation Technology. The term *mediation technology* has been coined to describe the kind of inheritance-based "wrapper" techniques available in an object-oriented environment for retrofitting non-conformant components. This can be used to integrate existing products that were not developed with the new object-oriented framework in mind, and it is an important vehicle for ensuring a smooth product evolutionary path and thereby good overall product economy. One approach to mediation relies on the message-based interaction of objects. The existing software component will then be seen as an object implementing its own message protocol, and one or more mediation objects can be implemented to translate between this object's messages and the messages used by other objects. Another approach is to use object-oriented programming techniques to wrap the existing code in a layer that reformulates the application's concepts in object-oriented terms and supports the inter-object message protocols. Both of these approaches rely heavily on object-oriented technology to achieve the desired wrapper effect.

Language Support. Many of the features of object-oriented distributed systems listed above will be reflected in languages that support such an architecture. The most obvious facility to be supported by a programming language in a distributed environment is *concurrency*. Concurrent languages (such as Concurrent Pascal [4], Ada, and ConcurrentSmalltalk [14]) support a process or task concept with intercommunication and synchronization facilities. Support for *distribution* in addition takes into consideration that the processes are executing in several geographically distinct locations. Languages supporting distribution (such as Argus [7], Emerald [3], and Distributed Smalltalk [2]), must cope with issues such as object persistence, protection, equipment failures, and extensibility.

Failure handling requires that the language contains an exception handling facility. Ada provides such a mechanism, and it is becoming available in object-oriented languages as well (Eiffel).

Extensibility means that it should be possible to add new processes or objects to the system while it is running. This has implications for the language's data typing system. It is no longer possible to have a compiler check all code interfaces before the system starts executing. It also means that objects generated by different tools will have to be integrated. This leads to *structural data type equivalence*, where type conformance is checked on the basis of an investigation of an object's attributes rather than on a type identifier guaranteed by a compiler. This notion has been explored in Emerald [3]. Note that the authentication aspect of typing is abolished in this scheme. Authentication has to be dealt with by the security system.

Another interesting language concept is *reflection*. A system that exhibits computational reflection can contain computational structures not only about the application domain, but also about itself [8]. Examples are self-modifying code and meta-classes as seen in Smalltalk-80. Reflective facilities are quite useful in distributed systems. For example, the message dispatch mechanism can readily be redefined in a reflective framework, allowing different addressing and synchronization mechanisms to be substituted without affecting application code. Structural type equivalence also depends on reflective capabilities. Tracing, debugging, and simulation are vastly simplified, activities that become complex in a distributed environment. Reflective facilities exist in several research languages [5, 13], but not yet systematically within production languages.

No production language exists today offering adequate support for distributed systems. However, with the momentum toward object-oriented distributed architectures, one should expect research in this area to influence products in the near future.

Operating System Support. The main problem with object-oriented distributed systems today regards performance, and the primary issue here is inter-object communication. With distributed objects, references that with traditional architectures would be within the same memory address space, end up in messages that may have to go through several layers of protocol translation, transfer data between address spaces, and even cross machine boundaries. Several different processes may be involved in name resolution. Object-oriented distributed operating systems solve some of the problems. For example, Mach [1] just sends memory references in messages and maps the corresponding pieces of data into the receiver's memory address space instead of copying the data. Another key area is security. Amoeba [10] has demonstrated how a capability-based approach can be used to guarantee protection in a distributed environment.

Database Support. In an object-oriented architecture, the classic notion of a monolithic database has been replaced by a scattered collection of persistent objects. This raises many questions. First, while the traditional database could be optimized with indices and tailored storage schemes, it is not obvious how to achieve performance with the additional high-level mapping from distributed objects. Object-oriented databases (e.g., GemStone [9]) use storage techniques such as clustering to optimize object storage. Mach also provides features such as memory-mapped files and user-defined paging that enable efficient object persistence. Second, a database can provide a centralized approach to security, integrity, transactions, etc., and none of the established solutions here transfer easily to the new architecture. These issues are being addressed by research in distributed databases, where techniques for distributed transactions, recovery, and integrity are being developed (e.g., TABS [12]).

Conclusion. While there is a performance penalty associated with object-oriented distributed systems implemented on traditional infrastructures, the momentum we are witnessing today in research and industry toward this new system paradigm suggests that the situation will soon be vastly improved. Thus, in planning next generation network management architectures we must already assume the existence of an OODS framework, a framework that provides significant solutions for the challenge of integrated network management.

3. THE OPEN SYSTEMS FRAMEWORK ARCHITECTURE

3.1 Model Overview

Goal. Several existing architectures (such as DEC's EMA, HP's OpenView, AT&T's UNMA, and IBM's NetView) address the needs of network management within a distributed environment. However, none of these architectures seem to attempt to fully utilize the recent advances in object-oriented technology. For example, while some of the systems cited above do allow multiple databases to exist on distributed machines, the finer granularity offered by objects, allowing increased flexibility and tailorability, is not taken advantage of. Likewise, many systems are designed around a central message router. The router easily becomes a bottleneck, and the problem can be alleviated by the kind of decentralized control natural to distributed objects. OSFA aims at fully realizing the benefits of OODS for network management.

Architectural support for the OSI model. OSFA follows the well-known OSI model of network management (see, e.g., [6]). However, a more implementation-oriented object taxonomy has been adopted to respond to the needs of management systems development. This taxonomy can be seen as a refinement of the ISO model. Figure 1 shows the class inheritance hierarchy of OSFA objects, and how some of the major classes defined by OSI network management can

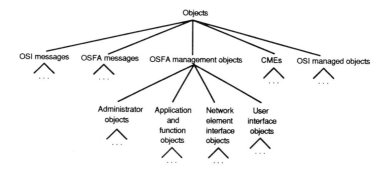

Figure 1: Object class inheritance hierarchy.

be seen to extend this hierarchy. The OSFA-defined object classes are:

- **OSFA management objects** constitute the superclass of all objects conforming to the OSFA architecture.

- **OSFA messages** describe the message protocols used for information interchange among OSFA management objects.

- **Application and function objects** contain all the levels of functionality needed for network management.

- **Administrator objects** perform name resolution, security, and other administrative tasks.

- **Network element interface objects** implement the link to physical network elements.

- **User interface objects** provide the human operator with a window into the management system.

The main OSI-related object classes are (cf. [11]):

- **CMEs**, or Conformant Management Entities, are objects foreign to OSFA that conform to OSI network management, and that an OSFA system therefore can communicate with. The role of a CME is to make those OSI managed objects it controls visible to other CMEs.

- **OSI managed objects** are OSI-compliant objects in the Management Information Base. These may be objects within an OSFA system or some other OSI-compliant system. They are made visible to OSFA systems or other CMEs through the CME that controls them.

- **OSI messages** are the messages defined in the context of OSI-based network management. These messages are exchanged among CMEs and OSI managed objects. Being OSI-compliant, OSFA management objects also understand these messages.

It is important to note that some, but not necessarily all, of the OSFA management objects will conform to OSI network management in terms of being CMEs or OSI managed objects. Many of the application and function objects will in general be CMEs. The network management

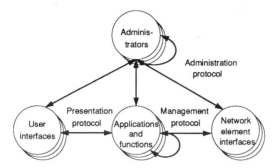

Figure 2: OSFA high-level structure in terms of objects and protocols.

interface objects will usually be OSI managed objects. Thus, there are multiple inheritance relationships not shown in Figure 1.

Figure 2 gives a high-level architectural view of how the different classes of OSFA management objects interact via the message protocols they support. There are three main protocols within OSFA:

1. The *management protocol* consists of the messages exchanged among OSFA application and function objects and network element interface objects. OSI network management messages are also supported by this protocol.

2. The *administration protocol* is used for communication with the administrators about OSFA object administrative matters.

3. The *presentation protocol* covers the interactions between user interface objects and other OSFA objects for the purpose of interacting with human operators.

Figure 3 shows OSFA objects in an OSI context. User interface and application/function objects on the management system side act as OSI managers. On the managed system side, network element interfaces are OSI managed objects, and applications/functions correspond to OSI agents. Each side needs administrator objects to set up connections, as illustrated by the dotted line. Note, however, that this need not be a static picture. Objects can migrate from one OSFA system to another if useful. The external CME interface is kept intact by the administrators' name resolution mechanism. From a system modelling perspective, system boundaries are not interesting, leading to the general picture in Figure 2.

Object structure. All OSFA management objects contain the definitions and mechanism necessary to operate within the OSFA framework. This involves both a common understanding of the classes of objects and messages that exist in such a system and a standard mode of operation that all objects adhere to. OSFA management objects can be structured in three layers (see Figure 4):

1. The *service template* embodies the mode of operation of management objects. It ensures that the objects will behave similarly with respect to message processing, exceptions, etc., and it is to be filled in with the specific services provided by each object.

2. The *system schema* supplies a view of the unified object world that OSFA is based upon. It merges facilities for persistent storage and message passing into a common notion of

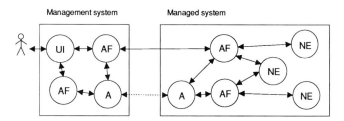

Figure 3: OSFA object categories in an OSI management model (*UI* = user interface objects; *AF* = application/function objects; *A* = administrator objects; *NE* = network element interface objects).

Figure 4: The structure of a management object.

object, and defines all the standard object classes available for use by the management objects. These classes form an extension of the OSI information model.

3. *Message and persistence support* perform all the low-level work needed to map different technologies for communication and permanent storage to a common model. The message and persistence components are exchangeable to accommodate a variety of communication protocols and storage schemes, without affecting the system schema definitions.

3.2 Administrator Objects

Managing the Management System. The OSI model of network management describes how a management system manages network resources. When designing a management system architecture, we also have to deal with the management of the system itself. This is the role of the OSFA administrator objects. A key observation here is that the relationship between an administrator and the other OSFA management objects is equivalent to the relationship between a manager and its agents in the OSI management model (Figure 5). Thus, when designing the OSFA administration protocol (see Figure 2) we can take advantage of the existing definitions and implementations of OSI management protocols. Indeed, the OSFA administration protocol is a CMIS-based protocol with added functionality for the specific tasks of administrators. This congruency between the two levels of management contributes to interoperability and economy. OSFA administration will appear simply as a proprietary extension to OSI network management protocols that will give added capabilities to those CMEs that support it.

Administrator Tasks. An administrator object is an ordinary management object, addressed

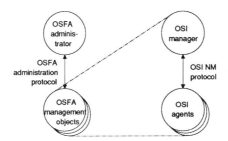

Figure 5: Two congruent levels of management.

just as any other object and containing the common parts described in Section 3.1, but its role is to provide specific administrative tasks for other objects. There will typically be at least one administrator per computing node, and the administrators exchange information to keep each other up-to-date about the state of the total system.

Among the responsibilities of the administrator objects are:

- Object creation, startup, and shutdown.
- Object dynamics. Administrators assist in operations such as object replacement, migration, and replication.
- Directory of services (name resolution). Objects announce their services to the administrators, which build an X.500-based directory.
- Security and accounting. Certain aspects of these functions are centralized in the administrators.

Domains and Name Resolution. An important concept supported by the administrators is *domains*. A domain is a collection of management objects that are managed by the same management system. Domains are central to the naming of objects and for viewing portions of the object space for various purposes. The OSFA domain concept described here is intended to comply with the domain concept emerging within OSI network management.

Administrator objects implement the domain concept through the *name resolution* mechanism. Each domain is managed by at least one administrator, and the name of a domain is normally the name of an administrator of that domain. Objects usually refer to other objects indirectly via their domain-based name when looking up a service. The administrators responsible for the domains that form parts of a name will map the name to a unique system-wide address. The address can subsequently be used for referring to an object directly and much more efficiently, without having to access the administrators. Name resolution via administrators is normally used only for setting up a connection, such as when initiating a logging connection.

The name resolution mechanism provides powerful abstraction. In Figure 6, the sender need not be aware of whether *B* is a subdomain or an object. Furthermore, the name *C* may just be a symbol that *B* uses to compute a receiver of the request. In this way a service can migrate transparently from one object to another, or it may be duplicated to more than one object.

Management Policies. The particular management policy to be used (for example, one central master administrator, hierarchical control, or a more "democratic" system with decentralized, peer-to-peer control) is not determined by the OSFA framework. Rather, while administrators have to conform to a common protocol, one is free to implement administrators with different

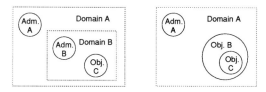

Figure 6: Two implementations of the name A.B.C.

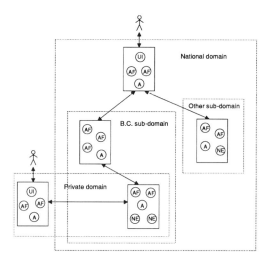

Figure 7: Network management example. The dashed rectangles are domains and the solid rectangles are management (and managed) systems.

management policies. The organization of the total network will be reflected in the algorithms embedded in the administrators. Security and accounting issues can also vary widely between installations.

4. MANAGEMENT EXAMPLE

Figure 7 shows how a nationwide telecommunications network could be managed in a distributed manner with an OSFA-based system. The figure also shows how the domain concept can be used to partition the network into management domains nested within, or overlapping, larger domains.

Each management system contains a collection of objects tailored to the specific task the system is to perform. For example, the systems actually interfacing network elements will contain network element interface objects, whereas only those systems that are directly accessed by a human operator contain user interface objects. Administrator objects in each system help locate the services needed by the objects. System boundaries (the solid rectangles in the figure) are invisible to the objects, and objects can migrate from one system to another without affecting other objects.

An application can be distributed among objects in the total system. For example, a performance analysis application may contain objects interfacing network elements, computing detailed statistics about their respective elements' performance. These objects may report to higher-level application objects that compute summary information, and, finally, a nationwide summary may be computed within an application object accessed by the top-level user interface object for presentation.

Due to the encapsulation and location transparency of distributed objects, explained in Section 2, this distributed application design means that knowledge about the local network characteristics is localized within objects close to the local networks. It further means that local networks and corresponding application support can be changed without affecting the management system at higher levels. New network elements and services can be supported by exchanging or adding objects, or moving objects around, all in a transparent manner that minimizes modifications outside the immediately affected area. Similarly, high-level objects can be exchanged without affecting lower levels, for example, a user interface object can be replaced by an application automating some previously manual task. A fixed framework with replaceable parts—this is the main contribution of the Open Systems Framework Architecture.

5. STATUS

A prototype of OSFA has been completed, mostly in Smalltalk-80 and C++, on networked UNIX^{TM} workstations. An example application (a decentralized, national performance evaluation system along the lines sketched in Section 4) has been implemented within the OSFA framework. Our future plans include the incorporation of supporting technology within operating systems and object-oriented databases, as they become available.

First experiences with OSFA confirm that there is considerable complexity involved in implementing a comprehensive framework supporting all the features we have listed in Section 2. Not surprisingly, though, it turned out to be much easier to develop dynamic aspects, such as object migration and replication, with a simple, typeless, interpreted language with reflective capabilities (Smalltalk-80), than with a more traditional object-oriented language (C++).

On the application side, we found it non-trivial to model applications to fully take advantage of the distributed object-oriented framework. Once in place, though, the resulting distributed application seems extremely flexible to use and to modify. Another point is that much of the framework nature will not be transparent to the application programmer (e.g., the needs for name resolution and deadlock analysis, the "almost" transparency of remote procedure calls). Much work clearly remains to be done both on application modelling and supporting technology.

6. CONCLUSIONS

The challenges posed by integrated network management are awesome. However, as we have shown in this paper, object-oriented distributed technology provides many powerful solutions that seem to fit well with the needs of distributed, heterogeneous, expandable management systems. We have focused on how, at a fairly detailed technical level, concurrency, persistence, object migration, location transparency, mediation technology, and language features, such as structural type equivalence and reflection, can support network management. The essence of the emerging type of architecture is that it is a *framework*, an infrastructure into which a changing set of components can be plugged while the base remains the same. This organization meets customers' changing needs for management services over evolving networks and improves product economy through the protection of long-term investments.

OSFA is an instance of an object-oriented distributed framework architecture for network management. The main contribution of OSFA is the mapping of the OSI model onto the OODS paradigm with the resulting object taxonomy, including administrator objects performing second order management in a manner congruent to OSI management. A prototype of OSFA has been implemented, along with a sample network management application. The prototype confirms that, while implementing an OODS framework is a complex task, the resulting system flexibility is extremely powerful.

ACKNOWLEDGEMENTS

The author would like to thank Paulin Laberge, the initiator of the OSFA project, for his enthusiasm and vision, and also the other participants in this and related projects, Richard Chow, Daryl Morse, Gail Murphy, Steve Ng, Shaun Smith, and Kevin Wittkopf.

REFERENCES

[1] M. Accetta, R. Baron, W. Bolosky, D. Golub, R. Rashid, A. Tevanian, and M. Young. Mach: A New Kernel Foundation for UNIX Development. In *Proc. of Summer USENIX*, 1986.

[2] J.K. Bennett. The Design and Implementation of Distributed Smalltalk. In *Proc. of the OOPSLA '87 Conf. on Object-oriented Programming Systems, Languages and Applications*. ACM, 1987.

[3] A. Black, N. Hutchinson, E. Jul, H. Levy, and L. Carter. Distribution and Abstract Types in Emerald. *IEEE Transactions on Software Engineering*, 13(1), January 1987.

[4] P. Brinch Hansen. The Programming Language Concurrent Pascal. *IEEE Transactions on Software Engineering*, 1(2), 1975.

[5] B. Foote and R.E. Johnson. Reflective Facilities in Smalltalk-80. In *Proc. of the OOPSLA '89 Conf. on Object-oriented Programming Systems, Languages and Applications*. ACM, 1989.

[6] S.M. Klerer. The OSI Management Architecture: an Overview. *IEEE Network*, 2(2), March 1988.

[7] B. Liskov. Overview of the Argus Language and System. Technical Report Programming Methodology Group Memo 40, Lab. Comput. Sci., Massachusetts Inst. Tech., February 1984.

[8] P. Maes. Concepts and Experiments in Computational Reflection. In *Proc. of the OOPSLA '87 Conf. on Object-oriented Programming Systems, Languages and Applications*. ACM, 1987.

[9] D. Maier, J. Stein, A. Otis, and A. Purdy. Development of an Object-oriented DBMS. In *Proc. of the OOPSLA '86 Conf. on Object-oriented Programming Systems, Languages and Applications*. ACM, 1986.

[10] S.J. Mullender and A.S. Tanenbaum. The Design of a Capability-based Distributed Operating System. *The Computer Journal*, 29(4), 1986.

[11] OSI/NM Forum. Forum Architecture. OSI/Network Management Forum, October 1989.

[12] A.Z. Spector, J. Butcher, D.S. Daniels, D.J. Duchamp, J.L. Eppinger, C.E. Fineman, A. Heddaya, and P.M. Schwarz. Support for Distributed Transactions in the TABS Prototype. In *Proc. of the 4th Symp. on Reliability in Distributed Software and Database Systems*, October 1984. (Also Carnegie-Mellon Report CMU-CS-84-132, July 1984).

[13] T. Watanabe and A. Yonezawa. Reflection in an Object-oriented Concurrent Language. In *Proc. of the OOPSLA '88 Conf. on Object-oriented Programming Systems, Languages and Applications*. ACM, 1988.

[14] Y. Yokote and M. Tokoro. Experience and Evolution of ConcurrentSmalltalk. In *Proc. of the OOPSLA '87 Conf. on Object-oriented Programming Systems, Languages and Applications*. ACM, 1987.

The LOTOS framework for OSI Systems Management

A.J. Kouyzer[a] and A.K. van den Boogaart[b]

[a]Netherland Railways Department of Engineering,
PO Box 2025, 3500 HA Utrecht, Netherlands
work done when employed by PTT Research Dr. Neher Laboratories

[b]PTT Research Dr. Neher Laboratories,
PO Box 421, 2260 AK Leidschendam, Netherlands
Email: AK_vdBoogaart@pttrnl.nl

Abstract

Since it is possible to make ambiguous specifications of Management Information Models using currently proposed methods, there is a need to develop a specification method for Management Information Models using a formal description technique.

A framework for management purposes is defined using LOTOS specifications in order to achieve this goal. The framework defines an environment in which it is easy to define a Management Information Model. This article describes the framework, introducing the main processes and the relations between them. It also gives the LOTOS specification of this part of the framework and a very small part of the more detailed specification of the framework.

1. Introduction

Methods currently proposed in the OSI/Network Management Forum and ISO OSI Management allow behavioural aspects of managed object classes to be defined by use of English text. Even if natural language descriptions are written very carefully, errors and omissions are difficult to prevent.

Formal Description Techniques (FDTs) enable much better quality descriptions to be produced which are unambiguous, clear and concise. Furthermore they act as a foundation for the analysis and verification of specifications.

The need therefore arises to develop a specification method for Management Information Models using a FDT. The use of a *standardised* FDT would speed up both acceptance and use of the method. Three standards for FDTs have full international status:

- SDL [1] (based on an extended finite state machine model).

- Estelle [2] (based on Pascal, with extensions to describe finite state machines).
- LOTOS [3] (based on process algebraic methods).

Considering the object-oriented environment in which Management Information Models have to be specified (CMIS [4], SMI [5]), the FDT must support concepts from the object-oriented paradigm. This support is present for LOTOS, but SDL and Estelle both need an extension of the language. It was therefore easy to choose LOTOS as a formal description technique for Management Information Models.

An object-oriented style for LOTOS has been developed within PTT Research [6], making it suitable for specifying Management Information Models. All data defined in the Management Information Model can be exchanged using CMIP [7], so the constructors of the LOTOS abstract data types (ADTs) should be based upon ASN.1[8] types. For this purpose an ASN.1 to LOTOS translator has also been defined by PTT Research [9].

It is necessary to make CMIS specific aspects such as *scoping, filtering, synchronisation* and *access control* invisible to the specifier of a Management Information Model, because these functions are not provided for in the specification of a Management Information Model. See SMI [5].

This article introduces a framework which specifies the service offered by CMIS. Example processes are given for the specification of managed object classes and managed objects to incorporate them into this framework. The framework describes the hierarchical relationship between a Management Information Service user (MIS user) in a manager's role (manager) and a MIS user in an agent's role (agent). However, the relationship between two peers is also defined with the framework. Such a peer relationship is in fact two relationships, where the manager in one relationship is agent in the other.

The framework only specifies how the manager *sees* the agent, so it is only an interface specification. However, LOTOS gates are defined to both the management and the managed system, so a LOTOS specification of these systems can easily be combined with the framework.

The description of the framework is in general terms and does not cover all aspects in full detail.

2. Main processes and their interaction

The framework specifies how the interface, as seen by the manager or the agent, is built up and how a managed system (Agent) can communicate with a management system (Manager). Communication with and within the managed system is beyond the scope of the framework, since it depends on the information model used. However, the fact that there will be communication with the Agent is obvious and provided for. Figure 1 shows the environment.

The Management Interface consists of two processes, an Interface and an CMISE. See Figure 2. These two processes represent the information model, which has several CMIS capabilities (Interface) and the communication between Manager and Agent (CMISE).

The interaction between CMISE and the other LOTOS processes consists of LOTOS gates which represent asap's (application service access points). The service of CMIS [4] is offered to the other processes on these gates.

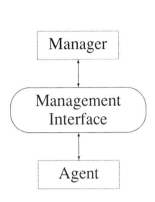
Figure 1: Management environment

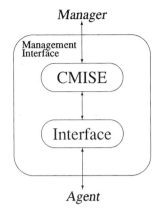
Figure 2: Management Interface process

The Manager and the Agent themselves are not part of the framework. CMISE is fully independent of the information model and its service can be defined easily in LOTOS. The Interface is the main subject of this article and it is further defined in the following subsection, and is also partly specified in section 3.3. and further.

2.1. The Interface process

The Interface process consists of three parts: the CMIS handler, a moderator (Interface Moderator) and processes representing the managed object classes. See Figure 3.

The CMIS handler process handles the following CMIS functions:

- Access control
- Scoping
- Filtering
- Synchronisation

The CMIS handler is independent of the information model used.

The Interface Moderator translates certain data (such as attribute lists, event data and others defined as ANY in CMIP [7]) from a generalized type to a class specific type. This means the Interface Moderator depends on the information model used. However, the structure is independent of the information model.

Each Class process represents one class of managed objects. The set of Class processes forms the Classes process. See Figure 4. The interaction between a class process and the CMIS handler is independent of the managed object class (except for those types which are translated by the Interface Moderator) and consists of a simplified CMIS and some additional control services (such as an existence check). The Class process depends heavily on the information model used, except for the generic part which handles creation, deletion and existence check and the part that handles the interface to the CMIS handler. For these parts the structure is independent of the information model.

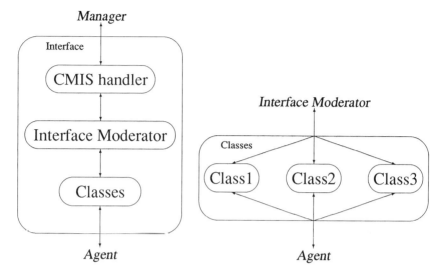

Figure 3: Interface process   Figure 4: Classes process

3. LOTOS specification

This section gives an incomplete specification of the behaviour of the agent (managed system) as seen by the manager (management system). The specification is only a framework, so a specific specification will have to be made for each managed system, or group of managed systems. In particular the *Agent* will have to be defined each time, as well as several ADTs.

The following conventions are used throughout the specification:
- Courier font is used for actual LOTOS text;
- Small Caps type style is used outside the LOTOS text for the names of processes.

3.1. ManagedSystem specification

As described in section 2, at the highest level the management framework can be viewed as an interface to a managed system; the MANAGEMENTINTERFACE and AGENT processes respectively. This general view is parametrised with the information model (managed object classes) used at a specific interface.

```
specification ManagedSystem [ manager ]
                            ( classes : classSet ) : noexit
behaviour
 hide agent in
  ManagementInterface [ manager, agent ] ( classes )
   |[ agent ]|
  Agent [ agent ] ( classes )
where
```

The AGENT is not specified in this article.

3.2. ManagementInterface process

The MANAGEMENTINTERFACE consists of two parallel processes, one specifying the communication between management system and managed system (CMISE), the other specifying the information representing the managed system as seen by the management system (INTERFACE).

```
process ManagementInterface [ manager, agent ]
                            ( classes : classSet ) : noexit
 :=
 hide asap in
  CMISE [ manager, asap ]
   |[ asap ]|
  Interface [ asap, agent ] ( classes )

where
```

The CMISE is not specified in this article.

3.3. Interface process

INTERFACE consists of three parts. First, the CMISHANDLER which handles the CMIS specific parts. Second, the INTERFACEMODERATOR which transforms several general CMIS constructs into class specific constructs and vice versa. Third, the CLASSES process which represents all classes of managed objects.

```
process Interface [ manager, agent ] ( Classes : ClassSet ) : noexit
:=
  hide generalized in
   CMIShandler [ manager, generalized ] ( Classes )
      |[ generalized ]|
   ( hide specific in
       InterfaceModerator [ generalized, specific ]
         |[ specific ]|
       Classes [ specific, agent ] ( Classes )
   )

where
```

The INTERFACEMODERATOR is not specified in this article.

3.4. CMIShandler process

The CMISHANDLER performs scoping, filtering, access control and synchronisation of the messages sent by the manager.

```
process CMIShandler [ manager, mibase ]
                    ( classes : ClassSet ) : noexit
:=
```

A message will be received from the manager.

```
manager ? class      : ClassId       ? object : ObjectId
        ? scope      : Scope         ? filter : Filter
        ? control    : AccessControl ? sync   : Bool
        ? operation  : Operation ;
```

First the managed objects for which the message is intended are selected. This means determining the scope of the message and applying the filter to the set of managed objects indicated by the scope. This will produce a set of managed objects to which the message has to be sent.

```
( SelectObjects ( class, object, scope, filter )
  >> accept selectedObjects : ObjectSet in
```

After having selected the managed objects where the message has to go, it has to be determined whether the manager actually has access to these managed objects. This will yield a set of managed objects which is a subset of the originally selected managed objects.

```
  ( AccessControl ( selectedObjects )
    >> accept objects : ObjectSet, reduced : Bool in
```

Depending on whether *atomic synchronisation* is present or *best effort synchronisation*, the selected managed objects will first have to be either claimed or not claimed. N.B.: if the manager does not have access to the full set of managed objects, atomic synchronisation will fail!

```
      (  [ sync and not ( reduced ) ] ->
            ClaimObjects [ mibase ] ( objects )
            >> accept claimed : Bool in
               [ not ( claimed ) ] -> ReleaseObjects [ mibase ]
                                                     ( objects )
               [] [ claimed ] -> Operate [ manager, mibase ]
                                         ( class, objects, operation )
      [] [ sync and reduced ] -> exit
      [] [   not ( sync )    ] -> Operate [ manager, mibase ]
                                          ( class, objects, operation )
      ) >>
      CMIShandler [ manager, mibase ] ( classes )
  )
)
where
```

A choice has to be made to determine which operation is to be performed on the set of managed objects selected. This is specified in the OPERATE process. In this article the choice is reduced to create, delete and action, where action represents the CMIS services Set, Get and Action. As it is the framework does not support Event Reports; this is, however, easy to add.

```
process Operate [ manager, mibase ]
                ( class     : ClassId,
                  objects   : ObjectSet,
                  operation : Operation
                ) : exit
:=
   [ operation = create ] -> Create [ mibase ] ( class, objects )
```

```
[] [ operation = delete ] -> Delete [ mibase ] ( class, objects )
[] [ operation = action ] -> Action [ mibase ] ( class, objects )
```

where

In the CREATE process one managed object can be created at a time, so the set of managed objects should contain one managed object. A message indicating the creation of a managed object, with the object identification in the set, is sent to the CLASS process, if no managed object with this identification already exists. If the set of managed objects is empty or contains more than one object identification, CREATE will terminate. CREATE has a *set* of managed objects as parameter, which allows for a general approach for scoping, filtering and access control.

```
process Create [ mibase ]
                ( class : ClassId, objects : ObjectSet ) : exit
:=
   [ Card ( objects ) ne succ ( 0 ) ] -> exit
[] [ Card ( objects ) eq succ ( 0 ) ] ->

        choice object : ObjectId []
          [ object IsIn objects ] ->
            mibase ! class ! object ! exist ? exists : Bool ;
            (  [         exists     ] -> exit
             [][ not ( exists ) ] -> mibase ! class ! object ! create ;
                                        exit
            )
endproc (* Create *)
```

In the DELETE process the message indicating that it has to be deleted is sent to one of the managed objects in the set (if it exists), as well as to the CLASS process, which keeps a record of all existing managed objects in its class. Then DELETE will call itself again, without the managed object which has just been deleted. If the set of managed objects is empty, DELETE will terminate.

```
process Delete [ mibase ]
                ( class : ClassId, objects : ObjectSet ) : exit
:=
   [ Card ( objects ) eq 0 ] -> exit
[] [ Card ( objects ) gt 0 ] ->

        choice object : ObjectId []
          [ object IsIn objects ] ->
            mibase ! class ! object ! exist ? exists : Bool ;
                ( [ not ( exists ) ] -> exit
                 [][ exists ] -> mibase ! class ! object ! delete ;
                                  mibase ! class ! object ! delete ;
                                  exit
                ) >>
                Delete [ mibase ] ( class, Remove ( object, objects ) )

endproc (* Delete *)
```

In the ACTION process the message indicating the action to be taken is sent to one of the managed objects in the set (if it exists). Then ACTION will call itself again, without the managed object, which has just received the message. If the set of managed objects is empty, ACTION will terminate gracefully.

```
process Action [ mibase ]
              ( class : ClassId, objects : ObjectSet ) : exit
:=
   [ Card ( objects ) eq 0 ] -> exit
[] [ Card ( objects ) gt 0 ] ->
       choice object : ObjectId []
         [ object IsIn objects ] ->
           mibase ! class ! object ! exist ? exists : Bool ;
               ( [ not ( exists ) ] -> exit
                [][ exists ] -> mibase ! class ! object ! action ;
                                exit
               ) >>
               Action [ mibase ] ( class, Remove ( object, objects ) )
endproc (* Action *)
endproc (* Operate *)
endproc (* CMIShandler *)
```

The SELECTOBJECTS, ACCESSCONTROL, CLAIMOBJECTS and RELEASEOBJECTS processes are not specified in this article.

3.5. Classes process

In the CLASSES process a separate process is started for each managed object class identified in the set of classes. These processes will run in parallel.

```
process Classes [ CMIShandler, agent ]
                ( classes : ClassSet ) : noexit
:=
  choice class : ClassId []
    [ class IsIn classes ] ->
        ( Class [ CMIShandler, agent ] ( class, EmptyObjectSet )
          |||
          Classes [ CMIShandler, agent ] ( Remove ( class, classes ) )
        )
endproc (* Classes *)
```

3.6. Class process

An instance of the CLASS process will be started for each managed object class. This process handles generic tasks for the managed object class, such as creating managed objects and administering existing managed objects of this class. Only the CLASS process can determine if a managed object does *not* exist!

```
process Class [ CMIShandler, agent ]
              ( me : ClassId, Objects : ObjectSet ) : noexit
```

```
:=
(
```

Check whether a managed object of this class exists with this identification.

```
    ( choice Object : ObjectId []
        CMIShandler ! me ! Object ! exists ! ( Object IsIn Objects ) ;
        Class [ CMIShandler, agent ] ( me, Objects )
    )
[]
```

Delete a managed object of this class.

```
    ( choice Object : ObjectId []
        [ Object IsIn Objects ] -> CMIShandler ! me ! Object ! delete ;
                                   Class [ CMIShandler, agent ]
                                        ( me, Remove ( Object, Objects ))
    )
[]
```

Create a managed object of this class.

```
    ( CMIShandler ! me ! Objects ! create ;
        ( Class [ CMIShandler, agent ] ( me, Add ( Object, Objects ) )
          |||
          Object [ CMIShandler, agent ] ( me, Object )
        )
    )
endproc (* Class *)
```

3.7. Object process

A separate instance of an OBJECT process will be started for each managed object in each class. The processes will have several aspects in common, such as creation, deletion, setting attributes and getting attributes. Other aspects, such as actions and event reports, will depend on the class to which the managed object belongs. The example only shows its behaviour in general terms for creation (starting the process!), deletion and an *action* representing the CMIS services Set, Get and Action. Only the basic principles are shown this way.

```
process Object [ CMIShandler, agent ]
                ( myClass : ClassId, me : ObjectId ) : noexit
:=
    CMIShandler ! myClass ! me ! delete ;
    stop
[]
    CMIShandler ! myClass ! me ! action ;
    i ;
    Object [ CMIShandler, agent ] ( myClass, me )
endproc (* Object *)
endproc (* Interface *)
endproc (* ManagementInterface *)
endspec (* ManagedSystem *)
```

4. Conclusion

As has become clear in this article, it is possible to create a framework to support the definition of a Management Information Model in LOTOS. This means the specifier of the Management Information Model can concentrate on the LOTOS specification of managed objects themselves, and is not concerned with CMIS specific aspects. However, the framework still needs to be elaborated in more detail. Work on this aspect is currently in progress at PTT Research.

References

[1] Blue book Fascicle X.1
Functional Specification and Description Language (SDL). Recommendation Z.100 and Annexes A, B, C, and E, Recommendation Z.110; CCITT, 1989

[2] International Standard ISO 9074: 1989
Information Processing Systems - Open Systems Interconnection - ESTELLE - A formal description technique based on extended state transition model

[3] International Standard ISO 8807: 1989
Information Processing Systems - Open Systems Interconnection - LOTOS - A formal description technique based on the temporal ordering of observational behaviour

[4] International Standard ISO/IEC 9595: 1990
Information Technology - Open Systems Interconnection - Common Management Information Service Definition

[5] Draft International Standard ISO/IEC 10165-1: 1990
Information Technology - Open Systems Interconnection - Management Information Services - Structure of Management Information - Part 1: Management Information Model

[6] PTT Research Technical Report 917/89
Object Oriented Specification Style in LOTOS
E.M. Dijkerman, W.H.P. van Hulzen, P.A.J. Tilanus

[7] International Standard ISO/IEC 9596: 1990
Information Technology - Open Systems Interconnection - Common Management Information Protocol Specification

[8] International Standard ISO 8824: 1987
Information Processing Systems - Open Systems Interconnection - Specification of Abstract Syntax Notation One (ASN.1)

[9] PTT Research Technical Report 665/90
Transforming ASN.1 into LOTOS
W.H.P. van Hulzen, A.J. Kouijzer, P.A.J. Tilanus

A Development Environment for OSI Systems Management

Shoichiro Nakai, Yoshiaki Kiriha, Yoshiko Ihara *, and
Satoshi Hasegawa

C&C Systems Laboratories, NEC Corporation
4-1-1, Miyazaki, Miyamae-ku, Kawasaki, Kanagawa 216, Japan

* Transmission Division, NEC Corporation
1753, Shimonumabe, Nakahara-ku, Kawasaki, Kanagawa 213, Japan

Abstract
A new development environment for the OSI systems management is proposed. This environment provides such functions as editing of management information models, processing of information models, and design of management information base(MIB). The editing function helps system designers to refer to and update the models described using object oriented templates which have been specified in the OSI systems management standard. The information model is processed so that any syntax errors included in the model can be detected. Through the processing function, a management information definition file is created. The MIB design function generates the relational database schema and access codes for the management information which is defined in the management information definition file. Each function together forms an integrated support tool to design and implement the management system, which enables the cost and time of the system development be drastically reduced.

1. INTRODUCTION

The OSI systems management has been actively discussed and standardized by ISO and CCITT as a target network management in the future multi-vendor environment. In the OSI systems management, key issues to be standardized are categorized in two areas, such as management services and protocols to exchange messages between open systems, and management information. The International Standard has been made for management services and protocols as Common Management Information Services/Protocols (CMIS/CMIP)[1]. Currently, the standard activities for specifying Systems Management Functions (SMF) are under processing, which employs CMIS/CMIP as application service elements. As for the management information, the Structure of Management Information (SMI)[2] has been defined.

In the OSI based network management systems (NMS)[3], it is essential to generate a management information model in a structured fashion, as well as implementation of CMIS/CMIP management protocols. The model, identifying the resources that exist in a network and their associated attributes, provides a foundation of the NMS design. Resources are modeled as *object* and the management view of a resource is referred to as a *managed object*.

In order to describe the model, the standard templates are used, where a plural number of attributes with operations, event notifications, etc., can be specified for each managed object. From an NMS development point of view, procedures to interpret event notifications and operations for managed objects shall be implemented, and the access mechanism to the management attributes defined in the model be implemented. The efficient and reliable development of NMS is one of the keys to deploy the OSI management in a wide variety of network. In this paper, an integrated development tool to support NMS implementation is proposed, which enables the cost and time of the system development be drastically reduced and the reliability of the system be improved.

2. CONFIGURATION OF OSI BASED NETWORK MANAGEMENT SYSTEM

Figure 1 shows an example of the OSI based network management system, where the manager-agent model which is defined in the OSI is employed. A manager initiates management operations to the agents through the OSI protocols. An agent performs the management functions upon receipt of a directive management operation on managed objects. The managed objects might reside on a real communication equipment, and hence, an agent has a local protocol which interfaces with the equipments, through which the management operations can be applied to the equipment. Agents may also forward directives to a manager to convey notifications such as alarm signals generated by equipments.

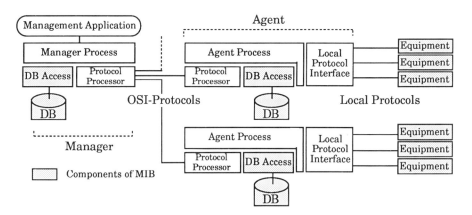

Figure 1: Network Management System based on OSI Systems Management

The management information specified in the OSI is stored in Management Information Base (MIB) which is regarded as a distributed virtual database in a network management system. The MIB is composed of entries, each of which contains information describing a single managed object. Each entry in the MIB is made up of attributes, each with a type and associated values. From an implementation viewpoint, the MIB might be database or file both in a manager and an agent, and memory (or register) in each communication equipment. The stored location of each attribute would be distributed according to whether the attribute belongs to a manager or an agent. Furthermore, the required management performance surely affects the stored location. In order to reduce the communication overhead, each attribute should be stored in a proper place. For instance, such attributes that have constant values should be stored in manager/agent database to achieve a fast retrieval time. An example of them is a serial number of the equipment. On the other hand, the attributes which have a dynamic behavior, such as a state of equipments, should be stored in the equipment itself in order to update values on a real time basis. From an application programmer's view point, it is desirable to provide such mechanisms that a manager or an agent can access any attributes without considering the stored location of them. To this end, the proposed tools maintain the information of the stored location for each managed object.

All the management information should be defined based on the management information model. In this sense, the creation of the model is one of the most important issues at the system design level. The proposed tool also gives a convenient mechanism for creation of the management information model to the designer. The following section explains the proposed tool in details.

3. DEVELOPMENT TOOLS FOR OSI BASED NMS

Figure 2 depicts an organization of a set of our proposed development tools, where the editing function of a management information model, processing functions of the model and a design support tool of the MIB are included. At the same time, the figure shows a processing flow of the management information model.

The editing function basically possesses the capability to update and retrieve the underlying information model. Through the editing function, the desired information model can be efficiently created or revised by referring to the standard models and templates. Then, the created information model is applied to the processing tool which is referred to as *Model Analyzer* in Figure 2, where potential errors in the created model can be detected. The management information dictionary is generated after completion of the desirable information model. By using the dictionary, a design support tool of the MIB, such as automatic generation of the MIB schema and access codes, can be provided.

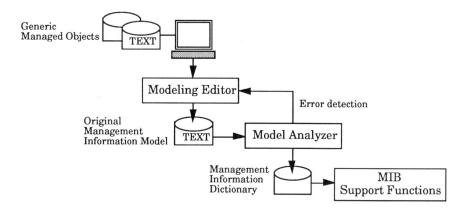

Figure 2: Proposed Development Environment

3.1. Management Information Model

The management information model describes the management information in terms of managed objects and a set of operations and notifications. The model is defined by using *templates*, specified in the Structure of Management Information in the OSI Systems Management. The *templates* provide the common formal format that represents the various aspects of the managed objects. The template examples are shown in Figure 3, where managed object class templates, attribute templates, a name binding template and ASN.1 syntaxes are shown. The template structure is briefly explained next by using the examples.

- **Managed object class template**

In Figure 3, *equipment* object class, which represents network equipments such as modem and PBX, is defined as the derived class of *top* object class. The equipment class includes its own attributes named *equipmentID*, *administrativeState* and *vendorName*, and inherits the attributes specified in the superior *top* class. Taking the *equipmentID* for instance, only the retrieval function (GET) can be allowed.

- **Attribute template**

The attribute template is used to define individual attribute types to which the before-mentioned class template refers. Other templates such as the name binding template would also refer to the attribute template. The definition of an attribute describes the attribute syntax label which shows a data type defined by ASN.1. Moreover, the template would indicate the possible operations such as *Equality* and *Ordering* which can be applied to the attribute value.

- **Name binding template**

The name binding template allows alternative naming structures to be defined for a given object class. The template should include one naming attribute to identify an object instance. Naming structure are based on the containment relationship among object classes. Figure 3 shows that the *equipment* is contained by *root* and the *equipmentID* attribute is specified as its naming attribute.

- ASN.1[4)] syntax

ASN.1 syntax specifies the data types referred to in the templates such as attributes or name bindings. *AdministrativeState* syntax as shown in Figure 3, is defined as ASN.1's ENUMERATED data type.

```
--Managed Object Class Templates          administrativeState  ATTRIBUTE
top         MANAGED OBJECT CLASS              WITH ATTRIBUTE SYNTAX   AdministrativeState
CHARACTERIZED BY:                             :
  ATTRIBUTES                              vendorName         ATTRIBUTE
    objectClass    GET;                       WITH ATTRIBUTE SYNTAX   VendorName
REGISTERED AS {mObjectClass 1}                :
                                          --Name Binding Template
equipment   MANAGED OBJECT CLASS          equipment-name      NAME BINDING
DERIVED FROM top                            SUBORDINATE OBJECT CLASS   equipment
CHARACTERIZED BY:                           NAMED BY
  ATTRIBUTES                                SUPERIOR OBJECT CLASS      root
    equipment ID    GET,                    WITH ATTRIBUTE             equipmentID
    administrativeState SET TO DEFAULT    REGISTERED AS {nameBinding 1}
                DEFAULT VALUE unlocked
                GET-REPLACE,              --Abstract Syntaxes
    vendorName    GET,                    EquipmentID       ::=   PrintableString
       :                                  AdministrativeState
REGISTERED AS {mObjectClassExample 2}                       ::=   ENUMERATED{
                                                                    locked       (0),
--Attribute Templates                                               unlocked     (1),
equipmentID     ATTRIBUTE                                           shuttingDown (2)}
   WITH ATTRIBUTE SYNTAX                  VendorName        ::=   PrintableString (48)
                EquipmentID
   MATCHES FOR   Equality
REGISTERED AS {attributeIDExample 1}
```

Figure 3: The examples of template definition

The object oriented approach is taken to describe the information model so that it is needed to refer to the inherited templates. Furthermore, since the templates have a nesting structure, the associated templates or ASN.1 syntaxes are necessary for the complete definition of management information. In order to get a consistent management information model, the mechanism to check whether the associated templates or ASN.1 syntaxes are defined or not, is crucial at the model design phase. Since the amount of the management information is supposed to be quite large, an automated checking mechanism is desirable. It should be noted that the before-mentioned templates do not reflect the latest DIS version in the OSI, but employs the DP version. Although the DP version template is used in the following discussion, the proposed development tools are applied to the DIS version template as well, because the key concept is quite the same.

3.2. Editing function of information model

In the proposed environment, the structured editor applied to the target information model is provided. The editor has knowledge on the structure of OSI templates, so that the information can easily be incorporated with the OSI templates. The editor features the following functions.

- Graphic display of the inheritance and the containment relationship

The inheritance tree and the containment tree of the information model are created by using the managed object class templates and name binding templates. The resultant relationships are graphically given to the operator as shown in Figure 4-(a). The graphic interface helps system designers to define the management information and look into the managed object definition in a user friendly way.

- Editing functions according to the template structure

The editor has knowledge on the template structure, so that it enables users to define the information model without difficulty. That is, template keywords, such as "MANAGED OBJECT" and "CHARACTERIZED BY," are initially displayed in order for users to create the model simply by fulfilling the appropriate items. Other desirable functions such as detection of syntax violation are provided in the editing function. The definition of the *modem* object template is depicted as shown in Figure 4-(b).

- Retrieval of the related template definitions

By using this function, users can retrieve all the templates, such as attributes, notifications and actions, which are derived from the superior managed objects. Figure 4-(c) shows the *modem* object which includes the derived attributes which are defined at its superior *equipment* and *top* objects. As shown in Figure 4-(c), the attribute templates and ASN.1 syntaxes which are associated with the selected *administrativeState* attribute, are retrieved and displayed.

- Consistency check of the information model

When a new template or ASN.1 syntax is defined to or deleted from the model, consistency check is automatically performed in the editor. The consistency check function prevents from defining the management information in duplicate or deleting information to which other templates refer. This function is closely related to the processing function explained in the next subsection.

3.3. Processing function of information model

The information model consisting of the template definitions and ASN.1 syntax definition is processed. The processing flow of the model is shown in Figure 5. First, the templates and ASN.1 syntaxes are parsed and the resultant data are stored in the intermediate files. Then, the intermediate files are checked in terms of the relationship among the definition of the template and the ASN.1 syntax. Finally, a management information dictionary, which includes all the information defined in the management information model, is created.

(a) Modeling window

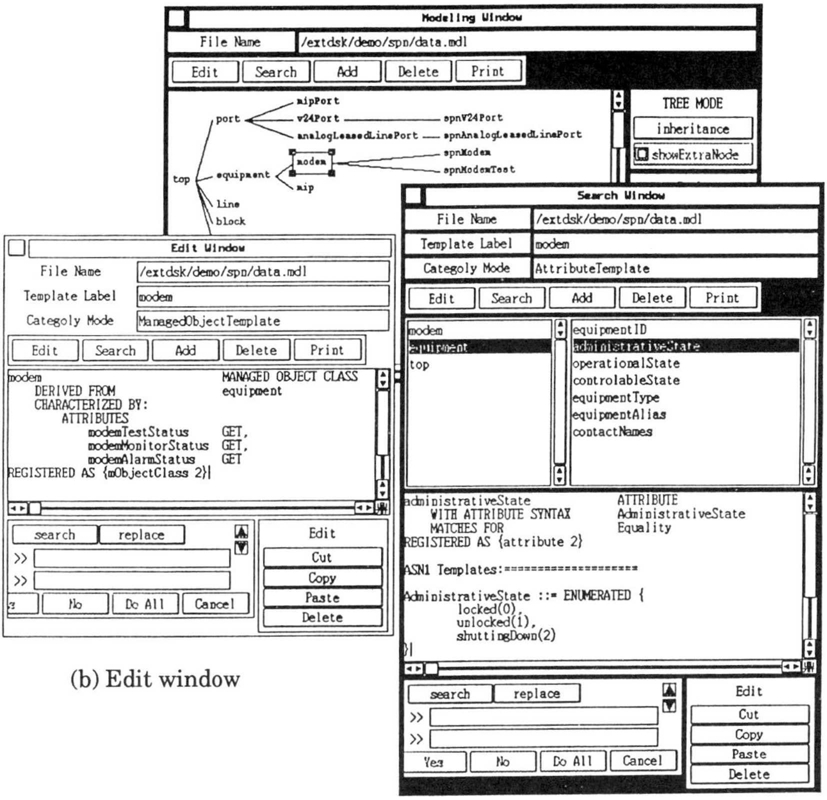

(b) Edit window

(c) Search window

Figure 4: User interface of Modeling Editor

Parsing and checking mechanisms, and the usage of the information dictionary are described in details in the following subsections.

3.3.1. Parsing and checking mechanisms of templates and ASN.1 syntaxes

After the template and the ASN.1 modules are parsed independently, the parsing results are stored in the separated intermediate files. Any syntax errors can be eliminated through the parsing process, and hence, the resultant files include no syntax errors. The required information could be obtained by referring to several templates which have such relations as inheritance, containment, etc. The information model should describe the complete relations in order for the required information to be correctly retrieved and updated. Therefore, it is essential to check the relations among the management information so that the duplicated or undefined templates can be detected.

After parsing and checking processes, the information model is converted to the information model dictionary file whose data can be used by the application programs. The application program refers to the management information by the query functions which are provided as the library functions coded by C language. The specification of each query function is described in next subsection.

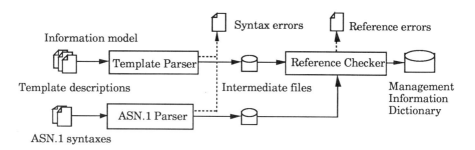

Figure 5: Processing Flow of Management Information Model

3.3.2. Query functions for information model dictionary

The management information defined in the templates and ASN.1 syntaxes are registered into the dictionary file. In the actual operation, the contents of the dictionary are transferred to the table on the memory for a quick look-up action. The following query functions are prepared for the dictionary.

- **Template identification**

This function examines whether the specified template such as the managed object template or the attribute template exists or not. A template label or identifier is used to specify the template to be examined.

- **Class modifier check**

This function examines whether the specified modifier such as CREATE or DELETE exists or not in the specified object class template.

- **Superior template retrieval**

Applying this function, the superior object or attribute template defined at the term of "DERIVED FROM" in the specified templates is retrieved.

- **Attribute Identification**

This function examines whether the specified attribute is defined in the specified managed object class or not. The examination is performed by searching and checking all the attributes specified in the superior managed object classes.

- **Validity check on the attribute operations**

This function examines whether the specified operation on the attribute, such as GET, REPLACE, ADD, etc., is permitted or not. Also the specified attribute value testing such as *Equality* and *Ordering* is examined whether it is allowed or not.

- **Attribute value retrieval**

The default value of the specified attribute is retrieved by using this function. It is possible to examine if a given value is within the range defined at its attribute template or ASN.1 subtype notation.

- **Name check**

This function examines whether an object instance name given when the object instance is created conforms to the rule defined in the name binding template.

By using above-mentioned query functions, the requested management operation from other open systems, which is provided through the OSI systems management protocols, can be examined whether it is executable or not. When the invalid operation, such as accessing the undefined attribute and undefined operation to the object, is detected, the error response is replied. Figure 6 shows an agent open system with the query function libraries in application programs. The same library functions can be applied to the management information dictionaries generating for the various types of managed networks, which potentially provides the environment of the efficient application code development.

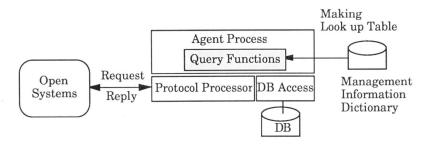

Figure 6: Protocols Implementation using Management Information Dictionary

3.4. Supporting tools for MIB

The management information which is defined using OSI templates, would be distributed over the manager system, a number of agent systems, and even communication equipments. Therefore, every attribute of managed objects must be specified the location where the attribute value is stored. In the development environment, the MIB supporting tool provides various functions. Among them, the following two function modules are specifically explained in this section.

- generating the relational database schema
- generating accessing codes for management information

3.4.1. Relational database schema generation

The relational database schema function has a capability to generate the DDL (Data Description Language) automatically for the SQL based database. Two types of database schema are considered such as the managed object class schema and the object directory schema. The managed object class schema is created for each managed object class, and only one object directory schema is created in NMS. The examples of both schema creation are shown below. In the creation of the managed object class schema, the definition of managed object class *equipment* (see Figure 3) is taken into consideration.

```
// Schema for the managed object "equipment"
  create table equipment {
  // internal identifier for object instance
    instanceID         NUMBER (32),
  // attribute derived from "top" object
    objectClass        CHARACTER (32),
  // attribute defined at "equipment" object
    administrativeState CHARACTER (32),
    vendorName         CHARACTER (48),
    :  }


// Schema for containment relationship
  create table objectDirectory {
  // internal identifier for object instance
    instanceID         NUMBER (32),
  // representing the object class
    objectClass        CHARACTER (32),
  // representing the name attribute label
    nameAttributeLabel CHARACTER (32),
  // identifier of superior object instance
    superiorInstanceID NUMBER (32)
    :  }
```

The tables of the relational database correspond to the managed object class definitions, and basically contain all attributes, where the inheritance from the superior managed object class definition is considered. The length of the data field (attribute of the relational database) of each attribute is specified as the default value or the specified value which is constrained by subtype notation in ASN.1 definition. Taking the *vendorName* attribute for instance, since the attribute is defined as "PrintableString (48)" in ASN.1 definition (see Figure 3), the length of the data field for the *vendorName* attribute is decided as 48 bytes. The creation of the directory schema specifies containment relations between managed objects. Hence, the name attribute definitions which is specified by the name binding templates, is processed and its schema is created as the object directory table. In both tables, *instanceID* field plays an internal role in relating between the managed object tables and the object directory table.

The attributes which are related to the ASN.1 syntax is stored in one data field of the relational database, the stored values of which are encoded based on the ASN.1 coding rule.

3.4.2. Generation of accessing codes to management information

The data structure of the managed object's attribute is based on ASN.1 syntax. The access code generation function first defines C++ base classes for a general purpose MIB access as shown in Table 1, each of which corresponds to the ASN.1 data type. In each C++ base class, several member functions are defined for encoding/decoding of ASN.1 code. The final access code for each attribute can be represented as a derived class of this base class. The example of the access code for the *AdministrativeState* attribute is shown next.

```
// "AdministrativeState" class is defined as derived class of asn1_ENUMERATED
// Value can be accessed by field name,
// such as "locked", "unlocked" and "shuttingDown"
class   AdministrativeState   public: asn1_ENUMERATED { };

class asn1_ENUMERATED   public: asn1_INT {
public:
        // member functions for accessing by field name
        virtual  char*   getValue ();
        virtual  char*   setValue (char* pt);
        };

class   asn1_INT {
public:
        // member functions for accessing integer value
        int      getValue ();
        int      setValue (int value);

        // member functions for encoding/decoding
        void     getPdu (int len, char* pt);
        void     setPdu (int len, char* pt);
        };
```

Table 1: ASN.1 base classes corresponds to ASN.1 data type

ASN.1 base classes	ASN.1 datat types
asn1_INT	INTEGER
asn1_ENUMERATED	ENUMERATED
asn1_REAL	REAL
asn1_STRING	Printable String, etc.
asn1_SEQUENCE	SEQUENCE

4. CONCLUSION

The development support tool to implement the OSI based NMS was proposed. The proposed tool aims at providing an integrated environment for the efficient NMS development, which covers a wide range of the development stages from the initial model creation phase through the MIB implementation phase. In addition to the efficient development, a high degree of reliability of the products is supposed to be achieved by using the debugging function of the information model and the capability of automatic generation of MIB schema and C++ access codes as well. The feasibility study of the tool has been completed, and the tool is now being upgraded to apply to the real NMS development.

ACKNOWLEDGMENT

The authors wish to thank Dr. B.Hirosaki and Mr. S.Sakata for their encouragement and support, and would like to give special thanks to Mr. H.Kuriyama and Ms. K.Arima for their support of the system development.

REFERENCES

1) ISO/IEC IS 9595/9596 (CMIS/CMIP)
2) ISO/IEC DP,DIS 10165 (SMI)
3) A.Kara, S.Nakai et al.,
   "An Architecture for Integrated Network Management",GLOBECOM'89
4) ISO/IEC IS 8824/8825 (ASN.1)

III

INTEGRATED MANAGEMENT

Global Commonality in User Requirements

Elizabeth K. ADAMS

Principal, Adams & Associates, Basking Ridge, NJ, USA
Director of Operations, OSI/Network Management Forum

Abstract

On the first anniversary of the founding of the Network Management Forum a new program was piloted. Called the President's Roundtable, it was the Forum's way of gathering user needs in network management. Now, one and a half years later, well over 100 users have spent the better part of a day with Forum leadership, describing frustrations with their current management environment, defining their needs, and influencing the Forum's direction.

The net of those discussions are user requirements for integrated network management, which are summarized in this paper. These requirements range from general comments on the preferred approach to multi-vendor management, to more specific statements of priorities among the various functional areas.

SETTING THE STAGE

The Growth of Information Networks

Information networks -- the collection of physical and logical elements that are the lifeblood of the modern corporation -- have grown tremendously in use, scope and complexity over the past decade. Several factors have contributed to that growth.

The deregulation of the telecommunications industry opened the door for innovation. Starting in the U.S. and extending to Europe and Japan, government action has hastened the development of new products and services. Companies have become more involved with the design of their networks in order to take advantage of potential improvements in performance and cost. The use of premises-based technology to fill the holes left by traditional service providers has put the user -- in this case the purchaser of multiplexers, modems and so on -- in more direct control.

As user involvement has grown, so has the strategic importance of information-sharing. The ability to link employees with critical systems has meant faster sales transactions, more accurate billing, and better service to customers at lower cost. At the same time, the pull of the world economy has caused a geographical branching-out, and companies have begun to rely on technology to bring their operations together. Over the space of a relatively few years, communications and computing have merged to make information networks a critical competitive factor.

The Changing Complexion of Network Management

Increased dependency on the timely delivery of networked information has brought a much greater focus on network management and control. The innovative corporations who first recognized the power of linking their employees and customers with information systems have pushed for the development of customer-controlled management tools, such that individual elements can now be monitored, tested, taken out of service and returned to service. But because the tools were developed piecemeal, they offer little help in dealing with management of the "total" network -- that which crosses product and service boundaries, vendor boundaries, geographic boundaries, and traditional management boundaries.

To compensate, companies have taken extraordinary measures to keep their information networks running smoothly. Some companies employ extra technicians in order to develop a high degree of specialized knowledge of individual equipment or service components. Others rely on redundancy, installing elaborate backup mechanisms to keep operations uninterrupted.

People and equipment substitute for the ability to easily summarize and interpret results, and to take positive action when faced with a performance problem that crosses vendor boundaries. As a result, the cost of complex information networks is high and growing steadily. Some estimates spanning a five-year period suggest that one-third of the cost is in purchase and installation, and two-thirds is in daily operation. Most of the operational cost is salaries.

The ability to analyze data, to isolate faults, to reconfigure the network, to take actions before service is disrupted, to identify chronic failures, and to gain maximum utilization from the most expensive resources are the keys to managing an information network effectively. Because information is not available across multiple management systems in common format, or using common terms, these tasks are nearly impossible for people to perform, and equally difficult to automate.

Without the means of obtaining management information in a common way, using consistent terms, neither the people nor the systems they employ can manage information networks effectively.

THE NETWORK MANAGEMENT FORUM

The Forum is a group of over 100 telecommunications and computing equipment vendors and service providers. It was formed to solve customers' needs in multi-vendor network management, spanning both the communications and computing environments. Its chief goal is to accelerate the availability of interoperable network management products and services.

It is accomplishing this goal by defining a common approach, common management structures and common terminology for the exchange of management information. The Forum provides specifications to its members and the means of testing products for conformance to those specifications. The Forum also works to educate the marketplace on the benefits of an interoperable approach. As a rallying point for network management, the Forum has pulled the diverging interests of computing and telecommunications together, and has defined a way forward.

The Forum does not deal with specific network management products or services. Rather, it defines the way in which management information can be exchanged to support a wide range of vendor-specified applications. The Forum's Release 1 Specifications, for example, define an architecture, data communications protocols, application services and a subset of management and managed objects which, when implemented by vendors, allow the exchange of management information among vendor products and services.

THE ROUNDTABLES -- WHAT WE LEARN

The users we've met aren't shy about describing their problems and their requirements. They are faced with extraordinary pressures to improve both the reliability and the cost-effectiveness of enormous and complex information networks. Because the problems in network management are so fundamental, Roundtable discussions center on fundamental needs -- specifically the growing need to exchange management information in a common way in order to manage the information network as a single resource.

Some Things Stay the Same...

Roundtables have been held in most parts of the world, including New York, Paris, Toronto, London, Rome, Tokyo, and San Francisco. But although the venues have been markedly different, user needs have reflected common experiences in network management, and a shared need to solve the problems caused by tying together multiple vendors, products and services.

Functional priorities have stayed remarkably constant over nine Roundtable sessions, with fault management leading the list of critical needs, followed closely by configuration and performance management. After one-and-a-half years, opinion is still divided on the relative importance of security management. And users are still frustrated to see how slowly products are emerging to help them solve their problems.

Some Things Change...

Users have grown in their understanding of the complexity of multi-vendor management. When asked to offer opinions on the relative priority of fault management and accounting management, users are now more likely to delve into details, suggesting specific fixes to problems in security or describing an application that would be possible with the exchange of performance data.

The configurations of the corporate networks managed by the Roundtable participants have changed, too, reflecting the growing importance of distributed LAN-based networks. With that shift have come even greater headaches for the managers who must try to keep end users happy without direct control over resources.

Some users are considering building their own "umbrella" management systems, tailored to fit their needs precisely. Convinced that no one vendor can satisfy their needs, some of these users view this approach as a regrettable but necessary last resort, while others conceive of the potential for selling their applications to other firms with similar needs.

DOWN TO DETAILS

Several key topics or areas of concern have been raised at nearly every Roundtable. They are:

- a total information network perspective;
- flexibility of approach;
- functional priorities;
- conformance testing;
- user education and involvement; and
- user interfaces.

More recently, additional topics have captured the interest of both the Forum and the participants, perhaps reflective of the shift from the Forum's "specifications" phase to the product delivery phase. These include:

- adherence to standards vs. product availability;
- common management information over alternative network types; and
- overall reduction in management systems.

Each of these is addressed in the following paragraphs.

"End-to-End" Scope

One thing all users agree on is the need for network management solutions to encompass the entire breadth of the information networks they manage. For some users the term "network management" connotes management of only the physical network. Considerable time has been invested at each Roundtable to assure users that the Forum's approach does address their multi-domain, end-to-end needs.

Because words to describe "end-to-end" don't seem to come easily, an example is often used to illustrate what is meant. Using a car rental agency as a model, the "end-to-end" information network includes all the communications and computing products and services necessary for the agent to receive a call from a customer, to access a database of available cars, to access current billing rates, to input order information, and to alert the car inventory system of the order. This car rental agency might have a network that consists of local area networks and workstations that are managed by each branch, with a corporate data network providing information on rate specials downloaded from the mainframe at company headquarters, billing data uploaded from the branch to the headquarters billing system, an "800" network provided by a carrier, and an ACD for routing customer calls to the appropriate agent.

Whatever the configuration, and whatever the functions performed by a company, be it a car rental agency, a brokerage firm, a retail chain or a manufacturer, it is clear that users need network management solutions which embrace the widest possible scope. Further, solutions must be able to be applied both to a distributed management environment and to a centralized environment, since individual companies do things differently.

Once the information network example is explained, users are ready to listen to the rest of the Forum's story. Because of their concern that we might only be addressing the physical characteristics of networks, this subject is often among the first to be raised in our discussions. One thing is clear: any multi-vendor solution that does not recognize the need for complete "end-to-end" management will fall short of the mark for the sophisticated user.

A Flexible Architecture

Virtually all users express their support for the Forum's approach to multi-vendor management, which provides the means for managing *any* network (open or proprietary) by exchanging common management information among management systems. They see in this approach the ability to take early steps to gain control without disrupting their current environment. Two points are raised by users in support of the Forum's architectural approach.

First, with the Forum's approach they are not faced with replacing existing systems by a monolithic "do-all" system, nor must they replace the (mostly proprietary) interfaces between existing management systems and the devices or services they manage. Modification of a management system is viewed as much simpler to accomplish than changing the various devices that comprise the network. Further, new technology can easily be accommodated by tying in new management systems with existing ones. Said one participant "What matters is that a vendor of a ten-year-old system and a vendor of a new umbrella system can talk together."

Second, the approach gives vendors broad leeway to develop applications in support of both centralized and distributed management environments. For those desiring an "umbrella" manager to collect, summarize, and display the status of the network, management information can be directed from many discrete systems to a single system. For those preferring a distributed approach, management information can be exchanged between systems such that a given system could act as manager to collect and process information in support of one function (e.g., accounting) and as agent sending information to support another function (e.g., performance management).

Functional Priorities

As both a calibrating tool and a means of getting to the issues faced by the Forum, a questionnaire has been employed at every Roundtable. Users are asked to rank the functional areas of network management in terms of the relative importance of multi-vendor interoperability to each function. Within the five major areas (fault, configuration, performance, accounting and security) a further breakdown is provided. For example, under fault management, users are asked the relative importance of interoperability for such functions as alarm reporting and logging, alarm management, testing and trouble ticketing.

Clear and away the most critical need is *fault management*, specifically the alarm reporting and logging functions. Less important are testing management and trouble ticketing. Many users have commented that until they can "stop the bleeding" they are unable to spend adequate time working other important issues, like performance management. Users see the ability to filter and correlate alarms as important time-saving aids, and they have commented that fault isolation is absolutely essential.

Configuration management, the next most important function, is viewed as a basic ingredient of successful fault management, because it establishes the inventory of resources to be managed, without which alarm management is not possible. Others view configuration management as most important for the reconfiguration function -- the ability to shift traffic to available resources when performance problems occur.

Performance management, ranked third in importance, is an essential function to ensure that the network continues to meet its design criteria and, ultimately, the business purpose for which it is operated. The most serious problem facing users is the lack of common metrics and common terms for such critical measurements as "response time". With common terminology and the ability to exchange information, users see the potential to reduce the need for redundant systems, and to increase the likelihood that end-user applications can be accommodated consistently. Users also see the relevance of performance data to overall cost management. By tying performance data to tariff-based routing, for example, organizations would be able to demonstrate significant value to the firm by avoiding blockages at least cost.

Among the interesting trends is that performance management has, over time, become more critical in the minds of users. In part, that may be due to the fact that the Forum's Release 1 specifications address the most pressing needs in fault and configuration . But it may also signal a growing recognition on the part of network managers that their real value to the corporation is in managing the use of corporate resources. Said one San Francisco participant, "I don't feel my job is putting out fires, although most of my time is spent that way. My job is to manage the utilization of resources."

Table 1
Summary of Responses From Participants of Seven Roundtable Sessions

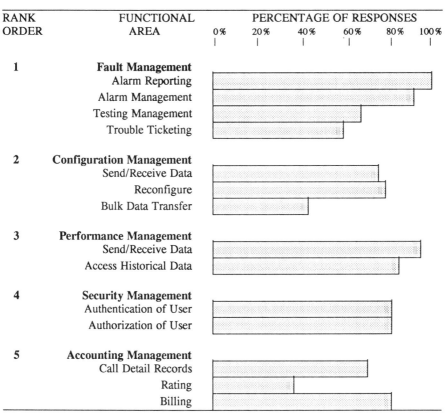

Participants were asked to rank the relative importance of the five functional areas, and to indicate, within each functional area, which items were most critical. This table shows the percentage of responses indicating a need for interoperability in each of fourteen sub-categories.

Security management is an area that consistently prompts divergent views, largely because the functional area encompasses very different dimensions. As a way of understanding user needs, the Forum has begun to use the terms "management of security" and "security of management" to help distinguish between the various needs.

"Management of security" connotes the protection of the network against unauthorized use. Using this definition, user views are divided, with some (such as resellers) viewing this as a top priority, and others viewing it as outside the scope of their network management responsibilities.

"Security of management" deals with the specific security issues that result when two or more management systems become interoperable. Here, most users agree that at least a basic solution is necessary, and some, reflecting corporate policy, view it as a mandatory requirement before purchases can be considered. A common authorization record was proposed by one group of participants as a way to provide a first-order of security without trying to solve the larger issues of preventing network fraud.

Within *accounting management,* the ability to manage costs based on call detail records in a common way across vendors would provide most users with what they need to determine end-user usage of the network. Some interest in rating and billing has been expressed, but the lack of depth of discussion on these topics indicates few users are yet able to turn their attentions to these problems.

Conformance Testing

In the course of providing an overview of Forum activities, we mention our conformance testing program with the Corporation for Open Systems International (COS) and the Standards Promotion and Application Group (SPAG). Invariably, users question this program extensively, and their views on conformance vary.

The level of sophistication of the users is often apparent in their understanding of conformance and interoperability testing. Initially, users were surprised and not a little dismayed to learn that conformance testing did not guarantee interoperability. Over time, their understanding of the whole testing discipline has grown, such that users are now more likely to debate the relative merits of conformance and interoperability testing.

Most users desire a conformance testing program as an important first step along the road to interoperability. Those who have had previous experience trying to make two products work together applaud any efforts the Forum makes in trying to lessen the extent of interoperability problems. In their view, conformance testing provides a level of quality assurance.

Others view conformance testing as "nice-to-have", but weigh that against the immediate need for interoperable products. They support early demonstrations of product interconnection, even if it means conformance testing has not been performed.

The way in which conformance testing is done is also of importance to users. Although many accept that vendor self-certification is necessary in the short term, their preference is for independent testing to be done by accredited testing houses, with criteria applied evenly across geographic and vendor boundaries.

Independent records of vendor interoperability are also desired by users, who are tired of hearing vendor X claim interoperability with vendor Y, only to find that vendor Y doesn't support the claim in quite the same way. These users have asked the Forum to consider establishing an industry matrix of interoperability, showing those cases where both vendor X and vendor Y claim interoperability.

User Education and Involvement

A key need of most users we meet is that of education and involvement in setting direction in network management. One London participant, when asked in a pre-meeting questionnaire his perception of the Forum's role, said "I hope that it will provide a framework for vendors to develop truly interoperable network management systems by utilizing standard interfaces. I also hope that real users needs will be voiced loudly, so that we are not forgotten, and get what we want out of the system, and not just what vendors want to give us."

Several users at one session lobbied for the Forum to allow them to help develop user requirements, and asked to be allowed to contact the technical team leaders to find out how the Forum is approaching specific problem solutions. Others have asked for tutorials on network management, for assistance in questioning vendors on their product plans, and for help in properly framing Requests For Proposal (RFPs). There is much more that can be done in this area, and the Forum is only now beginning to understand some of these needs.

User Interface

Although the Forum currently has no plans to define user interface specifications, this is a subject that often comes up in open discussion. Generally, the users would like to have a common user interface, since that would make their lives easier. They acknowledge, however, that it is unlikely vendors will ever agree on a single interface.

Some commonality, they feel, is both highly desirable and practical, including agreement on a common set of icons, and the association of colors with severity levels in a common way. Placement of critical fields in a common location would also help them operate more smoothly between management systems.

Standards Adherence versus Product Pragmatism

Users generally seem to view standards as the right way to go, but they are extremely frustrated by the lag in standards development and subsequent products. The Forum's mission has always stated that we will build to available standards, fill in any holes, and migrate non-standard specifications when final standards are approved. Yet, because that's easier said than done, and because we must wrestle with the difficulties caused when an approved standard, such as CMIP, is changed after our Forum specifications are frozen, we have, of late, asked users to tell us just how important it is for our specifications to be in synchronism with standards.

Their response, universally, is that what matters to them is getting products. If it is necessary to be non-standard for some period of time in order to freeze specifications and give the go-ahead to product developers, they're supportive. They have commented, though, that they wouldn't expect the Forum to persist in being non-standard. In fact, they suggested that an update every two years or so might be just about right, since that would coincide with normal product update cycles. The most satisfactory solution, it seems, is to provide a clear *path* to standards while making products available today.

Standards are the preferred route, but users would like to send a message to the standards bodies that a reasonable solution today is better than a perfect solution tomorrow. The advantage of an industry standard is that it frees vendors of all sizes to participate in providing network management solutions.

Common Management Information Over Alternative Network Types

One question asked of users is whether they feel it is necessary for the Forum to define other options for transporting management information in addition to the currently specified X.25 and 802.3. User views are mixed on this issue.

Most users believe the Forum should concentrate its efforts on defining the management information that is exchanged across the data communications protocols, since that is an area the Forum is uniquely able to address. Some have suggested that it would make their lives easier if the Forum were to endorse efforts by member companies to run Forum-defined management information over multiple protocols as an implementation alternative for those whose management networks are non-OSI based.

The disadvantage to extending the networking options is that interoperability problems may result. As ever, a balance needs to be struck. The Forum recommends OSI networking and provides conformance tools as a target. Migration to this target can be considerably eased by encouraging the use of Forum-defined management information transported over existing network types.

Reduced Number Of Management Systems

A goal of most users is to consolidate management functions in a smaller number of systems. Today, because each management system is a discrete island of information about specific network elements, they have been developed without regard to how the user fits each system into the larger whole.

Users would prefer an environment where basic device management is accomplished through systems that don't require a separate workstation. As one user explained, "My technicians are on call 24 hours a day. They take a terminal home with them, and when a problem comes up, they fix it from home. They deal in simple text 99% of the time." The need for sophisticated graphics, in the view of many users, can be reserved for an "umbrella" system that manages the end-to-end view.

At the very least, users would like to avoid paying for intelligent workstations to support every management system. They would like a common platform to be defined, so that multiple management applications could be run on a single workstation. Eventually, they'd like some of their smaller vendors to simplify their whole approach to network management by providing network management applications which run on a system common with other major network management applications.

SUMMARY

The network management users -- those individuals who are responsible on behalf of the network "owner" to gain the best performance and highest efficiency from a critical corporate resource -- face enormous challenges today. They must convince their suppliers to take a larger view of the problem, to recognize that the only viable solutions are the ones that recognize today's information networks for what they are: multi-domain, multi-vendor, multi-product collections of telecommunications and computing elements.

In the role of "director", users want all their suppliers to read from one script, enabling the user to assign the lead roles and the bit parts according to individual interpretation. An industry-accepted, standards-based approach is far preferable to fighting with individual vendors to cooperate, provided the approach is robust enough to handle the complexity of networks now in place. And an off-the-shelf solution is preferred over custom development.

The opportunity to state their needs is critical to the users. Almost without exception Roundtable participants have applauded the Forum's efforts to set up a direct link with users, rather than relying on vendor members to represent their needs. Said one user who has attended two Roundtables in the space of twelve months, "This is the only organization that listens to the users and acts on our input." Said another user, describing the value of the Roundtable session, "I know now what is reasonable to ask of my vendors. An approach that has the backing of the Forum and users is one I can insist that my vendors support."

The consistency of message from over 100 users worldwide over eighteen months tells the story quite effectively. Users want the ability to manage their information networks as a total entity, with the flexibility to tailor their management systems configuration to their own unique management environment. They see the Forum's approach as logical, flexible and practical, and they impatiently await products that will help them implement the Forum's approach.

III

INTEGRATED MANAGEMENT

A
Panel

Evolving User Requirements for Network Management

Moderator: *Janet Cohen*, Butler Cox PLC, UNITED KINGDOM

Panelists: *Hiawatha Cotton*, Department of Defense, USA,
Edwardo Duarto, European Space Agency, Germany,
Ashok Malik, American Express Travel Related Services, USA,
Chris Sluman, SEMA, on behalf of Ministry of Defense, UNITED KINGDOM.

The integrated network management systems that manufacturers are developing and delivering today are based on what has been believed to be well-understood requirements. However, recent work by both suppliers and standards bodies shows that our understanding of user needs is still evolving. Meanwhile, users are continuing to introduce new technologies and services into their networks that must be managed. This panel discussion will explore how user requirements for management systems are evolving, how they may change in future, and expose some of the differences between requirements in Europe and North America. Each speaker will make a short presentation on their view of user requirements, setting the scene for the following open debate.

III

INTEGRATED MANAGEMENT

B
Ways of Achieving Integration

Development of Integrated Network Management System NETM Based on OSI Standards

Michio Suzuki[*1] Ryoichi Sasaki[*1] Yasuhiko Nagai[*1]
Keizoh Mizuguchi[*2] Masato Saito[*2] Hideaki Kobayashi[*2]

*1 Systems Development Laboratory, Hitachi, Ltd.
1099 Ohzenji, AsaoKu, Kawasaki 215, Japan

*2 Software Works, Hitachi, Ltd.
5030 TotsukaCho, TotsukaKu, Yokohama 244, Japan

Key Words: OSI System Management, Network Management Architecture, Managing Network, CMIP, Managed Object Instance

Abstract

This paper describes the management architecture and the structure of the integrated network management system NETM developed by Hitachi on the basis of OSI standards (especially CMIP). The NETM management architecture has some unique characteristics such as a decentralized and centralized approach, a hierarchical structure of management systems, a managing network, and a NETM unique expression of managed object instance. In the decentralized and centralized approach, the network is divided into several subnetworks. Each subnetwork has an individual subnetwork management subsystem managing it's own subnetwork, and these SMSs are able to have hierarchical structure. NETM unique managed object instance expression, that is created by concatenating integer value given to each managed object instance within a containment tree, is able to reduce the length of managed object instance expression information and handling overhead of management information base. The managing network is introduced in order to manage the subnetwork management subsystems.

The NETM is aiming at a very large management system that will manage almost all of Hitachi's network facilities, and will be able to include other vendors' network management systems.

1. Introduction

Rapid advancement in new communication technology, such as optical fiber and digital transmission, has accelerated the enlargement and integration of computers and communication networks in many companies. These days, it seems that the greater part of many companies' activities depend on the "network". [4, 5, 6, 7]

The greater the dependence of their activities on their networks, the greater the damage by network faults, the ratios of network costs within companies' expenditures increase year by year. The needs for network management are apparent today to reduce damage and save money by reducing downtime and solving performance problems.

The many companies' networks in almost all cases include various kinds of facilities from multi-vendors, and these networks gradually grow and become more varied day by day. Therefore, it is very difficult to make up network management systems for company networks having these characteristics.

This paper will discuss some network management requirements, Hitachi's approach to developing network management systems, and Hitachi's integrated network management system, NETM, developed on the basis of OSI standards. [1, 2, 3]

2. Needs and goals of the network management system

When a new system is developed, it is very important to analyze the background that makes system necessary, to clarify the current needs of the system, and to define the system's future goals.

The background that make network management systems necessary is considered as follows.

(1) Once a fault occurs in a network, company activity will suffer very serious damage from that fault. Therefore, network management system which can reduce the damage is necessary.

(2) Since the ratio of network cost to company expenditure has increased, the network management system is necessary to reduce cost and to save money.

(3) Although many companies' networks have become worldwide, it seems difficult to employ the necessary number of skilled network operators having the requisite expert knowledge of the network.

Recognizing the above, we defined the future goals of Hitachi's network management system as follows.
 -maximum usage of network
 -high reliability of network
 -easy operation of network management system

In order to achieve the above goals, we have developed the integrated network management system called NETM.

3. Architecture of Hitachi's integrated network management system NETM

In all cases of developing systems, an assured architecture is the most important issue. If a system is developed on the basis of an uncertain architecture, it may not be completely adequate in the future.

This section will discuss the architecture of Hitachi's integrated network management system

3.1 Essential requirements
The essential requirements that the management architecture must satisfy are:
(1) Network management systems must be able to handle the various configurations and scales of networks.
(2) The network management system has to guarantee it's own reliability.
(3) The network management system must be capable of interfacing with other network management systems.

3.2 Decentralized and centralized management approach (DCMA)
As illustrated in Fig. 1, the network is divided into several subnetworks that are classified as follows.
(a) Physical network
 The physical network is constructed with physical facilities which execute the functions of the physical layer of the OSI model. Examples in this category are a TDM network and a modem network.
(b) Switching network
 The switching network consists of switching devices deriving their functions from the 1st to the 3rd layers of the OSI model.
There is a PBX subnetwork or a packet switching subnetwork in this category.
(c) Application network
 The application network is constructed with facilities that execute functions from the 1st to the 7th layers of the OSI Model.
One of the application networks is a network that includes host computers and workstations.

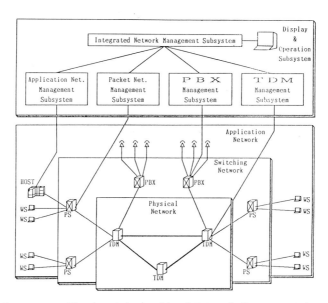

Fig. 1 Decentralized and Centralized Network Management Approach

The switching network uses the transmission services of the physical network and the application network uses the switching services of the switching network. Some actual company networks do not need to include all three kinds of subnetworks. When a company network does not need the switching function, the network may be constructed with just the physical and application networks. Since the network is constructed by combining these subnetworks in accordance with the purpose of the network, there are many kind of network. Network management system must cope with this characteristic of network.

The management function of each subnetwork management is different, for the function of each subnetwork is different as mentioned above. As several subnetwork management systems have been already developed respectively up to the present, it is important to make use of the resources realized by these subnetwork management systems.

It is generally said that the network continues to grow up day by day. In some cases, even one subnetwork which may be provided by many kinds of vendors is added to the network. Considering these matters, it is important that each subnetwork is able to be managed respectively, consequently the decentralized management approach is thought to be a basic one for realizing network management system.

As mentioned already, each subnetwork has some relationships which are very important objects for managing the network as a whole. In order to manage these relationships, the subnetworks' boundary information must be concentrated, in results the centralized approach is thought to be necessary for managing these information.

We have thought that the decentralized and centralized approach (DCMA) is most suitable to network management system. In the DCMA, each subnetwork has an individual subnetwork management subsystem (SMS) managing it's own subnetwork, and the whole network is managed by the integrated network management subsystem (INMS) which is connected to all of the SMSs. If the network management system is based on DCMA, the problem that there are many kinds of network is able to be solved by combining several SMSs in accordance with the structure of the network. Therefore, Hitachi's integrated network management system NETM have adopted the DCMA.

The DCMA has the following advantages.
(a) the ability to meet the configuration complexities and the scale varieties of the different kinds of networks.
(b) the ability to continue managing its own subnetwork even if trouble occurres in the INMS

3.3 Relationship between OSI management and DCMA

The OSI system management model, illustrated in Fig. 2, consists of a managing process (MGR), an agent process (AGT), and some managed objects. The managing and agent processes use the Common Management Information Protocol(CMIP) to exchange management information between them.

As shown in Fig. 3, the decentralized and centralized management system has a hierarchical configuration including a human-machine interface (HMI). Each management subsystem mentioned at the clause of 3.2 can be mapped to the MGR or AGT.

HMI: INMS = MGR: AGT
INMS: SMS = MGR: AGT
SMS: network facilities = MGR: AGT or AGT: Managed Object Instances

In this approach, the HMI is the superior MGR, and there is no limitation on the number of hierarchical levels of MGR: AGT subsystems. A management subsystem (MGR) can recognize only those subordinate management systems such as AGTs directly below the MGR. Conversely, a management subsystem (AGT) can recognize only the superior MGR management system, directly above it.

Fig. 2 OSI System Management Model

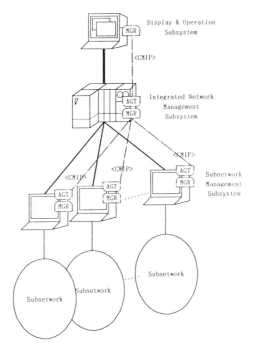

Fig. 3 Hierarchy of Management Subsystems

3.4 Introduction of managing network

Since it is very important to manage the management subsystems themselves, we have introduced the managing network into the NETM architecture. This architecture contains all of the subnetwork management subsystems. The concept is illustrated in Fig. 4.

Although the two networks, the managing network and the managed network, are independent, they are integrated into just one instance tree as shown in Fig. 5. The management subsystems are merely contained in the managing network instance and are not contained in the other subnetwork management subsystems. The general network resources (managed by the subnetwork management subsystem) are contained in the managed network.

The managed objects within the managed network, which are managed by the SMSs within the managing network, are defined in the common attribute that all the SMSs contain.

One major advantage of introducing the managing network into the NETM architecture is that we can distinguish the states of the SMSs from those of the (general) managed objects within the managed network.

In order to manage the managing network, we have introduced a management agent of which the managed objects are the SMSs. This agent is logically independent of the already described agents, but may physically exist in the INMS of those agents.

The display and operation subsystem (DOS) that provides the HMI is the manager of both the managing and the managed networks. That is, DOS can display not only the status of each managed object in the managed network, but also the status of each management subsystem in managing network. Network operators can operate both networks with DOS. The management protocol between the INMS and the DOS is based on CMIP.

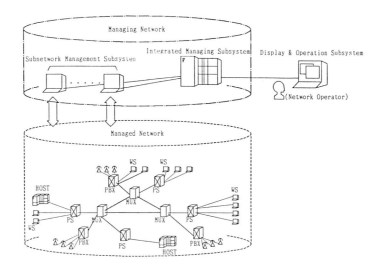

Fig. 4 Concept of Managing and Managed Networks

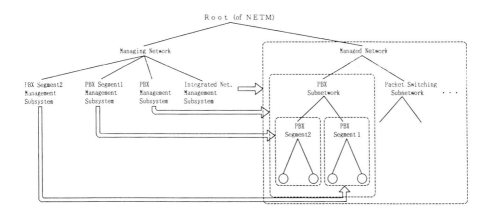

Fig. 5 Instance Tree of NETM

3.5 Representation of managed object instance

When management information is exchanged on CMIP, each managed object instance is represented by D.N. (Distinguished Name), R.D.N. (Relative Distinguished Name), or integer. However, since D.N. and R.D.N. are represented by using a sequence of the pair of a name attributetype and its value, the length of the information to be represented is too long, and the integer representation cannot show the containment relationship within the managed object instance. Besides, in D.N. and R.D.N., it is thought that the searching of managed object instance within MIB (Management Information Base) needs complicated procedures because of handling with attributetype and its value.

As for the length of the managed object instance representation, R.D.N. is shorter than D.N. However, in the hierarchical system, the intermediate management system, that send events from subordinate management systems to superior management system and operations from superior to subordinates, must modify managed object instance representation.

In the NETM architecture, we have introduced a new representation form called HOS (Hierarchical Octet String) to represent managed object instances in order to solve these problems, in addition to the standard representations. The HOS representation form has both advantages of R.D.N. and integer representation, and is identified by a private tag from ASN.1.

The HOS is similar to the R.D.N. sequence, except that the R.D.N sequence uses the name attributetype of each objectclass. In the HOS representation, each subordinate managed object that a superior managed object instance contains, is given a different integer value. The HOS is created by concatenating these given values in a

containment tree. The difference between the HOS and the R.D.N. is that the HOS does not use name attributes and the R.D.N. uses them as follows.
   R.D.N. sequence of {attributetype, attributevalue}
   HOS sequence of {integervalue}
The encoding rule of the HOS is similar to that of the ASN.1 object identifier, except that first and second components are calculated in the case of object identifier encoding.
Fig. 6 shows examples of the HOSs exchanged between management subsystems. As shown in Fig. 6, the representation of the same managed object instance will vary with management subsystems as it is exchanged between subsystems. For example, as for the managed object instance A, its HOS is "1.1" when it is exchanged between the PBX segment 1 management subsystem and the PBX management subsystem, but it is "1.1.1" when exchanged between the PBX management subsystem and the INMS. For example, in the case of representing a managed object instance 5 levels below the top managed object instance, the HOS uses only 5 bytes while the R.D.N. uses 50. This supposes that the length of the name attributetype identifier is 5 bytes and the name attributevalue is 1 byte. Additionally, each management subsystem is responsible for the modification of HOS representation forms.
The HOS has the following advantages.
(a) the ability to shorten the length of managed object instance, compared with D.N. and R.D.N.
(b) the ability to show the containment relationship, compared with integer.
(c) the ability to simplify the searching within management information base.

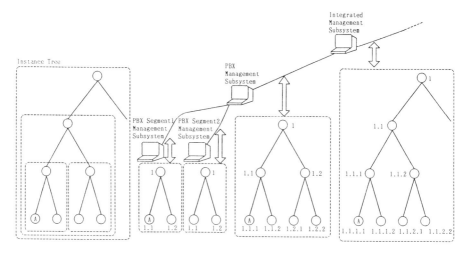

Fig. 6 Examples of Managed Object Instances in HOS

4. Development of Hitachi's integrated network management system
4.1 Management information

In order to realize the Hitachi's integrated network management system NETM, we have defined objects identifier tree for itself. Using this tree, object classes, attributes, notifications, and actions are developed based on the concept of OSI management information.
Hitachi's object identifiers are as follows.

NETM OBJECT IDENTIFIER::={iso(0)member-body(2)Japan(392)
                        Hitachi(99107010) NETM(1)}
NETMCommonInfo OBJECT IDENTIFIER::={NETM 100}
NETM-objectClass OBJECT IDENTIFIER::={NETMCommonInfo 1}
NETM-attribute OBJECT IDENTIFIER::={NETMCommonInfo 2}
NETM-notification OBJECT IDENTIFIER::={NETMCommonInfo 3}
NETM-action OBJECT IDENTIFIER::={NETMCommonInfo 4}

In this object identifier tree, we have registered about 70 objectclasses, 130 attributes, 30 notifications, and 15 actions.

4.2 Construction and functions

Fig. 7 shows the integrated network management system NETM that has been developed in accordance with the architecture mentioned in section 3.
The NETM is able to manage not only almost all network facilities provided by Hitachi, but also other vendors' network facilities. The NETM includes many subsystems such as NETM/EYE (display & operation subsystem), NETM/MGR (integrated network management subsystem), NETM/AGT (application network management subsystem), NETM/MUX (multiplexer subnetwork management subsystem), NETM/HIPA (packet switching subnetwork management subsystem), NETM/PBX PBX subnetwork management subsystem), and NETM/TADI (TADI subnetwork management subsystem).
The NETM/EYE and all the other subnetwork management subsystems are used with Hitachi's workstation 2050/32 and NETM/MGR on Hitachi's host computer.
The functions that have been developed so far by the NETM are :
    (a) Configuration management functions
        object management
        state management
    (b) Fault management functions
        alarm management
        error statistic management
    (c) Performance management functions
        gathering performance information
In the future, the other management functions will be continuously developed.

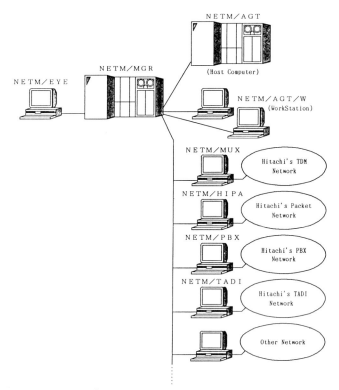

Fig. 7 Structure of NETM

4.3 Development of management subsystems

(1) Development of the NETM/MGR

The program module structure of the NETM/MGR is illustrated in Fig. 8. The CMISE has achieved the OSI standard. As to SMASE, we have defined the SMAS for the NETM considering the movement of OSI. The MIC (Management Information Control) manages and controls all management information such as objectclasses, attributes, notifications, and the information concerned with managed object instance. The objectclasses, the attributes, and the notifications have been individually defined by Hitachi. Changes in this information do not influence any other program module except the MIC.

(2) Development of subnetwork management subsystem

As previously described, NETM/***(subnetwork management subsystem) has been developed using Hitachi's workstation 2050/32. Each NETM/*** uses the same CMIP control program and individual functions to manage each subnetwork have been developed for each subnetwork.

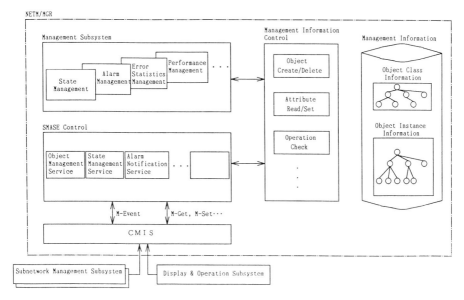

Fig. 8 Components of NETM/MGR

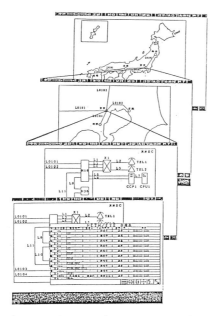

Fig. 9 Examples of Display of NETM/EYE

4.4 Display and operation subsystem

The human interface is a very important issue for large systems needing various operations like in the network management systems.
The human interface of the NETM/EYE has been achieved by using GUI (Graphical User Interface). Examples of human interfaces are shown in Fig. 9. Network operators can perform the operations using the functions provided by NETM/EYE such as zooming.

5. Summary

We have described some goals, the management architecture, and the subsystem developed for Hitachi's integrated network management system NETM.
The NETM architecture is characterized by several aspects such as the decentralized and centralized management approach, the pursuing of OSI system management, the managing network for managing the subnetwork management subsystems themselves, and the new expression form for the managed object instance. The advantages of the NETM architecture are thought of as that it is now possible for management systems to cope with any of the network varieties, to guarantee reliable management itself, and to be open to other management systems.
Finally, we have showed the developed system NETM to be a very large system intended to manage almost all of Hitachi's network facilities and able to manage others as well.

Acknowledgements

We are grateful to Shingi Dohmen, Dr. Koichi Haruna, Dr. Tatsuo Mitsumaki, Kazuhiko Ohmachi, Yasuo Ueda, Morio Iino, all of Hitachi for their continuous help.
We are also grateful to all of our colleagues with whom we have developed the network management system NETM.

Reference

[1] ISO/IEC IS 7498-4: 1989 Information Processing Systems - Open Systems Interconnection - Basic Reference Model - Part4: Management Framework
[2] ISO/IEC DIS 9596-2: Information Processing Systems - Open Systems Interconnection - Management Information Protocol Specification - Part2: Common Management Information Protocol
[3] ISO/IEC DP 10164-1 Information Processing Systems - Open Systems Interconnection - Management Part 1: Object Management Function
[4] David J. Millham and Keith J. Zwilletts,"BT's Communications Management Architecture," Proceedings of the IFIP TC 6/WG 6.6 Symposium on Integrated Network Management, '89
[5] Paul J.Brusil and Daniel P.Stokesberry,"Integrated Network Management," Proceedings of the IFIP TC 6/WG 6.6 Symposium on Integrated Network Management, '89
[6] Steven Goldsmith, Umberto Vizcaino,"Enterprise Network Management, Integrated Network Management," Proceedings of the IFIP TC 6/WG 6.6 Symposium on Integrated Network Management, '89
[7] Lev Feldkhun,"INTEGRATED NETWORK MANAGEMENT SYSTEMS(A Global Perspective on the Issue," Proceedings of the IFIP TC 6/WG 6.6 Symposium on Integrated Network Management, '89

SNMP for non-TCP/IP sub-networks: an implementation

E. Duato[a] and B.Lemercier[b]

[a]European Space Operation Centre, Robert-Bossche Strasse 5, D-6100 Darmstadt, Germany, Tel: (49) 6151 886 860, e-mail: eduato@esoc.bitnet

[b]BIM s.a/n.v. Kwikstraat 4, B-3078 Everberg, Belgium, Tel: (32) 2 759 59 25, e-mail: bl@sunbim.be

Abstract

The paper reflects the experience learned during the implementation of an Integrated Network Management System (INMS) for the European Space Agency (ESA) general purpose network (ESAnet). It discusses the architecture of the system and in particular the implementation of SNMP proxies and MIB in large heterogeneous multi-vendor networks, including non-TCP/IP equipment. The discussion emphasizes the solutions developed, the problems encountered and special interests. Finally, the INMS project is compared with a CMIP proxy and MIB implementation.

Keywords

SNMP, proxy, INMS, MIB, CMIP

1. THE EUROPEAN SPACE AGENCY GENERAL PURPOSE NETWORK

The ESA operates a private communication network, known as ESAnet. It includes Local Area Networks (LAN) as well as Wide Area Networks (WAN). The ESAnet supports communication between ESA establishments, space industry, research institutes and National Space Agencies all around the world [1]. ESAnet provides "transport services", in accordance with ISO/OSI model. ESAnet is formed of digital and analog terrestrial and satellite circuits being permanently dedicated by PT&T administrations [2]. The link capacity is subdivided by various multiplexing techniques. For analogue lines, time division multiplexers are integrated into modems, whilst for digital links, high speed multiplexers are used. On top of this backbone, ESA operates special services, as compressed voice, and two virtual sub-networks, SNA and ESApac. ESApac is ESA's X.25 network and carries asynchronous DECnet, SNA and Facsimile traffic. ESApac also supports TCP/IP services through multi-protocol routers. As illustrated in figure 1, ESAnet is composed of the following elements:
- Time Division Multiplexers (TDM) from Timeplex;
- Time Division Modems (MODEMS) from AT&T-Paradyne;
- X.25 Network switches (ESAPAC) from Hughes Network Systems;

- Multi-protocol Routers (ROUTERS) from cisco;
- Packet Assembler/Disassemblers (PAD) from Timeplex;
- Ethernet and Token-Ring LANs (LAN)..

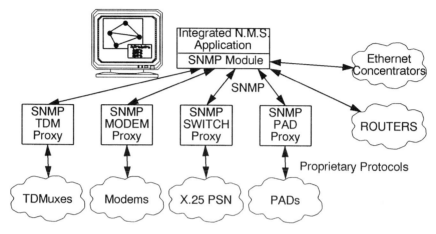

Figure 1. The ESANet Integrated Network Management System.

1.1. ESAnet network management

In the past, the whole ESAnet was monitored and controlled from the European Space Operation Centre (ESOC, Darmstadt, Germany) using several proprietary network management systems. This approach allows the Management and Operations (M&O) Team to detect alarms as soon they occur, and to reduce the expertise and manpower required. This is due to the fact that network analysis and reconfiguration can be done from a central site, where the expertise resides. Nevertheless, the result was a wild proliferation of screens. A single failure of a digital network is reported to three different screens, using different messages. Congregated statistics collection is rather difficult. The use of disparate NMS requires highly trained networks operators, which implies additional costs. The solution to this problem is to provide the network operator with an integrated view of the network on one single screen. A unique man-machine interface should allow access to all data needed for the supervision and control of the different sub-networks, using the same "look-and-feel". This results in less training and a clear view of the network and its interconnections.

1.2. The ESAnet Integrated Network Management System

To unify the management of all sub-networks under a single focal point, ESA started in January 89 a project called INMS (Integrated Network Management System).

Up to 20 proprietary NMS were analyzed, but no suitable off-the-shelf tool was found. The integration of multi-vendor sub-networks into a proprietary NMS was found expensive and very complex, when possible. It was decided therefore to make a development based on a standard protocol. CMIP was considered as the long term goal, but due to the immaturity of standards, the lack of products, and the complexity of CMIP, a more pragmatic solution was urgently required [3]. It was concluded to implement an evolutive solution based on SNMP [4][5][6][7], one of the TCP/IP network management protocols. SNMP was a simple, mature, and well-known network management protocol [8]. Agents implementations and off-the-shelf libraries were available, and the trend showed that more vendors would provide SNMP agents in the short term [9]. SNMP also supports the concept of proxy. A proxy provides the ability to manage network elements that either are not addressable by means of an Internet address or use a network management protocol other than SNMP.

SNMP also allows the definition of enterprise private objects. Using this facility, it is possible to extend the scope of the standard MIB, defining objects for multiplexers, modems, and X.25 switches

1.3. The INMS architecture

From the network operator point of view, the INMS is a window based application, fully mouse-driven and graphic oriented (see figure 2). The main screen is a map representing ESAnet, in which the operator can recursively decompose an object according to the MIB by clicking on it. The map also represents the current status of the network, using a color code. Each time an event occurs, the status of the map changes, and an alarm window is opened showing the message sent by the node that produced the event. The basic platform is composed of Unix, C, X11R4 and a RDBMS. On top of this platform, INMS has the following functional modules:
- MMI: responsible for the graphical representation of the objects, the dialog boxes, and the operator interaction;
- INMS Kernel: the heart of the INMS: it concentrates the knowledge and control of the application. This knowledge resides in the configuration files (containing the object class information) and the database (containing the objects instance information and the event log file);
- Alarms Reception: receives the SNMP traps and processes them;
- Log and Trouble Ticket Tool: two separated applications for off-line events management;
- Proxies: translates SNMP PDU to each sub-network proprietary protocol and vice-versa.

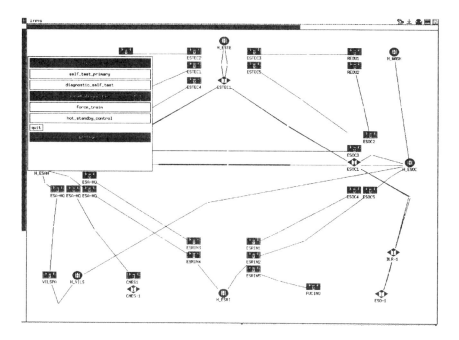

Figure 2. The INMS top level window.

2. PROXY ARCHITECTURE

This chapter describes the SNMP proxies built for the INMS. Three proxies were developed: one for the Paradyne modems sub-network, one for the Timeplex Link/2 multiplexers sub-network and one for the Hughes X.25 switches. These proxies hide all the peculiarities of the managed equipment. However, they were constructed according to the same architecture presented below [10].

2.1. General architecture of the proxies

Figure3 shows the proxy architecture. The normal flow of information is as follows: an SNMP request is received on UDP port 161 by the dispatcher (its exact role will be discussed in the next chapter) and placed in the message queue. The request is then analyzed, i.e. decompiled, and translated according to local MIB information into one or more proprietary commands, which are sent through the equipment-dependent driver. The answer is then processed and an SNMP response is generated. For SNMP trap generation, the proxy makes use of a polling unit detailed in the next section.

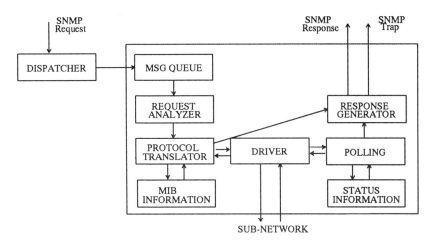

Figure 3. Proxy architecture.

2.2. The proxy functions

2.2.1. Drivers to a proprietary NMS

The drivers implement the proprietary network management protocols. They act as if they were the proprietary consoles. Four specific drivers were written for the INMS:
- the modems driver is connected to the RS232 control port of one of the modems. The asynchronous protocol is based on a "request-response" model. All the modems constituting the sub-network are connected to each other either through a daisy-chain link for local modems or through a 110 baud channel allocated in the line bandwidth for remote modems.
- the multiplexers driver is connected to the RS232 computer port of one of the Link/2 multiplexers. The asynchronous protocol is based on a "request-response" model. The management information is exchanged between muxes through a special channel allocated in the total bandwidth.
- the ESApac driver takes the place of one of the Network Operator Consoles (NOC). This NOC connects to the Network Control Processor (NCP) through an X.25 virtual circuit. This particular driver is the most complex: it implements a TP2-like transport layer and the functions of this NMS are really sophisticated (more than 150 different screens on the NOC).
- the PAD driver can be attached to any asynchronous PAD port. It emulates a VT-100 terminal connected to the control port of the PAD via a triple-X protocol. The asynchronous protocol is based on a "request-response" model.

2.2.2. Protocol translation

The protocol translation function is far from straight-forward since many SNMP variables can be present in a single request. Furthermore, the retrieval of the value of one variable can lead to a rather complex sequence of proprietary commands. Therefore, a generic skeleton for the protocol translator was developed. It is a finite state automaton that performs the actual translation using the MIB and proprietary information tables as input.

The SNMP SetRequest caused an extra problem: in most of the cases, the equipment commands do not allow to set only one value but a whole set of them in one command. In these cases it is necessary to fetch the set of values from the sub-network, modify the appropriate variables and then write everything back to the equipment.

All this can lead to a considerable management traffic, so a special attention has been paid to minimize the interaction with the sub-network by a proper grouping and ordering of the proprietary commands. For that purpose, logical groups of MIB variables were identified, that divide standard MIB groups in sub-groups and tables. Requests are always issued on an entire logical group.

2.2.3. The polling unit

SNMP assumes that the equipment generates Traps spontaneously. Unfortunately this is not the case for most equipments. This function is performed by the polling unit: each node is polled in turn and the status of the equipment is retrieved and compared to the previous status kept inside of the proxy (the Status Information table on figure 3). In case of change, an SNMP Trap is generated and sent to the INMS application on UDP port 162.

2.2.4. Alarm shield

It is important to note that the proxy sends Traps to the INMS application only when a node status change has occurred. It thus shields the application from redundant alarm storms, reducing the SNMP traffic. The SNMP trap numbers are specific traps. Two trap numbers are allocated per kind of problem: one to express the fact that the equipment goes from a normal state to fault and one for the reverse transition.

The shielding mechanism is tunable on-line by the application, which can select relevant Traps to be forwarded.

2.3. MIB design

The managed objects of the INMS were created and added in the SNMP MIB under the enterprises(1) part of the private(4) MIB sub-tree (1.3.6.1). At that time the vendors had not registered their private trees, so four sub-trees were added under the registration number of ESA (162): modem(50), mux(51), ESApac(52)and PAD(53).

These sub-trees can be relatively complex and are structured in order to include the notion of containment rules. Here is a small part of the MIB illustrating this concept:

"mux_nest",	"1.3.6.1.4.1.162.51.2",
"mux_slot",	"1.3.6.1.4.1.162.51.2.1",
"NCM_slot",	"1.3.6.1.4.1.162.51.2.1.1",
"NCM_hex_code",	"1.3.6.1.4.1.162.51.2.1.1.1",
"BPM_slot",	"1.3.6.1.4.1.162.51.2.1.2",
"BPM_hex_code",	"1.3.6.1.4.1.162.51.2.1.2.1",
"ILC_slot",	"1.3.6.1.4.1.162.51.2.1.3",
"ILC_hex_code",	"1.3.6.1.4.1.162.51.2.1.3.1",
"ILC_redundancy_flag",	"1.3.6.1.4.1.162.51.2.1.3.2",
"ILC_primary_flag",	"1.3.6.1.4.1.162.51.2.1.3.3",
"ILC_modem_type",	"1.3.6.1.4.1.162.51.2.1.3.4",

The MIB was implemented on the proxies as sequential lists contained in "include" files that are compiled with the proxy code. This is realistic since the MIB part known by the proxy is rather small. Any change to the MIB would automatically imply modifications to the code of the protocol translator or the driver so that the proxy has to be recompiled anyway.

However, the MIB on the application part of the INMS was implemented as separate ASCII files that are loaded at start up, in order to allow changes without any recompilation. It is also possible to modify the MIB on-line, so that new equipment (supporting SNMP) can be added smoothly.

3. SPECIAL POINTS OF INTEREST

This chapter emphasizes some difficult points encountered during the design, implementation or operation of the SNMP proxies. It describes solutions envisaged during the INMS project as well as suggestions for future work.

Experience has shown that writing proxies is a far from trivial task. First, a proxy managing an entire sub-network must contain some intelligence to be efficient. Secondly, the process of gathering and analyzing proprietary information is long and painful, mostly because equipment vendors are in general reluctant to provide documentation and support.

3.1. SNMP limitations

The limitation to 484 bytes per PDU in the INMS SNMP implementation (from SNMP Research, Incorporated) does not allow the retrieval of long attribute lists with a single Request PDU. One could be tempted to increase the size, but it should be done in an adaptive manner in order to remain inter-operable with other implementations.

Another problem with SNMP is that it does not support an Action service. In order to perform tests and resets on the ESAnet equipment, special variables were created, which, when set, do not result in any parameter setting; instead, the action is performed by the proxy as a side effect.

3.2. Discussion on the polling

The polling mechanism is necessary to simulate spontaneous network alarms and generate SNMP Traps. Nevertheless, experience has shown that a fine tuning of the polling is a mandatory issue.

3.2.1. When to poll?

Many factors have to be taken into account. The proprietary protocol driver is shared by the protocol translator and the polling module. In order to minimize the response time of the application, priority has to be given to the SNMP requests. A synchronization mechanism was thus developed to inhibit the polling while requests are processed.

The polling frequency was made configurable on a node-by-node basis, in order to focus on problems of major nodes. This is done using a configuration file read at proxy start-up. It can be changed on-line.

The time-out value for one poll is configurable (if possible it should be adaptive), because the time needed to obtain a response depends on the distance (in hops) from the proxy. This is especially true when working with low speed control channels.

3.2.2. Polling frequency and bandwidth

A compromise has to be found between up-to-date status information and network management traffic. A too high frequency can lead to complete saturation of the management channels. Experience shows that polling 60 nodes in a round-robin fashion at 1 second intervals occupies a good portion of the bandwidth in some cases and gives an updated equipment status after only one minute.

3.3. Distribution of proxies

This section discusses different aspects where the centralization or decentralization of proxies is involved.

3.3.1. CPU power problems

A proxy is a rather complex program which performs heavy data processing in real time. It requires a lot of CPU power when the amount of managed nodes is important. In the early stages of the INMS prototype, both applications and the proxies resided on the same machine. Tests have shown that the proxies used an important part of the computer resources. Other tests were conducted, separating the applications on one machine from the proxies on another one. These tests were much more convincing.

3.3.2. Multiple proxies on one machine: the dispatcher

A proxy is identified by an IP address and UDP port 161 [7]. If more than one proxy resides on a single machine, it is impossible to distinguish between them. This is the reason why a dispatcher module was added to the architecture, as a front-end to the proxies. The dispatcher examines the community name included in the SNMP packet and transfers the integral packet to the appropriate proxy using UNIX IPC mechanisms.

3.3.3. Duplication of proxies in the network

When the amount of nodes to be managed by one proxy becomes important (over 50), it is wise to duplicate the proxies, each of them managing one part of the sub-network. The INMS was designed to take this possibility into account. Another reason for duplicating proxies of one type is to place proxies closer (in terms of network hops) to the equipment each proxy manages. In this case it is assumed that a (fast) IP link exists between the proxy and the applications.

3.3.4. The network management information transport

If SNMP was designed to minimize interference of the management traffic with the normal network traffic, it is, in most of the cases, impossible to guarantee minimal interference while designing proxies since they are based on proprietary protocols. In critical conditions, it could be necessary to have this information transit through other links, possibly a "meta" network dedicated to network management. (Naturally, that "meta" network would need a "meta" network management system...)

3.4. Need for management support objects in the proxies

As stated above, the proxy is a rather complex program. Not only does it keep network status information, but it should also embed some management support objects and functions. This allows the application to perform on-line tuning on the proxy functions, like polling parameters, but also improve the synchronization between application and proxy. Other functions can then be envisaged like obtaining information on the proxy load for example. A proxy should definitely implement the SNMP group defined in MIB II [6].

4. COMPARISON WITH CMIS/CMIP

This chapter makes a comparison between CMIS/CMIP and SNMP, as far as the proxy and MIB concepts are concerned. The purpose is to identify:
- what problems found using SNMP are solved by CMIP;
- what new problems can appear using CMIP.

4.1. Management information transport

SNMP was designed to use connection-less UDP. This mechanism has the advantage that it does not introduce a big overhead on the network. The drawback of UDP is that there is no end-to-end error correction. In the INMS, the network status is updated each time a SNMP trap is received from a proxy. If the SNMP trap is lost, then the INMS looses its synchronization with the proxy. There is also no guarantee that Get, GetNext, SetRequest and GetResponse will arrive at their destination [11].

CMIP uses a connection-oriented transport. This guarantees a very reliable communication between the manager and the proxy. The drawback is the overhead that this type of transmission mechanism can produce, both in the network and in the applications. In a network near to congestion this can have terrible consequences. Another problem in CMIP could be the number of open connections a manger has to handle. Proxies can provide a shelling mechanism: CMIP can be used between the manager and the proxy. The proxy itself can then use a more efficient mechanism to reach the network element within its sub-network. Some works suggest to use SNMP as a rather low level protocol between manager and equipment and CMIP for manager-to-manager communication [12].

4.2. Services

SNMP primitives are a subset of the primitives offered by CMIS [13]. Table 1 presents a comparison between both protocols.

Table 1.
Comparison of CMIP and SNMP primitives.

CMIP	SNMP
M-Initialize (c)	
M-Terminate (c)	
M_Abort (nc)	
M-Event-Report (c/**nc**)	Trap
M-Get (**c**/nc)	GetRequest, GetNextRequest, GetResponse
M-Set (**c**/nc)	SetRequest, GetResponse
M-Action (c/nc)	
M-Create (c)	
M-Delete (c)	

One of the problems faced during the INMS development was the reduced number of primitives offered by SNMP. As explained above, the lack of an SNMP Action primitive had to be turned around.

Furthermore, CMIP offers three additional mechanisms: scoping, filtering and synchronization. They allow one, for example, to retrieve all attributes in a sub-tree in one CMIP request. However, the practical implementation of these mechanisms in a proxy agent will not always be possible, due to the restrictions imposed by the proprietary protocol behind the proxy (see section 2.2.2.).

For alarm reporting, both CMIP and SNMP have an open-ended syntax, which allows vendors to support their own events. CMIP proxies shall also support an event sieve and event log file. CMIP generic notifications include a lot of useful information, like severity, trend, problem type, etc.

Dynamic creation and deletion of objects is only possible with CMIP. This is an important feature when managing a large network, with several management domains. New objects can be introduced on-line in a management domain. Nevertheless, this interesting feature from the conceptual point of view requires powerful and sophisticated mechanisms in the application. The manager and the proxy agent have to be able to dynamically change the instance representation at any level. Furthermore, the management application shall be able to reflect the change in all views including the object affected by the change.

4.3. Security

SNMP does not offer real security mechanisms. There is an authentication process, which makes use of the community name. Checks can also be made on the IP addresses. To use SNMP in open networks, additional security mechanisms are required.

4.4. The Structure of Management Information

Both SNMP and CMIP use the ASN.1 and Basic Encoding Rules to represent the information. Only a subset of the basic types are permitted in the SNMP SMI, which does not allow the definition of complex attributes. CMIP supports the full richness of the ASN.1 definitions; the major drawback is that the complexity is not hidden in lower protocol layers: the massive use of the type "ANY DEFINED BY" has the consequence that the upper application layer must implement ASN.1 parsers and formatters, that must be written for each new type.

4.5. Conclusions

As stated above, SNMP proxies require important computing resources to perform basic functions.

In addition to these functions, CMIP proxies have to support a connection-oriented mechanism, a complex set of services and a powerful set of data structures. They also need to control two additional kinds of objects: Support Objects like complex discriminators or event logs and Aggregated Managed Objects for computing statistics on information gathered from many different managed systems. CMIP proxies can also be requested to establish CMIP connections simultaneously with several managers at the same time. Finally, objects can be created and deleted dynamically.

The implementation of CMIP in currently installed networks can only be done via proxy agents. Although some of the functions offered by CMIP are urgently requested by network managers. The development and installation of CMIP proxies will be a costly exercise.

5. REFERENCES

[1] ESAnet Development Plan. ESOC/ECD/SPD/COM. Oct.90
[2] Final Report on the Study on Space Data Network Communication Specification and Signalling System. July 89. Departamento de Ingenieria Telematica
[3] Network Management of TCP/IP Networks: Present and Future. A. Ben-Artzi a.o. IEEE Network Magazine. Jul. 90.
[4] rfc 1155: Structure and Identification of Management Information for TCP/IP based Internets. May90
[5] rfc 1156: Management Information Base for Network Management of TCP/IP based Internets. May90
[6] rfc 1157: Management Information Base for Network Management of TCP/IP based Internets: MIB II. May90
[7] rfc 1158: A Simple Network Management Protocol. May90
[8] Connexions. The Interoperability Report March 89 Vol3 No3
[9] rfc1147: Tools for Monitoring and Debugging TCP/IP Internets and Interconnected Devices
[10] Study on Integrated Network Management Facilities. Contract No 7917/88/D/IM, Several issues, BIM
[11] Considering CMIP. March90. Data Communications
[12] OSI/Network Management Forum General Meeting notes. Oct. 90. San Francisco.
[13] ISO IS 9595-2 and 9596-2 Common Management Information Service and Protocol

Evaluating Network Management Systems: Criteria and Observations

H. Elston Carter and **Januario P. Dia**

The MITRE Corporation
Burlington Road
Bedford, MA 01730

Over the past two years, the number of standards-based network management system (NMS) product announcements has grown considerably. To assist in choosing an appropriate NMS from this growing list of products, a methodology for evaluating them is required. This paper describes a set of evaluation criteria that could be used to develop such a methodology. In addition, it also presents the authors' observations on installing and operating four first-generation NMS products in the context of these criteria.

INTRODUCTION

During the past few months, the authors have surveyed various commercial off-the-shelf (COTS) network management systems for networks based on the Transmission Control Protocol/Internet Protocol (TCP/IP). Of these products, one Release 1.0 and three beta version systems were procured for testing in a laboratory environment. This paper documents our observations on installing and operating these NMS products.

Because the concept of standards-based, integrated network management is still in its infancy, few (if any) methodologies exist for evaluating the capabilities of available products. In the absence of any documented methodologies or procedures, the authors have developed evaluation criteria using the work performed by the various network management standards bodies (The National Institute of Standards and Technology, The Internet Netman Working Group, etc.). As such, the criteria are not static and will evolve as the network management standards evolve. Since the criteria only serve as guidelines for evaluating NMS products, the decision to select a particular NMS will ultimately depend on the user's requirements.

Unfortunately, user requirements in the area of network management are vague because traditionally, network management has never been accorded the importance it deserved in designing, implementing, and operating user networks. For these reasons, users frequently look upon the service provider (or systems engineer) to define the requirements and dictate the scope of network management to be provided.

Determining the user's network management requirements is but one hurdle that the network service providers must face when recommending a suitable network management system for the user. A bigger problem is finding the one product that provides all the important network management functions and capabilities.

AREAS OF EVALUATION

In conducting the product evaluations, the efforts were primarily focused on three areas demarcated by the evaluation criteria defined for them. The criteria were arbitrarily grouped as follows: functional, operational, and other. Functional criteria refer to the product capabilities that provide some level of standards-based, integrated network management. Operational criteria are those features that the authors believe should be incorporated into the product implementation to facilitate use by network operators. The other criteria refer to various factors that could also influence the recommendation or rejection of a product, but are not by themselves specific to network management.

As our primary area of interest was on TCP/IP-based NMS products, our evaluation made extensive use of TCP/IP-specific criteria. This is especially true of the functional criteria. The operational and other criteria are more general in nature and do not rely specifically on the underlying network technology. The following sections discuss the evaluation criteria.

Functional Criteria

Because standards play an integral part in ensuring transparent interoperability and integration among different vendor products, it is only natural that they occupy a central role in our functional criteria. As mentioned, the authors have limited the scope of the functional criteria to the standards that are defined for networks using TCP/IP. For TCP/IP networks, three network management standards are defined in the Internet Request For Comment (RFC) documents: The Management Information Base (MIB) for TCP/IP networks in RFC 1156; the Simple Network Management Protocol (SNMP) in RFC 1157; and the Common Management Information Services/Protocol over TCP/IP (CMOT) in RFC 1095. In addition to complying with the network management standards defined in the RFCs, the systems must also support the five system management functional areas as defined by the standards bodies. It should be noted that the evaluation criteria used could easily be extended to other types of networks, provided some well-defined standards exist for those networks.

Support of the Internet MIB (RFC 1156)

The MIB defines the required variables (or managed objects) that can be accessed and modified by the network manager to monitor and control specific network components. The MIB contains only generic, or non-implementation specific, objects that the Internet working group has determined as essential to managing a network component. From this minimal set of objects, all other variables specific to an implementation can be derived. The Internet MIB can be extended to support proprietary as well as any future standard protocols. The current Internet MIB defined in RFC 1156 provides support for 112 objects. RFC 1158 defines a new MIB, called MIB-II, which extends the number of manageable objects supported to 171.

In all cases, the managers advertising compliance to SNMP did in fact provide support for the entire set of Internet MIB objects that are applicable to their implementation. This means that the devices only provided information for the MIB objects that made sense for that particular device. This is exactly what should be expected - for example, devices which do not implement the Exterior Gateway Protocol (EGP) should not be expected to provide any information for this object type. All the NMS products examined claimed to provide a means of extending the SNMP MIB, although this could not be verified thoroughly during the product evaluations. One manager supported both a proprietary MIB and the Internet MIB. This manager used a proprietary protocol to manage the vendor-specific objects, and provided a SNMP proxy implementation to manage RFC 1156 MIB objects.

Provided the proper encoding techniques are followed, other non-standard variables can be added to the MIB and can be accessed or modified by the network manager. Few, if any, changes can be expected to this current practice, although the advent of Open Systems Interconnection (OSI) standards could bring with it a set of new objects. It is expected that a new MIB (incorporating the Internet MIB) will be developed for the Common Management Information Services/Common Management Information Protocol (CMIS/ CMIP) for OSI network management. In fact, work has been done to cast the Internet MIB in terms of the OSI protocol stack by the OSI Internet Management Working Group.

Support for SNMP (RFC 1157)

The SNMP defines the protocol through which a network manager can inspect, set, or modify attributes of a network element in a TCP/IP-based network. By manipulating the network device's attributes, a network management station can monitor or control the device remotely. These attributes are the managed objects defined in the Internet MIB.

All the network managers evaluated supported SNMP and implemented the "get" command Protocol Data Units (PDUs). On the other hand, the SNMP "set" command PDU was not implemented in any of the vendors' products. Some vendors implemented the SNMP Trap PDUs; however, in many cases this could not be ascertained.

Certainly, one of the future developments in this area would be the support of the entire SNMP suite, including the traps. In addition, it is likely that some products will eventually address manager-to-manager communications such as proprietary-to-SNMP, or SNMP-to-SNMP, or even SNMP-to-CMOT.

Support for CMOT (RFC 1095)

The CMOT network management architecture defines a protocol that uses the ISO-defined CMIS/CMIP to manage a TCP/IP network. Like SNMP, CMOT provides the means through which a network manager can inspect, set, or modify attributes of a network element in a TCP/IP-based network. By manipulating the network device's attributes, a network management station can monitor or control the device remotely. The attributes are the managed objects defined in the MIB.

Only one of the network management systems evaluated supported the CMOT protocol. However, this product could not be tested thoroughly since there were no stable CMOT agent implementations at the time of the evaluation,.

As OSI applications proliferate, it can be expected that interest in CMIS/CMIP, and thus CMOT, will grow. The authors also expect that as approaches for transitioning to OSI-based networks become a reality, more manufacturers will attach importance to CMOT and begin to introduce products which support it.

Configuration Management Capabilities

Configuration management provides the means to identify, configure, and control devices on a network. Mechanisms are required to define network resources and associate attributes with them. The storage of this configuration information is important for status and inventory reporting. In order to configure and control devices, the ability to set and modify device parameters is required. Inherent to configuring and controlling devices is the ability to remotely access devices on the network.

The products evaluated provided remote access through "telnet" connections or "rlogin" sessions. As mentioned above, none of the products supported the SNMP "set" command. This command would have provided the capability to remotely modify managed objects. Each of the products evaluated stored the configuration management information (including IP addresses, locations, serial numbers, etc.) using a flat file implementation.

The major component of the configuration management functionality is the storage mechanism for the management information. Storing the information in a database avoids several disadvantages of the flat file implementation. A database implementation would allow a master database to reside on a server rather than at the management station. This in turn would provide ease of maintainability (one copy of configuration information), larger data storage capacity, quicker access to the stored information, and data access to other applications (eg., a spreadsheet for inventory description). In a manager-of-managers hierarchy, the database implementation would allow information retrieval with a query instead of a request to the management station. Although none of the evaluated products had an underlying database, the authors believe this will be the trend in the future.

Performance Management Capabilities

Performance management deals with observing and ensuring the quality of service of the communication system. To this end, performance management is separated into the two broad functions of monitoring and tuning system parameters. A monitoring capability provides access to the attributes which affect system performance. Performance tuning is done by modifying those attributes. Documenting system performance requires the collection, storage, and reporting of performance statistics.

The products evaluated were for the most part passive and only provided a monitoring capability. Tuning capabilities have not yet been provided due to problems with NMS security (primarily verifying user access and authority). As mentioned, the monitoring capabilities provided access to all the RFC 1156 MIB attributes for devices that used SNMP. Other system attributes (proprietary information) were accessed depending on the NMS being used. Mechanisms were provided to log this information, but applications for data analysis (e.g., spreadsheets) had not been provided with the products. Presentation of the data was done with real-time plots, bar graphs, and strip charts that could not be stored or regenerated. In all cases, the network manager had to assimilate the information gathered through the monitoring of the various system elements. No applications were provided to assist the network manager in this task.

Along with the ability to tune system parameters, future NMS products should allow artificial conditions to be generated on the network. This capability will allow the generation of an increased load on the network and network devices in an attempt to determine performance impacts. This information can then be used for capacity planning.

Fault Management Capabilities

Fault management provides for the detection, isolation, and correction of faults on a network. Diagnostic capabilities are required to isolate faults and to ensure faults are corrected. Error condition logging is necessary for record keeping and diagnostic purposes, and for reviewing events that occurred while the NMS was unattended.

The products provided the capabilities to generate and log alarms. From there, the "human" network manager had full responsibility for diagnosing (possibly using existing diagnostic packages) and correcting faults. The generation of alarms in three of the systems evaluated was based on the ability to set thresholds on the monitored attributes. Faults were then detected based on a collection of alarms or on a single alarm. Notification of the alarms was done via auditory and visual means and through electronic mail. SNMP traps were also used to report changes in state and problems to the network management console. Two of the products examined provided an attempt at fault isolation by automatically polling all devices with links to a device that appeared to be down.

Fault management capabilities of future products will need to provide automatic isolation of faults. Expert system applications to perform fault determination based on a collection of alarms will also be necessary. This may be extended to initiating fault diagnostics once the location of an error has been determined. Any applications that can off-load the responsibilities of the network manager to the NMS will obviously be of great utility.

Security Management Capabilities

Security management deals with monitoring the usage of secure devices and controlling access to network resources. It requires capabilities that provide for encryption of information, authentication of users, controlling access to resources, maintaining audit trails, and logging all security events.

No security management capabilities were provided with any of the products evaluated. Standards for security management are still evolving and vendors have yet to concentrate on this portion of network management functionality. The vendors have stressed the areas of configuration, performance, and fault management in an attempt to provide some basic and useful management functionality in a timely manner. It is expected that both security and accounting management capabilities will be built into future products once the basic capabilities have matured.

Accounting Management Capabilities

Accounting management allows for the creation and monitoring of network accounts and associates charges with resource usage. Therefore, it requires the capability to define and specify the accounting information to be monitored and collected. Additionally, accounting management requires the ability to generate reports on the collected data.

None of the products evaluated provided any form of accounting management. As mentioned above, capabilities in this area should be available once the configuration, performance, and fault management capabilities have matured.

Operational Criteria

The operational criteria focus on the products' installation and use. These criteria provide a means of judging a product's suitability to the user's environment apart from its network management functionality.

Ease of Installation

Almost all NMS vendors market their products as user-installed software, meaning that users (rather than the vendors) are the ones primarily responsible for its installation. For this reason, the ease of installation of a NMS is an important consideration when evaluating them.

All of the evaluated products required superuser (root account) privileges to install their software. In most cases, the software installation script automatically created or updated the files and directory links required to use the software. Implicit in the installation requirements is the existence of coherent and organized installation manuals. Strangely enough, the authors have observed that those vendors that required the user to manually create/update system files or directory links were the ones that had the least documentation. It is hoped that in the future, vendors will provide an automated means for installing their products and will supply the user with the manuals that document it.

Ease of Use

In discussing the ease of use of a NMS, it is important to establish the level of sophistication/experience of the NMS user. The most desirable (and logical) qualification for a network manager is a person with solid experience with the equipment/system to be managed. Someone with an understanding of the protocols, operating system, applications, etc. is best qualified to make management decisions. In addition, because of the technical nature of the interfaces provided, it is especially critical to have a knowledgeable and experienced user.

Since there are vast numbers of opinions on the features that make one system easier to use than another (eg., pull-down menus vs. list menus), this discussion is limited to features that we think are lacking in the products we have evaluated. The addition of these features will, without a doubt, improve ease of use. The area that needs the most improvement is the presentation of the attributes of the managed devices.

Examining ease of use from the viewpoint of the presentation of the attributes uncovered several deficiencies in the products. First, most of the products provided no amplifying text to assist the network manager. The network manager was assumed to have an understanding of the attributes by name. With no amplifying text, remembering the MIB attribute names in a homogeneous system using the RFC 1156 MIB (112 defined objects) was difficult, even if some MIB names were somewhat self-explanatory. This will be impossible in a heterogeneous system with an expanded MIB (with an even larger number of attributes). Second, the products did not display the units of measure for each of the attributes. Admittedly, the units were obvious for some items, but for others the choice of seconds, milliseconds, and microseconds, for example, was not obvious. Also, whether the value of an attribute had been tallied since system boot-time or was given on a per interval basis was not indicated. Finally, for most of the products, the collected data was not formatted and no tools were provided for data analysis and report generation.

Obviously, the products are collecting useful data. However, the authors think that the data should be presented in a more useful manner. This will help the on-line real time monitoring done by the network manager. Reports generated for off-line analysis would also be more readable and, therefore, used more efficiently.

Man-Machine Interface

One of the major aspects of the man-machine interface for a NMS is the presentation of the managed system. Ideally, the man-machine interface provides a graphical representation of the network. The following are the four important features of the graphical representation: 1) the ability to have a hierarchical map, 2) the ability to represent unmanageable devices in the map, 3) the ability to show both physical and logical maps, and 4) the ability to tailor the icons that represent the network and network devices.

Two of the four products evaluated used a hierarchical map to represent the network. A hierarchical map was a necessity when dealing with a large network with many devices to be managed and displayed in the graphical representation. With a hierarchical map, different views were used to show different levels or branches of the network (excellent for representing backbone/branch architecture). It was impossible to represent each device of a large network in an easily usable format when using the products that had the network map constrained to one window or one screen.

The capability to display unmanageable devices (in addition to the manageable devices) on the network map was a helpful tool provided by some of the products. In the products evaluated, network device icons were displayed when descriptive information for a device (e.g., IP address and device type) was entered into the system. When both types of devices were displayed, the network manager was able to visually determine all elements (manageable or otherwise) on the path to a failure. Using this information, it was easier to isolate the source of a problem.

Another visual aid for the network manager was the ability to display both a physical and logical representation of the network. One of the evaluated products provided this feature. Showing devices on a particular branch, while at the same time showing the physical location of these devices (eg., using floor plans as backgrounds) can aid in quickly locating a network fault and minimizing network down-time.

The final important feature for the network manager was the capability to tailor the icons representing the network devices. The packages that represented different types of devices (e.g., bridges, workstations, etc.) with different icons were much easier to use. This capability, in conjunction with the ability to display both manageable and unmanageable objects, provided an easily understood display of the network.

The evaluated products provided a mixture of the above capabilities. The hierarchical structure, combining both physical and logical representations, provided the most concise and easy to use representation of the network. In the future, products should combine each of the above features in the graphical representation of the network. Also, to lessen the burden on the network manager, automatic generation of the network map should be an added feature of future NMSs.

Performance

Performance of a network management system can be defined as the system's ability to efficiently process large amounts of requests and responses in an acceptable manner. This means that the NMS should provide information to the network manager at a reasonable rate without noticeably degrading the response times of other users on the network. Performance also applies to the fact that the network management system should use a relatively small portion of the available bandwidth for management activities.

One product provided a feature which if enhanced could provide increased performance. The product allowed distribution of management functionality by permitting a proxy system to make management requests to some of the agents provided with the package. More specifically, a machine other than the main management console could be designated to poll other devices with agent software. The requested information was then sent to the main management console. An enhancement to this would be to allow the main management console to send one request to the proxy system, which in turn would poll (at specified intervals) the agent device. The information collected during a larger time frame (ie., more than one proxy system to agent polling interval) would then be returned to the manager in one response. This would generate less traffic on the network and decrease the burden on the management station. This type of functionality seems desirable and should be considered when evaluating a NMS. Because the network being used in the evaluations was active, it could not be overrun with requests or repeatedly exposed to the potential problems caused by the beta software bugs. This restricted the authors' efforts to quantify NMS performance.

Reliability

Because a network management system provides the primary means of monitoring and controlling network operations, it is imperative that reliability be established as one of the criteria for evaluating a network management system. In this case, reliability refers not only to the product but also to its effect on total network reliability.

Ideally, the network management station should be robust enough to minimize system crashes. If the system does fail, it should not affect the network being managed. The beta version software did fail on occasion and in some cases, even created problems on the nodes being managed. The vendors have assured us that such problems will eventually be resolved in future releases of the software.

Security

The operational criteria of security focuses on access control and authentication for using the management station and management data, and the security of manager-to-agent communications. Authentication is determining the identity of a requestor, while access control is determining if the requestor has sufficient authority to make a given request. Access and use of the management station can be controlled by user accounts and passwords. Different levels of access are a desirable feature of a NMS since different types of users may need access to management information (e.g., accounting personnel for reports on resource charges). The communication between manager and agent applications should be protected so that management information cannot be maliciously modified. Also, any attempts to access the management station should be logged.

The products evaluated afforded one level of security by using features of the underlying operating system. The extent of this security was the ability to control directory/file access via the UNIX file system. One of the products provided further security mechanisms based on its architecture. This product was based on the proprietary Remote Procedure Call (RPC) protocol. This protocol combined with the "netgroups" feature of Sun OS 4.0 provided for both authentication and access control between the manager and agents. The authors expect that the security of the management station will become an important issue with the vendors once their management architectures have been established and tested (i.e., once the products have matured).

Scalability

The scalability of a network management system refers to its potential for use in larger and more complex networks than the test environment. Among other things, it means that the NMS should be capable of handling the magnitude of management information that could be generated in a large network. It also requires that the NMS be capable of representing large networks on a map. In addition, it requires that the manager be capable of supporting multiple management domains, where the domains may represent geographical, administrative, or functional boundaries. Besides the capability to define multiple management domains, the NMS should have the ability to support manager-to-manager interactions, such as trap forwarding between different domain managers.

In all the cases examined, the NMS products were capable of multitasking and appeared to have the ability to support multiple manager-agent interactions. In addition, all the systems were capable of creating large network maps, although some managers could only have single-level maps. The ability to define multiple management domains and support manager-manager interactions could not be tested, as guidelines for organizing and sizing network management domains were missing. This is one area where vendors should devote their future research and development efforts.

Other Criteria

Besides the functional and operational criteria, other more general criteria exist for judging the merits of a network management system. These other criteria are not specific to network management and could easily be adapted for evaluating generic software systems.

Documentation

As mentioned above, the existence of organized and coherent documentation is essential to installing and using a NMS. With this in mind, it is obvious that one of the criteria to be used for recommending or rejecting a product should be its documentation.

In three of the vendor products examined, an installation guide and a user's manual were the only documents supplied. Only one of the products provided documentation on the architecture of the NMS. For organizations that expect to develop applications for their NMS, it is imperative to choose products that have adequate documentation on their architecture.

User Training Requirements

Because the skill level of network managers varies from organization to organization, part of the decision to purchase a NMS should depend on its user training requirements. Among the questions that should be addressed are the following: What is the type of network manager for which the product is targeted? What facilities (if any) exist for tailoring the system for other types of users? What sort of information is given to, or required from, the operator of the network management station?

All of the products examined required that the network manager/operator have some experience with analyzing network data and have some knowledge on interpreting MIB variable information. Future products should have the capability to customize and filter the type of information queried or received from the managed object to tailor the system to a specific type of operator/user.

Cost

As with all product evaluations, the cost of the NMS is a factor that must be considered when recommending/rejecting a system. For a NMS product, the cost may include both the cost of the manager software (and the agents, if included), as well as the cost of maintenance and license agreements.

The list prices for the systems ranged from $7,500 to $15,000. These prices were for the NMS software only and did not include the hardware platform. In some cases, the NMS included agent software. Product maintenance and license fees varied.

It is expected that as the technology matures, the prices for the network management software will drop. It is also expected that as SNMP and CMOT become more readily available, more managers will either include device-specific agents with their system or offer them as options.

Vendor Customer Support

Once a NMS has been purchased, a major concern is the level of vendor support. This support should include installation assistance, bug fixes, new releases, etc. for the purchased product. With the network management field rapidly changing, a vendor with a strong commitment to continuing customer and product support is a necessity. Since the standards are still evolving, the products will need to change in order to comply with the standards. The consumer must look to the future and determine if the vendor will be supporting the current product, enhancing the product, and developing new products to keep pace with the rapidly changing environment. This determination should be made based on corporate experience with a given vendor.

The products evaluated were all first generation software packages, and contained some bugs. All the vendors provided adequate support in diagnosing and fixing problems experienced with their products.

Experience Base

The best advertisement for a vendor's NMS is its installation record and efficient use on many networks. Hopefully, the product will have been used on a network similar in size and architecture to the customer's target network or has been proven sufficiently flexible by providing the necessary management capabilities in many different environments.

Many of the products evaluated were beta software and therefore the experience base was limited to trade show demonstrations and beta site evaluations. For this reason, this element of the evaluation criteria was not applicable. However, this criteria will prove very useful when products have been released and used on a variety of networks.

SUMMARY

In this paper, we have described a set of criteria that could serve as a basis for evaluating network management systems. In addition, we have presented our observations in using these criteria to evaluate a number of NMS products. Finally, we have also identified some areas where these products could use some improvement.

The criteria we have developed provide a means of assessing the capabilities and features of a NMS, but are by no means comprehensive. They only provide some general guidelines for judging the merits of a product and do not take into account the environment in which the NMS will ultimately be deployed. A weighting system should be applied to the criteria to accurately reflect the user's requirements when evaluating a NMS.

While our results show that the first-generation systems meet most of the important evaluation criteria, they still contain some deficiencies. Some of the deficiencies are due to the fact that these products are beta or early customer release versions. Others can be attributed to the immaturity of the standards and technologies that encompass network management. We can expect continued enhancements in future NMS products as they incorporate new versions of the evolving standards and integrate applications such as database management systems and expert systems.

REFERENCES

Aronoff, R., Chernick, M., Hsing, K., Mills, K., and Stokesberry, D., "Management of Networks Based on Open Systems Interconnection (OSI) Standards: Functional Requirements and Analysis," National Institute of Standards and Technology, NIST SP 500-175, November 1989.

Brusil, P. J. and LaBarre, C. E., "Standards-Based Network Management," M89-47, The MITRE Corporation, August 1989.

RFC 1095, "The Common Management Information Services and Protocol over TCP/IP (CMOT)," U. Warrier, April 1989.

RFC 1155, "Structure and Identification of Management Information for TCP/IP-based Internets," M. Rose and K. McCloghrie, May 1990.

RFC 1156, "Management Information Base for Network Management of TCP/IP-based Internets," K. McCloghrie and M. Rose, May 1990.

RFC 1157, "A Simple Network Management Protocol," J. Case, M. Fedor, M. Schofferstall, and J. Davin, May 1990.

RFC 1158, "Management Information Base for Network Management of TCP/IP-based Internets: MIB-II," M. Rose, May 1990.

IV

IMPLEMENTATION

AND CASE STUDIES

Management By Exception: OSI Event Generation, Reporting, and Logging

Lee LaBarre

The MITRE Corporation, Burlington Road, Bedford MA 01730

1. Introduction

Networks require near real-time monitoring and control of network resources to provide optimum use of available equipment and communication facilities and to provide the required performance in terms such as availability, bandwidth, delay, and error rates. However, the scale and complexity of today's networks, the stringent command and response time requirements, and the volume of status and traffic data are increasing to the point where network operators are unable to adequately handle the near real-time control of the networks. At the same time, the number and skill level of available network management personnel often is decreasing. Methods must be developed to reduce the volume of raw data offered to the operators, improve the information available for decision making, and assist the operator in making those decisions.

Most current network management systems overload network operators with raw data that is redundant and difficult to assimilate for problem diagnosis and determination of proper real-time control actions. They are usually based on polling for the transfer of significant volumes of raw data from managed systems to a central manager station. The information content relative to the total data transferred is usually very low, and is difficult to extract for the average network operator. This approach can put a significant drain on network resources, in terms of link and cpu bandwidth and does not provide network operators the timely information needed for near real-time control.

An alternate approach is to use management by exception as described below:

- Use infrequent polling to determine connectivity between managed systems and the management station. Such polling could be done using efficient lower layer protocol facilities, e.g., network layer echo.

- Maintain information about the network configuration, fault history, and "normal" levels of component and system performance as indicated by metrics derived from raw data sample measurements made at the managed system.

- Transfer information - not raw data - from the managed system to the manager station in the form of timely events containing notifications about significant occurrences in the network.

- Use the timely exception reports (event reports) about faults, configuration changes, and deviations from "normal" or desired performance, along with information about the network configuration, fault history and "normal" component and system performance levels to diagnose problems and make decisions to effect control actions. One such decision may be to request additional information from the affected devices to assist in the decision process.

Management by exception provides the information in the proper form needed for applying expert systems technology to network management. Expert systems can use the timely event and historical information to diagnose problems, recommend decisions, and in some cases automatically take control actions to effect those decisions.

This article concentrates on the functions developed by the International Standards Organization (ISO) to provide tools for managing the timely transfer and manipulation of information in the form of events. It provides a tutorial on ISO mechanisms for notification generation (section 5), the ISO system management functions related to reporting of notifications within event reports (section 6) and controlling their logging (section 7), and the services related to each (section 8). It also identifies the contents of standard events and log records defined in these and other system management functions (section 9), and describes ongoing work in defining metrics that may be used to characterize performance (section 10). Readers are expected to reference the ISO documents for complete definitions of the functions and associated ASN.1 syntax.

2. Management by Exception Approach

Figure 1 illustrates the management by exception approach to network management, including use of an expert system. Network components report events related to configuration changes, status changes, faults, security, and performance deviations to the network management control system (NMCS). Performance metrics are calculated within components from raw data measurements made on component resources. They characterize individual component and system performance in terms such as: availability, throughput, error rates, error ratios, and response time.

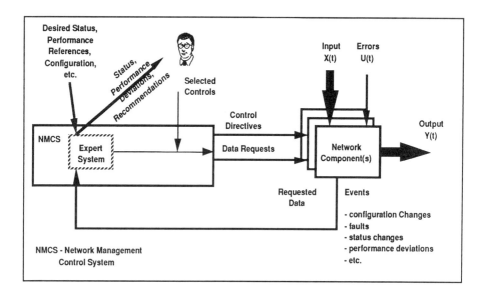

Figure 1: Management by Exception

229

These raw data measurements are samples taken at instances in time on random variables related to the component's status, traffic input $X(t)$, traffic output $Y(t)$, random inputs $U(t)$ due to errors. However, performance related events are generated only when the performance exceeds specified bounds.

The NMCS may display events directly to the operator, who must then use his knowledge of the desired status, performance bounds, network configuration, and experience to diagnose problems and determine appropriate control actions. For example, this may involve NMCS requests to the managed systems for detailed raw data to assist in the diagnosis. Alternatively, the events may be reported to an expert system in the NMCS that may diagnose problems and recommend control actions to the operator, and possibly even make those control actions automatically.

3. The OSI Structure of Management Information

Some knowledge of the OSI structure of management information (SMI) [1] is essential to understanding how notifications are generated, and their reporting and logging are controlled. The collection of management information that characterizes the resources in a managed system is called its management information base (MIB). The SMI uses an object-oriented approach to model the structure of management information. Physical and logical resources that may be contained within managed systems are modeled abstractly as managed object classes. Characteristics of the resource are modeled as attributes within a managed object class. A specific resource within a managed system is represented an an instantiation of an object class and is called a managed object. Managed objects may be created and deleted as resources are created and deleted within a managed system. Attribute values of a managed object may be retrieved and modified.

Managed objects may be designed to generate notifications when significant changes happen to the related resource. The notifications are transferred from the managed system to manager stations using management protocol event type protocol data units. Since event reports contain notifications, the two terms are often interchanged.

Two special managed object classes have been defined for controlling event reporting (Event Forwarding Discriminator) and logging (Log). They are described in sections 6 and 7.

4. OSI System Management Functions

The management functionality needed for the management by exception approach is being developed within the framework of the Open Systems Interconnection (OSI) model. Event reporting, logging, and generation are problems that have received extensive investigation in the ISO community and the International Telegraph and Telephone Consultative Committee (CCITT) community. ISO has developed models and specified objects for OSI event management and log control, and specified many useful events, their content, and the content of related log records. These models, objects and events have been documented in a multi-part series of related Draft International Standards, termed System Management Functions [2, 3, 4, 5, 6, 7, 8, 9]. ISO has also defined specific management information related to these functions, including notification contents and generation mechanisms [10]. In addition, ISO is defining mechanisms for calculating metrics that characterize performance [11, 12].

5. Notification Generation

ISO has defined several mechanisms for controlling the generation of notifications from managed objects if significant changes happen to the related resource. Some mechanisms are defined such that the change itself is the significant occurrence. Examples of such mechanisms are to emit a notification when:

- an object is created or deleted,

- a software or hardware processing failure happens that is peculiar to a resource,

- a state type attribute changes value, e.g., attributes indicating the operational or administrative state of the resource,

- a settable non-state type attribute value is changed, e.g., an attribute indicating an operational parameter such as the transport retransmission parameter.

However, changes to the most prevalent type of attribute in any MIB, counters and gauges, should not cause the generation of events. Such behavior would flood the network with traffic. The significant occurrence is not that the counter or gauge changed value, but that the value it changed to exceeded some specified bound. For example, when the ratio of the number of corrupted PDUs to the total number of received PDUs exceeds an acceptable value, a notification may be emitted as a fault indication. Or, when the count of users of a resource exceeds a defined number, the manager is informed so that some users may be shifted to another resource.

ISO [10] has defined generic mechanisms to generate notifications when a related counter or gauge exceeds a specified value.

5.1 Counter Thresholds

Counters may only increase until they reach their maximum value, and then they wrap to zero. Some counters may be reset to zero (though use of such counters is not recommended in multiple manager situations). The operation of the counter threshold is illustrated in figure 2.

A counter threshold attribute type has three associated values that must be specified by the manager. Those values and their associated semantics are as follows:

- comparison level: A notification is emitted when the value of the associated counter becomes equal to this value.

- offset value: If the value is not zero, then when a notification is generated, add the offset value to the comparison level. When the associated counter again equals the new comparison level, the process repeats. If the offset value is zero, the result is a "one shot", i.e., no more notifications are generated.

- notification switch: This determines whether notification generation is enabled.

The above three values define a single threshold level. Multiple threshold levels may be defined for a single counter threshold attribute.

5.2 Gauge Thresholds

Gauges associated with thresholds may increase and decrease in value. Consequently, if a gauge value oscillated about a comparison level, multiple notifications would be emitted. To prevent this, gauge threshold attribute types are defined with a hysteresis mechanism to prevent such repeated notifications. Figure 2 illustrates the operation of the gauge threshold attribute.

Gauge thresholds are defined as a pair of comparison levels: "notify high" and "notify low". The "hysteresis interval" is the difference between the values associated with the the notify high and notify low levels. A notification may be generated when the gauge value increases above the notify high value, or when it decreases below the notify low value. A notification switch defined for each level determines whether or not a notification is actually emitted.

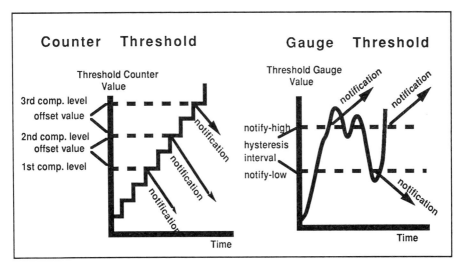

Figure 2: Counter and Gauge Threshold Operation

After generating a notify high notification, another one may not be generated until the gauge has decreased below the notify low level. Thus in figure 2, the second crossing of the notify high value does not cause a notification to be generated. Similarly, after generating a notify low notification, another one may not be generated until the gauge has increased above the notify high level.

The above two comparison levels and the associated enabling switches define a single threshold level. Multiple threshold levels may be defined for a single gauge threshold attribute.

6. Event Report Management Function

ISO has defined the Event Report Management Function [6] to schedule the distribution of event reports to one or more destinations, including logs. The destination selection is based on information contained in each event report. The function provides for the following:

- selection of (or filtering) event reports to be forwarded,
- distribution of the selected event reports to chosen destinations,

- ability to suspend and resume event reporting on a per destination basis,
- ability to create/destroy/modify selection criteria and destination indicators,
- ability to control the event report scheduling on a daily, weekly, or other basis.

The Event Report Management Function provides services by which event reports can be distributed. Event selections are done by a filtering process using the Event Forwarding Discriminator (EFD) managed object. Event Reporting Services are provided to initiate, terminate, suspend, resume, modify, and schedule event reporting through the manipulation of Event Forwarding Discriminator managed objects. These services are described in section 8.

6.1 Event Report Management Model

The Event Report Management Function conceptually behaves as follows (figure 3).

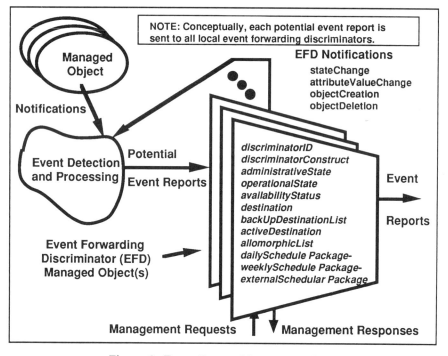

Figure 3: Event Report Management Model

- At least one managed object is capable of generating notifications that may be forwarded within an event report.

- A detection and processing module receives the locally generated notifications and forms potential event reports. Potential event reports contain all the information in the notification, plus information added by the detection and processing module, such as: event time, object class and instance.

- EFD managed objects provide the scheduling of event reporting and select the potential event reports that are forwarded to destinations as real event reports.

- EFDs actively forwarding events match attribute values in the potential event reports against criteria specified within the EFD by a manager. Potential event reports with attribute values that match those criteria are reported to the specified destination.

- Potential event reports are sent to all Event Forwarding Discriminators.

- Event reporting can be managed remotely by manipulation of the EFDs.

6.2 Event Forwarding Discriminator (EFD) Managed Object

The Event Forwarding Discriminator is a managed object that specifies criteria for selecting the events that are to be forwarded, the destinations to which they must be reported, and the schedule when processing of potential event reports is allowed. The destinations may be local or remote managing processes. The managing process may be one that logs event reports.

6.2.1 EFD Attributes

The attributes of the event forwarding discriminator object are described below.

discriminatorID - Contains a value that is used to uniquely name the EFD object.

discriminatorConstruct - Changed by the manager to define the criteria that specify whether the associated potential event report should be forwarded. Only those potential event reports that pass (match) the discriminator construct criteria will result in the associated potential event report being forwarded. The attribute is a logical expression on attributes within the potential event reports, e.g., {NOT {(A=100) and (B>50)}}, where A and B are attributes.

administrativeState - Changed by the manager to suspend or resume forwarding of events. Its value indicates whether the EFD object is "locked" by the manager, i.e., the associated potential event reports are not allowed to proceed, or is "unlocked", i.e., the associated potential event reports are allowed to proceed.

operationalState - Changed by the EFD object itself to indicate whether the EFD object is operationally able to perform its function. Its value may be "disabled", i.e., the associated potential event reports cannot proceed, or "enabled", i.e., the associated potential event reports can proceed.

availabilityStatus - Changed by the EFD itself to qualify the operational state. Its value may be "off-duty", indicating that the EFD is not processing potential event reports since it is "off-duty", i.e., the schedule currently inhibits processing of potential event reports. If the *availabilityStatus* is "off-duty" then the *operationalState* is "disabled".

destination - Identifies the title of the primary application to which the event reports are to be forwarded. This usually indicates an application in the managing system, but it may be the title of a local application or log.

backUpDestinationList - Is an ordered list of destinations to be tried in case the primary destination fails. This list is in priority order.

activeDestination - Identifies the title of the application to which the event reports are currently being forwarded. This is either the primary title specified in *destination* or one of the alternate application titles specified in *backUpDestinationList*.

allomorphicList - A managed object may behave as (i.e., be allomorphic to) one of its superclasses in the inheritance hierarchy [1]. Each managed object has knowledge of classes to which it is allomorphic, including itself, as an attribute called *allomorphs*. The *allomorphicList* attribute in the EFD contains an ordered list of managed object classes, one or more of which may be allomorphic to the object that emitted the notification. The object class associated with the event report will be the first object class encountered in EFD *allomorphicList* which is also a member of the *allomorphs* attribute of the object that emitted the notification.

6.2.2 EFD Scheduling Packages

Scheduling packages specify attributes and behavior used to control the automatic switching of the EFD between reporting-on, and reporting-off, i.e., "off-duty", conditions. Three conditional packages are specified to control reporting on a daily, weekly, or arbitrary basis.

dailyScheduling Package - Specifies a list of time intervals during a 24 hour time period when the EFD will be in the reporting-on condition.

weeklyScheduling Package - Specifies the date and time at which the EFD will start and stop the periodic weekly schedule, and the selected days of the week, relative to the starting date, and a list of intervals of the day, when the EFD may be allowed to be in the reporting-on condition.

externalSchedular Package - Specifies the name of an external schedular managed object that will control the reporting-on and reporting-off conditions of the EFD.

6.2.3 EFD Notifications

The EFD object may emit the following notifications, which are then themselves processed by an EFD object.

objectCreation - emitted when the EFD object is created.

objectDeletion - emitted when the EFD object is deleted.

stateChange - emitted by the EFD when the *administrativeState, operationalState, or availabilityStatus* attributes change value.

attributeValueChange - emitted by the EFD when settable non-state type attributes change.

7. Log Control Function

The Log Control Function [7] provides services by which records associated with event reports can be logged. Records can be logged according to a schedule. Record selection is done by a filtering process using mechanisms similar to those used by the EFD managed object. Log control provides the services to initiate, terminate, suspend, resume, modify, and schedule event logging through the manipulation of Log managed objects. These services are described in section 8.

7.1 Logging Model

The Log managed object (figure 4) contains storage for records. It adds information (e.g., unique record identifiers, logging time) to received event reports and transforms that information into potential log record objects. Attributes (shaded area of figure 4) associated with the log provide control over which potential log records are selected for storage, and control the availability of the logging activity. A subset of these attributes, as indicated in the trapezoid in figure 4, control the filtering of potential log records in the same way as the EFD object controls the forwarding of potential event reports. The filtered log records are stored and may be available for retrieval either via the management protocol or via other services such as file transfer.

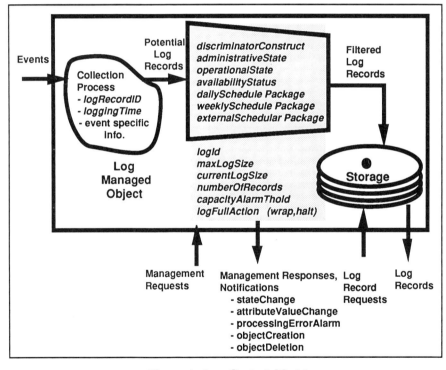

Figure 4: Log Control Model

7.2 Log Managed Object

The log's behavior is determined by its state attributes, discriminator construct attribute, and schedule attributes.

The log stores records in the order in which they are presented for logging. New records will be added to the log only if the log is:

- not in the "off-duty" condition due to scheduling,
- not in the "locked" administrative state,
- not in the "log-full" (for a log that halts) availability status,
- not in the "disabled" operational state, and
- the record passes the discriminator construct test criteria.

When the log has the "log-full" availability status, new records will not be entered into the log, but records currently in the log may be retrieved. Placing the Log object in the "locked" administrative state will prevent entry of new records into the log.

The log operational state and availability status reflect the internal activity of the log and cannot be changed by direct management action. Two options have been defined for the behavior of the log when it reaches the maximum log size. The log may either halt logging (discard new records) or wrap (discard old records). A log that halts will always generate a capacity threshold event. A log that wraps will discard an integral number of records, and may also generate a capacity threshold event indicating that a new wrap has occurred. Every log must be able to support the halt behavior; support of the wrap behavior is optional.

7.2.1 Log Attributes

The attributes of the Log object are described below.

logID - Uniquely identifies an instance of a Log managed object

maxLogSize - The maximum size in octets that the log can contain, exclusive of the system overhead involved in establishing the log.

currentLogSize - The current size of the log measured in octets.

numberOfRecords - The current number of records in the log.

logFullAction - Specifies the action to be taken when the maximum size of the log is reached. Options are: (a) Wrap - the oldest set of records in the log will be deleted to make room for new records. (b) Halt - no more record will be logged. Records already in the log will remain. All logs must support the Halt behavior.

capacityAlarmThreshold - See *logFullAction*. As a percentage of *maxLogSize*, the condition for which an event will be generated to indicate that a log full or log wrap condition is approaching. Support is mandatory for the Halt behavior.

The *discriminatorConstruct, administrativeState, operationalState,* and *availabilityStatus* function as defined for the event forwarding discriminator object, except that they control the logging of potential log records instead of the forwarding of potential event reports.

7.2.2 Log Scheduling Packages

The scheduling packages for the Log object are the same ones defined for the Event Forwarding Discriminator object. The scheduling packages control the times when the Log object may be active.

7.2.3 Log Notifications
ProcessingAlarm - emitted by the Log object when the *capacityAlarmThreshold* is generated with a probable cause of "storageCapacityProblem".

The *attributeValueChange, stateChange, objectCreation, and objectDeletion* notifications function as described for the Event Forwarding Discriminator object, except that they apply to the Log object.

8. Event Report Management and Log Control Services

Many of the attributes of the EFD and Log objects are the same, and the services used to control event reporting and logging are similar. Basic services have been defined to initiate and terminate, to temporarily suspend and resume, and to modify conditions for event forwarding or logging. The Event Forwarding Discriminator and Log attributes may be retrieved at any time through the normal use of the management protocol.

8.1 Initiation
A user at a managing system may desire that particular events generated at a managed system be reported to particular destination systems, or that logging should be initiated. This is accomplished by the creation of Event Forwarding Discriminator or Log managed objects at the managed system.

8.2 Termination
A user in a managing system can use this service to turn off the forwarding of events or logging at a specific managed system. This is accomplished by the deletion of the EFDs of the unwanted events, or Log objects, at the managed system. The absence of an EFD will not stop the generation of notifications from the managed objects; it simply prevents those particular notifications from being forwarded as event reports. Controlling the event generation mechanism, e.g., by setting to "off" the notify high and notify low switches on gauge thresholds, will prevent the generation of notifications.

8.3 Suspension/Resumption/Modification
This service enables the manager to temporarily suspend, and subsequently resume, the forwarding of event reports normally passed by an EFD, or logging of log records, and to modify conditions for event forwarding or logging.

To suspend event reporting, or logging, the EFD or Log *administrativeState* attribute is changed to "locked"; and to resume event reporting, or logging, it is changed back to "unlocked".

A managing system can modify the conditions of event forwarding for selected events, or logging of selected log records, by changing the values of the attributes which are used in the processing associated with event distribution and control, or with logging. For example, the user may want to modify the reporting of a specific type of event to a different destination, or change the schedule of the event reporting. To achieve such results, a managing system will need to modify the value of the EFD *destination* and/or *discriminatorConstruct* attributes and the appropriate scheduling attributes to reflect the new needs. Similarly, changes would be made to the *discriminatorConstruct* attribute and scheduling attributes in the Log object to control the selection of records to be logged.

9. Standard Event and Log Record Contents

The OSI system management functions have defined the contents for some standard event reports and log records. The contents, in addition to providing needed information about the event or log record, are used to select which events are to be forwarded or logged and for log record retrieval.

9.1 Standard Event Reports

The Event Report Management function uses the standardized information contained within an event report to select event reports that should be forwarded to specified destinations.

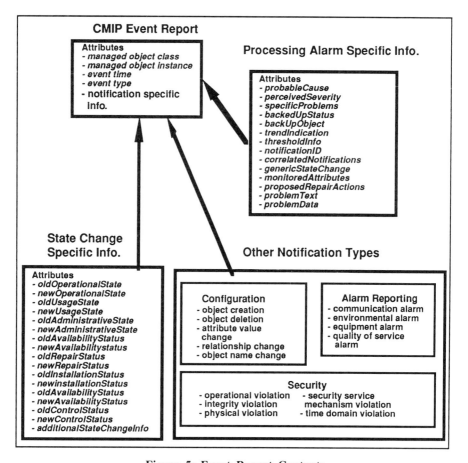

Figure 5: Event Report Contents

Several system management functions define standard information for notifications. This information is contained in the event report. For example the Alarm Reporting Function [5] capabilities, whether the condition is getting more or less severe, threshold related information, defines information to help with understanding the cause of the alarm, its severity, backup and other information for correlating the alarm to other alarms or providing additional diagnostic information.

Similarly the State Management Function [3] defines the information to be included in State Change notifications, and the Object Management Function [2] defines information to be included in notifications related to object creation and deletion.

All event reports also contain the additional information specified by CMIS [13] for event reports (event time is optional). Figure 5 illustrates the event (notification) specific information contained in the Processing Alarm event, and the State Change event.

Other notification types thus far defined in the standards for configuration management [2, 3, 4], alarm reporting [4], and security alarms [8], are also indicated in figure 5.

Attributes corresponding to the parameters of an event may be defined. These attributes can be referenced by a *discriminatorConstruct* attribute of an EFD object for determining if and where event reports should be sent, or referenced by the *discriminatorConstruct* attribute of a Log object to select which records should be stored in the log.

The standards allow some of the attributes defined for a notification to be conditionally present. Therefore, all events for a notification type may not contain the same attributes.

9.2 Standard Log Records

Log records are managed objects that represent information stored in logs. The records may contain information related to management defined events, or events related to protocol data units (PDUs) (e.g., transport connection PDUs). The Event Log Record (figure 6) managed object class contains attributes common to all log records (*logRecordID* and *loggingTime*), CMIS defined attributes common to all events, plus information specific to each notification

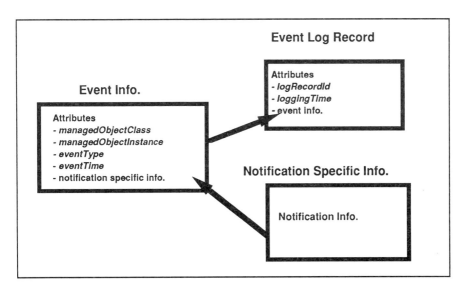

Figure 6: Log Record Contents

type. Standard information related to ISO notifications are defined in the system management functions, with their syntax defined in [10]. For example, the Alarm Record contains the attributes defined for the Event Log Record, including information specific to Alarm type notifications.

10. Performance Metrics

Classical performance models characterize system performance in terms of metrics that include: input traffic (offered load), output traffic (throughput), delays (e.g., service time), error rates, and fraction of system capacity in use (utilization). These performance metrics may be calculated from samples on the raw data contained in counters and gauges. For example, dividing the counter value that represent the number of input messages by the time between samples provides the metric for the offered traffic rate. Thresholds can be applied to the performance metrics and cause events to be reported to the manager station when acceptable performance bounds are exceeded.

10.1 Statistics on Performance Metrics

All of these performance metrics change over time, perhaps very rapidly from one short time interval to the next. These "instantaneous" performance metrics characterize the transient performance of the system components. But "instantaneous" metrics are often inadequate to characterize the performance of network components for the purpose of remote control across a communications network. By the time either the raw data samples, or the calculated "instantaneous" metrics, have been transmitted to the management station the component or system performance may have changed due to random fluctuations in the input traffic, or due to self-correcting algorithms in the individual components and protocols. Control actions based on the transient performance characterization may therefore be unnecessary or inappropriate to the changed conditions, and may even push the system into unstable operation.

Experience with control systems indicate that control decisions should be based on performance metrics derived from the weighted time-averaged distribution parameters of the performance metrics, which are random variables in time, calculated from many samples taken over a significant period of time. The weighted time-averaged distribution parameters, including the mean, variance, inter-quartile range, etc., of the metrics, tend to change slowly and filter out indications of transient behavior, thus adjusting to variations in traffic and the self-correcting behavior of components. Significant changes in these distribution parameters are more likely to be a true indication that a component's performance has altered. Thresholds applied to the time-averaged parameters of the performance metrics will result in fewer event reports being generated due to transient conditions that need no management action to correct.

ISO is developing standards that specify mechanisms for calculating the time-averaged statistics on counter type time differentials and gauge type attributes. The Workload Monitoring Function [11] defines algorithms for calculating the time-average statistics (mean, variance, percentiles) of counter differentials and gauges, defines thresholds on these statistics, and indicates how they may be applied to estimations of resource requests, resource rejections, and resource utilization that characterize the resource workload. The Summarization Function [12] specifies a method for scheduling the reporting of events that carry this summary statistical information and dynamically defining their contents.

10.2 Metric Objects

Managed objects are usually defined without specific consideration of the metrics that characterize the performance of the related resource, although their definers often define the counters and gauges that are the raw data measurements needed to calculate the performance metrics. Moreover, in most cases they should not include the metric calculations in the managed object, since which resources should be monitored for performance is usually a policy or system (or experimental) design decision.

A "meter", or metric object, that can be "attached" to the managed object to be monitored is needed to monitor the performance of resources represented by objects not designed with a requirement for performance monitoring - in much the same way that a voltmeter is attached to points in an electronic circuit (figure 7). The concept of metric objects defined in [11] fulfills that need. A metric object makes observations on one or more attributes in a monitored object according to a periodic schedule determined by the sample time. The observations may be used to calculate a value. If the calculated value represents a gauge value, as is the case with the metric objects defined in [11], then a gauge threshold may be associated with the calculated metric value.

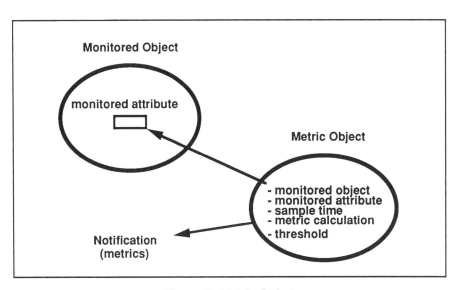

Figure 7: Metric Objects

11. Summary

One approach to network management is management by exception, where exceptions are reported via events. This paper has provided a tutorial on the standard tools being developed by ISO for event generation, event reporting, event logging, and performance monitoring. These tools are expected to become International Standards in 1991. Managed objects are currently being defined in the standards and implementors agreements communities that will use these tools for management of events. Implementors agreements on the system management function standards that define the event report and log control tools are now in progress within several communities. Future expert systems will use these tools to improve the timeliness and quality of information available to network operators and assist them in effecting control decisions.

12. References

1. ISO/IEC DIS 10165-1, Information Processing Systems - Open Systems Interconnection - Structure of Management Information - Part 1: Management Information Model.
2. ISO/IEC DIS 10164-1, Information Processing Systems - Open Systems Interconnection - Systems Management - Part 1: Object Management Function.
3. DIS 10164-2, Information Processing Systems - Open Systems Interconnection - Systems Management - Part 2: State management Function.
4. DIS 10164-3, Information Processing Systems - Open Systems Interconnection - Systems Management - Part 3: Attributes for representing Relationships.
5. ISO/IEC DIS 10164-4, Information Processing Systems - Open Systems Interconnection - Systems Management - Part 4: Alarm Reporting Function.
6. ISO/IEC DIS 10164-5, Information Processing Systems - Open Systems Interconnection - Systems Management - Part 5: Event Report Management Function.
7. DIS 10164-6, Information Processing Systems - Open Systems Interconnection - Systems Management - Part 6: Log Control Function.
8. ISO/IEC DIS 10164-7, Information Processing Systems - Open Systems Interconnection - Systems Management - Part 7: Security Alarm Reporting Function.
9. ISO/IEC DIS 10164-8, Information Processing Systems - Open Systems Interconnection - Systems Management - Part 8: Security Audit Trail Function.
10. ISO/IEC DIS 10165-2, Information Processing Systems - Open Systems Interconnection - Structure of Management Information - Part 2: Definition of Management Information.
11. ISO/IEC CD 10165-12, Information Processing Systems - Open Systems Interconnection - Systems Management - Part 12: Workload Monitoring Function.
12. ISO/IEC WP 10165-13, Information Processing Systems - Open Systems Interconnection - Structure of Management Information - Part 13 Summarization Function.
13. ISO 9595, Information Processing Systems - Open Systems Interconnection - Management Information Service Definition - Common Management Information Service.
14. ISO 9596, Information Processing Systems - Open Systems Interconnection - Management Information Protocol Specification - Common Management Information Protocol.

IV

IMPLEMENTATION

AND CASE STUDIES

A
Implementing Network Management

Communications Network for Manufacturing Applications (CNMA)

A European Initiative for Network Management in Industrial Communication

Dr. W. Kiesel and K.H. Deiretsbacher
Siemens AG, Automation Group, Department AUT 961
8520 Erlangen, Federal Republic of Germany

Abstract

CNMA is an initiative of the European Communities to promote international standards for factory automation (Computer Integrated Manufacturing - CIM) as the European complement to MAP, however based on more advanced standards. This project is part of ESPRIT, the European Strategic Programme for Research and Development in Information Technology.

CNMA's activities are to specify and implement industrial communication, and to demonstrate the results. Network Management (NMT) is one of the key areas within CNMA Phase 4 and the currently planned Phase 5. Most standards for Network Management are still immature and their promotion has therefore highest importance.

The current CNMA-implementations support one Manager System and several Managed Systems. Network management is performed through communication between System Management Application Processes (SMAP's) executing on these systems. Agent SMAPs collect or modify management information under control of the Manager SMAP.

1. CNMA AND ITS GOALS

For cost effective Computer Integrated Manufacturing (CIM) it is essential that equipment from many different vendors is capable of simple and cheap integration. The model of OSI (Open Systems Interconnection) provides the framework for this integration.

The first major OSI based initiatives are the General Motors MAP (Manufacturing Automation Protocol) and Boeing TOP (Technical and Office Protocol). CNMA (Communications Network for Manufacturing Applications) is a complementary European initiative, supported by the European Communities under its ESPRIT programme.

The CNMA programme is performed in several phases. It has been initiated in 1986 (CNMA Phase 1). Since 1989 it is progressing phase 4. This phase will be completed at the end of 1990. The next phase (January 1991 - December 1992) is currently in the planning stage.

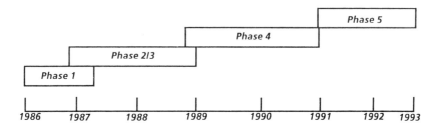

Fig.1: CNMA-Project Phases

One of the most important goals of CNMA is the promotion of ISO activities in the area of industrial communication. The area of Network Management is in an early standardization state, but users require solutions now.

Network Management has been defined as a key area within CNMA phase 4 with the ultimate objective to support and accelerate the standardization process. For example representatives of the CNMA members are active within national and international working groups for standardization.

The final outcome of each CNMA phase is the specification and implementation of selected standards which are at last demonstrated within pilots. The participants which contribute to the work therefore range from leading manufacturing companies (including vendors and users of local area networks, LANs) through software houses, universities and test centers.

CNMA's main focus is industrial communication (factory automation) based on Local Area Networks (LAN). Therefore, its activities for Network Management are oriented towards LANs (IEEE-802.3, IEEE-802.4), too.

2. CNMA ACTIVITIES

The key topics may be summarized as follows:
- Specification of user requirements.
- Selection of existing and emerging standards.
- Specification for implementations based of the selected standards (standard profiles) in a so-called Implementation Guide.
- Implementation of these standards on CNC/PLC controllers, minicomputers and PC's.
- Verification of the implementations by Conformance and Interoperability Testing-Tools.
- Validation and demonstration of the implementations by use in real production cells.
- Promotion of these standards to encourage their widespread acceptance.

3. CNMA AND NETWORK MANAGEMENT

3.1 Functional Overview

CNMA Network Management provides tools for the purpose of:
- problem detection and diagnosis
- installation and checkout
- performance monitoring.

The basis of CNMA Network Management is the OSI Management Framework (OSI 7498-4). Due to the most important user requirements for industrial communication it is currently concentrated on the Functional Areas (FA) of Configuration Management, Performance Management and Fault Management. The current chosen System Management Functions are:
- Object Management Function (Configruation Management)
- Confidence and Diagnostic Testing Function (Fault Management)
- Workload Monitoring Function (Performance Management)
- Management Association Control Function (common for all FAs).

Within the Funtional Areas these functions have reached the most advanced state in the standardization process and have been identified as most useful. Most of these standards are currently at Committee Draft or Draft Standard status.

The selected Network Management Application Protocol is the ISO-OSI Common Management Information Service and Protocol (CMIS/CMIP) which is meanwhile International Standard.

3.2 MAP V 3.0

CNMA specifications have also been largely influenced by **MAP V3.0**. MAP V3.0 is based on standards which existed approximately end of 1987. Since CNMA focus on the use of most recent ISO Standards, and since several selected standards are much more stable in the meantime, there are some differences between CNMA and MAP V3.0. However, these are mostly enhancements, so that (from a general point of view) CNMA is a superset of MAP V3.0. Major vendors of Network Management systems plan to marketize manager systems which are able to manage both CNMA and MAP systems.

CNMA is only concerned with OSI systems, therefore currently special MAP nodes like MAP/EPA nodes or Mini-MAP nodes are not addressed.

4. Implementation Guide

One of the major work items within CNMA is the production of a public availiable Implementation Guide. This guide specifies the funtional standards (profiles) and the requirements for implementations of Manager Systems (applications) and Managed Systems (agents). In addition to protocol profiles, the Implementation Guide specifies the needed management information and rules for the agent behaviour.

An addendum to the Implementation Guide specifies the architecture and functionality of the Network Management applications. In general, the content is based on the funtionality identified as a requirement by user representatives.

5. NETWORK MANAGEMENT ARCHITECTURE - OVERVIEW

Currently (for Phase 4) CNMA supports one Manager System (the NM-Console) and several managed systems.
Network Management is performed through communication between System Management Application Processes (SMAPs) residing on the various nodes. The SMAP on the manager system is called Manager-SMAP. The SMAPs on the managed systems are called Agent-SMAPs.
Agents collect or modify management information under control of the Manager. They also have the ability so send events related to predefined event conditions set by the Manager-SMAP.
Communication between SMAPs is performed by use of the Common Management Information Protocol (CMIP).
The Network Manager application processes all collected information in order to achieve Configuration Management, Performance Management and Fault Management.

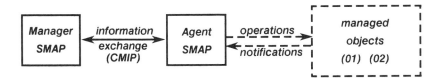

Fig.2: Interactions between Management SMAPs

6. Network Management Components

6.1 Management Information

Management information is defined in terms of managed objects. A managed object is the management view of a system resource (e.g., a layer entitiy, a computer system, a scheduling queue, or a printer). The set of managed objects in a system, controlled by a single Agent-SMAP is called that system's Management Information Base (MIB).

A managed object is an abstracted view of a system resource. Each managed object is defined by:
- the attributes visible at its boundary,
- the management operations which are applied to it,
- its behaviour to management operations, and
- the notifications emitted by the object.

Within CNMA, the definition of the Management Information was an essential work item. Main focus was thereby put on the selected "System **Management Functions**" and on MAP V3.0 upward compatibility.

CNMA Phase 4 has selected more that 20 Object Classes (Layer Entities, Test Performer, Connections, Service Access Points) covering all important communication aspects. A total number of more than 350 attributes, actions and events has been defined, half of which is mandatory within the CNMA project.

6.2 Management Information Transfer

The interactions between management SMAPs is realized through the use of OSI's Common Management Information Services and Protocol (CMIS/CMIP). The current CNMA-phase 4 specification is based on the Draft International Standard (DIS) and will be updated in the next project phase.

The following CMISE (CMIS Element) services have been chosen by CNMA:

    o **M-INITIALISE** service: to establish an association with another CMISE service user for the purpose of exchanging management information.

    o **M-TERMINATE** and **M-ABORT** serivce: to release an association orderly or abrupt

- o **M-EVENT-REPORT** service: to report an event about a managed object to another CMISE service user.
- o **M-GET** serivce: to request retrieval of management information (attribute values of managed objects).
- o **M-SET** service: to request modification of management information (attribute values of managed objects).
- o **M-ACTION** serivce: to request another service user to perform an anction (e.g., start - test, de-enroll) upon a certain object.

6.3 Management Architecture

The Network Management Applications provide means for the human network administrator to read or alter data, control the data and access reports. It is part of a manager process which is outside the OSI standardization activities of ISO.

CNMA specifies and implements also Management Applications. A subgroup of CNMA (the Netweork Management Kernel Group) has defined a common architecture and has developed the components of the Management Applications in a joint effort.

6.3.1 Decomposition of Activities

The basis for the architectural structuring is the decomposition of the application into four levels of activity:

Level 0: Information display

Level 0 deals with the user interface. It contains all necessary functions to give the user access to administrative information, access to perform administrative operations, and access to signals coming from the managed objects.

Level 1: Functional Area Specific Information Processing

Level 1 contains management application services specific for the functional areas: configuration management, performance management and fault management.

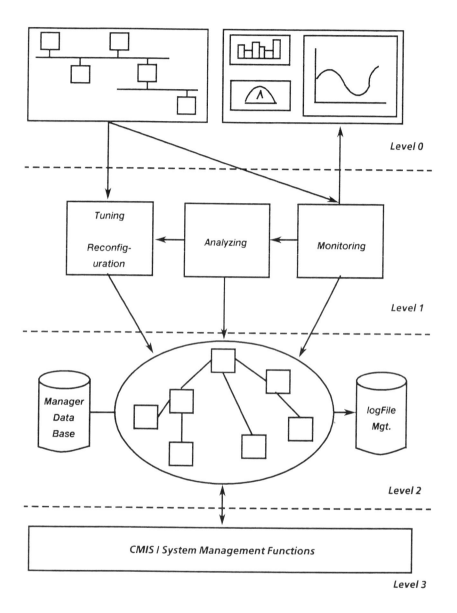

Fig.3: Manager Architecture

Level 2: Common Information Processing

Level 2 contains services common to one or more management applications. It also manages a Manager Data Base, containing all administrative information not contained in the Management Information Base (MIB).

Level 3: Information Transfer

Level 3 contains all the standardized functions giving access to the Management Information Base (MIB).

These four levels have been defined to achieve modularity, extensibility and openness for all modules within the manager architecture.

6.3.2 Management Activity Levels

Each level is defined to be independent of the level built on top of it. This means in particular, that Level 2 modules are indepent of User Interface issues. The advantage of this separation is manyfold:

o it is a method of modularization (parallel development, maintainability, extensibility), and

o it makes large parts of the application independent of the user interface. This allows, for instance, to use the modules underneath by different user interface implementations (command level, semi-graphic, full graphic).
This in particular is an important requirement for CNMA, where Level 2 modules will be ported between the vendors in order to produce a complete Network Management System.

6.3.3 Management Applications

For the Manager System several applications have been implemented, which primarily integrate the following management aspects:

o **Configuration Management** identifies and describes the network topology and the network components, and changes and monitors configuration characteristica during installation phases and operation phases.
Two types of objects are handled by configuration mangement services: hardware components like controllers, computers, network equipments and software compo-

nents like application packages, support tools. Configuration Management offers support for the installation, modification, removement and observation of hardware and software components.

Installation includes services to set up the manager database as well as services to ensure that the installed component is up and running well.

Observation means the monitoring of the operational status (e.g., MIB of managed systems, transport connections, associations).

o **Performance Management** is concerned with system and communication performance aspects.

Performance Management provides functions for the administrator to monitor the communication quality at runtime (workload, throughput, reliability), and to change certain parameters for performance improvement (tuning).

Serveral jobs may run in parallel, monitoring different entities of the network. Diagrams and gauges are used to display statistics or show a system's or network component's performance. Monitoring results are usually written to log files. The recorded data can be interpreted for longterm statistics or trend analysis.

o **Fault Management** identifies and localizies faulty components, and provides diagnostics and necessary corrective actions.

Fault Management includes a knowledge based system in order to facilitate diagnosis and to support the operator in repair and configuration tasks. These functions will be performed by the Network Diagnosis System. An inference engine carries out deductive and abductive reasoning in both forward and backward chaining.

Static information about the network (real world) is derived from the manager database; dynamic state information is received via event notifications reported by agent-SMAPs in various fault situations; additional data may be obtained by asking the user.

Diagnosis concentrates currently on hardware faults. The diagnostic mechanisms are based on associational knowledge of the meaning of symptoms and about the verification of suspicions and is supported by a structural model of the network system.

The quality of the Diagnosis System depends on the number of fault situations recognized and reported by managed systems and the knowledge availiable in the knowledge base. This knowledge increases when experience with operating networks has been gained, and thus it is only partially available at the beginning of pilot installations.

Therefore the Fault Manager is designed as system which can be easy extended during the operating phase of the pilot.

Wherever possible and reasonable, user interactions and presentation of monitoring results are performed in a fully graphical manner.

7. CONFORMANCE AND INTEROPERABILITY TESTS

Before their integration in Pilots, the vendor's implementations of the communications protocols will be tested, against a suite of Conformance and Interoperability Test Tools. Experience from previous phases of CNMA has shown that conformance testing is vital when installing complex systems to tight timescales. The Test Tools for Phase 4 are being produced by a complementary ESPRIT Project, known as TT-CNMA.

8. CNMA PILOTS

The profile defined in the Implementation Guide will then be validated and demonstrated by use in a number of real production cells, known as pilots. The use of the communications software in these pilots ensures that the project addresses genuine CIM communication problems, and implements practical solutions.

The implementations of Network Mangement will be tested and demonstrated in CNMA phase 4 specifically at the Renault Experimental Demonstrator in Boulogne Billancourt, France.

In addition to the features mentioned above, this Pilot will provide a fault injector to input simulated faults into the network. With that the ability of Network Management to tune and reconfigure the network, to diagnose and manage fault conditions will be illustrated.

Once the Network Management Applications have been proven on the Renault Experimental Demonstrator, it will be installed in the Aerospatiale factory and in the Magnetti Marelli (FIAT) factory in Italy, thereby showing its value in a real production environments.

9. PROMOTION

Throughout the project, the partners are promoting the use of OSI communications. CNMA has hold a workshop at the University of Stuttgart Demonstrator site in September 1990. The CNMA pilots are all be used for demonstrations to visiting members of the public.

The CNMA partners are liaising with national and international standards bodies, as well as other related organizations, for example EMUG, NM Forum, X/Open, and the NIST-OSI workshop.

10. OUTLOOK-CNMA PHASE 5

The planning for CNMA Phase 5 is in progress. Phase 5 work will be based on the results of Phase 4. Newest, then available OSI standards for **Network Management** will be used.

It is expected, that additional System Management Functions (e.g. Response Time Monitoring) will have reached a state where they are usable.

Work in Phase 5 is planned to focus on
- Integrating LAN Analyzers into Network Management
- Distributed Network Mangers,
- Cooperation with non-CNMA NMT systems, e.g. Internet Network Management
- Security for Network Mangement and,
- Knowledge aquisition for Fault and Performance Management.

REFERENCES

IG 4.0: ESPRIT Project 2617 - CNMA Implementation Guide - Volume 2, A-Profile (Sep. 89), Addendum I (March 90)

Mgt. Framework: (ISO/IEC 7498-4) Information Processing Systems - Open Systems Interconnection - Basic Reference Model Part 4: Management Framework (1989)

Mgt. Overview: (ISO/IEC DIS 10040) Information Processing Systems - Open Systems Interconnection - System Management Overview (June 1990)

CMIS: (ISO/IEC DIS 9595-2) Information Processing Systems - Open Systems Interconnection - Mangement Information Service Definition Part 2: Common Mangement Information Service (1989)

CMIP: (ISO/IEC DIS 9596-2) Information Processing Systems - Open Systems Interconnection - Management Information Protocol Specification Part 2: Common Management Information Protocol (1989)

SMI: (ISO/IEC DP 10165) Information Processing Systems - Open Systems Interconnection - Structure of Mangement Information (Parts 1-4)

Object Mgt.: (ISO/IEC CD 10164-1) Information Processing Systems - Open Systems Interconnection - Systems Management - Part 1: Object Management Function (May 1990)

PMFA: (ISO/IEC JTC1/SC 21 N4981) Information Processing Systems - Open Systems Interconnection - Performance Management Working Document (July 1990)

FMFA: (ISO/IEC JTC1/SC 21 N4077) Information Processing Systems - Open Systems Interconnection - Fault Mangement Working Document (December 1989)

Experience of Implementing OSI Management Facilities

Graham Knight , George Pavlou and Simon Walton

Department of Computer Science, University College London, Gower Street London, WC1E 6BT, United Kingdom

ABSTRACT

The Computer Science department at UCL has experimented with OSI management systems for several years and has implemented a pilot management system on a Unix workstation. A second version of this system is now being implemented. This paper briefly reviews our experience with the pilot system and outlines the capabilities that were felt desirable in its successor. The architecture of the successor system is then described.

1. INTRODUCTION

The Computer Science department at UCL has experimented with OSI management systems for several years and has built a pilot system. We have been interested in investigating how practical is the OSI approach when compared with that of (say) SNMP[1]. In particular, we have wanted to experiment with the complex filtering and event control facilities that OSI management provides.

Section 2 describes the key features of the pilot system and Section 3 outlines the areas in which development was felt to be necessary. Section 4 gives an overview of the architecture of the updated system whilst Sections 5-7 provide a detailed description of its internal operation. It is assumed that the reader is familiar with the basic concepts of OSI management. A good tutorial introduction to these may be found in [2].

2. THE PILOT SYSTEM

Our first implementation "Osimis" [3] was developed under the ESPRIT INCA project[4] and was designed to provide OSI Management facilities for a Unix workstation. It included a CMIP implementation to the DP of 1988 built upon ISODE[5]. This system is illustrated in Figure 1; it provided communication between a Unix "Systems Management Agent" (SMA) and three clients; an event logger ("Osilog"), a status monitor ("Osimon") and an MIB browser ("Osimic"). The MIB was restricted solely to the ISODE Transport Layer.

Some of the key features of this system are described below.

2.1. Managed System Internal Communication

The ISODE Transport protocol code that was used ran in user space. At any given moment, there could be several instances of the protocol active within several Unix processes. In order to de-couple management protocol operation from communications protocol operation and to collect management services in one place, it was decided to implement the management services in a single process - the "Systems Management Agent" (SMA).

The problem then arose of how and when management information should pass between the communications protocol processes and the SMA. In the event, a UDP socket was chosen, with all communication being initiated by the Transport protocol processes. The Transport protocol code was liberally seeded with entry points to the management code that would trigger the dumping of management information to the SMA at significant moments. Triggering events included T-Connects, T-Disconnects, a fixed data quantity transferred etc. This avoided the problem of having to interrupt Transport protocol operation in order to deal with asynchronous requests from the SMA - a procedure that was held likely to have unpredictable effects on Transport (and other) protocol operation. Unfortunately, this arrangement also made it difficult to pass information from the SMA to the protocol processes. Since our main interest at the time was in event-driven management we decided that one-way communication was acceptable.

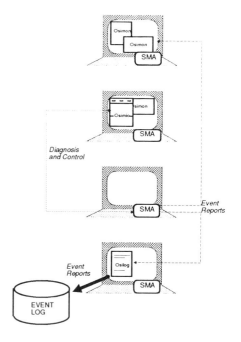

Figure 1. UCL "OSIMIS" version 1.

2.2. Event Report Control

At the time of the original implementation, the management of event reporting was not very complete in the standards. There was a notion of a "Defined Event" - an attribute of a Managed Object (MO) which corresponded to some event in the real world, a Transport Disconnect for example or a management event such as a threshold being exceeded. There was also a "Report Control" MO which was tied to a Defined Event and specified the information to go in the report, together with a list of recipients.

Having Report Control and Thresholds as MOs in their own right meant that, in principle, these could be created and deleted dynamically and that many-to-many relationships could exist between Defined Events, and Threshold and Report Control MOs. This suited the way in which we proposed to use event reports which were seen as a general purpose tool for driving loggers, status displays and diagnostic tools.

Although the main purpose of the system was to support human "managers", it also needed to cater for the requirements of our relatively sophisticated users who like to know what is happening on the department's systems. If performance seems poor, they like to be able to find out why by getting an up-to-date picture of activity. The "netstat" program on Berkeley Unix systems is an example of one which addresses this need. We wished to satisfy a similar requirement for ISODE. If facilities like these are to be event-driven, then it must be possible for several remote systems to receive event reports from a target system simultaneously and for the event reporting criteria (thresholds etc.) to be individually tailored. This requirement is in contrast to the more conventional one in which is only necessary to report events to a single remote sink.

2.3. The MIB Browser

The aim of the browser was to provide a tool that would be useful to system developers and maintainers who needed to focus in on some aspect of the operation of an OSI component and view its operation in detail. In OSI Management terms this means providing a detailed view of a single MO. The problem we faced was in the naming of transient MOs such as those representing Transport Connections. These are named by identifiers (usually integers) allocated by the parent system at "create" time. Their scope is purely local and there is no way that an external management process can know of these *a priori*. The browser provided a graphical interface to the tree of MOs on a remote system. From any object, it was possible to list the MO classes and Relative Distinguished Names of the subordinates. In the case of transient objects, the subordinates could be examined in turn until the required one was found.

3. REQUIREMENTS FOR THE SUCCESSOR SYSTEM

When the time came to update the pilot system we had several things in mind:

i) The standards had developed further, this was particularly noticeable in the area of event control, "discriminators" having been introduced in order to enhance this. These had some of the properties of our Report Control MOs but with much more sophisticated filtering.

ii) Our original CMIS/P implementation was not quite a full one. The M-CREATE, M-DELETE and M-ACTION services were omitted as was filtering, scoping and wildcarding. CMIS and CMIP had now stabilised at the final DIS stage and differed substantially from the versions we had used. An update was essential.

iii) It was clear, even from our small-scale pilot scheme, that OSI management systems were inclined to be large - too large for many small but high-performance network components. We were keen to investigate the use of proxy systems in these circumstances.

iv) Though it was clear that the MIB we managed would need to be expanded to include additional components, we could not say in advance precisely what these should be.

v) Our own MO definitions were "interim" at best, skeletal at worst! We needed to be able to adapt our MO definitions as the standards process proceeded, with a minimal disruption to existing code.

vi) The proprietary management facilities inherent in network components are rarely designed with OSI management in mind - nevertheless we wanted to incorporate these. Quite complex data mappings are sometimes needed in order to massage native management information into an OSI-like form. Further, proprietary management protocols vary in the message set they offer and in fundamentals such as which party initiates communication. We needed to construct systems which were flexible enough to allow new components to be incorporated no matter what proprietary facilities were on offer.

vii) Notwithstanding the variety of components alluded to above, many features of OSI management are common to all components. We wished to extract the common facilities into a few tightly-defined modules that could be used as a basis for building a variety of systems by a variety of people.

4. GENERIC MANAGED SYSTEM OVERVIEW

To date, in our design of a successor system, we have concentrated on the Managed System and how this can be structured with extensibility in mind. We have attempted to provide the generic features of an OSI Managed System together with a support framework which may be used by the implementors of MOs. The target environment is a Unix workstation requiring OSI management facilities and which may also act as a proxy for another machine. The result we call the "Generic Managed System" (GMS).

The internal structure of GMS is shown in Figure 2. To a large extent, the structure reflects the OSI model of a Managed System so that the major software interfaces correspond to those of the OSI model. For example, the external CMIS interface specified in [6] and the internal "object boundary" interface outlined in [7] are both represented by software interfaces. However,

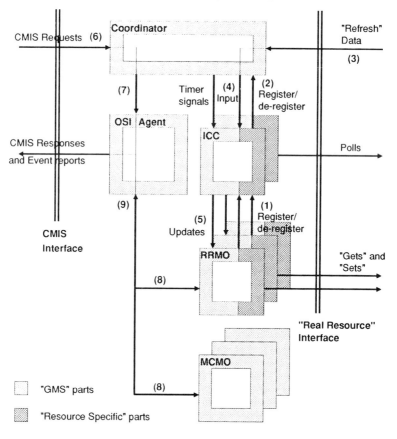

Figure 2. The UCL Generic Managed System

the OSI model was never intended to be an implementation model and significant divergencies have been made in order to arrive at a practical design.

The GMS is implemented as a single Unix process; the implementation language being C++. There are five major software components, each realised as a C++ object or set of objects:

i) Real Resource Managed Objects
ii) Management Control Managed Objects
iii) Internal Communication Control Objects
iv) The Coordinator.
v) The OSI Management Agent

These are described in the sections below.

5. INTERNAL STRUCTURE AND FUNCTIONALITY

5.1. Managed Objects

From the implementation point of view, two sorts of MO classes may be identified. The first are abstractions of "real resources" (Transport connections for example) which need to be managed; these we call Real Resource MOs (RRMO). The second relate to features of the management system itself and exist so as to allow the operation of the management system to be controlled via standard management operations; we call these Management Control MOs (MCMO). One example of a MCMO is an Event Report Forwarding Discriminator (ERFD). This contains information specifying which events should be reported and to where. Event reporting behaviour can be modified by changing this information through the use of CMIS M-SET operations.

5.1.1. Real Resource MOs

Implementations of "Real Resource" MO classes (RRMO) may be considered to have two parts (see Figure 2):

i) A part common to all RRMO classes. This is provided by the GMS and includes:

- A C++ object class for a generic MO. This has methods corresponding to the MO boundary interface plus some additional ones to assist with maintenance. Specialised subclasses of this may be derived as required.
- C++ object classes for commonly occurring attributes such as counters, gauges, thresholds and tidemarks. In fact, all the attribute types in [8] are supported in this way.
- A support environment to assist with and coordinate communication with the real resource. This environment is described more fully in Section 5.2

ii) A part specific to a particular RRMO. This must be tailored not only to the real resource type but also to the means the real resource uses to present management information. These "resource specific" parts are not provided by the GMS and must be supplied by the individual implementors of RRMO classes.

5.1.2. Management Control MOs

MCMO classes are common to all management systems no matter what real resources are being managed. Hence, GMS provides implementations of MCMO classes in their entirety.

At present, the only MCMO class provided is the ERFD one. ERFD objects may be created, destroyed and updated as a result of CMIS messages from a remote manager.

5.2. Communication with Real Resources

We now consider the ways in which management information may be obtained from real resources. Real resources may reside in the operating system's kernel, on communications boards, in user-space processes or even at remote systems which are managed via proxy management. The information they contain may be accessed by reading the kernel's virtual memory, talking to a device driver, communicating with another user-space process using an IPC mechanism or - in the case of proxy management - with a remote system using a communications protocol. In general, communication needs to be two-way as it should be possible to perform "intrusive" management by setting management information values in the real resources.

From the point of view of the GMS, information flow may be triggered:

i) asynchronously as a result of some activity on the real resource.

ii) by a timeout indicating that a real resource should be polled.

iii) as a result of a CMIS request from a remote manager process.

Each of these embodies a trade-off between the timeliness of the management information that the GMS can make available and its responsiveness. Used exclusively, iii) makes event reporting infeasible and implies the operation of a pure polling regime by the manager such as is favoured for SNMP.

5.2.1. Internal Communications Control

If generality is to be achieved, the GMS must support all the communications methods above, maintaining at the same time well-defined and uniform interfaces between the RRMOs and the rest of the system. It must be remembered that several RRMOs may be associated with a single real resource; ideally such a "family" of RRMOs should share a single communications path to the real resource. In order to achieve this, the notion of an Internal Communications Control (ICC) object is introduced. An ICC object coordinates the updates of a family of RRMOs that are realised in a similar fashion. ICC objects are repositories for information about the mode of communication to be employed, they initialise this communication and understand conventions such as the nature and structure of the messages exchanged, i.e. the protocol used. ICCs are created at system start-up time for each RRMO family that is to be managed.

As there are no real resources associated with MCMOs these do not have corresponding ICC objects.

5.2.2. The Coordinator

Given that several real resources are being managed and that messages are also being sent and received across the CMIS interface, it can be seen that some organisation is necessary to ensure that incoming messages are delivered to the correct objects and that no object can do a blocking read thus disabling the whole system. This is achieved by ensuring that all incoming messages are delivered first to a "Coordinator" object which then distributes them.

When the first RRMO in a family is created (either as a result of a CMIS **M-CREATE** request or of some activity on the real resource), its ICC interacts with the Coordinator in order to register an endpoint of communication (typically a Berkeley socket) to the real resource through which asynchronous messages may be expected.

An ICC may ask the Coordinator to call one of its methods at regular intervals so that it may poll the real resource. Alternatively, it may ask that whenever data becomes available at the communication endpoint a method should be called. Typically, this method will read the incoming data and pass this to the correct RRMO. The only case when RRMOs interact directly with the real resources is when they set management information.

5.3. The OSI Agent

The other major component of the OSI managed system model is the "OSI Agent". This too is represented by a C++ object and handles wild-card naming, scoping, filtering and (eventually) access control.

The OSI Agent services the messages it is handed by the Coordinator. These may be either association establishment/release requests or CMIS operation requests. In the latter case it first performs access control functions and then synchronises the potentially multiple replies according to the scoping and filtering parameters. In order to perform CMIS requests, it interacts with the selected MOs to get, set, etc. management information.

The OSI Agent may also receive event "notifications" from the RRMOs. According to the OSI management model, MOs issue notifications to the agent which then checks with the event filtering information in the ERFDs to determine whether the notification should result in a CMIS **M-EVENT-REPORT.** Unfortunately, if an implementation follows this model it results in a great deal of wasted processing in the case that no remote manager is interested in the event in question. There are also some logical problems; for example, the filtering expression may reference the MO that issued the notification but this may, by now, have been destroyed. Within the GMS, ERFD filtering information is applied in advance and notifications are only issued by RRMOs if it is known that **M-EVENT-REPORT**s will result. Although we have implemented the full generality of the filtering mechanism specified in the draft standards[9], we can see that certain filters will be extremely expensive to process. We expect that, in practice, only quite simple filtering expressions will be used.

6. METHODS

As an aid to understanding the information flow within the GMS we now summarise the methods applicable to the objects above. The C++ **obj.method** notation is used (somewhat loosely) to indicate a method **method** being applied to an object with id **obj**.

6.1. The OSI Agent

Three methods are used by the Coordinator to report incoming messages from the CMIS interface:

```
assoc_id = agent.cmis_connect (connect_ parameters)
agent.cmis_work (assoc_id)
agent.cmis_lose (assoc_id)
```

The first is used to notify a request for the establishment of a CMIS association, the second to indicate that a message has arrived on an existing association, (including a disconnect request), and the third to indicate that an existing association has been abnormally released.

A further method is used by the RRMOs to notify events:

```
agent.notify (my_class, my_name, event_type, event_report_info,
                            destination_address_list)
```

6.2. Managed Objects

The OSI agent interacts with the RRMOs and MCMOs to perform requested CMIS operations. The procedures and methods used are:

```
result = Create (parent_MO, rdn, init_info)

result = mo.Get (attribute_ids)
result = mo.Set (attribute_id/value pairs)
result = mo.Action (action_type, action information)
result = mo.Delete ()
```

Create is a static method which checks whether a create request is valid and, if it is, calls the constructor for the appropriate C++ class. The identity of the parent MO in the containment hierarchy and the Relative Distinguished Name (RDN) of the new MO are supplied as parameters. The four methods shown then embody the interface defined in [7]. The

Four other methods are provided to assist the OSI agent in locating the required MO:

```
target_mo = mo.find (name)
mo_list = mo.scope (scope_info)
answer = mo.filter (filter)
answer = mo.check_class (my_class)
```

The first searches the subtree below mo for a MO called name, the second returns a list of MOs which are "in scope" according to scope_info, the third applies a filter to a MO and returns a boolean value, the fourth checks that the class my_class is appropriate for the MO in question.

An ICC may need to create or delete transient RRMOs and to refresh the RRMOs it controls with new management information according to activity in the real resources. RRMOs are created by the ICC calling the constructor directly. The methods used by ICCs are:

```
mo.do_update (management information)
mo.destructor ()
```

6.3. The Coordinator

The ICCs tell the coordinator to register or de-register endpoints of communication to the real resource and are subsequently informed of activity on these. They also tell the coordinator to schedule and cancel periodic polling signals. The methods used are:

```
coord.register_cep (icc, cep_id)
coord.deregister_cep (icc, cep_id)

coord.schedule_poll (icc, interval, MOclass)
coord.cancel_poll (icc, MOclass)
```

6.4. ICCs

RRMOs in a family register and de-register themselves with their ICC as they are created and deleted:

```
icc.register_object (mo, mo_class, rdn)
icc.deregister_object (mo)
```

Note that the first RRMO to register triggers the establishment of communication to the real resource.

RRMOs may optionally request a special polling regime for a particular MO class and these requests are passed to the Coordinator via the ICC:

```
icc.schedule_poll (interval, MOclass)
icc.cancel_poll (MOclass)
```

A RRMO may need to talk directly to the relevant real resource - for example when the setting of some attribute value should rapidly be reflected in system operation. In this case, the RRMO must ask for its communication end-point from the ICC:

```
cep_id = icc.get_cep ()
```

The Coordinator needs to inform ICCs of the arrival of a message from a real resource or of the necessity to issue a poll. These two methods are used:

```
icc.do_cepread (cep_id)
icc.do_poll (MOclass)
```

The object class parameter in the poll method is only used when the polling takes place for a single MO class within a family rather than for the family as a whole.

7. THE GMS IN USE - AN EXAMPLE

The first RRMO to be implemented was a port of the ISODE TP0 management functions from the old OSIMIS system. An important test of the GMS structure was the ease with which this could be done.

In the case of ISO TP0, we identified two MO classes: the T-Entity class and the T-Connection class. There may be one and only one (static) instance of the T-Entity class. This summarises activity for all incarnations of the protocol and contains information such as the number of current and previous connections, the amount of data transferred, and error counters. Instances of the T-Connection class are transient - existing only during the lifetime of a connection. They contain information such as creation time, source and destination TSAP addresses and traffic counters. These are subordinate to the T-Entity instance in the MIB containment hierarchy.

ISODE TP0 is implemented as a set of library routines that are linked with the applications, this means that it runs in user space - the "real resource" in this case is effectively a Unix process. The IPC method used to communicate management information is a UDP socket - communication is only possible from the real resource to the GMS at present.

Implementation was straightforward. C++ object classes for the T-Entity and T-Connection RRMO classes were derived from the generic C++ MO class. Many of the additional attribute types required were instances of C++ classes already available within the GMS. An ICC object class was written (again, derived from the generic C++ one). This registers a socket bound to a well-known port with the Coordinator. An ICC method (**icc.do_cepread()**) is then called by the Coordinator each time a message arrives at the socket. Detailed operation is as follows (the numbers in the text are references to Figure 2).

When the T-Entity RRMO is created, (which happens either at initialisation time or through a CMIS M-CREATE request), it registers itself (**icc.register_object ()**) with the "ISODE" ICC object (1). If it is the first ISODE-related RRMO to register, the ISODE ICC object initialises the UDP socket, so that it may be contacted by active ISODE processes. It also registers this "communication endpoint" with the Coordinator (**coord.register_cep ()**) so that it will be notified in case of activity (2). After this, T-Connection RRMOs may be created and these too will be registered with the ISODE ICC.

When a message arrives at the UDP socket from an ISODE process (3), this is fielded by the Coordinator which recognises the socket as being managed by the ISODE ICC which it then informs (**icc.do_cepread ()**) (4). The ISODE ICC then passes the incoming information to the relevant RRMOs (5) (**mo.do_update ()**).

A CMIS M-GET request will also be fielded by the Coordinator (6) and, in this case, will be passed to the OSI Agent (7) (**agent.cmis_work ()**) which will perform the scoping and filtering tasks in order to select the relevant MOs. It then performs the requested operation (8) (**mo.xxx()**) on these.

As a result of processing information received from an ISODE process, an RRMO may determine that a threshold has been exceeded and that a notification may be required. If the RRMO determines that an ERFD filter is set to forward such a notification, it informs the OSI Agent (9) (**agent.notify ()**).

Finally, when an ISODE RRMO is deleted, (which happens, for example, when a T-Connection closes or as a result of a CMIS M-DELETE request, it deregisters itself with the ISODE

ICC object. If it was the last RRMO, the UDP socket is closed and the Coordinator is notified accordingly, so that future protocol instances will not talk to the agent.

8. CURRENT STATUS AND FUTURE PLANS

We have, at present, an implementation of the GMS as described. The only RRMOs supported so far are those related to the ISODE implementation of ISO TP0 described above.

The next step will be to add RRMOs related to further classes of real resource. One of the first of these will be the Berkeley Unix TCP/IP implementation. Although it might seem odd to manage a non-OSI protocol suite in this way, its extensive use in our environment means that it will exercise the GMS in a realistic way.

The OSI management work is being undertaken as part of the ESPRIT project "PROOF"[1]. This project is building a connection oriented Ethernet ISDN gateway, UCL is also building a connectionless version. Both of these gateways will be managed by using the GMS as a proxy; ISDN, X.25 and Ethernet RRMOs will be needed.

Another possibility we will look at is the implementation of RRMOs with SNMP back-ends. These would enable the GMS to operate as a proxy system for SNMP agents, enabling these to be managed by OSI manager processes.

The standards continue to develop, one [8] has been superceded [10] and no doubt others will suffer the same fate. We intend to modify the GMS as soon as these changes seem to be fairly stable.

Finally, we will investigate the use of a managed system specification language with a compiler to generate code for the generic parts of RRMOs, to define a MO schema and to provide initialisation data for static MOs. In this way, only the "back-end" code for interacting with the real resources will need to be hand written.

We do not claim that this system provides all the answers for OSI management. The GMS was built as an experimental tool with flexibility rather than performance and compactness in mind. However, we do hope that it will give us some practical insights into the problems of OSI management which will be valuable in the future, and that our experience will be useful to others facing similar problems.

---

1 The PROOF partners are: 3Net (UK - Prime Contractor), SNI (Germany), System Wizards (Italy), University College London (UK).

REFERENCES

1. Case, J.D., Fedor, M., Schoffstall, M.L., Davin, C. "Simple Network Management Protocol (SNMP)", (DARPA RFC 1098), April 1989

2. Sluman C., "A tutorial on OSI Management", "Computer Networks and ISDN Systems" vol 17 pp 270-278, 1989

3. Knight G. J., "The INCA Network Management System", Connexions - The Interoperability Report, Vol. 3, no. 3, pp. 27-32, March 1989

4. Knight G.J., Kirstein P.T., "Project INCA - An OSI Approach to Office Communications", Proceedings of the OSI87 Conference, pp 245-254, Online, London 1987.

5. Marshall T. Rose, "The ISO Development Environment User's Manual", U. Delaware, 1990

6. ISO/DIS 9595 (Final Text N3874), "Information Technology - Open Systems Interconnection - Common Management Information Service Definition", 2nd DP N 3070, January 1990

7. ISO/DP 10165-1 (Proposed DP Text), "Information Processing Systems - Open Systems Interconnection - Management Information Services - Structure of Management Information - Part 1: Management Information Model", June 1990

8. ISO/DP 10165-3 (Proposed DP Text N 3302), Information Processing Systems - Open Systems Interconnection - Management Information Services - Structure of Management Information - Part 3: Definitions of Management Attributes, January 1989

9. ISO/DP 10164-5 (N3845), Information Processing Systems - Open Systems Interconnection - Systems Management - Part 5: Event Management Function, September 1989

10. ISO/DIS 10165-2, Information Processing Systems - Open Systems Interconnection - Management Information Services - Structure of Management Information - Part 2: Definition of Management Information, June 1990

The Architecture of LANCE:
A Simple Network Management System

Michael A. Erlinger, PhD

ABSTRACT

The architecture of *LANCE*, a simple network management system for the management of heterogeneous Ethernet networks, is presented. *LANCE* uses a network management station communicating with various network agents via SNMP, the Simple Network Management Protocol. The architecture of *LANCE* is described at several levels: the network management agent, the network management station user interface, and the management station functional models. The *LANCE* architecture matches the current operational paradigm for network management, yet the implementation is flexible enough to change both in terms of overall architecture and in terms of individual management features.

1. Introduction

LANCE, the Local Area Network Control Executive, is a simple network management system for the management of heterogeneous Ethernet networks. The *LANCE* architecture is presented from several views — current network management taxonomy, user, and software system architecture. *LANCE* subscribes to the current operational approach to network management architecture — Internet-standard Network Management Framework [Rose 91], but is also flexible enough to change both in terms of overall architecture and in terms of individual management features. *LANCE* is considered to be a simple network management system for two related reasons. First, *LANCE* does not attempt to be *all things to all people*. Currently, the system provides specific support for one network management agent and generic support for other agents. *LANCE* was designed so that future management agents and management activities can be easily added. Network management is rapidly evolving, driven by changes in technology. Thus, network management tools must be capable of rapid change. Second, *LANCE* subscribes to the approach "We will learn what [Inter]Network Management is by doing it" [RFC 1052]. The developers of *LANCE* do not pretend to understand all the issues of network management. *LANCE* was designed to act as a core network

---
Author's Address: Micro Technology, 5065 East Hunter Ave., Anaheim, CA 92807, 714-870-2117 mike@mti.com
LANCE group includes: Mike Peterson, Robert Curley, Adrian McCarthy, Dave Stuit, and John Murakawa
LANCE, LANCE/Tap, and LANCE/NMS are trademarks of Micro Technology Inc.
SunOS is a trademark of Sun Microsystems Inc.
DEC, DECstation, VAXstation, Ultrix, and VMS are trademarks of Digital Equipment Corp.
The X Window System is a trademark of the Massachusetts Institute of Technology.
Motif is a trademark of The Open Software Foundation, Inc.
Novell and LANtern are registered trademarks of Novell, Inc.
Multinet is a trademark of TGV.
DV-Tools is a trademark of V.I. Corporation
UNIX is a registered trademark of AT&T Bell Laboratories.

management system that can be easily augmented in the future. In its current state, *LANCE* is a valuable tool in the morass of network management, but the inherent flux associated with network management will require constant growth within *LANCE*.

Standards were a guiding principle for *LANCE*. The development group decided early on to make all choices in the direction of existing standards and, in particular, standards that had been thoroughly tested by implementation. Another guiding principle was *don't reinvent the wheel* – the group searched out existing tools that could be integrated into the *LANCE* product. This approach allowed a much shorter development period.

The obvious goal of the *LANCE* team was to develop the best architecture that would allow for the creation of the premier network management station. But the important part of the *LANCE* project was the development of an architecture that provides user support for current technology and allows for rapid growth and change.

2. Network Management Architecture

In the general parlance of networking, various terms related to network management have evolved: network elements, network agents, network management station, and network management protocol. Network elements (NE) are devices that exist within the network, e.g., workstations, printers, bridges, routers, PC's, terminal servers, etc., while a network agent (NA) is a network element that has network management functionality.

A network management station (NMS) is a network element that has the functionality to communicate with network agents and that provides a user interface to the agents – allowing users to view network agent data and/or to control network agents and thus the network elements. A network management protocol is used to communicate between network management stations and network agents.

LANCE follows the general network management model and can be divided into a network management station, LANCE/NMS, and network agents that use SNMP (the Simple Network Management Protocol, [RFC 1157]). The LANCE/NMS can communicate with any NA that supports SNMP, but it has a special relationship with its primary network agent, the LANCE/Tap or *Tap*. A *LANCE* environment can have a number of *Tap*s and one or more LANCE/NMSs communicating with one or more of the *Tap*s – Figure 1.

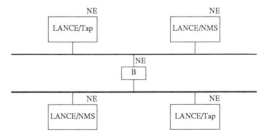

Figure 1. Example LANCE network management architecture

Normally such communication is via the monitored network, but each *Tap* also has a secondary communication channel, a serial port, that can be used when the network is unavailable or to avoid affecting the network being monitored.

3. LANCE Architecture

3.1. LANCE/NMS Architecture Overview

LANCE/NMS was developed to be ported easily to a number of different operating system and machine environments. In particular, it runs on Sun 4's (SunOS), DECstations (DEC RISC Ultrix), and VAXstations (VMS). The user interface was developed under the X Window System and Motif, i.e., the LANCE/NMS was designed to run in the X server/client environment. The LANCE/NMS user interface was created using the *object oriented* paradigm: users select an object and then perform operations associated with that object.

The LANCE/NMS supports three types of network management agents. First is the functionality to communicate with any SNMP agent supporting MIB I [RFC 1156] or MIB II [RFC 1158]). Second, the LANCE/NMS supports proprietary SNMP agents with proprietary MIBs. This support is of two flavors; the LANCE/NMS incorporates a directory of such known MIBs and the LANCE/NMS can support any correctly entered MIB OID. Finally, the LANCE/NMS provides detailed support for a particular network management agent, the LANCE/Tap – a derivative of the Novell LANtern [Novell 90].

3.2. LANCE/Tap Architecture

The LANCE/Tap is a stand-alone network element that acts as a remote network management agent [Novell 90]. Once installed it gathers network traffic information by listening to all packets on the network. The *Tap* uses a private enterprise management information base (MIB) as its virtual storage area for its data objects and communicates to an NMS via SNMP.

By monitoring all traffic on its network segment the *Tap* gathers statistics on general network events, individual station activity, station pair activity, and monitors the segment for common network errors. The general network statistics include total events, total good packets, total errors, local collisions, collision fragments, total bytes, and a packet size distribution. These general statistics are also logged within the *Tap* over a number of discrete equal time intervals to provide a history of network utilization.

Statistics (total packets transmitted, total bytes transmitted, total errors transmitted, protocols used, and date and time of first transmission and of most recent transmission) are kept on traffic into, and out of, each station on the segment, and for traffic between each source/destination pair of stations on the segment. The SNMP trap – PDU (Protocol Data Unit) mechanism [RFC 1157] is used by the *Tap* to asynchronously notify specified LANCE/NMSs (each *Tap* has a list of such stations) of significant network events — e.g., *Tap* power-on or reset and failure of SNMP message to pass the authentication procedure. The *Tap* also uses SNMP traps to support a set of segment threshold (high and low) exceeded alarms, e.g., average percent utilization and broadcast packets as percentage of total good packets.

4. LANCE/NMS Structural Architecture

At the network layer LANCE uses TCP and UDP riding on IP to provide the networking protocol (using the native system implementation for UNIX and Multinet [TGV] in the VMS world) – Figure 2. A commercial version of SNMP [PSI] was chosen as the network management protocol (support issues outweighed the free cost of the public domain versions). Because SNMP is well defined [RFC 1157], there were no protocol problems between the SNMP code in the *Tap* and that used within the LANCE/NMS. SNMP provides only the means (protocol) to extract information from the *Tap* (management mechanism [Rose 91]), the choice of data objects, the period over which to extract such objects, etc., is a function of the *LANCE* Utilities (management policy [Rose 91]). The functional architecture of these programs is delineated below.

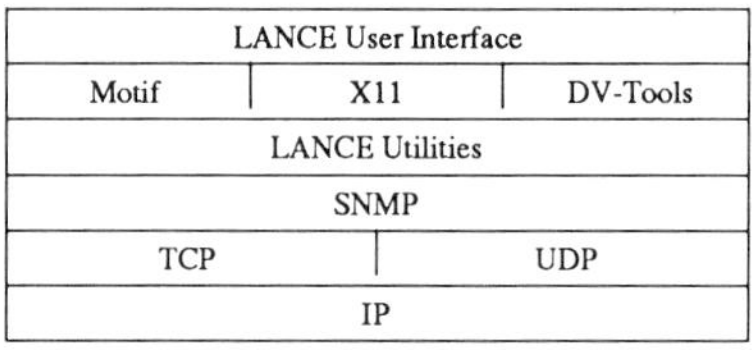

LANCE User Interface		
Motif	X11	DV-Tools
LANCE Utilities		
SNMP		
TCP		UDP
IP		

Figure 2. LANCE structure

The topmost structure is the LANCE/NMS user interface [MT 90]. By means of the *LANCE* User Interface routines, the LANCE/NMS user manages his network. The user interface uses the X Window System and Motif [Motif]. X11 was an obvious choice, especially given the desire to follow standards and to facilitate portability to other systems. Motif, on the other hand, was a choice of convenience. It was a given that LANCE/NMS would run in the VMS arena and Motif development had progressed further than other GUIs when the user interface decisions were made. DV-Tools provides a library of graphics routines that are used to display various graphs created from data acquired from the *Tap*.

5. User Interface Architecture

The LANCE/NMS user interface was developed knowing "It is not clear that we know what data should be collected, how to analyze it when we get it, or how to structure our collection systems." [RFC 1147]

The *LANCE* interface for the network administrator (user interface) had the following basic requirements:

- It must be simple and easy for new users to operate, with little or no need to refer to documentation.

- It must provide support for more advanced users to navigate easily and quickly to the desired tool.

- Whenever possible, a graphical interface should be provided instead of a textual interface. Text should be used as necessary to complement graphical elements.

- Every window should offer help to the user; help information should pertain to the context of the current window.

- For any given function the number of mouse and keyboard operations should be kept to a minimum.

- Whenever possible, conform to the conventions specified by the *OSF/Motif Style Guide* [OSF 90].

LANCE presents the user with a two-level graphical view of the network topology. The top level is an overview of the network (network view), showing all segments and intervening

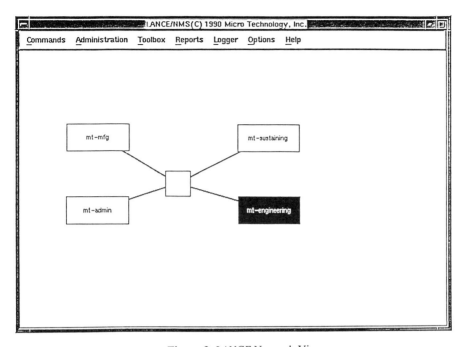

Figure 3. *LANCE* Network View

bridges, routers, and gateways – Figure 3. The lower-level view (segment view) shows a particular segment in detail – Figure 4.

The general paradigm for performing functions is to select first a network object from the graphical display (network or segment view) and then the tool to perform the desired function on the selected object. (This provides a consistent interface and conforms to the OSF/Motif conventions.) The "Toolbox" menu option at each level contains the tools appropriate to the context of the current view.

Each view and each function occupies a separate window, which is placed and sized "reasonably" when created, so intervention by the user usually is not required. However, the user may change the window positioning and sizing if so desired. Although the user interface is intended to be easy to use, help is available in every window to explain how to use *LANCE*'s functions and interpret its displays.

5.1. Network View

The top view is intended to give an overview of the network – Figure 3. At this level, network segments and intervening bridges, routers, and gateways are visible. Initial placement of the graphical objects on this screen is done by the user. The menu strip at this level contains six options:

| Commands | Administration | Toolbox | Reports | Logger | Options | Help |

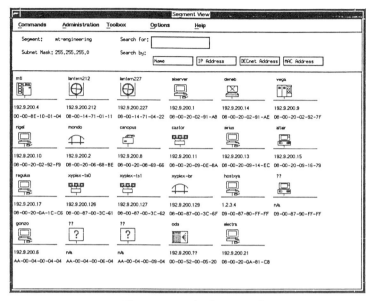

Figure 4. *LANCE* Segment View

Each of the options is a pull-down menu. The selectable objects at this level are the segments, so the tools available via the network view are segment-oriented. When a LANCE/Tap is attached to each segment, selecting a segment implicitly selects the *Tap* on that segment, though the user is not burdened with these details.

For example, the "Toolbox" option is a pull-down menu, containing the following options:

Toolbox
Zoom to Segment
Set Segment Alarms
Display Traffic Statistics
Profile Segment Traffic
Start MIBulator
Ping IP host

Each option will open a new window. "Zoom to Segment" takes the user to the segment view level for the selected segment and is described below. The threshold alarms supported by the LANCE/Tap are managed via the "Set Segment Alarms" window. The "Display Traffic Statistics" window allows traffic monitoring on the selected network segment. Network traffic statistics are collected by the *Tap* in a set of cumulative counters which may be read by LANCE/NMS. In addition to the counters, a *Tap* maintains a log of these statistics over 100 equal time intervals. Thus a *Tap* can provide the LANCE/NMS with a picture of network traffic over some recent time period. Whether current or historic statistics are chosen, the data is displayed in a standard graph window. Strip charts or bar charts may be displayed – Figure 5. The "Profile Segment Traffic" option generates a network traffic profile for the specified segment and identifies the stations that are consuming the most network capacity.

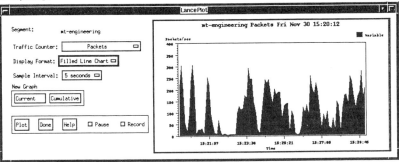

Figure 5. Segment Performance

5.2. Segment View

The "segment view" window (Figure 4) gives a graphical view of a particular segment, including all hosts, *Tap*, and other devices on the segment. The segment view includes a menu strip with the following options, each of which is a pull-down menu.

| Commands | Administration | Toolbox | Options |

The "Toolbox" at the segment view contains tools for the types of objects visible at the current level — the segment and its stations. The segment functions are the same as those at the top view, and apply to the current segment. The station functions apply to selected network elements, following the standard object/function selection paradigm. As before, the toolbox option is a pull-down menu. It contains the following options:

Toolbox
Set Segment Alarms
Display Traffic Statistics
Open Notebook
Monitor station
Monitor station–station
Start MIBulator
Ping IP host

Each tool in this toolbox opens a new window. Note that these tools are divided into three groups. Tools in the first group act on the segment. In fact, they are the same as the tools available in the top-level toolbox, except that these act only on the current segment. The tools in the second group act on stations and are described below. The third group are special cases and are explained further below.

The "Open Notebook" function manages a notebook entry for each network element visible in the segment map and also provides access to station-specific alarm capabilities. The entries have the same structure as that provided at the Network View level. The "Monitor station" window allows monitoring of traffic in and out of a selected station. Graphs in the usual format — the user's choice of strip or bar chart — are displayed for total packets, total bytes, and error packets in and out. The display also shows basic host information, plus the times of the first and last transmissions by the host and the last transmission to the host. A list of recognized protocols is displayed, with an indication of all protocols seen in a packet sent to or from this host. The statistics for the station monitor display are obtained from the *Tap*'s station table.

The "Monitor station-station" tool works in much the same way as the station monitor, except that in this case two stations must be selected — a source and a destination – Figure 6. Only traffic with the selected source and destination is monitored with the display showing total packets, total bytes, and error packets in each direction between the two stations.

The "Start MIBulator" tool operates on any node on the current segment that has an SNMP agent. Its function is to monitor a MIB variable selected by the user from a scrolling list of variables in standard MIBs, experimental MIBs, and any known private MIB. This tool is intended to allow the network administrator who is familiar with the MIBs to perform any special monitoring functions not otherwise supported by *LANCE*. The user may obtain either a single reading of the selected variable or a graph (with the usual options) of values obtained at a specified interval. The "Ping IP host" tool allows the user to ping any host that supports TCP/IP.

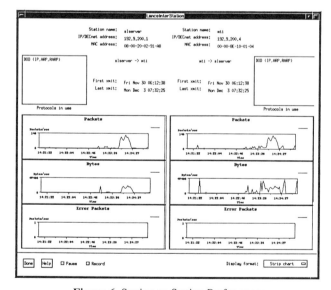

Figure 6. Station-to-Station Performance

6. LANCE/NMS Functional Architecture

The software architecture of *LANCE* is very simple; it is diagramed in Figure 7. A user interface, *lance*, provides the primary means of running the *LANCE* family of applications. Each of the applications was written as a separate utility that can be executed independently from *lance*. Thus, users could ignore the *lance* interface and directly execute the various utilities. Figure 7 illustrates relationships between some of those utilities — *lancestn*, *lanceinterstn*, *lancelogd*, and *snmptrapd*.

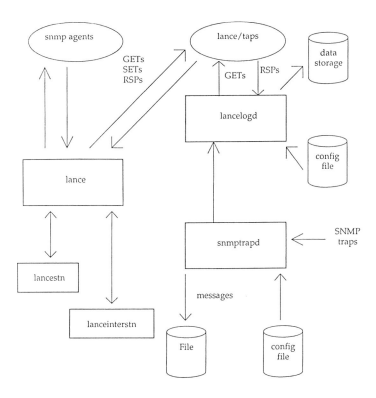

Figure 7. *LANCE* Functional Architecture

snmptrapd is a trap daemon designed to sit and await SNMP—based traps directed at the host on which it is running. A configuration file is used to indict whether a trap should be received and what action to take upon trap reception. In *LANCE* traps are logged to their own file and passed to *lancelogd*, which also saves the traps for later incorporation into the report program.

The *lancelogd* program is the *LANCE* software that polls the *Tap*s at fixed intervals, generating raw data files for use by the *report* program. The polling configuration is read from *configfile* which contains entries indicating the probes to be polled and the interval at which to poll them. The polling configuration changes infrequently. If the configuration file is modified, *lancelogd* will note that the modification time of the file has changed, and will reread the file and reconfigure itself.

The control flow of *lancelogd* is straightforward. After reading the configuration file and setting up the communication ports and the current log file, it aligns its master clock to the hour.

---

snmptrapd is part of the PSI suite of SNMP tools.

On the first hour following its startup, *lancelogd* polls all of the probes and waits in a timing loop until the next minute (up to 60 seconds). Data from *snmptrapd* can interrupt this timing loop. If such data is available, *lancelogd* logs it and resumes its timing loop for the remainder of the 60-second basic timing cycle. At the end of a cycle, a determination is made of which, if any, *Taps* are to be polled. If there are none, the 60-second timing cycle is repeated. Otherwise, the polling cycle is entered. After the polling cycle is finished, the captured data (or an entry noting a failed poll) is written to the current log file, and the main 60-second cycle, or what remains of it after data polling and processing, is re-entered. *lancelogd* closes the present day's log file and starts a new log file each day at 12 midnight.

lancestn monitors the traffic to or from a selected Ethernet station. The program needs 3 parameters: the agent to be queried, the MAC address of the station to be monitored, and the interval between queries. Two additional parameters, the IP address and the name, can be provided and are used for display purposes. The program creates an X window and displays the station's traffic, protocols in use, and times of transmission and reception.

lanceinterstn monitor traffic between two Ethernet stations. The program creates an X window that displays the stations' traffic, protocols in use, and times of transmissions (Figure 6). *lanceinterstn* accepts 8 parameters: agent to be queried, interval of the queries, name, IP address, and MAC address of each of the stations.

The approach taken to the *LANCE* functional architecture allows for rapid change in *LANCE* functionality. New programs can be added and modifications quickly implemented. In fact, the MIBulator tool has been rewritten recently to take advantage of an X11 tree widget.

7. Future LANCE

Network management is in its infant stages. The current tools, while providing service, are helping to determine the scope and functionality of network management. *LANCE* is no different than other such tools: each *LANCE* feature motivates the users and developers to define additional features. For example, when the hierarchical approach to *LANCE* use (i.e., choose a segment, zoom to that segment, then choose a node) was presented a potential user replied: *"LANCE assumes that I know where the network nodes reside – an ideal situation. In a network of 10 segments and 4,000 nodes, the only node whose location I am sure of is the one that I am currently using – users connect and disconnect stations at will. Before choosing any network segment, LANCE needs to provide a feature where I can search its databases for a particular node..."* Thus a top level function to search (by name, IP address, or MAC address) all the network configuration files is being added.

In analyzing *LANCE* various future development areas have been identified – Figure 8. The current LANCE/NMS together with the LANCE/Tap is viewed as the *core*. Expansion in terms of core functionality (Figure 8 – Core enhancement) is an obvious direction. The *Tap* can be augmented in a number of ways. First, it currently provides only protocol recognition. Expansion should include the concept of percentages based on filters, i.e., filtering particular protocols or packet types. Second, the *Tap*'s MIB could be modified to support monitoring of token ring networks. This would expand the purview of *LANCE* and could be accomplished by creating a separate MIB subtree.

Currently *LANCE* provides special support for a particular network management agent, the LANCE/Tap. Additional specialized support must be provided for other network agents, e.g., bridges and routers. The obvious direction is to provide such support for devices which support SNMP. Because most such devices rely on a private MIB, such development requires a close relationship with device manufacturers. With the current number of private MIB positions well over 100, *LANCE* core development will be selective in providing specialized agent support. A parallel core development will involve expanded MIB capability in LANCE/NMS. Given a particular MIB variable (generic or private) this capability will provide the user with an option list of possible display formats.

The LANCE/NMS could also be modified by the installation of an underlying database system (Figure 8 – Core database). Currently, because of cost and performance considerations, the LANCE/NMS uses flat files for its internal configuration files, records, and statistical data. A database system will provide functionality that would be difficult to implement in the current flat file structures.

Applications provide another direction for *LANCE* development. Each area of the OSI network management architecture (performance, configuration management, security, faults, and accounting) delineates an area of possible applications. Such applications would probably exist outside of the LANCE/NMS, but would use the data captured and stored by the LANCE/NMS.

There exist, or are being developed, several large network management systems: IBM's NetView, DEC's EMA, AT&T's Acumaster, Sun's SunNet Manager, etc. *LANCE* is not meant to compete with these products. Rather (Figure 8 – Integration), *LANCE* needs to be integrated into, or exist in parallel, with such large systems. For example, DEC's EMA claims to provide an interface for other network management systems. *LANCE* development needs to include such interface development.

Finally, *LANCE* must be poised to join the eventual implementation of the OSI network management architecture (Figure 8 – ISO). It is still unclear how *LANCE* would integrate into such an environment – move to CMIP or use SNMP as an OSI application.

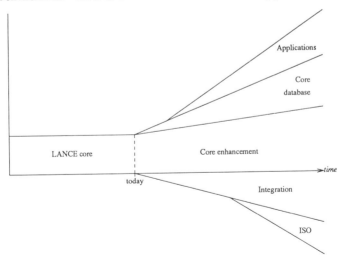

Figure 8. LANCE enhancement strategy

8. Summary

The architecture of *LANCE*, a simple network management system for the management of heterogeneous Ethernet networks has been presented. The *LANCE* architecture has been shown to match the current operational paradigm for network management, but the important part of the *LANCE* architecture is its flexibility in terms of overall management tool architecture and in terms of individual management features.

Bibliography

[DV]   Data Views Reference Manual, V.I. Corporation, 1988.

[Klerer]   Klerer, S. Mark, "The OSI Management Architecture: An Overview" *IEEE Network*, Vol. 2, No. 2 March 1988, pg. 20-29.

[Motif]   OSF/Motif Programmer's Guide, Open Software Foundation, 1989.

[MT 90]   *LANCE* Network Management System User Guide, Micro Technology, 740012-001, Rev. A, November 30, 1990.

[Novell 90] Novell, Inc., LANtern Network Monitor Technical Reference, June, 1990.

[OSF 90   The Open Software Foundation, OSF/Motif User's Guide, Prentice Hall, 1990.

[PSI]   SNMP, Network Management Station (NMS) and Agent Implementation, Performance Systems International, Inc., 1989.

[RFC 1109]   V. Cerf, Report of the Second Ad Hoc Network Management Review Group, RFC 1109, August 1989.

[RFC 1147]   R. Stine, Editor, FYI on a Network Management Tool Catalog: Tools for Monitoring and Debugging TCP/IP Internets and Interconnected Devices, RFC 1147, April 1990.

[RFC 1052]   V. Cerf, IAB Recommendations for the Development of Internet Network Management Standards, RFC 1052, April 1988.

[RFC 1155]   M. Rose and K. McCloghrie, Structure and Identification of Management Information for TCP/IP-based Internets, RFC 1155, May 1990.

[RFC 1156]   K. McCloghrie and M. Rose, Management Information Base for Network Management of TCP/IP-based Internets, RFC 1156, May 1990.

[RFC 1157]   J. Case, M. Fedor, M. Schoffstall, and J. Davin, A Simple Network Management Protocol (SNMP), RFC 1157, May 1990.

[RFC 1158]   M. Rose, Management Information Base for Network Management of TCP/IP-based Internets: MIB-II, RFC 1158, May 1990.

[Rose 91]   M. Rose, The Simple Book, Prentice Hall, 1991.

[TGV]   Multinet Programmer's Reference Manual, TGV, 1989.

[X11]   X11, Massachusetts Institute of Technology.

IV

IMPLEMENTATION

AND CASE STUDIES

B
Case Studies

HOW WELL DO SNMP AND CMOT MEET IP ROUTER MANAGEMENT NEEDS?

Sharad Sanghi
Asheem Chandna
Gregory Wetzel
Soumitra Sengupta

Abstract

This paper provides a comparison of the extent to which SNMP and CMOT (the two TCP/IP network management protocols) meet IP router management needs. We have classified IP router management needs into three broad categories: functional, operational, and implementation. We have further broken the functional needs into seven categories: fault, configuration, performance, access control, accounting, network planning, and data management. Wherever applicable, we have used the ISO Management Framework to provide a comprehensive statement of management needs. To what extent SNMP and CMOT meet these needs is illustrated by means of a case study of the Columbia University/Columbia Presbyterian Medical Center network.

Keywords: *TCP/IP network management protocols, IP routers, SNMP, CMOT, CMIP/S, network management requirements, fault, configuration, accounting, performance, access control, network planning, data management.*

Sharad Sanghi works in the UNMA Department at AT&T Bell Laboratories in Holmdel, NJ. He is also a Graduate Research Assistant for NetMATE, a network management project at Columbia University. He holds a B.Tech. in Electrical Engineering from the Indian Institute of Technology, Bombay and an M.S. in Electrical Engineering from Columbia University.

Asheem Chandna works at AT&T Bell Laboratories in Middletown, NJ. His work responsibilities include CMIP and SNMP related planning and systems engineering, for AT&T's ACCUMASTER Integrator network management product. He holds B.S. and M.S. degrees in Electrical and Computer Engineering, respectively, from Case Western Reserve University.

Gregory Wetzel works at AT&T Bell Laboratories in Naperville, IL. He is involved in planning and co-ordinating SNMP based network management for the TCP/IP network at AT&T Bell Labs.

Soumitra Sengupta is a Research Scientist at the Computer Science Department and at the Center for Medical Informatics at Columbia University. He holds a Ph.D in Computer Science from Stony Brook University.

1. INTRODUCTION

The last couple of years have witnessed a large upward trend in the deployment of IP routers. In many cases, these routers are being used as the fundamental building blocks for the design and integration of enterprise-wide data networks. A recent Infonetics Research Institute study surveyed 150 information managers from Fortune 1000 companies [1]. The study showed that the annual cumulative growth rate for IP routers will continue to be over 50% through 1993.

Given the above facts it is clear that network management of IP routers is a topic that will gain considerable importance in the coming future. This paper provides a comparison of the extent to which SNMP and CMOT meet IP router management needs. It is presumed that the reader is familiar with SNMP and CMOT fundamentals. (Readers are referred to [9] for an overview of SNMP and CMOT.)

The basis for comparison is as follows:

• *IP Router Management Needs:* We have used the ISO Basic Reference Model Management Framework document [2] to provide a comprehensive statement of management needs. Using Columbia University's TCP/IP network management as a case study, we have then further interpreted these needs in the context of IP router management.

• *Functionality provided by SNMP:* Most of the current IP router SNMP agent implementations are centered around a core set of SNMP specifications: the SNMP protocol [3], the rules for the structure of management information [4], and the objects in the common management information base (MIB-II) [5]. We have used the functionality provided by these three standards as a basis for comparison. (Except for a couple of *cisco* specific examples, we have not included the additional functionality provided by the technology specific and proprietary MIBs implemented by several vendor SNMP agents.)

• *Functionality provided by CMOT:* CMOT (CMIS over TCP/IP) is an implementors' agreement for providing ISO CMIS [14] functionality over TCP/IP networks. To date, CMOT agents have been implemented on IP routers by fewer than five vendors. These implementations have been based on [6], which covers the CMOT protocol implementation agreements, and also provides an ISO translation of the original common SNMP management information base (MIB-I). Since then, the more recent SNMP common management information base (MIB-II) has also been translated into ISO format [7]. We have used the functionality provided by the latter MIB translation, along with the CMOT protocol implementation agreements as a basis for comparison.

It is worth noting that the SNMP comparisons are based on early but real experiences in utilizing SNMP to manage IP routers. In contrast, due to a lack of deployment, the comparison to CMOT is based on theory.

The rest of this paper is organized as follows: Section Two provides an overview of the network configuration at Colmbia University and Columbia Presbyterian Medical Center. Sections Three and Four discuss functional and operational requirements, respectively. Section Five provides a brief discussion of vendor implementation requirements. Section Six provides a summary of conclusions.

2. COLUMBIA UNIVERSITY/COLUMBIA PRESBYTERIAN MEDICAL CENTER NETWORK

The Columbia University/Columbia Presbyterian Medical Center (CU/CPMC) has a geographically dispersed network that contains equipment from many different vendors [Figure 1]. Two independent organizations, Columbia University and the Presbyterian Hospital, jointly use and maintain this network. The network connects the hospital computing center at Tarrytown, New York and three major locations in New York City (Manhattan) - Allen Pavilion Hospital at 210 St, CPMC Main Campus at 168 St, and CU Main Campus at 116 St. All these locations use *cisco* multi-protocol routers. The following is a brief overview of network configurations at each of these locations.

Computing Center, Tarrytown, NY: The network at this location has one router (CR1), with two serial interfaces (to Allen and CPMC Main) and one token ring (local) interface. Protocols used in this router include IBM token ring source routing, Novell IPX and AppleTalk routing in addition to IP routing. The local token ring connects to two IBM mainframes (through IP-SNA and Token Ring-SNA gateways) and a small number of PCs (<50). Primary IP traffic is the access to mainframes from remote locations; most SNA traffic from PCs at other hospital locations, however, are handled by SNA controllers and dedicated remote links that are separate from the router and its serial links.

Allen Pavilion, Manhattan, 210 St, NY: The network at this location has one router (CR2), with two serial (to Computing Center and CPMC Main) and one token ring (local) interfaces. Additional protocols used in this router include token ring source routing, IPX routing, and AppleTalk routing. The local token ring (comprising of 6 physical rings and IBM source routing bridges) connects to over 100 PCs and workstations. Primary IP, IPX and AppleTalk traffic through the router is the access to hosts and servers at CPMC and the CU Main Campus. There is some SNA traffic (on source routing) through the router for access to IBM AS400 at the CPMC Main Campus.

CPMC Main Campus, Manhattan, 168 St, NY: The network at this location has one router (CR3) with two serial interfaces (to Computing Center, Allen Pavilion) and one token ring interface, another router (CR4) with one serial interface (to CU main campus), one token ring interface (local) and one ethernet (local) interface. These routers support IPX and AppleTalk routing among all links other than the CU Main Campus link, DECNet routing between local ethernet and CU Main Campus, token ring source routing among local token ring and links to Computing Center and Allen Pavilion. In addition to the cisco routers, there are multiple Kinetics AppleTalk gateways connecting Apple LocalTalk networks to the ethernet. These gateways support both AppleTalk and IP routing.

The local token ring (comprising of over 25 physical rings and IBM source routing bridges) connects more than 500 PCs. The local ethernet (comprising of 4 segments and DEC bridges) connects more than 30 hosts and PCs. The LocalTalk networks connect over 50 Apple PCs. The cisco router is primarily used for IP and IPX communication. Major SNA traffic to the computing center travels through dedicated SNA controllers and gateways on token ring and ethernet, and hence, bypasses the router. The token ring source routing support in the routers is used only as redundant backup access to the mainframes. The majority of IP traffic to the CU Main Campus is related to electronic mail, research, and CU administrative activities, as compared with other traffic that is related to hospital patient care and administration.

<u>CU Main Campus, Manhattan 116 St, NY</u>: The network at this location has one router (CR5) with one serial interface (to CPMC Main Campus) and one ethernet (local) interface on the CU backbone network. Additionally, there are routers (mostly ethernet to ethernet) to individual departments such as Computer Science, Philosophy, Chemistry, etc., and two routers to the NYSERNET backbone. IP packets form the primary traffic through these routers, and are related to mail and research.

The NetMATE project represents a real-world management example of a heterogeneous, geographically dispersed network in a multi-vendor environment [8]. The NetMATE system uses SNMP to manage the IP routers in the CPMC network.

3. FUNCTIONAL REQUIREMENTS

Network management functionality has been broadly classified into six major areas. Requirements pertaining to each of these functional areas are discussed below. It must be mentioned however, that the boundaries of these areas are not well defined. They overlap and are intended to complement and supplement each other.

3.1 Fault Management
Fault management encompasses fault detection, fault isolation, fault recovery and fault prevention. It includes the ability to perform the following functions [2]:

a. Maintain and examine error logs.
Both CMOT and SNMP provide information which can be used to populate error logs. CMOT provides more detailed object-specific error information than does SNMP. A CMOT message provides information about the problem type. It also provides a textual description of problem data. In addition, CMOT can give different levels of fault severity. In general, however, the maintenance and examination of error logs are more of an application issue than a protocol issue.

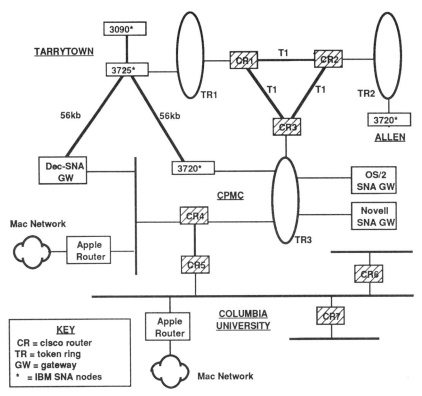

Fig. 1   CPMC/CU NETWORK CONFIGURATION

b. Generate error detection notifications based on thresholds and object status.
The standard SNMP MIB provides six pre-defined threshold/object status - triggered events at the agent level. This requirement can also be met locally, with a manager application. CMOT provides the capability to generate threshold and object status based events at both the agent and manager levels.

c. Accept and act upon any error detection notifications.
The ability to accept and act upon error detection notification is more of a management application issue than a protocol issue. Both protocols provide supporting information for the use of an application. CMOT, however, presents a more comprehensive view of network error conditions as described in (a) above.

d. Trace and identify faults and carry out sequences of diagnostic tests.
Currently, fault diagnosis and testing are within the domain of applications. For instance, Performance Systems International provides an *snmproute* application to debug routing problems. Also, at CU/CPMC, protocol-independent applications, such as *ping* (a connectivity test), are currently in widespread use.

e. Correct faults.
Both protocols provide information which can be used to correct faults. CMOT provides this information in a more timely fashion since it can send threshold-based alarms, whereas SNMP must wait for the next poll. (The SNMP standard only provides a limited number of alarms.) A drawback of CMOT however, is that it depends on the establishment and maintenance of associations between the management station and the managed object. During times of network trouble (e.g., periods of high errors on serial lines, routing problems, etc.), the underlying transport level associations are prone to disconnection due to time-out. For instance, if only one of the 25 datagrams sent can get through, then retransmission for reliable transport (required by associations) will take a long time and may take longer than the association may tolerate. Since SNMP uses only connection-less communication, theoretically it has a much higher probability of getting through. (Simply send the same datagram over and over again until a response is received.) Thus, when router management is needed the most (during times of trouble), the CMOT paradigm may be less effective than SNMP. In both cases, network management requests and replies are carried over the managed network, thus making management in the face of network trouble dependent on the very network having the trouble (a drawback that is inherent in both SNMP and CMOT).

f. Prevent faults by acting upon alerts triggered by thresholds on critical parameters.
Automation and customization features are not supported directly by the protocol. However, some vendor-specific implementations provide the ability to users to specify actions that are automatically executed to prevent faults.

To detect whether or not a fault has occurred, one would have to know the operational and administrative status of the link and/or the interface, for example, whether the link and/or the interface is up or down. Once a fault has been detected, the fault needs to be isolated either through diagnostic tests (as mentioned earlier CPMC uses *ping,* a connectivity diagnostic which is independent of SNMP and CMOT), or by polling for and retrieving information about certain critical parameters before corrective action can be taken. CMOT has the advantage of a more extensive message set than does SNMP, and hence may be able to provide more timely notification.

Fault recovery is usually done off-line, although a fault isolation algorithm may trigger an automatic fault restoration service. An example of fault recovery would be the rebooting of a system which has crashed. Another aspect of fault management is the prevention of a fault. When certain critical parameters reach or exceed their thresholds, the management station must be alerted so that the fault can be prevented. At CPMC for instance, link-down transmission error variables are checked for the indication of an impending total failure; which is prevented by doing performance management on or replacing the marginal components. It may be noted here that the ability to do real-time trend analysis or history analysis would definitely help the process of fault prevention.

At CPMC there is a need to retrieve and correlate many dynamic SNMP objects from managed systems. Sometimes there is a need to retrieve a portion of the MIB (for instance, an entire routing table or a portion of it) when a fault occurs. The only way a correct analyses can be done is if all the values are obtained at the same time so that interdependencies can be observed. Unlike CMOT, SNMP does not support linked replies. Therefore the crucial interval between the replies causes inconsistencies in data thereby preventing analyses. The inability of SNMP to efficiently retrieve large amounts of data from the MIB, is in our opinion, an inherent shortcoming of the protocol. (When compared to the scoping and filtering mechanism of CMOT, the *GetNext* feature provided by SNMP is an inefficient way of performing MIB browsing [9].)

3.2 Configuration Management
Configuration management identifies, exercises control over, collects data from and provides data to open systems for the purpose of preparing for initializing, starting, providing for the continuous operation of, and terminating interconnection services [2]. Configuration management includes functions to [2]:

a. Set the parameters that control the routine operation of the open system.
Ideally, one would like to be able to set all the parameters in the MIB. But parameter setting is only partially supported by SNMP. For example, the choice of the routing protocol is not accessible via the MIB. This is also true in the case of CMOT. The lack of authentication and authorization in both SNMP and CMOT restricts the use of the *set* feature in general.

b. Associate names with managed objects and sets of managed objects.
Names are associated with managed objects and sets of managed objects in both SNMP and CMOT. CMOT, however, provides a more flexible, powerful and network protocol-independent naming mechanism. The advantage of the CMOT naming approach is that it is independent of the networking architecture and address. For example using the same naming scheme, the fact that one is

operating using CMOT over TCP/IP (with an IP address) or in future using CMIP over OSI (with an OSI address) or over the data link layer (with a MAC level address) is hidden. In SNMP however, the application is dependent on IP addresses since SNMP was designed specifically to manage TCP/IP networks.

c. Initialize and close down managed objects.
Managed objects are initialized by setting them to a default set of values and closed down by setting them to an off-line state in both SNMP and CMOT. CMOT also provides extensive support for the creation (during initialization) and deletion (during close down) of objects.

d. Collect information on demand about the current condition of the open system.
Both SNMP and CMOT can collect information on demand by polling for variables about the current condition of the open system.

e. Obtain notifications of significant changes in the condition of the open system.
Currently, SNMP supports six pre-defined *traps* (alarms). CMOT supports a more comprehensive set of alarms via event messages. The existing CMOT specification [6] covers the six traps present in SNMP. In addition, the management implementation agreements in the emerging CMOT document [10], based on the OSI Implementors Workshop - Network Management Special Interests Group (OIW-NMSIG) implementation agreements, support object-sieves which allow users to specify filters for alarm conditions thereby allowing them to be highly selective about the nature of alarms received, based on their needs.

f. Change the configuration of the open system.
Both SNMP and CMOT have the capability to change the configuration of a managed system via parameter setting, but due to security reasons (lack of an authentication/authorization mechanism), these parameters are manually updated (for instance, via remote login to the router).

3.3 Performance Management

Performance management enables the behavior of resources and the effectiveness of communication activities to be evaluated [2]. It includes functions to [2]:

a. Gather statistical information.
The gathering of statistical information is well supported by SNMP and CMOT.

b. Maintain and examine logs of system state histories.
Both protocols provide information which can be used to populate logs of system-state histories. For instance, the *sys_up_time* variable present in the MIB indicates the time since the network management portion of the system was last re-initialized. In general, however, the maintenance and examination of these logs is more of a data management application issue than a protocol issue.

c. Determine system performance under natural and artificial conditions.
To determine system performance under natural and artificial conditions, the management system should have the ability to re-configure snap-shot intervals

(intervals at which critical performance parameters are sampled) and to do trend analysis on the basis of this information - both in real time. Currently this function cannot be invoked via SNMP or CMOT and one has to depend on remote logins to the router.

d. Alter system modes of operation for the purpose of conducting performance management activities.
Though the system requirement of altering modes of operation of the system is supported to some extent by both SNMP and CMOT, it is not currently allowed because of lack of a suitable authentication mechanism.

e. Allow customized data collection and reporting.
The management station should be flexible enough to allow customization of performance data collection. Both SNMP and CMOT provide performance data (such as number and size of packets in and out, number of packets discarded due to various types of errors etc).

Statistical performance information obtained through the cisco router - CR4 (Figure 1) at CPMC includes the number of packets and bytes sent and received, and snap-shot errors (for example, cyclic redundancy code (CRC) errors, errors caused by the packet size being too big or too small, etc). An important performance management requirement for the management station at CPMC is the ability to produce standard and customized trend reports on a daily, weekly, monthly, quarterly and yearly basis.

At CPMC today, there is a need to be able to determine the system performance under natural and abnormal conditions. For instance, there is a need to change the standard five minute snap-shot error interval to one minute during abnormal conditions.

3.4 Access Control
Access control to a network would be required for various reasons. One of the main reasons is security. Other reasons include the restriction on the use of remote services (as this might prove expensive), the allocation of higher priority to local stations for the use of local services, and the placing of restrictions on the use of these services to remote stations. Another reason for restricting the use of local services to remote stations would be to control the resulting traffic congestion.

a. Creation, deletion and control of security services and mechanisms.
Currently, the creation, deletion and control of security services and mechanisms are not supported by SNMP or CMOT and can only be done via remote login to routers.

b. Distribution of security relevant information.
The distribution of security-related information (for instance, the distribution of access lists) is not supported by either SNMP or CMOT. Currently, this distribution is achieved by using file transfer protocols.

c. Notification of security relevant events.
The existing notifications of security-related events are limited, both in SNMP and CMOT. An example of such an event would be the *SNMP authentication failure*. This is achieved by an event report in CMOT.

d. Allowing or disallowing network access based on policy data.
The ability to permit or deny access to the network on the basis of policy data (for instance, on the basis of accounting limits, policy domain, affiliation, time of usage, etc.) is not supported by either SNMP or CMOT.

e. Authentication and authorization of management control request, including access to sensitive information by customers.
The SNMP standard only provides "trivial authentication" (passwords). However, significant work is underway in this area. CMOT provides an "access control" field, the use of which is currently an implementation issue.

The lack of a suitable authentication/authorization mechanism is a serious drawback, and this is the reason for not allowing parameter setting at CPMC. It may be noted here that it is possible to allow SNMP and CMOT *sets* and some of the CMOT imperatives and they will work. However, this may be done at the risk that some user may also decide to send *sets* and imperatives to the router too, and change its configuration (accidently or purposefully). Since the agent has no way to authenticate requests (i.e to determine who really sent it), most people specify that the agent honors only read (*get*) access requests. Also, an unauthorized user could reduce the efficiency of cisco routers (for example CR4 and CR5) at CPMC if he wrote a program doing a series of SNMP MIB retrievals, by merely knowing its IP address and community string (which is transmitted in clear text with the normal management requests - and so is vulnerable to interception).

3.5 Accounting Management

Accounting management enables charges to be established and costs to be identified for the use of resources. The functional requirements for accounting management are [2]:

a. To inform users of costs incurred or resources consumed.
The MIBs supported by SNMP and CMOT provide interface-related information (number of packets in and out, number of packets discarded due to various errors etc) that can be used for accounting purposes. Certain vendor-specific implementations provide source-destination related information on the basis of which the cost incurred or the resources consumed, can be computed. For instance, the cisco-specific MIB provides an IP accounting table which gives the number and size of packets between a source-destination pair. However, computing the actual cost is left to an application.

b. To enable accounting limits to set tariff schedules to be associated with the use of resources.
Currently, neither SNMP nor CMOT address the enabling of accounting limits to set tariff schedules that are associated with the use of resources.

c. To enable costs to be combined where multiple resources are invoked to achieve a given communication objective.
Currently, neither SNMP nor CMOT can combine costs that are incurred when multiple resources are invoked to achieve a given communication objective.

d. The ability to dynamically change the fee structure.
Neither SNMP nor CMOT currently support a dynamic change in the fee structure.

For accounting purposes, Columbia University and Columbia Presbyterian Medical Center are two different entities. Accounting problems arise while considering applications or resources that can be used by both the university and the hospital. There are some university applications for hospital use (an example of such an application would be an electronic text-book for taking care of patients). A need exists to ensure that this facility is being used for the hospital rather than the university. Also, accounting needs to be done on the basis of the number and size of packets sent, time of usage, etc). Currently to simplify matters, it is done on the basis of a flat fee structure (monthly charge).

3.6 Network Planning
One of the primary reasons for doing IP router management is network planning. Network planning requirements include:

a. Allowing customization of data collection for network planning.
The management station should be flexible enough to allow customization of data collection for network planning. Both SNMP and CMOT provide this data although the customization of this data for network planning is left to the application. Further, the emerging CMOT specification will allow the user to specify filters and define thresholds which will further ease the customization of data by the application.

b. Classifying data according to sender or receiver.
The network planning function needs end-to-end traffic data to properly engineer interlocation trunks and even to engineer data balance between different LANs at the same location (especially among high-use servers and their clients). The cisco router maintains an "IP Accounting Table", available via SNMP, that keeps IP source and destination pairs together with the total number and size of packets sent from the source to the corresponding destination. This facilitates the retrieval of information on traffic that passes through the router. Most of the other information available via the MIB is on an interface basis. For instance, consider a source X and a destination Y. All traffic between X and Y goes via station A. Assume that the trunks from X to A, and from A to Y begin to saturate. Without being able to classify the data on the basis of source-destination pairs, one cannot determine how much of the traffic starting at X is really headed towards Y. If it is a substantial amount of traffic, then one might decide to put in a new trunk between X and Y. If it is not substantial then the network manager might decide to upgrade bandwidths on

existing trunks. A similar scenario can show the need for end-to-end data for balancing and/or positioning servers in an internetwork.

3.7 Data Management

Data management is essentially independent of the network management protocol. However, the large volume of network management-related data transfers taking place within the network, make it a very important aspect of IP router management. Some of the functional requirements for Data Management are:

- *generating reports that log important events and alarms.*

- *storing this information in a database for future retrieval by an application process.*

- *generating sub-network (in the OSI sense) utilization reports.*

- *storing summary information instead of details.*

- *providing automatic removal of old information.*

4. OPERATIONAL REQUIREMENTS

One of the main operational requirements is to keep the cost of management operations to the minimum. The two significant operational costs are those associated with the consumption of network resources and the degree of skilled human intervention.

a. Minimize operational cost incurred due to consumption of resources .
SNMP uses a connection-less, datagram delivery transport service; CMOT requires the establishment and maintenance of sessions. This makes SNMP cost-effective (especially for smaller management transactions) when compared to CMOT. However, when retrieval of large amounts of data from the MIB is considered, SNMP proves to be in-efficient and expensive [9].

b. Minimize operational cost incurred due to the degree of skilled human intervention.
The services of a skilled network manager should only be needed on the occurence of pre-defined threshold or object-status triggered alerts. However, applications based on both SNMP and CMOT currently require a greater degree of skilled intervention. Certain variables (such as the routing protocol to be used, the password to authorize an inbound telnet, etc.) are absent from the MIBs of CMOT and SNMP. In order to set the configuration of the router, it becomes necessary for the skilled network manager to manually log in to the router.

At CPMC, one would like to be able to separate day-to-day monitoring functionality from operation monitor functionality. Only alerts and other vital information which require intervention by an experienced network manager need be reported on the operation monitor. The day-to-day monitoring of the network can be left at the hands of a relatively unskilled person. (Presently, network monitoring and management is achieved through a graphic user-interface which combines both these functionalities).

5. IMPLEMENTATION REQUIREMENTS

Industry vendor experience has shown that SNMP is significantly easier to implement than CMOT. Indeed, this is one of the key reasons behind SNMP's success in the marketplace!

In particular, the implementation experience of Netlabs (a network management OEM for AT&T), has shown that CMOT managers and agents are at least twice as complex to design and implement than SNMP managers and agents [9]. The implementation requirements can be classified into the following categories:

a. Memory requirements:
Excluding the code required for TCP/IP, SNMP occupies approximately one-third the total code-space required by CMOT [9]. In addition, CMOT requires more dynamic memory than SNMP.

b. Processing requirements:
Limited benchmarking indicates that SNMP *Get* requests execute about three times faster than CMOT *M-Get* requests [9]. (However, it is worth noting that unlike SNMP *Get* requests, a small number of CMOT *M-Get* requests can selectively retrieve large volumes of management information.)

c. Application requirements:
The application programming interface (API) for SNMP is much simpler than that for CMOT. As a result, basic network management query-type SNMP manager and agent applications can be designed and built a lot faster than similar CMOT based applications.

6. CONCLUSIONS

Table One illustrates the level of management support provided by SNMP and CMOT, by providing numeric rankings for each of the major categories discussed in sections three, four, and five. In addition, the following points summarize the previous discussions:

• *Functional requirements:* The Columbia case study shows that SNMP provides a fair amount of support for fault management, configuration management, and network planning, but falls short in all the other categories. In theory, CMOT provides better support (than SNMP) for configuration management, performance management and access control.

Protocols Mgmt. Needs	SNMP	CMOT
Functional - Fault Management: - Configuration Management: - Performance Management: - Access Control: - Accounting: - Network Planning: - Data Management:	 2 2 1 1 1 2 -	 2 3 2 2 1 2 -
Operational - Efficiency and Cost-Effectiveness: - Personnel Related Costs:	 2 2	 1 1
Implementation - Memory Cost: - Processing Cost: - Applications Cost:	 4 3 3	 1 1 1

SNMP: Protocol, SMI, MIB-II
CMOT: CMIS, OIM MIB-II

↑ 5 - Excellent
4 - Very Good
3 - Good (Meets Mgmt. Needs)
2 - Fair
1 - Poor
↓ 0 - Lacking

Table 1 Level of Protocol Support for IP Router Management Needs

• *Operational requirements:* The Columbia case study shows that SNMP is efficient and cost-effective for network transactions involving simple retrievals. For larger transactions such as MIB browsing, the case study shows that SNMP is inefficient and not cost-effective. In theory, CMOT is more efficient, especially for larger transactions, but is also less cost-effective. Both SNMP and CMOT fall short in that they require frequent intervention by network administration personnel who need to be very well versed in the details of network management.

• *Implementation requirements:* CMOT is signicantly more complex to implement than SNMP. CMOT requires more memory, more processing power, and more complex network management applications. As a result, CMOT is more difficult to implement in resource constrained IP routers.

• *SNMP and CMOT:* Finally, looking ahead, we believe that the future end-to-end enterprise management of IP routers will involve both SNMP and CMOT [11]. Despite the current growing popularity of SNMP, we believe that over time, CMIP and CMOT will be successfully deployed in the Global Internet. Just as both RISC and CISC processor technologies co-exist today (with somewhat overlapping applications), in the future both SNMP and CMOT will co-exist within the global Internet [12].

7. ACKNOWLEDGEMENTS

We offer special thanks to Nirali Khanderia, without whom this paper would not have been possible. Also, several of our network management colleagues provided us with valuable comments. We would like to thank Rich Bantel, Subodh Bapat, Beverly Bernoske, Matt Bush, Tom Capotosto, Jayant Kadambi,Steven Russell, and Stephen Wilkowski.

8. REFERENCES

1. *LAN Interconnection Products: User Requirements and Buying Plans,* Infonetics Research Institute, October 1990.

2. ISO/IEC 7498-4, International Standard, *Information processing systems - Open Systems Interconnection -Basic Reference Model- Part 4: Management Framework* 1989.

3. Case Jeffrey D., Mark S. Fedor, Martin L. Schoffstall, and James R. Davin, *A Simple Network Management Protocol (SNMP),* RFC 1157, May 1990.

4. Rose, Marshall, *Structure and Identification of Management Information for "TCP/IP -based internets,"* RFC 1155, May 1990.

5. McCloghrie, Keith, and Marshall Rose, *Management Information Base for Network Management of "TCP/IP - based internets",* RFC 1158, May 1990.

6   Warrier, Unni, and Larry Besaw, *Common Management Information Services and Protocol over TCP/IP (CMOT)*, RFC 1095, April 1989.

7   LaBarre, Lee, *OSI Internet Management: Management Information Base II*, Internet Draft, OIM Working Group, July 1990.

8   Dupuy, A., S. Sengupta, O. Wolfson and Y.Yemini, *Design of the Netmate Network Management System,* Second International Symposium on Integrated Network Management, April 1991.

9   Ben Artzi, Amatzia, Asheem Chandna,and Unni Warrier, *Network Management of TCP/IP Networks: Present and Future*, IEEE Network Magazine, July 1990.

10 Warrier, Unni , Larry Besaw, L. LaBarre, B.D. Handspicker, *The Common Management Information Services and Protocols for the Internet (CMOT and CMIP)*, RFC Draft,July 1990.

11 Bush, Matt and Asheem Chandna, *Network Management: Present and Future,* Invited Presentation, New Jersey Chapter of the ACM, Monthly Meeting, December 1990.

12 Private Communication with Subodh Bapat, Racal Milgo, December 1990.

13 ISO/IEC DIS 10165-4 (ISO/IEC JTC/SC21 N 4065), *Information Processing Systems - Open Systems Interconnection- Structure of Management Information - Part 4: Guidelines for the Definition of Managed Objects*, June 1990.

14 ISO - 9595, *Information Processing Systems - Open Systems Interconnection - Management Information Services Definition - Common Management Information Service, CMIS*, Dec 1989.

15 ISO - 9596, *Information Processing Systems - Open Systems Interconnection - Management Information Protocol Specification, CMIP,* Dec 1989.

Experience in Network Management: The Merit Network Operations Center

Kraig R. MEYER[1]
Computer Science Department MC0782, University of Southern California, Los Angeles, CA, 90089-0782. Phone +1 213-740-7287. Electronic mail: kmeyer@usc.edu

Dale S. JOHNSON
Merit/NSFNET Network Operations Center, Merit Computer Network, 1075 Beal Avenue, Ann Arbor, MI, 48109-2112. Phone +1 313-763-3448. Electronic mail: dsj@merit.edu

Abstract

Over the last twenty years the TCP/IP research internetwork, generally referred to as *the Internet*, has grown from being a loosely connected set of experiments in wide area networking protocols to a production-level national resource. Management of all portions of this Internet, and in particular management of the NSFNET backbone, has become essential in providing consistent, reliable service to its users in the U.S. research community. In this paper we describe our experiences in the Merit Network Operations Center (NOC) as we designed technical and human systems to manage a rapidly expanding and changing network. We also examine how organizations in general can cope with the increased importance of, and demands placed on their NOCs. The emphasis of this case study is on how human and technical systems interact in a network management environment, and not on technical or implementation details.

Key Words and Phrases

Networks, Internets, Internetworking, Operations, Management, Monitoring, Human Factors, Network Operations Centers, Phases, TCP/IP

1.0 Introduction

The evolution of the TCP/IP Internet into a national production-level network has required the academic networking community to take a more serious look at the task of managing TCP/IP networks. From a technical standpoint there has been a lack of adequate tools to allow for distributed management [CLAR88]; the high degree of heterogeneity in the Internet differs substantially from that of corporate and commercial networks--presenting new challenges in network management. A variety of management systems and protocols [CASS89] have been developed. At the same time, a number of new human systems have been developed to operate in the framework of this new technology. Regardless of the technical differences, the primary goal of TCP/IP network managers is the same as other network managers--to provide consistent, reliable services to the network users.

The next section of this paper briefly describes how the management of the TCP/IP research internet has evolved to its present form. In Section 3, we describe the role that a NOC

---

[1]Part of Mr. Meyer's work was completed while he was with the Merit Computer Network.

plays in network management. Section 4 describes a model we use to describe the stages through which a NOC passes, and in Section 5 the Merit NOC's history is described as it relates to this model. Section 6 discusses what conclusions can be reached about the different phases in a NOC's life cycle. Finally in Section 7, the future of the Merit NOC and NOCs in general is examined. Section 8 contains concluding remarks.

2.0 The Evolution of the TCP/IP Internet

The TCP/IP Internet first grew out of the ARPAnet, a network supported by the Defense Advanced Research Projects Agency (DARPA). ARPAnet was first administered by Bolt, Beranek, and Newman, Inc. starting with its design in 1969 [McKE75]. Later, in 1983, this network was split into two networks--a production network called the MILNET, and an experimental network which retained the ARPAnet name. This experimental network remained the primary backbone for researchers in academia until 1984, when the National Science Foundation decided to lease high-speed data lines between its major supercomputing centers [NART89].

2.1 Early Internet Maintenance

The maintenance of the research network was problematic; its practical maintenance was informally distributed among a relatively small number of networking professionals who all watched the network from their respective points of view. The monitoring was done in a variety of formal and informal ways, with networking gurus noticing and fixing many problems on their own. When they were unable to fix problems themselves, they generally depended on a human network of personal contacts to resolve problems. This informal management depended on a great deal of cooperation as well as a willingness on many peoples' parts to be woken up in the middle of the night.

2.2 NSFNET Redesign

In June of 1988, a new backbone network sponsored by the National Science Foundation was put into place and its management was centralized at the Merit Computer Network in Ann Arbor, Michigan. Merit had previously had over 20 years experience operating the Merit statewide regional network in Michigan. With the new backbone, Merit was entirely responsible for monitoring the T1 links and packet switches, resolving problems in a timely fashion, and providing a central point of contact for the regional networks which were connected to the NSFNET backbone.

One convenient way to model this present-day Internet is as a hierarchy. The NSFNET backbone is one of a number of backbone networks which are at the highest level of this hierarchy. The backbones span the United States and provide moderate bandwidth (1.544-44 Mbps) between major access points--analogous to the way that airlines provide transportation between regions of the country via their hub and spoke systems of airports. Just as each of the various airlines provide service to different airports, each backbone provides network connectivity to different access points, with geographic, economic, and political factors determining which points get served. At each of these access points there are typically regional networks that provide low to moderate bandwidth (9600bps-1.544Mbps) between the backbone networks and a variety of smaller networks, which are generally within a contiguous geographic region of the country. The lowest level of this network hierarchy is comprised of campus, corporate research, and local networks. These typically range from moderate to high bandwidth (1.544 Mbps to 80Mbps and higher) and take the form of networks based on router, bridge, and LAN technologies.

In contrast to a corporate network, or the public telephone company networks, not all levels of the TCP/IP Internet hierarchy have consistent, centralized, and formalized network management and maintenance. This is due to a number of reasons. First, a large degree of technical heterogeneity exists at lower levels in the hierarchy. Before the advent of standard network management protocols, management of such a network was nearly impossible. Historically computer networks have not been considered high priority resources. Each regional network and its members have different expectations and provide different funding levels in support of networking. As such, a number of regionals are non-profit organizations and may be trying to limit expenses for their members. The result of all this is that some portions of the Internet are supported with 24 hour monitoring and problem resolution while others are not watched at all.

3.0 The Role of a NOC

As previously mentioned, the general goal of any Network Operations Center (NOC) is to provide a level of consistent network service to its user community. There are four major aspects to providing consistent service: fault detection, fault recovery, fault notification, and future planning. A NOC's primary focus should be on the daily activities of *fault detection*, and on *management* of the fault recovery process. In some organizations the same people who staff the NOC may have a secondary responsibility of actually fixing problems that they discover. Under the Merit model of network operations these secondary activities are not considered prototypical NOC activities.

3.1 Fault Detection

Fault detection in a NOC should be accomplished primarily through the use of automatic monitoring tools. This allows problem resolution to begin at the earliest possible time-- potentially before a service disruption has occurred, and before it has been noticed by a user. Users must also have ways available to directly report network problems to the NOC-- preferably via both in-band communications, such as electronic mail, and out-of-band, such as the telephone. Users often need or want to know that the cause of their problem has been identified and is being resolved. This action is the conclusive evidence to the user community that the network is being watched and kept in good repair. In addition, some problems will always occur that monitoring tools cannot identify, and users provide the only feedback available to resolve these problems.

A NOC should never rely *solely* on reports from users to identify network problems. This causes significant delay in the resolution of problems, especially on those sections of the network that may not be regularly used. Also this does not allow the identification of problems for which there are no obvious user-visible symptoms, such as degradation of service and redundant link failures. Technical staff members need to know there is a place they can contact to report problems and determine status. But user reports should be used to supplement, not replace, the information provided by monitoring tools.

3.2 NOC Tools

It is necessary for a NOC to have tools that explicitly identify faults and trigger operator-visible alarms. An alarm may take many forms, such as a bell on a terminal, a green icon that turns red, a light bulb or a text list of outstanding problems. All of these alarms have something in common--they all convey a boolean state which explicitly tells the network operator that a problem exists. For maximum efficiency, an alarm should appear when an operator needs to take an action, and not otherwise. Tools, for example, that solely display raw packet traces or error counts should *not* be the primary monitoring tools for a NOC because they alone do not identify specific problems to an operator. The output from such tools may, however, be fed into other network tools which in turn analyze the data and trigger alarms. These tools may also be useful on their own to aid in the related tasks of fault recovery and network planning.

3.3 Network Problem Resolution

Once a problem has been identified, the NOC is responsible for seeing that it is resolved in a timely and efficient fashion. This does not mean that Network Operators should be required to have the technical expertise to solve all problems they are faced with in a given day. In fact, we have found that it is preferable for Network Operators to simply have a basic knowledge of networking and have resources available to draw upon in order to resolve major problems. This is a situation analogous to a high school nurse--who knows how to treat cuts and bruises, and knows how to get trauma cases to the hospital. Most people can immediately recognize that having a full time M.D. on a high school staff would be unnecessary, expensive, and inefficient use of the doctor's time. Yet many do not realize that the same is true of having a full-time engineer work as a network operator.

The network manager, then, should hire operators who are computer-literate, enthusiastic, and willing to learn, but who are not engineers or experts in networking. The network operators can learn a few basics of networking and be trained to treat the network's cuts and bruises via cookbook-style procedures. In following these procedures the operator may have to "dial 911" when an engineer is needed. Under the Merit model of network operations, a network operator's knowledge of networking can theoretically be quite minimal--but in reality we strongly encourage our operators to learn as much as they can about networking. This prevents the operators from getting bored, and also allows them to increasingly take work away from the engineers. Our success in managing the NSFNET backbone has been due largely to the network operators' continually expanding base of knowledge. All of this allows the engineers' time to be used more effectively--on real engineering tasks such as designing monitoring tools and planning network expansion.

3.4 The Importance of Record-Keeping

The NOC's most important role in problem resolution is coordinating what is done and ensuring that problems don't fall between the cracks even if they must be referred to personnel outside of the NOC. This requires some type of record keeping, or "trouble ticket" system. Trouble tickets act like hospital charts: they allow any Network Operator to pick up any open problem and to work on it. Most network problems cannot be solved in the first few minutes that an operator works on them--and many problems will extend across one or more operator-shift boundaries. The problem may need to be referred out to a phone company, or for help from someone at a remote site, or to a consulting engineer. In any of these cases, a note needs to be made that the problem has been passed on, and that further action will be necessary later at a later time.

Each record should provide the date and time of each action which was taken to resolve a problem, as well as a description of the action itself and who in the NOC coordinated it. Such a record allows the NOC staff to make sure that problems are resolved efficiently. It also allows them to answer questions from the user community and the personnel who are assisting in the problem resolution such as telephone linemen, technicians, and engineers. This record allows network planners to calculate reliability parameters of the network such as the mean time between failures and mean time to repair. Finally, the record is useful during funding reviews, to document the amount of work involved in maintaining a transparent, reliable network service.

A very small NOC can operate without trouble tickets, at the expense of losing some problems because things are not written down. A larger NOC risks losing even more incidents, given the increasing numbers of people who can be involved in any particular incident, and the increasing number of hours per week that are covered. A simple multi-user database system can be made into a reasonable trouble ticket system; or even a paper system will work in a pinch.

Managing the present day Internet, or any distributed heterogeneous internetwork, requires addressing a few additional issues which do not apply to homogeneous network architectures. The Internet is administered in a distributed, semi-coordinated fashion, and the gateways run a variety of software available to different groups. No single NOC is able to get complete information about all gateways in the Internet. As a result, a connectivity problem between two parts of the Internet might require the involvement of 3 or more major organizations before it can be resolved. In order to resolve problems in those portions of the Internet that do not have significant management resources, NOCs still count on the culture of volunteerism and home phone numbers described earlier in this paper.

3.5 Sample Outage: A Fiber Cut

We now examine the series of steps which would be taken to resolve a typical network outage. Suppose a backhoe cuts a piece of fiber in the NSFNET backbone between Lincoln, NE and Boulder, CO. Operators in the Merit NOC may be notified in a number of ways. First, the monitoring software[2] running on the packet switches at each end of the link will notice that traffic is no longer being passed on the link. Two tools in the NOC will separately poll this monitoring software and will display a red line on a graphical display and add a new line of text to a list of problems. MCI, who provides the T1 lines for the NSFnet backbone, also has monitoring equipment which will notice the outage and will indicate an alarm on an MCI station in the NOC. Finally, a contact person in Lincoln may notice a performance degradation and report the problem either via electronic mail or a telephone call to the NOC.

One of the network operators on duty will open an electronic "trouble ticket" noting the first symptom of the problem and the time of its occurrence. For example, they might note that the graphical line between the Lincoln icon and the Boulder icon turned red at 11:30 EDT. The trouble ticketing system will automatically note the time the report was made and the person making it. A different operator may receive the phone call or the electronic mail message (both of which are automatically distributed among the on-duty operators) from the staff member in Lincoln. The trouble ticket will be updated with these additional reports of problem.

Assuming that the operators on duty do not know that a fiber breakage has occurred, they might follow the cookbook procedure for resolving this problem. The first step might be to check whether the channel service units on each end of the link are reporting alarms. The operators might also check to see that interfaces involved are also operating properly. When these steps fail to resolve the problem, the operator would likely call the MCI National Accounts Service Center (NASC) to report a problems with a given circuit. The NASC would then open their own trouble ticket, inform the Merit operators that indeed a piece of fiber had been cut and provide an estimated time of repair. Based on the severity of the outage and the estimated time to repair, the operators also must make a decision who to inform about the problem. For example, mail may be sent to contact people at each regional network.

The Merit operator would note all of these events in the trouble ticket, and would allow the ticket to remain idle for no more than two hours once MCI has acknowledged the problem. Every two hours or less, MCI calls Merit back to provide a status update on the problem, and each of these interactions is noted in the trouble ticket. In some cases, the NOC operator might contact MCI to further determine the extent of the problem. The problem is considered resolved when the alerts have disappeared from the operators' screens, and when MCI has verified they have fixed the problem. The Merit operator would put the final resolution data into the trouble ticket and mark it closed--meaning that no further action need be taken on the problem. If the outage was of a severe nature, the Merit staff member would also send electronic mail to regional contacts indicating that the problem had been resolved.

---

[2]SNMP daemons

4.0 Phases in a NOC's Life Cycle

From our work at Merit we have observed that as a network and a Network Operations Center grow in size, there is a series of phases through which a typical NOC passes. Not every NOC will necessarily go through every part of every phase. The borders between phases are inherently points of discontinuous growths, and some NOCs may completely skip some phases.

At the beginning of each phase, the staffing level is more likely to be adequate for the amount of work involved. The closer the NOC gets to the end of each phase, the busier the staff becomes. If adequate funding is provided, new staff members are added and a new phase begins. Along with a more relaxed pace at the beginning of each phase, staff members are also increasing their breadth of skills--taking on a wider variety of tasks. As they become busier, they narrow the focus of their daily tasks and may actually develop very specialized skills. This phase repeats itself throughout the life cycle of a NOC.[3]

4.1 The Startup Phase

Many networks start without a Network Operations Center. A network may initially be watched only by a single engineer or programmer, with a NOC slowly being formed around that person's desk. This lone staff member has no choice but to be a generalist. S/he is responsible for watching everything that goes on in the network, setting up monitoring tools, debugging tools, and keeping track of what was down and when and why it was down. It is not uncommon for this first staff member to be quite busy, seemingly conflicting with our claim that the beginning of each phase is a time of relaxation. In reality, this staff member may have built the network as part of another job. Alternatively s/he may have been given these responsibilities long after the network had grown too large for one person to watch, and the staff member is busy due to lack of adequate resources.

The second half of the startup phase begins when a few additional people are hired to assist the soloist. Almost immediately the responsibilities will be split up into workable portions. The original staff member will likely begin expanding the monitoring and debugging tools while the newer employees learn how to use the existing tools and take over responsibility for record keeping.

4.2 The Second Phase

The number of staff members in a NOC should dramatically increase either when significant new responsibilities are taken on or when the NOC begins to provide 24-hour per day service. The beginning of the second phase is characterized by this sudden increase in staffing, as the number of staff members--and the amount of money--required for providing around-the-clock staffing is significant. Yet staff members who are on non-peak shifts will likely have a lot of spare time. It is natural for the network manager to take on more responsibilities to keep the employees busy and interested in their job, to solidify administrative justification for the new budget, as well as to generate revenue from new sources to cover the added cost. These responsibilities may include monitoring additional/different types of networks, providing better statistics, and performing unrelated tasks such as running backup tapes for unrelated systems.

---

[3]Each employee's role in the first and second halves of a phase might be compared to an employee's role in a small or large company, respectively. In a small company there are fewer employees, so each employee is individually responsible for a larger portion of the company's well being--and thus a wider variety of tasks. In a large corporation each individual is responsible for only a marginal portion of the company's well being and generally focuses on a very specialized set of tasks.

The second half of this phase is entered when the NOC has taken on sufficient new responsibilities such that staff members on all shifts are busy. As the NOC continues to add employees and responsibilities, it no longer is prudent for every employee to learn every task. Instead, it is sufficient for any given task to be completed by some employee on each shift. The focus in this part of the cycle has shifted from generalization to specialization of employee skills.

4.3 Additional Phases

The third and subsequent phases of this model are essentially repeats of the second phase. For example, a single NOC may wish to split into two physically separate NOCs for redundancy purposes (see Section 7). In this case, each of the two NOCs would need to have a staff with a complete repertoire of skills.

5.0 The Role and History of the Merit NOC

The Merit Network Operations Control Center (NOCC), as it was originally called, was formed in early 1987 to manage and control the Merit Computer Network, a Michigan statewide regional network consisting of over 150 custom-built nodes--and the University of Michigan campus network, called UMnet. This NOCC illustrates the resources typical of startup NOCs: The staff consisted of one full time engineer, and a half time liaison provided by the computing center operations staff, who together answered phone calls on one phone line, tracked problems on notecards, and did troubleshooting by typing operator commands on simple terminals.

As NOC workload grew quickly it became necessary to increase staffing and other resources. Merit hired three full time network operators and a part time engineer to cope with the increasing responsibilities of managing a rapidly growing network. The operators began to focus more on tracking problems and following the few standardized procedures. This allowed the two engineers to focus on improving monitoring tools, developing better record keeping systems, and assisting in network planning. With 6 employees the notecard trouble tracking system quickly proved to be inadequate and plans were made to replace it with a simple PC distributed database system so that simultaneous access by multiple operators was possible.

5.1 The New Backbone

In November 1987, the National Science Foundation announced that Merit would be taking on responsibility for managing the new NSFNET backbone. This proved to be an onerous task; simply providing adequate staffing to monitor the NSFNET, Merit, and the University of Michigan campus network required hiring eight new operators for 24-hour per day, 7 day per week coverage. Training the operators for the highly specialized environment exposed another difficulty of managing TCP/IP networks: most classes that teach the basics of TCP/IP networking are taught as senior undergraduate engineering courses--certainly not appropriate for the majority of operator candidates, who had little university-level computer experience. Through a combination of formal lecture series and an apprenticeship "hands-on" program, we managed to train the eight new operators well enough in a month to put them in pairs on every shift of the week. Many phone calls were made to the engineering staff during the first few months.

Moving to such a large staff changed many of the organizational dynamics of the NOC, as it was no longer a small offshoot of another group in the organization. The NOC became a major group in its own right, with different personnel and management requirements than were previously needed. The large increase in staff size required an internal structure to develop, as well as the formalization of procedures which had been very ad-hoc with a staff of five. A new management position was created, with several support positions. Engineers were moved away from the daily NOC work to concentrate on network monitoring tools, problem resolution, and network planning.

It was decided early in this process that it was very desirable for every operator to be prepared to respond to every major task in a similar manner. This would assure that satisfactory resolution of a problem was not dependent on who happened to be working on a particular shift. At the same time, we did not want to establish procedures that prevented the more advanced operators from making use of their skills--thus we viewed procedures as a minimal set of guidelines to be used in problem resolution and not a lowest common denominator which all operators were required to follow. A new group of senior operators worked with the engineering staff and the NOC manager to establish this formal set of procedures for the operators to follow under many different circumstances. A theme of using many technical, automated solutions to assist the operators with their standard procedures proved to be one of the most important ones in the developing NOC.

5.2 Merit Takes on a New Role

In taking on the management of the NSFNET backbone, the Merit NOC found itself in a new role in the Internet community. Merit expanded its NOC staffing to provide 24/7 ("every minute, forever" as we fondly call it) staffing. Our corporate partners in the NSFNET effort-- IBM and MCI--also provided "every minute" service and made it possible for our NOC to both observe problems 24 hours per day and dispatch service technicians for T1 line and packet switch problems any time of the day or night. While Merit has often relied on the excellent cooperation from the NSFNET sites to reboot machines and provide minor adjustments as necessary, the additional flexibility of being able to call on our corporate partners at any time has always remained available. In addition, Merit's position as a national backbone provider has also allowed it to play a continuous facilitating role in resolving problems that involve multiple autonomous systems. All of these factors have contributed to the reliability of the NSFNET backbone, and the robustness of the Internet as a whole.

The Merit NOC also played an important role in defining the quality of service expected of a national TCP/IP network. Modeling its service commitments after its corporate partners, Merit strived to meet commercial production-level standards rather than the traditional research standards which had previously been accepted as status quo by the TCP/IP community. These high standards have served as a catalyst for some regional and campus networks. The goals of high bandwidth backbones, 24 hour service, off-hours maintenance, continuous network monitoring, and speedy problem resolution are slowly becoming the norm for academic computer networks.

5.3 Employee Stress in the NOC

An additional challenge that quickly surfaced for network management was the question of how to deal with employee stress. It is important to recognize that there are factors in a NOC that potentially can cause great stress on NOC employees: an environment of rapid technological change, occasional barrages of angry clients, and poorly designed work areas. The NOC manager must be aware of and regularly discuss stress and stress management. These discussions could be facilitated by an outside expert in an employee seminar. Second, by providing proper technical backup and training for new tasks, the employees will not feel overwhelmed with the new technology that shows up frequently in a NOC. Finally, it is important to create good work environments--by providing ergonomic furniture, proper lighting, and taking into account other human factors.

The same technological change that frustrates network operators affects the NOC managers as well. They must use anticipatory styles of management, so they can foresee what affect each new network will have on network management. To further address this problem NOC staff should be intimately involved in the planning of new networks, not only so that they are prepared to monitor them when they are implemented but also to assure that adequate network monitoring is included in the network design. Simply being able to monitor a new network may not be adequate. Ideally a NOC would like to be able to monitor each new network with a common set of tools. While this is dependent on a choice of good, "integrateable" monitoring tools, it also plays a role in the selection of technology and vendors for new networks.

6.0 Addressing the Needs of the NOC

The life cycle model is useful because it allows the identification of what stage a NOC has entered, and further, what the immediate goals and needs are of that NOC. When a NOC is starting up, it needs someone who has good skills not only in dealing with technical issues, but also dealing with people. S/he not only must monitor the networks but must also design an entire strategy for monitoring and resolving network problems. It is typical for this first person to be hired (or assigned these duties) when the network is already in place, causing this first person to be overloaded with limited resources. In these situations it is necessary for a great deal of support to be provided by other engineers and/or programming staff in assisting the lone NOC member.

Typically the only solution to lack of resources is to provide additional staff members. Automated procedures and technical solutions can help some, but they can not generally be counted on to answer phones or call service technicians.[4] When the second and third staff members join the NOC, the original staff member can be encouraged to spend his/her newfound time to either develop new monitoring tools, or research publicly available tools. The new staff members should try to learn, and document, the procedures that the original staff member has been using to isolate and resolve network problems. In later stages these procedures will become more formalized. Record keeping is still somewhat informal, but its importance is increasing.

6.1 Twenty-four Hour Service

Providing 24 hour NOC service is difficult, in part simply because there are so many hours in a week. It takes a minimum of five 40-hour workers to staff a 168-hour week, even if no one gets sick, takes holidays off, or goes on vacation. With six employees, there is enough flexibility to begin to work out sane shifts (three eight-hour shifts, with two people covering a given eight hours). With more than that, you can begin to eliminate single staffing and to concentrate extra coverage at midday, when NOC demands are at their highest.

One common way to extend the hours of NOC operations is to combine the NOC with other similar operations, such as Computing Center operations. In this case, NOC functions become only a part of each employee's responsibilities. In order to be able to perform the NOC duties as well as the other responsibilities, all the employees' functions must be well defined and well supported with training, documentation, and concise tools and procedures. Another approach, once the NOC is in place, is to pick up other campus responsibilities that may come with their own funding. University of Michigan, for instance, has a separate Electronic Mail group charged with providing reliable electronic mail service for campus. The Merit NOC was able to add one Network Operator in return for providing UM with 24-hour mail system monitoring and support. If you find that the nightshift is not always busy, backups and other support projects may make the NOC more useful to the organization. Beware, though, of combining

---

[4]The interested reader is referred to [PEAC88] which describes "Big Brother: A Network Services Expert." "Big Brother" is a prototype network monitoring system in the College of Engineering at the University of Michigan that includes both a telephone audio response unit and an automatic paging system for contacting staff members when network problems occur.

excessively menial tasks with network operations support. Network operations is not yet a menial job!

As the NOC grows, its need for internal communication and organization grows. Training by "apprenticeship" works well when the staff is small and stable and when the apprentices work the same shifts as their mentors. But as the number of new employees grows, the need for both "cookbook" and theoretical documentation also grows. Also, the more employees are expected to work alone, the more vital it is for them to have references to help them with unexpected problems, and "on call" resources to help out in emergencies.

Once the funding for 24 hour service is stable, and enough work is provided for employees on each shift, it may be possible for some of the operators to begin specializing in particular troubleshooting tasks. This should be particularly encouraged for the advanced operators, who may be able to learn techniques from the network engineers and programmers. This not only reduces the amount of work on the engineers, but also makes the operator's job more interesting. With each additional operator who is hired, it is possible to take on additional specialties.

7.0 Future NOCs

As the Internet becomes an ever-more ubiquitous part of our national research and education infrastructure, it will become necessary to further increase the reliability and survivability of network operations. Merit has already installed basic network management software on site at Princeton, New Jersey, and San Diego, California. If the Merit NOC in Ann Arbor goes down for any reason--be it a line or power problem, natural disaster or building evacuation--the network can be run by NSFNET development personnel from any point on the network. For further reliability, Merit is working on plans to automatically coordinate these "warm" backup facilities and to extend this automatic backup service to more of its statistics collection, database services, and other NOC functions.

It is clearly desirable from a redundancy and reliability standpoint to have either "warm" backup facilities or multiple NOCs. A second NOC, or alternate monitoring point, would provide a second vantage point from which to view the network. Particularly if this were combined with an out of band exchange of information--such as a dialup modem connection--it would provide the network operations staff with a more complete and accurate picture of the state of the network.

A second NOC has the advantage that it is guaranteed to always be staffed with properly trained personnel and up-to-date resources, ready to take over if the first NOC has to go offline. It has the distinct disadvantage that extensive coordination would have to occur between the two NOCs. For example, normally each NOC might be responsible for monitoring one geographic half of the Internet. If one of the NOCs becomes unreachable, there must be a series of human protocols whereby one NOC assumes responsibility for the entire Internet, and transfers responsibility back when the other NOC becomes reachable.

Another service that might be provided by future national NOCs is a national network trouble center. Such a center could provide an easy way for users to request assistance and report problems via an 800 number and an electronic mail address. Such a center would work closely with the various regional and campus NOCs, and could be ultimately responsible for seeing that any network problem, anywhere on the Internet, was resolved.

8.0 Concluding remarks

The Internetwork has gradually been driven from an experimental collection of networks into a production level national resource for research and academia. As the importance of consistent service increased in the research community, so did the need for appropriate technical and human systems for managing the network.

The Merit NOC has played an important part in developing the role of network management in the TCP/IP environment, as well as in helping establish a set of service standards by which many internets are now judged. Staffing and user interface issues have been continually re-examined and addressed throughout the growth of the NOC. Finally, the NOC has fulfilled its goal of providing consistent network service by using a variety of techniques. Fault detection is provided via automatic monitoring tools, and speedy problem resolution has been aided through the careful tracking of network problems in an online system.

9.0 Acknowledgements

The authors would especially like to thank Laura Kelleher for her extensive editorial assistance in preparing this paper. A number of others provided useful comments on various drafts, including Lee Breslau, Deborah Estrin, Abhijit Khale, Jeff Mogul, and Glenn Trewitt.

9.1 References

[CASS89] L. Cassel, C. Partridge, and J. Westcott, "Network Management Architectures and Protocols: Problems and Approaches," *IEEE Journal on Selected Areas in Communications*, Vol. 7, No. 7, September 1989, pp. 1104-1114.

[CLAR88] David Clark, "The Design Philosophy of the DARPA Internet Protocols," *Proceedings of the ACM Sigcomm '88, Stanford, CA, Aug 16-19, 1988*, pp. 106-114.

[McCL89] K. McCloghrie, M. Rose, and C. Partridge, "Defining a Protocol-Independent Management Information Base," *Proceedings of the IFIP TC6/WG6.6 Symposium on Integrated Network Management, Boston, MA, 16-17 May 1989*, pp. 185-195.

[McKE75] A. A. McKenzie, "The ARPA Network Control Center," *Proceedings of 4th Data Communications Symposium, Oct. 1975.*, pp. 5-1 to 5-6.

[NART89] T. Narten, "Internet Routing," *Proceedings of the ACM Sigcomm '89, Austin, TX, Sept 19-22, 1989*, pp. 271-282.

[PEAC88] D. Peacock and M. Giuffrida, "Big Brother: A Network Services Expert," *Proceedings of the Summer 1988 USENIX Conference, Jun 20-24, 1988, San Francisco, CA*, pp. 393-398.

9.2 Biographical Information

Kraig Meyer is a Ph.D. student in the Computer Science Department at the University of Southern California in Los Angeles, CA. His research interests include management, routing and protocols in large internets, and human interface design. Between 1985 and 1989 he worked for the Merit Computer Network, directing much of the creation of the Network Operations Center as an engineer and programmer. Meyer received a Bachelor of Science in Engineering (Computer Engineering), from the University of Michigan in December 1988. His current address is Computer Science Department MC0782, University of Southern California, Los Angeles, CA, 90089-0782.

Dale Johnson has been the manager of the Network Operations Center of the Merit Computer Network in Ann Arbor, Michigan since June 1988. His research interests include integrated alert and problem management approaches for operational environments. Johnson's background includes fifteen years of automated process control and Unix† product development. He received Bachelor (1973) and Master (1982) of Science degrees in Computer and Communications Science from the University of Michigan, and a Master of Science in Public Systems Engineering (1981) from U of M's Institute for Public Policy Studies. His current address is Merit/NSFNET Network Operations Center, Merit Computer Network, 1075 Beal Avenue, Ann Arbor, MI, 48109-2112.

†Unix is a trademark of AT&T.

Providing CMIS Services in $DECmcc^{TM}$

Mike Densmore

Digital Equipment Corporation, 550 King Street LKG2-2/T2, Littleton MA 01460-1289, USA

Abstract

Digital's DECmcc implementation of the Enterprise Management Architecture (EMA) includes an access module which supports management of DECnet/OSI. This module provides, among other services, CMIS services which are mapped to the DECmcc synchronous call interface. This paper describes the relationship between CMIS and the services provided by the DECmcc DECnet/OSI Phase V Access Module.

Digital Equipment Corporation's Enterprise Management Architecture (EMA) and the DECmcc implementation of EMA were developed to provide an extensible management system for use in a multi-vendor, distributed environment. One major component of DECmcc is the Access Module (AM) which provides basic management services for a specific technology to other components through a generic, synchronous call interface [1][4]. One such Access Module is the DECnet/OSI AM which provides services for DECnet/OSI Phase V based implementations. These services include CMIS management operation services (e.g.; M-SET). The CMIS services are mapped to the DECmcc call interface by the AM in order to take advantage of the features of that call interface while maintaining conformance to CMIS/CMIP services and protocols.

This paper will briefly describe the DECmcc AM call interface, the overall design of the DECnet/OSI AM and the general mapping between the call interface and CMIS [2][3].

Brief Overview of DECmcc Access Module Call Interface

AM services are invoked using a call interface consisting of a single procedure call, MCC_CALL_ACCESS, which causes the DECmcc Dispatcher to dispatch to the appropriate entry point in an AM. The entry point is determined by the *verb* (or directive[1]), *in_entity*[2] and *attribute*[3] arguments of the call. (The *time_spec* argument is also used in the process but may be ignored for the purposes of this paper.) Table 1 lists the arguments for MCC_CALL_ACCESS along with a brief description. The key for the Req (request) and RSP (response) columns is equivalent to that used for CMIS [2] : M = mandatory, U = user option, C = conditional, - = not present.

This call interface applies to all AMs regardless of the underlying protocols used in any particular AM to deal with managed objects. This allows higher level applications (Functional Modules) to deal with information, not protocol, since all requests are made independent of managed object protocol with the AM supplying any translation. The call interface is synchronous to ease development of Functional Modules (FMs).

Access Module Implementation

To help understand the relationship between the DECmcc call interface and CMIS services, it is useful to examine the structure of the AM. Figure 1 is a general view of the AM structure.

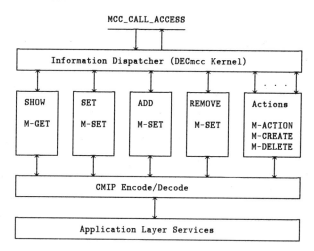

Figure 1: AM Structure

The DECmcc Dispatcher, upon receipt of a call, uses the arguments to determine an entry point into the DECnet/OSI AM. (The AM has entered an entry point for

---

[1] A directive is a management operation initiated by the manager to the managed object that performs the operation.

[2] An entity is a managed object.

[3] An attribute is a single piece of information that describes a part of an entity. It can usually be examined, and possibly modified.

Parameter	Req	Rsp	Description
verb	M	-	Specifies the directive to be performed.
in_entity	M	-	Specifies the complete entity specification.
attribute	M	-	Specifies the attributes for an examine or modify directive. NULL for action directives.
time_spec	M	-	Specifies scheduling and scope of interest to apply to the directive.
in_q	U	-	Specifies any qualifiers to the directive.
in_p	U	-	Specifies any input arguments to the directive.
handle	M	M	Specifies context and state information used to identify and control related requests.
out_entity	U	C	Specifies the complete entity specification of the entity on which the directive was performed. Required if the caller provides this argument or if wildcarding is used in the in_entity argument.
time_stamp	-	M	Specifies the time associated with the response to a directive.
out_p	-	C	Specifies any output arguments associated with the directive.
out_q	-	-	Reserved for future use.
CVR	-	M	Specifies the status of the call. If an error occurred as part of the service, the CVR will specify a service error and any additional, specific error will be returned in out_p.

Table 1: MCC_CALL_ACCESS Parameters

every directive when it was enrolled[4] into the DECmcc director.) The AM validates the arguments in the call and performs some directive-dependent preprocessing (e.g.; mapping an action directive to the appropriate actionID for the M-ACTION service). The resultant data is then passed to CMIP Encode/Decode services where a CMIP PDU is built. The PDU is then passed to Application Layer Services where associations are established, maintained and released and where context is maintained for multiple requests and responses over a single association.

Event services are shown in Figure 2. Since the AM has been developed for use in a director, or managing system, role only, the AM does not generate M-EVENT-REPORT PDUs but does accept event reports from managed objects. The AM uses DECmcc Kernel services to synchronize and distribute events among FMs which have made requests for events.

Each request for an event is passed to the queue. The requesting thread is blocked until the event is sent to the queue by the AM putEvent module. When an event is sent to the queue, all threads waiting for that event are unblocked and given a copy of the event. The AM then returned to the caller.

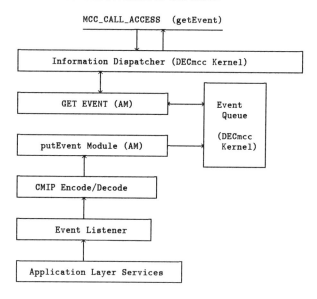

Figure 2: AM Event Handling

Because the call interface has inherent scoping on all calls (see Table 1), the getEvent call used by clients of the AM may invoke scoping for events by using wildcard specifications in the *in_entity* argument.

---

[4]Enrollment is the process by which a new management module is added to a director.

These figures do not show the control of the AM processes (e.g.; enabling and disabling event listening) or services not related to CMIS. These are not germane to this paper however.

General Mapping Between CMIS and the Access Module Interface

This section describes the general mapping between common CMIS parameters and the AM call interface. (Detailed mapping for specific CMIS services are in an appendix to this paper.) Some of the mapping is not a one-for-one translation between a call argument and a CMIS parameter. This is a consequence of requirements placed on the call interface. The AM mediates any such differences to maintain conformance to the DECmcc interface on the one hand and to the CMIS/CMIP specifications on the other. The following paragraphs describe mapping for the CMIS parameter indicated.

Invoke Identifier The Invoke and Linked IDs are not visible at the AM call interface except implicitly in the *handle* argument where context is maintained between successive responses. The AM maintains the IDs on behalf of the caller.

Linked Identifier See Invoke Identifier.

Mode The Mode is supplied by the AM. All services, except events[5], in DECmcc are confirmed. If necessary, the AM provides the confirmation on behalf of the managed object.

Base Object Class The Base Object, both Class and Instance, is encoded in the *in_entity* argument using the DECmcc *Abstract Entity Specifier*. The AM translates this into the proper encoding for CMIP. DECmcc does not support the full range of entity specification allowed however. See Scope.

Base Object Instance See Base Object Class.

Managed Object Class The Managed Object, both Class and Instance, is encoded in the *out_entity* argument using the DECmcc *Abstract Entity Specifier*. (Managed Object is encoded in the *in_p* argument for M-CREATE.) The AM translates the CMIP encoding into the DECmcc encoding. The *out_entity* argument is conditional and is used under the same conditions as the CMIS parameters with one additional condition. If the caller provides the argument, the AM returns the Managed Object Class and Instance even if the Base Object is the only object specified in the call.

Managed Object Instance See Managed Object Class.

---

[5] DECmcc, as a managing system does not use ROIV-m-EventReport or ROIV-m-EventReport-Confirmed.

Access Control Access Control information is optional and, if supplied, is specified in the in_q argument in the form of three qualifiers: *BY_USER, BY_ACCOUNT, BY_PASSWORD*.

Current Time A timestamp is provided for all operations in the *time_stamp* argument. The AM provides a timestamp on behalf of the managed object if necessary.

Scope The Scope is not directly visible at the call interface. The AM derives Scope from the in_entity argument and supplies it on behalf of the caller. If no wildcarded instance specifications are used in the in_entity argument, Scope is baseObject. If a wildcarded instance is used, Scope is individualLevel, where the level is the level which has a wildcarded instance. (DECmcc allows a single wildcarded level.)

Filter Filtering is specified in the in_entity argument using the WITH qualifier to specify the filter.

Synchronization BestEffort synchronization is always used. The AM provides bestEffort on behalf of the managed object if the object uses atomic synchronization only.

Attribute Identifier List The *attribute* argument specifies the attribute partition[6] to which the attributes belong. The list of attribute ids are passed in in_p as a set of IDs. Note: all of the attributes specified in in_p must be in the partition specified in *attribute*.

Attribute List Attribute List is passed in in_p where a list is to be passed to the managed object (modify, create or action directives) or in out_p where a list is to be returned (examine, modify or action directives). M-SET is always confirmed in DECmcc and there will always be an attribute list returned unless there was a failure. The list is encoded as a set of attribute ids with values or status codes (equivalent of CMIP attributeList, GetInfoStatus, SetInfoStatus).

Action Type The value of the *verb* argument indicates the action type.

Errors Errors are reported using two arguments. The *CVR* (condition value or status) indicates success or failure and, in the latter case, the specific failure. Additional information may be returned in out_p. The *CVR* is equivalent to the CMIS error parameter while the information in out_p represents a DECmcc extension.

---

[6]The attribute partition is a collection of attributes defined by the architecture for a managed object. An attribute belongs to one and only one partition.

Conclusion

We have seen how CMIS services are provided in the DECmcc environment, maintaining most features of CMIS without losing the advantages of the DECmcc call interface (see Reference [4]). While some of the more complex filtering and scoping options allowed in CMIS are not currently supported in the DECmcc interface (most notably the *wholeSubtree* and *baseToNthLevel* scoping choices), this is balanced by some key advantages of the DECmcc call interface:

- protocol independence to allow universal, consistent access to different protocols and different protocol versions by higher level applications

- comprehensive scheduling and scope of interest specification, including current, past and future time, repetitive scheduling and time ranges

- expanded error reporting using both status and argument lists

The AM also insulates the client from CMIS behavior and options in order to simplify the client's view of the managed objects:

- synchronization is always bestEffort

- timestamp is always returned

- event confirmation is handled by the AM

- SET and ACTION are always confirmed

- context is maintained by the AM

This allows clients (FMs) to be more generic and easier to develop.

References

1. DECmcc System Reference Manual, Order Number EK-DMCC1-RM-001, Digital Equipment Corporation

2. ISO/IEC 9595: *Information technology - Open Systems Interconnection - Common Management Interface Service Definition*, 6 December 1989

3. *Stable Implementation Agreements for Open Systems Interconnection Protocols*, National Institute of Standards and Technology, 1990

4. Strutt, Colin and Shurtleff, David G., Architecture for an Integrated Management Director, *Integrated Network Management, I*, Proceedings of the IFIP TC 6/WG 6.6 Symposium on Integrated Network Management, B. Meandzija and J. Westcott (Editors), North-Holland, 1989

Acknowledgments

The author would like to recognize the work of Arundati Sankar, Jim Halpin and Ramasamy Jesuraj, the DEC engineers who built the DECnet/OSI Access Module.

Author

Mike Densmore is a Principal Engineer in Digital Equipment Corporation and is a project leader for DECmcc DECnet and OSI management products. Prior to this assignment, Mike was Network Operations Engineer for DEC's internal data and electronic mail networks. Among his current assignments, Mike is editor of the Management Communications section of NIST OIW network management implementor agreements and Rapporteur to the EWOS-NIST joint CMIP ISP project.

Appendix A - AM Directive to CMISE PDU Mapping

CMISE Param	AM Param	Comment
Invoke ID	none	The ID is not visible to the AM service user.
Linked ID	handle	The AM provides for the linking in the Handle argument.
Base Object Class	in_entity	
Base Object Instance	in_entity	
Scope	in_entity	Scope is handled using wildcarded instances.
Filter	in_entity	The WITH qualifier specifies any filtering.
Access Control	in_q	Specified using BY_USER, BY_ACCOUNT, BY_PASSWORD qualifiers.
Synchronization	none	BestEffort is always used.
Attribute Identifier List	attribute, in_entity	
Managed Object Class	out_entity	
Managed Object Instance	out_entity	
Current Time	time_stamp	Always provided.
Attribute List	out_p	
Errors	CVR and out_p	
	time_spec	DECmcc only.

Table 2: M-GET Mapping

CMISE Param	AM Param	Comment
Invoke ID	none	The ID is not visible to the AM service user.
Managed Object Class	in_entity	
Managed Object Instance	in_entity	
Superior Object Instance	in_entity	
Access Control	in_q	Specified using BY_USER, BY_ACCOUNT, BY_PASSWORD qualifiers.
Reference Object Instance	in_p	
Attribute List	in_p	
Current Time	time_stamp	Always provided.
Errors	CVR and out_p	
	time_spec	DECmcc only.

Table 3: M-CREATE Mapping

CMISE Param	AM Param	Comment
Invoke ID	none	The ID is not visible to the AM service user.
Linked ID	handle	The AM provides for the linking in the Handle argument.
Mode	none	All modify operations are confirmed.
Base Object Class	in_entity	
Base Object Instance	in_entity	
Scope	in_entity	Scope is handled using wildcarded instances.
Filter	in_entity	The WITH qualifier specifies any filtering.
Access Control	in_q	Specified using BY_USER, BY_ACCOUNT, BY_PASSWORD qualifiers.
Synchronization	none	BestEffort is always used.
Managed Object Class	out_entity	
Managed Object Instance	out_entity	
Modification List	in_p	
Attribute List	out_p	SET, ADD and REMOVE are always confirmed services.
Current Time	time_stamp	Always provided.
Errors	CVR and out_p	
	time_spec	DECmcc only.

Table 4: M-SET Mapping

CMISE Param	AM Param	Comment
Invoke ID	none	The ID is not visible to the AM service user.
Linked ID	handle	The AM provides for the linking in the Handle argument.
Mode	none	All action operations are confirmed.
Base Object Class	in_entity	
Base Object Instance	in_entity	
Scope	in_entity	Scope is handled using wildcarded instances.
Filter	in_entity	The WITH qualifier specifies any filtering.
Managed Object Class	out_entity	
Managed Object Instance	out_entity	
Access Control	in_q	Specified using BY_USER, BY_ACCOUNT, BY_PASSWORD qualifiers.
Synchronization	none	BestEffort is always used.
Action Type	verb	
Action Information	in_p	
Current Time	time_stamp	Always provided.
Action Reply	CVR and out_p	CVR indicates success and out_p contains any information to be returned by managed object.
Errors	CVR and out_p	
	time_spec	

Table 5: M-ACTION Mapping

CMISE Param	AM Param	Comment
Invoke ID	none	The ID is not visible to the AM service user.
Linked ID	handle	The AM provides for the linking in the Handle argument.
Base Object Class	in_entity	
Base Object Instance	in_entity	
Scope	in_entity	Scope is handled using wildcarded instances.
Filter	in_entity	The WITH qualifier specifies any filtering.
Access Control	in_q	Specified using BY_USER, BY_ACCOUNT, BY_PASSWORD qualifiers.
Synchronization	none	BestEffort is always used.
Managed Object Class	out_entity	
Managed Object Instance	out_entity	
Current Time	time_stamp	Always provided.
Errors	CVR and out_p	
	time_spec	DECmcc only.

Table 6: M-DELETE Mapping

CMISE Param	AM Param	Comment
Invoke ID	none	The ID is not visible to the AM service user.
Mode	none	The AM provides confirmation if required. The caller is responsible for logging an event to permanent storage.
Managed Object Class	in_entity, out_entity	
Managed Object Instance	in_entity, out_entity	The caller requests the events from an entity, or set of entities, using *in_entity*. The AM returns the event source using *out_entity*.
Event Type	out_p	
Event Time	time_stamp	Always provided.
Event Information	out_p	
Event Reply		Any replies required would be handled in the AM.
Errors	CVR and out_p	
	time_spec	DECmcc only.

Table 7: M-EVENT-REPORT Mapping

Network Management In the Space Transportation System Mission Control Center: Lessons Learned

by
Robert C. Durst
The MITRE Corporation

Abstract

This paper discusses some of the lessons learned from NASA's implementation of a network manager for the Space Transportation System Mission Control Center. The lessons are directed toward implementors of network managers for in-house use. These lessons address such topics as selection of a host machine for the network manager, the trade-offs between standards-based and custom network management, and appropriate levels of experience for users and implementors.

1. Introduction

This paper discusses some of the lessons learned from NASA's implementation of a network manager for the Space Transportation System (STS) Mission Control Center (MCC) at the Johnson Space Center in Houston, Texas. (The Space Transportation System is the formal name for the Space Shuttle and its supporting ground systems.)

The lessons put forth in this paper are directed toward implementors of network managers for in-house use. These lessons address such topics as selection of a host machine for the network manager, the trade-offs between standards-based and custom network management, and appropriate levels of experience for users and implementors.

The document is organized with a background section, followed by a section for each lesson and its explanatory text, and a brief summary at the end of the paper.

2. Background

The MCC network manager was part of an upgrade to the control center that included implementation of a distributed, workstation-based environment for some aspects of Shuttle mission control. This upgrade constituted the first use of Local Area Networks (LANs) in the MCC for Shuttle mission support, and the first need for centralized network management. LANs had been used in the MCC before, but not for mission support. Previously, network administrators had no centralized management capabilities, and in general, management activities were performed using discrete dedicated equipment. The network manager implementation began with requirements definition in early 1989, with design completion near the end of 1989. Implementation followed, with initial delivery scheduled for late 1990. Refer to figure 1 for a high-level view of the MCC network architecture. The following paragraphs describe the main elements of the MCC's networks: workstations and LANs.

The workstations selected for the MCC all share identical architectures. This architecture is relatively unique: the workstations have multiple central processing units (CPUs), multiple graphics processors, and an implementation of System V Unix™ with real-time extensions. The architecture was selected for applicability to the tasks performed by the flight controllers, but the requirement for commonality also applies to the network manager workstation. The workstations share a common architecture for two reasons: first, this improves the probability of a close, available spare in the event of a workstation failure; second, the restriction limits the number of development environments that must be maintained.

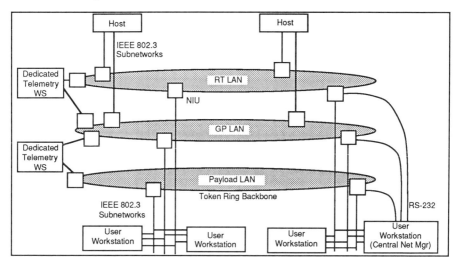

Figure 1. Block Diagram of MCC Network Architecture

The LANs in the MCC provide data distribution among the MCC host computers, some dedicated workstations that process and ship spacecraft telemetry data, and the users' workstations. The MCC uses three LANs to support a mission: a Real Time LAN (RT LAN), a payload LAN, and a General Purpose LAN (GP LAN). The RT LAN carries cyclic information from the host computers and the dedicated workstations to the user workstations. This LAN uses a custom protocol, called the Real Time Message Protocol (RTMP), above the IEEE 802.2 (IEEE 1984b) Logical Link Control (LLC) protocol and IEEE 802.3 (IEEE 1984a) Media Access Control (MAC) protocol. The payload LAN carries payload information to a subset of the user workstations, and uses the same protocols as the RT LAN. The GP LAN carries non-cyclic data that one would consider "typical" workstation to workstation traffic: file transfers, electronic mail, application-to-application communication, and so forth. The GP LAN uses Open System Interconnect (OSI) protocols that conform to the Technical and Office Protocols (TOP) 3.0 specification (SME 1987b), with a few exceptions. (Notable exceptions include the use of the inactive subset of ISO 8473, the connectionless network layer protocol (ISO 1988), and use of the International Standard version of ISO 9594, the Directory Services protocol (ISO 1990).) The GP LAN protocols were purchased from Retix, and ported to the host and workstation environments. All network management traffic is carried on the GP LAN, using the MAP/TOP 3.0 Common Management Information Protocol (CMIP) (SME 1987a). Retix's protocol products had TOP 3.0-conformant layer management entities included with each protocol layer. Further, the MAP/TOP 3.0 CMIP implementation included example code for an agent that could interact with these layer management entities. This software formed the basis for the MCC agent applications.

329

Each of the MCC LANs is implemented with a "backbone" architecture that uses a proprietary token ring LAN as a backbone, and IEEE 802.3 "tail circuits." The selected IEEE 802.3-conformant interface boards were chosen on the basis of their ability to meet the MCC throughput requirements; the statistics maintained by these interface boards do not map one-for-one with the attributes specified in the TOP 3.0 MAC ISO 8802/3 mechanism. The token ring LAN has network interface units (NIUs) that act as bridges between the backbone and the tail circuits. One of the NIUs is designated as "master" for control and monitoring of the backbone ring. The master NIU control software is currently personal computer-based with an RS-232 interface to the NIU, but is being ported to the central network manager workstation.

As a final note of background information, the implementors of the MCC network manager were under intense schedule pressure related to contract issues. The existing contract was due to expire, and the winner of the new contract had not been selected. This forced NASA to hold to an accelerated schedule throughout the design and development cycle.

3. The Lessons

This section presents seven lessons that we learned as a result of the definition and implementation of the network manager. Each lesson presents the ideal case, our experiences, and suggestions for other implementors.

Lesson 1: Seek help from experienced network administrators.

Ideally, when one sets out to build a network management system, network administrators who are experienced with the selected network equipment and the operational environment are available to assist in requirements definition and design review. These users can assist in at least three key areas: definition of the user interface, definition of appropriate statistics and methods for collection, and identification of areas for productivity enhancement tools. Users should be involved in display definition, definition of display-to-display transitions, and the identification of appropriate levels of flexibility and "streamlining" in the user interface. Experienced users can lend expertise in defining appropriate defaults for statistics collection and storage, and appropriate "tunability" of the statistics collection functions. Finally, experienced users can identify key areas for productivity enhancement tools, such as automation of frequently-used diagnostic procedures, simplification of LAN configuration management and change procedures, and definition of programs that provide expert assistance appropriate to the given network environment.

In the MCC, the designers of the network manager did not have access to experienced network administrators who were familiar with the types of LANs being installed in the MCC. As a result, the MCC network manager is not as easy to use as it might be, and does not make the best use of the system resources. For example, when the MCC network manager software polls the end systems, it collects and stores a great deal of information about each one for subsequent analysis. The network administrator has the ability to control the frequency at which end systems are polled, but not the information gathered on each polling cycle - all available information is reported. Current estimates indicate that our disk allocation of 40M bytes for management statistics will be consumed within approximately eight hours at the maximum polling rate. A review by experienced users would have allowed specification of a better default set of statistics that could be periodically gathered and provided the flexibility to easily revise that set.

Suggestions: When developing the requirements for a network manager, get help from experienced network administrators. Ask them to review the network manager design, with particular emphasis on the user interface design. If the local user community has no *applicable* experience with network administration, seek outside assistance.

Lesson 2: Host your network manager on a machine that vendors care about. -or- Don't buy support products that you'll end up debugging later.

In an ideal situation, the designers of the network manager pick an execution platform in response to the network manager design and the selection of necessary support products. If execution platform selection is restricted, and if those restrictions affect the availability or maturity of selected support products, designers should schedule time for reliability demonstrations of the selected support products. These actions will ensure that the system on which the network manager is built is capable and stable.

In the MCC, the execution platform was selected in advance of the network manager design. Few of the necessary support products, such as the database manager and the human interface development tool, had already been ported to this platform. The schedule was such that the software ports had to be performed in parallel with system development, with no time for reliability demonstrations. As a result, some of the key capabilities were not successfully ported. For example, the port of the human interface development tool didn't include either a color or mouse capability. Further, the support products did not exploit the architecture of the MCC workstation to improve their own performance - they were as close to "plain vanilla" System V ports as possible. This is understandable from the vendor's perspective, but the architectural advantages of the workstation are wasted.

Suggestion: Pick an execution platform in response to the design of the network manager and availability of required support tools. If software ports of necessary support products are required, implementors should develop test scenarios for reliability demonstrations. These test scenarios should represent how the product will be used during the operation of the network manager, and support products should successfully complete these reliability demonstrations before being procured.

Lesson 3: Understand network management concepts, applicable standards, and applicable products.

In an ideal situation, the implementors have detailed knowledge of all aspects of the system they are designing. The implementors understand the equipment being managed and its most likely failure modes. Further, the implementors understand the way that the users intend to operate the system. Finally, the implementation team understands the process of network management, the standards related to network management, and the products that are available to simplify their implementation.

In the MCC, the implementors had good experience with the LAN equipment, since they had been involved with its specification, procurement, and installation. As previously mentioned, there were no experienced network administrators available, so the concepts of operation for the network management system were not well documented. The implementors had no prior experience in the development of a network management application, and were generally unaware of the current state of network management practices. The implementation team did attend commercially-offered classes in network management, but this occurred late in the design cycle. As a result the implementation effort was underestimated, which has caused a schedule slip and the deferral of some requirements.

Suggestions: Have experienced users document the concept of operations. If possible, the implementation team should "get a feel" for network administration by sitting in with administrators while they manage a similar network. Implementors should seek adequate levels of training early in the requirements definition phase of the development. Finally, the implementors should locate existing network managers that meet similar requirements and evaluate the good and bad points of these as part of the network manager design.

Lesson 4: If you conform to emerging standards, be prepared to take some lumps.

If the world were perfect, standards would be stable and well-supported from their inception. Changes to the standards that occurred over time would never require changes to existing software that was built upon that code.

The MCC is not being implemented in a perfect world. NASA adopted TOP 3.0 as the protocol specification for its General Purpose LAN, including the TOP 3.0 specification of the Common Management Information Protocol (CMIP). CMIP was an emerging standard at the time, and TOP 3.0 was clearly an interim step. This selection presented both advantages and disadvantages.

There are two significant advantages. First, TOP 3.0 CMIP was purchasable, which saved development costs associated with the protocol implementation and management resources supported by the product. Second, for the resources not supported directly by the product, the TOP 3.0 CMIP specification and the example code accompanying the protocol product have provided a reasonable model for the interaction between managers and managed resources and the software basis for extension.

There are, however, some disadvantages with selection of an emerging standard such as TOP 3.0's CMIP. First, since the specification was relatively new at the time the MCC network manager was designed, not all the equipment to be managed had TOP 3.0 management interfaces available. This has necessitated some development that probably would not have been required had the standard been more mature. Second, since the selection of TOP 3.0 network management, the state of network management has evolved. Eventually, some equipment procured for the MCC will no longer support TOP 3.0 interfaces. At that point, the MCC network manager will be faced with a choice of upgrading the existing manager-agent protocol, or supporting a second set of management protocols. When the situation is encountered, a trade off analysis must be performed to weigh the additional development costs of an upgrade to the existing equipment against the additional sustaining cost of a dual-protocol solution.

The result to the MCC of the selection of an emerging standard has been a reduction of development costs when compared to the selection of no standard at all. Due to evolution of the standard, there may be some additional development or sustaining costs in the future, but these may be avoided if vendors provide backward-compatibility in their products.

Suggestions: If an emerging standard is well-designed and mature enough to be available as a product, consider it as an alternative to designing and implementing similar capabilities from scratch. Be aware of other users' experiences with the standard and the products, and examine possible future requirements for retrofit or design expansion of the network manager if the standards change.

Lesson 5: You don't necessarily have to use all the management capabilities you bought.

In the ideal case, implementors can tailor their purchase of management capabilities to buy only what is required, or integrate the additional management capabilities without additional cost.

In the MCC, the MAP/TOP 3.0 network management protocol implementation was purchased from Retix. In addition to the protocol software, this implementation included resource classes and layer management entity interface code for each layer of the protocol stack. However, there are costs associated with incorporating this software into the agent and the manager applications. As a result of a cost-benefit trade, the network manager for the MCC does not manage the upper layer protocols, because the benefits of doing so did not appear to justify the costs. The following paragraphs describe the costs of upper-layer management, then the benefits to MCC.

To provide upper-layer management, the network manager application must incorporate display and control mechanisms appropriate to those layers. However, a more difficult task is incorporating upper-layer statistics gathering into the agent. In the MCC's network implementation, the upper-layer protocols are linked into applications that use network services. These applications use an inter-process communication (IPC) interface to communicate with the (single) process providing the local lower-layer network services. The agent software exists as its own process, and uses IPC mechanisms to communicate management information to the lower-layer protocols. In order to manage all instances of the upper-layer protocols in this environment, a means of registering operational upper-layer entities to the agent would have to be defined. Each of the operational upper-layer entities would have to communicate with the agent via an IPC interface.

The benefit of upper-layer management is that it provides the network administrator information about upper-layer failures. There are two main types of failures that are reported by the upper-layer protocols in a TOP environment: failures due to incompatible end-system configurations and failures due to protocol errors. In the MCC's relatively homogeneous environment, the capabilities of the end systems are known, and negotiation of upper-layer connections should not fail. Again, in a relatively homogeneous environment, protocol errors should not occur. It is possible that, in a highly-congested end-system, data could be lost at the interface between the application (with the upper-layer protocols) and lower-layer protocols. This data loss would most probably manifest itself as an upper-layer protocol error.

Without an interface between the agent and the upper-layer protocols, the MCC is at some risk of failing to communicate protocol error information to the central network manager. The complexity of gathering the upper-layer statistical information was considered to be substantial enough to justify this risk. Omitting this information simplified both the agent design and the manager design.

Suggestion: Examine the costs associated with managing each resource, and weigh those costs against the benefits of having that centralized management capability. Be sure to take into account the influences of the system operating environment and the network management requirements when evaluating these costs and benefits.

Lesson 6: Not all resource management has to conform to a single standard.

Ideally, all resources that are to be managed would be supported by the selected management standard. In a slightly-less-than-ideal world, the cost of bringing all resources into conformance with the selected standard would be low enough to make this an easily-justifiable decision.

In the MCC, there is a combination of OSI-conformant communication resources and non-OSI resources. A dogmatic approach to network management would have dictated that TOP 3.0-conformant agent interfaces be developed for all of the non-conformant resources. The MCC employed what appears to have been a more pragmatic approach of implementing some agent interfaces for their non-conformant resources, while leaving custom interfaces to other non-conformant resources, based on the costs and benefits associated with each.

The Real-Time Messaging Protocol (RTMP) is an example of a non-TOP 3.0-conformant resource that is managed using the TOP 3.0 agent. RTMP is supported on the same end systems that support the OSI protocols, and uses the same agent software as its interface to the central network manager. Incorporating RTMP into the agent software was a relatively simple undertaking, requiring the development of a layer management entity for RTMP, the definition of a custom resource class to identify the RTMP-specific information, and development of parsers and formatters for the information.

Some non-conformant resources in the MCC do not use the TOP agent for management. One example of this type of resource is Network Interface Unit that is the bridge between the backbone LAN and the IEEE 802.3 tail circuits. As mentioned in the Background section, the backbone LAN uses a proprietary token ring protocol. The backbone LAN is controlled and monitored with PC-resident software that communicates with one (arbitrarily selected) of the backbone's NIUs via RS-232. In order to integrate management of the backbone into the workstation hosting the rest of the management software, this control and monitoring scheme had to be changed.

Three alternatives existed for integrating the backbone LAN management with the other network management capabilities: retain the control PC and port the network management agent software to the PC, adding a GP LAN interface to the PC in the process; do away with the control PC by porting the agent software directly onto the NIU; or do away with the control PC, port the control software to the manager workstation, and integrate the management functions at the user interface level. Supporting the agent on the PC would have involved creating more new resource classes, porting the agent and a supporting protocol stack to the PC, and adding hardware to the PC for a LAN interface. This was judged to be a very high cost approach, both in terms of development cost and sustaining cost. The second approach, porting the agent software to the NIU, was not possible because the NIUs are not programmable (by purchasers). The final alternative, which was selected, required porting the control software to the network manager workstation and using the one of the workstation's RS-232 ports to control the master NIU of each backbone LAN. This approach minimizes the required software development and allows removal of the PCs.

Using cost-benefit analyses to decide how to manage non-conformant resources should result in an acceptable trade between development costs and life-cycle costs in the MCC.

Suggestions: In deciding whether to implement a standard management interface for equipment that does not already provide such an interface, implementors should compare the cost to the benefits. There are three major benefits:

- simpler incorporation of management information into the management application,
- possible simplification of upgrade path in the future (for upgrade of managed resource or manager), and
- for resources that have no current management capabilities, a well-defined specification for control and monitoring.

The corresponding costs of implementing a standard management interface are as follows:

- possibly porting the network management agent to a new execution platform,
- specification, implementation, and test of a mapping between the managed resource's native management information and actions and those specified in the standard, and
- development of a layer management entity.

Lesson 7: If you can buy a network manager that meets your requirements, buy it!

Ideally, system implementors could buy a network management application that exactly matches their requirements, and at a reasonable price.

In the MCC, the execution platform was dictated to the implementors of the network manager. While a source-code implementation of a MAP/TOP 3.0 network management application was available, the implementors thought the port of this code would be too difficult, since both the supporting protocol stack interface and the window system interface would have to change. As a result, the network manager was designed and implemented from scratch. In retrospect, procurement of this available implementation could have served as a requirements-definition tool, a learning aid for the implementors, and possibly as a basis for the delivered network manager. A cost-benefit analysis comparing use of the available implementation in its native environment versus porting it or developing the application from scratch may have justified a waiver of the requirement dictating the execution environment.

Suggestions: Implementing a network manager is expensive enough to warrant a search of available network management products. Consider using a purchased network manager as a requirements definition tool and user training tool, even if it is not used in the operational system. Even if a network management product does not meet all requirements for a particular project, there may be value in adding external capabilities to the purchased product, or sacrificing some less-important requirements.

4. Summary

This paper has presented a few of the lessons learned during the implementation of the network manager for the Space Shuttle's Mission Control Center. The following paragraphs summarize the key points.

Success in developing a network manager begins with the definition of the manager's requirements. If the local user community does not have a great deal of experience with network administration, seek outside assistance in defining and reviewing the requirements.

Careful consideration given to the appropriateness of the network manager's host system and to the experience of the implementors can prevent unexpected delays in the implementation and unmet requirements.

Using network management standards appropriately can reduce development costs. Depending upon the maturity of the standards and the changes expected in the operating environment, network management standards can also be effective tools in reducing life cycle costs.

Possibly the most important lesson learned in the development of the MCC network manager is "If you can buy a network manager that meets your requirements, buy it!" If such an implementation is unavailable, and the requirements cannot be changed, perhaps these lessons can help make the implementation proceed more smoothly.

Acknowledgement

The author would like to thank John W. Linn, III for the application of his unique motivational methods, which contributed directly to the preparation of this manuscript. This work was performed under NASA contract NAS9-18057.

References

IEEE Standards for Local Area Networks: Carrier Sense Multiple Access with Collision Detection (CSMA/CD) Access Method and Physical Layer Specifications, ANSI/IEEE 802.3-1985, 1984, New York, New York: Institute for Electrical and Electronics Engineers

IEEE Standards for Local Area Networks: Logical Link Control, ANSI/IEEE 802.2-1985, 1984, New York, New York: Institute for Electrical and Electronics Engineers

Information Processing Systems-Data Communications-Protocol for Providing the Connectionless-mode Network Service, ISO 8473, 1988, Geneva, Switzerland: International Organization for Standardization.

Information Processing Systems-Open Systems Interconnection-The Directory, ISO 9594, 1990, Geneva, Switzerland: International Organization for Standardization.

Manufacturing Automation Protocols, Version 3.0 Implementation Release, MAP 3.0, 1987, Dearborn, MI: Society of Manufacturing Engineers

Technical and Office Protocols, Version 3.0 Implementation Release, TOP 3.0, 1987, Dearborn, MI: Society of Manufacturing Engineers

Acronyms and Abbreviations

CMIP	Common Management Information Protocol
CPU	Central Processing Unit
GP LAN	General Purpose Local Area Network
IPC	Inter-process Communication
LAN	Local Area Network
LLC	Logical Link Control
MAC	Media Access Control
MAP	Manufacturing Automation Protocols
MCC	Mission Control Center
NASA	National Aeronautics and Space Administration
OSI	Open Systems Interconnect
PC	Personal Computer
RT LAN	Real Time Local Area Network
RTMP	Real Time Messaging Protocol
STS	Space Transportation System
TOP	Technical and Office Protocols

IV

IMPLEMENTATION
AND CASE STUDIES

C
Panel

Polling vs. Event Reporting:
What are the Tradeoffs ?

Moderator:
Keith McCloghrie
Hughes LAN Systems, Inc.

Almost every Network Management architecture includes the concept of asynchronously generated messages being sent to the Network Operations Centre as a means of alerting a Network Manager to error or unusual conditions. However, there is large variance in the use of, and reliance placed upon these asynchronous messages by today's major Network Management architectures. In particular, Telecommunications management has traditionally placed a very high reliance upon them. In contrast, TCP/IP network management (SNMP) places very little reliance upon them. OSI network management tends to fall somewhere between these extremes. This panel will explore the advantages and disadvantages, costs and tradeoffs of placing reliance on these asynchronous messages, including discussion of the following:

- The amount of network traffic required.
- The robustness in critical situations.
- The latency of notifying the Network Manager.
- The amount of processing in managed devices.
- The tradeoffs of reliable versus unreliable transfer.
- The suitability for particular types of network management communication.
- The contingencies required in case a notifying device dies before sending an asynchronously generated message.

V
TELECOMMUNICATIONS
NETWORK MANAGEMENT

Management of Telecommunications Services Provided by Multiple Carriers

Wladyslaw (Walter) J. Buga

AT&T Bell Laboratories
480 Red Hill Road
Middletown, NJ 07748
USA

This paper discusses management concepts for multi carrier[1] telecommunications services with a focus on international services and recommends approaches to be taken to meet customer service management needs. Recommendations are made on implementation of a management interface between service providers, and potential benefits of it are highlighted. The recommended interface is modeled upon the OSI layered model. All major blocks of the interface are addressed, with a focus on application layer and management processes blocks.

1. INTRODUCTION

The technological progress in communications, computing and transportation is making geographical boundaries of the world invisible. Many national corporations are expanding to become, or currently are, multinational. Their presence in foreign countries ranges from a few locations in one country to many locations throughout the world. As these companies become global, their communications and information movement and management needs are becoming global. Successful implementation and operation of the international networks and services require cooperation between Administrations and RPOAs to share and exchange service and network related data during network and

---

1. The term *carrier* used in this paper refers to telecommunication services providers such us Telecommunication Authorities (TAs); Post, Telephone, and Telegraph (PTT) Administrations; Registered Private Operating Agencies (RPOAs); Interexchange and Local Exchange Carriers.

344

service design/expansion, provisioning and operations. This paper's focus is on aspects of the management planning for these international services required to ensure service quality and manageability.

In the past (prior to introduction of direct dialing), the telephone operator served as the interface between different carriers' networks. Introduction of direct dialing for the international long distance service added a new capability, the International Switching Center (ISC), providing a mechanized gateway between switched telecommunications networks. As Common Channel Signalling No. 7 (CCITT#7) networks, and Intelligent Networks become a reality, sophisticated new multicarrier services will be offered in the 1990s. Some of the more advanced services may implement intelligent database to database communications using CCITT SS7 protocols. Such implementation is based upon the assumption that call processing databases will be distributed across the networks of all involved carriers. These new service architectures will pose a challenge to service planners and managers.

Unique aspects of international services are: significant time differences, language barriers, different technologies and standards, different network management methods and different market needs.

2. MULTICARRIER SERVICE ARCHITECTURE AND SUPPORT

A multicarrier service architecture includes customer premise equipment located in multiple countries, local exchange carrier networks, domestic carrier networks international networks (including international facilities using terrestrial links, satellite links and undersea cables), foreign international carrier networks and foreign interexchange carrier networks. In other words, customer end-to-end services will be provided over a complex multi-carrier network configuration.

The customer's satisfaction will be determined by the quality of the offered service as characterized by the combined aspects of service support performance, service operability performance, service integrity, service availability and other factors,[1] on an end-to-end basis.

Customers may choose to work directly with each carrier for the portion of the overall service provided by that carrier, or may contract service management services to one of the involved carriers, or may contract such services to an independent contractor. Therefore, the service manager, responsible for such service, must obtain and maintain information on the operation, performance and repair of the customer's service from all involved carriers.

Service management functions to be supported include all five standard OSI Management Functional Areas (MFAs) as defined by International Standards Organization (ISO) and CCITT: Configuration, Fault, Performance, Accounting and Security Management. However, telecommunications services management requires additional functionalities such as planning, operations etc. Based upon the experience of managing telecommunications network and services, AT&T included four additional functional areas[2] in its Unified Network Management Architecture (UNMA). Other vendors and carriers recognized the need for the additional functionality in their management architectures, as well. As an example, British Telcom (BT) modified 5 standard MFAs and added new areas such as Financial Management.[2] IBM offers their view of network management functions which includes such areas as Change Management and Operations Services.[3]

Multicarrier service may be implemented on a set of network resources distributed across multiple and different types of networks. Global Software Defined Network (GSDN)^3 is an example of a such complex multicarrier service. Currently, GSDN service is offered to numerous countries based upon arrangements made with France Telcom in France, Regie des Telegraphes et Telephones (RTT) in Belgium, Mercury Communications LTD. in the United Kindom, Kokusai Denshin Denwa Co. Ltd. of Japan and Unitel of Canada. AT&T also has GSDN connections with British Telcom's International CityDirect and International FeatureNet, and with Call Plan service offered by OTC of Australia.

It is clear that in a such multicarrier arrangement service quality depends upon the ability of each carrier to meet the specified performance objectives and Operations, Administration and Maintenance (OA&M) agreements. Each carrier's network is controlled by that carrier and represents its own management domain. Each involved carrier must provide quality support within its own domain, and its management domain will have to interact with management domains of other involved carriers. Therefore, service management must hide from the customer the technical and administrative aspects of service implementation over multiple networks and to take the burden of service management from the customer. To fully understand the interactions between management domains, the relationships between service support processes, customer services and carrier networks must be also understood.

3. MANAGEMENT PROCESSES

In general, networks are made up from two parts: (1) physical equipment (such as transmission systems, switching exchanges, signalling equipment, supporting equipment,etc.) and (2) circuits, paths, links, etc. that are derived from them. These networks then carry the various types of telecommunications services to meet the needs of the customers and other users. Networks and services need to be provisioned, maintained, reconfigured, etc., to meet user requirements. These activities, performed by the carrier's technical personnel and with Management Systems (MS) assistance, are defined as OA&M or management processes.

The management processes require interactive information flows between them and all aspects of the network and services they provide. These processes should be considered from both physical and logical viewpoints in order to take into account all of the required interactions between services and networks. Figure 1 shows such relationships. Analysis

---

2. These additional MFAs are: Planning, Operations, Programmability and Integrated Control.
3. GSDN is AT&T's international service offering and is also known as an International Virtual Private Network (IVPN).

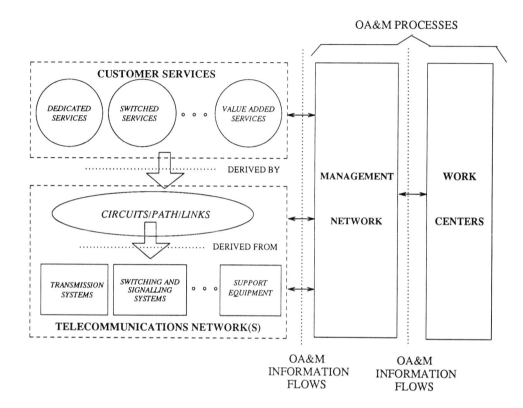

Figure 1. Relationships Between Network, Service and Service Management

of management processes indicates that they are hierarchical in their nature, and can be divided into layers.

Different carriers may view this management hierarchical structure their own way. As an example, AT&T's UMNA recognizes three layers: network elements, element management systems and integration systems. BT's ONA Management Architecture[2] recognizes five structural layers: network elements, element management, network management, service management and business management. IBM's network management structure defines[3]: an entry point, a service point and a focal point, and their management roles. Also, the layered approach is used to model network and service management processes by various groups defining network management standards[4] and service maintenance principles[5]

Generally, in such layered structure the lower layers will support the higher ones and the higher layers are not concerned with the lower layers' details but with its functions. Not all layers are required to exist within these processes. It is logical to make an observation that management requirements and methods may be different for each layer. The tasks performed at each layer are also different. However, these layers interact and pass information between them. Higher layers rely on lower layers to provide the needed resources for services. Also, the lower layers will pass to the higher layers only that management information which is relevant to its functions and services.

Without arguing which or whose layered architecture is the best, let us focus on processes which deal with service management.

4. SERVICE MANAGEMENT PRINCIPLES

The service management process is concerned with, and responsible for, the contractual aspects of services that are being provided to customers or available to potential new customers. It has the following functionalities:

- Interfacing with the customer,
- Interfacing with other carriers/Administrations/RPOAs,
- Interactions with layers above and below, and
- Performing service management functions.

Service management functions will be driven by the customer requirements. In order to meet customer needs and expectations, AT&T adopted the following service management principles:[6]

- Total service support - *Global end-to-end service mangement, including single point of contact.*
- Unified end-to-end performance objectives - *Defined service quality across the global network which consists of networks of all involved carriers, including allocations to each network.*
- End-to-end provisioning and maintenance responsibilities - *Coordinated service installation by all involved carriers and service quality assurance, including allocations to each carrier.*
- Coordinated service design - *Mutual agreements and definition of global service architecture and operations requirements.*
- Communications methods - *OA&M information exchange capability between carriers.*
- Security of information - *To protect vital customer and network information.*

The above principles were adopted by AT&T for their international offerings such as GSDN service. All management processes (on an end-to-end basis) were defined and implemented by all involved carriers prior to service introduction. Within AT&T, the Software Defined Network Control Center (SDNCC) was designated as a single point of contact for all GSDN service related activities. Also, its detailed responsibilities and requirements for the AT&T service portion were defined.

5. END-TO-END SERVICE SUPPORT

In order to provide end-to-end service, service management processes in one carrier's management domain[4] have to communicate with service management processes in other carriers' domains. These communications, in most cases, will be done on a peer to peer basis. Interaction may also occur in layers above and below service management functions (e.g., facility maintenance functions). However, the specific arrangements on the level of the interaction are the policy issues and must be negotiated between interested parties, as they are subject to the following constraints:

- Technological
- Competitive
- Political
- Operational

Service management processes within the carrier management domain may require human action or assistance or they may be automated. Therefore, interaction between management domains will involve both human assisted and automated processes. The degree of their involvement (human vs automated) will vary from carrier to carrier as the technological and operational capabilities will vary from carrier to carrier.

6. INTERDOMAIN INTERFACE REQUIREMENTS

The success of multicarrier service support will rely heavily: (1) on the ability to communicate with management domains of all involved administrations and carriers and (2) on the ability to exchange in a timely manner service and network OA&M information that directly affects the customer. The communications between these domains may be accomplished by: mail, telephone, telex, facsimile, electronic mail, electronic transaction processing, electronic file transfer, etc.) The first four communications methods are available within most carrier domains and therefore are the most commonly used in today's OA&M environment. However, these methods do not lend themselves to automation and are error prone.

In addition, multi-carrier service management will introduce needs for near-real time and highly accurate OA&M information flows between work centers and OA&M data bases of the involved carriers. To meet such requirements, electronic mail, electronic transaction processing and electronic file transfer should be considered by the service providers. Since OA&M networks will carry vital customer and network information, security is of paramount importance.

---

4. The term management domain is used in this paper for the purpose of describing the area or scope of management control. In this particular instance "domain" is used in sense of administrative or ownership control. An interesting discussion of management domain concepts can be found in reference.[7]

It is required that the interdomain interface should possesses the following characteristics:

- provide interoperability between management domains,
- provide a uniform definition of application functions,
- minimize ambiguities during information interpretation,
- automate information transfer, storage and retrieval, and
- provide access control and protection of customer and network information.

7. INTERDOMAIN INTERFACE ARCHITECTURE

ISO and CCITT are in the process of defining the international standards for management of open systems, telecommunications networks and telecommunications services. Both these organizations have adopted object oriented techniques in definitions of their standards. Therefore, the telecommunications networks and services can be viewed as a set of managed objects (see Figure 2).

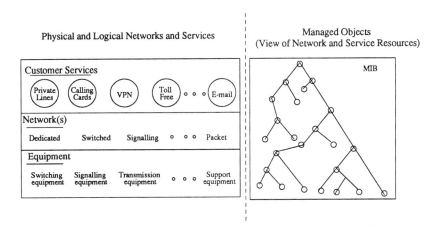

Figure 2. Object Oriented View of the Network and Services

The managed objects including their attributes will constitute the Management Information Base (MIB). It is expected that the managed objects will follow the hierarchical structure of the network/services and management processes. Management activities will be performed through manipulation of managed objects, by using standard management services. In order to define an architecture for the interdomain interface, based upon the requirements specified in Section 6, this interface will be modeled upon ISO's OSI layered structure. Therefore, the general structure of the interface may be defined as in Figure 3. Five major building blocks are identified:

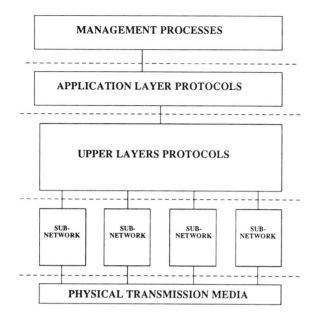

Figure 3. Interdomain Interface Structure

- Management processes,
- Application layer (7),
- Upper layers (4-6)[5],
- Sub-networks (1-3), and
- Physical transmission media.

A brief discussion on each block's characteristics will be provided in the following subsections.

7.1 Management Processes

An application in the sense that is used in OSI is an application which is implemented in the computing environment (hardware and software). As an example, such implementation of the management processes discussed in Section 3 of this paper would

---

5. Usually OSI transport layer is considered to be a part of the lower layer infrastructure. The reason for including it in the upper layer structure is to contrast end-to-end functionalities provided by the transport layer vs functionalities provided by subnetworks.

consist of application process(es). OSI does not standardize applications; however, CCITT may standardize the network applications that are used to support the various management functions. The OSI approach is to standardize the interactions that can occur between those parts that reside in different end systems. The part of an application that resides in a single end system is called an application process. The part of an application process that takes part in the interaction with another application process is called an application entity. Therefore, an application entity represents the communications functions of a management process. The process of defining an interface will begin with the selection of the management applications for the interface. The major management application for the multicarrier services is the exchange of service management information between involved carriers. Such an application may be a subject for standardization by CCITT.

7.2 Application Layer

The application layer must be able to perform the functions necessary to communicate any information the management processes need to convey to a remote peer. Unlike the other layers in the OSI stack, the application layer must provide functions that are application specific. Therefore, the form and content of the functions in the application layer are fully dependent on the needs of the application process using these functions. Figure 4 shows the application layer structure as it may be relevant to telecommunications applications.

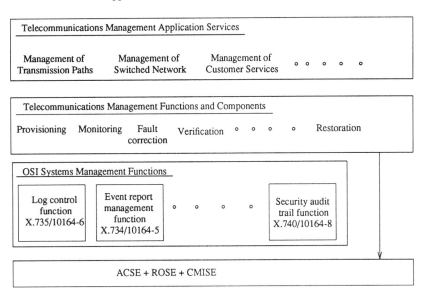

Figure 4. Application Layer Structure

The functionalities that have been identified in this layer include: communications protocols and services, OSI management services, telecommunications management functions/components and telecommunications management application services.

Communications protocols and services included here are these defined by ISO/CCITT and they include ACSE,[8] [9] ROSE,[10] [11] and CMISE.[12] [13] ISO and CCITT also address various systems management functions such as Object Management function,[14] Alarm Reporting Function,[15] Test Management Function[16] and others. The reason for the standardization of these functions is to provide general management functionalities, such as alarm reporting or test management, for the specific applications.

Telecommunication management functions are currently being studied by CCITT, and relevance of systems management functions is also under study as well. The objective is that these functions should provide the generic functionalities for all telecommunications activities such as circuit testing, provisioning, restoration, etc. The top layer specifies telecommunication management application services which are applicable for specific telecommunications areas such as exchange maintenance, traffic management, etc.

7.3 Upper Layers (4-6)

Upper layers (4-6) are positioned between subnetworks and the application layer. These layers provide a set of well defined functions and services that are contained in a monolithic layered structure of the presentation and session layers. In addition, the transport layer ensures reliable data transfer between end systems. The issue here is the selection of a specific subset of functions and services to be implemented for the interdomain interface.

7.4 Sub-networks

A vital component between remote communicating systems is a real communications network (eg. PSTN, packet switched, etc.). In order to accommodate different network technologies without redesign application over any given real network, the network layer has been introduced. In addition to being real-network independent, the network layer has to be able to organize the interconnection of many different real networks in such a way as to provide the appearance of a single network to the user of end-to-end services.

Currently, telecommunications carrier networks consist of multiple type subnetworks. The major ones are listed below.

- PSTN (POTS)
- Private Lines
- Packet switched
- SS 7 signalling links

Early consideration should be given to the proper selection of the sub-networks to ensure interworking. It is recommended that the following criteria be used during the selection process.

- Reliability requirements

- Availability requirements
- Security requirements
- Traffic requirements
- Performance requirements
- Cost

7.5 Physical Transmission Media

The objective of the physical layer is to mask, as far as possible, the characteristics of physical media. Some consideration should be given to the physical medium as it may affect transmission performance.

8. RECOMMENDATIONS

To meet customer expectations for service management, the whole telecommunications world is moving from monopolistic arrangements to business driven arrangements reflecting a competitive environment, and is becoming more customer oriented. Also, as stated in this paper, management of telecommunications services provided by multiple carriers requires interactions between involved carriers' management domains. Therefore, it is expected that the majority of telecommunications carriers will be implementing interdomain interfaces and operating agreements to maintain a competitive position.

It is recommended that interdomain interfaces be implemented based upon TMN[6] principles. The TMN concept is recommended due to its attributes, which are:

- International standard - *this should provide interoperability between and various carrier OA&M networks.*

- Predefined application functions - *this should provide a uniform shopping list of possible TMN application functions to perform an application service.*

- Messages based on objects and its attributes - *this will provide a uniform object definition and naming convention, therefore, minimizing ambiguities during information interpretation in different languages.*

- Object oriented design - *this will allow for easy control of the access to information (hiding/ disclosing) by classifying information as private or public.*

- Automated information transfer, storage and retrieval - *this will minimize impacts of time differences.*

---

6. Telecommunication Management Network (TMN) is a concept specified in CCITT Recommendation M.30. The idea behind a TMN is to provide an organized structure to achieve interconnection of the various types of management systems and network elements using agreed upon architecture and interfaces.

- Security management - *this should provide access control and protection of customer and network information.*

Currently CCITT is still working on the definition of TMN and its interfaces. However, it is expected that in the next study period the specific standards will be defined for most of the management functional areas.

It is understood that the implementation of TMN will take many years and not all carriers may wish to implement it. This implies that a carrier will exchange the same service management information using different technologies (TMN and non-TMN) with different carriers. Therefore, the objective of the implementation should be such that the interdomain interface be applicable to all or majority of interconnections. In other words, such an interface should provide a capability to communicate via different transport networks without the need for changing the application (e.g., sending the same trouble ticket via Fax, E-mail or TMN).

9. CONCLUSION

Implementation of an international interdomain interface, compatible with TMN standards, by telecommunications carriers will make invisible geographical boundaries. It will provide network and service management capabilities to allow telecommunications carriers to offer fully global (end-to-end) telecommunications services for a growing number of multinational business corporations with global communications needs. It also will allow full implementation of service management principles outlined in this paper, which will ensure that the customer service will not be impaired when provided by a multinetwork configuration.

10. ACKNOWLEDGMENTS

The author would like to express appreciation to his colleagues in AT&T, in particular, to Joel Boroff, Ken Hanson, Dave Sidor, Vern Werth and Doug Zuckerman for their reviews of this paper. The author wants also acknowledge that some of the ideas presented here were shaped up during his discussions with members of T1M1.5 Committee and CCITT SG IV.

REFERENCES

1. CCITT Recommendation G.106, "Terms and definitions related to quality of service, availability and reliability."

2. David J. Milham and Keith J. Willetts, "BT's Communications Management Architecture", First International Symposium on Integrated Network Management, Elsevier Science Publisher B.V. (North-Holland), May, 1989.

3. Randall Cambell et al., "IBM's Network Management Approach", IEEE Communications Magazine, March 1990, Vol.28 No.3.

4. CCITT Draft Recommendation M.30, ANNEX B, Q.23/IV Expert Group meeting, Paris, October 1-5, 1990.
5. CCITT Draft Recommendation M.1x, Q.6/IV Expert Group meeting, Geneva, April 3-4, 1990.
6. W.J. Buga, K.L. Hanson, "Considerations on operations planning for new global telecommunications services", GLOBECOM '89, Dallas, Texas, November 27-30, 1989.
7. M. Sloman and J. Moffett, "Domain Management for Distributed Systems", First International Symposium on Integrated Network Management, Elsevier Science Publisher B.V. (North-Holland), May, 1989.
8. ISO 8649 : 1988, "Information processing systems - Open Systems Interconnection - Service definition for the Association Control Service Element" (CCITT X.217)
9. ISO 8650 : 1988, "Information processing systems - Open Systems Interconnection - Protocol specification for the Association Control Service Element" (CCITT X.227)
10. ISO/IEC 9072-1 : 1989, "Information processing systems - Open Systems Interconnection - Remote Operations - Part 1 : Model, Notation and Service Definition" (CCITT X.219)
11. ISO/IEC 9072-2 : 1989, "Information processing systems - Open Systems Interconnection - Remote Operations - Part 2 : Protocol Specification" (CCITT X.229)
12. ISO/IEC 9595 : 1990, "Information technology - Open Systems Interconnection - Common Management Information Service definition"
13. ISO/IEC 9596 : 1990, "Information technology - Open Systems Interconnection - Common Management Information Protocol specification"
14. ISO/IEC DIS 10164-1 "Information technology - Open System Interconnection - System Management - Part 1: Object Management Function" (CCITT Draft X.730)
15. ISO/IEC DIS 10164-4 "Information technology - Open System Interconnection - System Management - Part 4: Alarm Reporting Function" (CCITT Draft X.733)
16. ISO/IEC Draft 10164-? "Information technology - Open System Interconnection - System Management - Part ?: Test Management Function" (CCITT Draft X.tmf)

V

TELECOMMUNICATIONS

NETWORK MANAGEMENT

A
Architecture and Design

An Architecture for the Implementation of an Integrated Management System.

Paul Senior.　　　　　　　　Bell Northern Research, Maidenhead, UK
　　　　　　　　　　　　　　(formerly Roke Manor Research Ltd.)
Steve Harris.　　　　　　　　Roke Manor Research Ltd.
Declan O'Sullivan.　　　　　Braodcom.
Yiannis Trodullies .　　　　　British Telecommunications.

Abstract:

The ADVANCE Project R1009 is part of the European RACE Programme (Research into Advanced Communications for Europe). ADVANCE is a project investigating the applicability of Advanced Information Processing (AIP) techniques to Network and Customer Administration Systems (NCAS) for the Telecommunications Management Network (TMN). The project is involved the development of prototypes that demonstrate the benefits of these new techniques so that they can be assessed for industrial applicability. ADVANCE has developed an implementation architecture that is compatible with the emerging TMN systems architecture and components currently being standardised by CCITT and ISO.

The motivation for Integrated Management.

A major criticism of existing network management systems is their lack of integration. Many stand-alone management applications have been developed in an ad hoc and incremental fashion and have dedicated and non-integrated information resources. Integrated information resources will be one of the key results of ADVANCE prototyping.

Some typical deficiencies in existing network management systems are:

1) Duplication of functionality within applications. For example, user interface services, which are common to a number of applications, are embedded separately within each application, rather than being available as a common service to all applications. (Furthermore, home-grown user interface implementations are often of poorer standard and may not be designed or developed by MMI specialists).

2) Inefficiency because no automatic means of communication between applications is provided.

3) Concentration of management functionality at lower management layers eg existence of element management tools, but no tools to manage a network of elements.

4) Systems tend to be passive when applications cannot communicate with each other.

5) Operators have to cope with a multiplicity of terminals and MMIs.

6) There will be no control on system security if responsibility resides within individual management applications.

7) Lack of generally agreed terms and concepts leading to common information bases.

8) One of the key requirements on ADVANCE is to demonstrate the benefits of an integrated management system, and to demonstrate how such an integrated system might be constructed.

9) The implementation architecture for the integration of prototypes will explicitly address the deficiencies highlighted above.

Development of the Architecture.

The ADVANCE project which started in 1988 included an activity called Prototype Planning. Its goal was the development of an initial prototyping architecture. This architecture has now been updated in the light of initial prototype experiments in order to support the definition of Prototype Version One (PV1).

The developers of the architecture had two main goals. The first is to provide a framework which supports design implementation and test of network management functions in the form of management applications. The second is the longer term goal of defining the Telecommunications Management Network Operating System (TMNOS). The TMNOS is the infrastructure for NCAS and this is not to be confused with the TMN Operations Systems from CCITTs M.30 recommendations.

One of the key challenges facing the architects is the need to provide a fusion between computer industry standards and network management standards and architectures. Typically this means that there needs to be support for use of industry standard components such as relational databases and networking software in addition to the ADVANCE research requirement to support the application of AIP techniques within the same architecture.

External Influences.

Today there are numerous bodies that influence the design and implementation of management systems. World standards bodies such as CCITT and ISO are supplemented by region or country based organisations like ETSI and ANSI. These together with initiatives such as the OSI Network Management Forum and British Telecom's Open Network Architecture - Management (ONA-M) help to provide a common architectural understanding as well as interface and interworking mechanisms. It is clear that any system design and implementation must utilise the broad thrust of these recommendations.

CCITT's M.30

The CCITT Recommendation M.30, "Principles for a Telecommunications Management Network"[1], plays a central role in the architecture of any Telecommunications Management System. M.30 presents the general principles for planning, operating and maintaining a TMN. It states that the purpose of a TMN is to support (the) management of (a) telecommunications network. A TMN provides management functions to the telecommunications network and offers communications between a TMN and a telecommunication network.

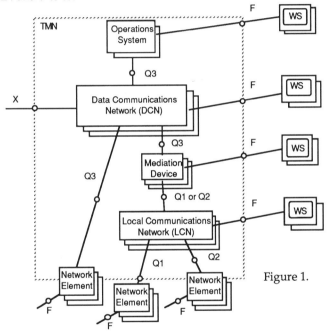

Figure 1.

M.30 details a general physical model for the TMN which is shown in Figure 1.

The basic concept behind a TMN is, therefore, to provide an organized network structure to achieve the interconnection of the various types of Operations Systems (OSs), which carry out the management tasks, and telecommunications equipment in the underlying managed network.

British Telecomms Open Network Architecture (ONA-M).

Before any design and construction of a Management System can take place, a further refinement of the Operations System (OS) is required. The functionality that resides within the OS as stated by CCITT M.30 has no structure. One useful suggestion, of how this mass of functions can be better treated has been developed by British Telecom. Called ONA-M[2], this approach attempts to partition management into different roles. ONA-M partitions management systems in four layers, see Figure 2.

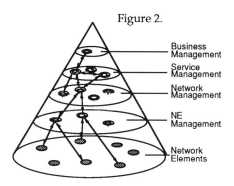

Figure 2.

"Management of Elements" or Element Management, is a type of management concerned and dedicated to the management of particular equipment. This type of management is highly individual, and specific to equipment types. "Management of Networks" or Network Management, can only take place when a management process has a view across the network. This type of management can be equipment independent, perhaps managing Q3 compliant systems.

The next layer, the "Service Layer" is one of the most important concepts. ONA-M splits the management of networks away from management of services. It assumes that services are provided across network(s) and that networks can, and should be managed independently from the services they supply. Therefore the "Management of Services", builds upon the underlying support provided by Element and Network Management.

Finally Business Management, or the "Management of a Business" addresses the decisions about the levels of service provision, priority etc... These are best separated and classified as aspects of management which are dedicated to a particular business organisation.

All these layers of management stack one upon another, and help to resolve responsibilities and provide a further classification which aides the construction of OSs for management systems.

Each of the different types (layers) of management each have distinguishing characteristics. Element management is concerned with, and may have to respond to, stimuli from hardware. This means that performance, and response times become important. Element management software is analogous to real-time software which controls exchanges. Network, Service and Business management layers are more "Data Processing" oriented. Accounting, planning etc... are less time critical and accordingly are generally analogous with current DP systems technology. The ADVANCE domain, ie. Network and Customer Administration exists primarily at the Network, Service and Business layers and is therefore more off-line in nature.

ISO Systems Management.

The first concept utilised from ISO is that of the Management Information Base (MIB). Conceptually the MIB[3] is the repository of all the data held in the management system about the network and the management system itself. The information base is required to store information on the network and system configuration, customers and services, current and historic performance and trouble logs, security parameters and accounting information etc... The MIB is seen as the heart of an integrated management system, since most of the information held within the MIB is shared between two or more functions.

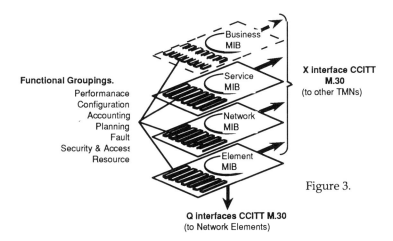

Figure 3.

The second major architectual concept adopted from ISO is that of object management. ISOs DP 10040 defines how using the CMISE application layer service managing processes, via agents, access the managed objects.

The management of a network by defining and using managed objects is now widely accepted and proposed by almost all network management standards bodies.

The ADVANCE view of a TMN Architecture.

By synthesising the architectures, models and mechanisms given above, ADVANCE has developed a view of the most important elements of a standardised TMN[4]. Figure 3 shows the layering of an OS, and the MIBs that exist at the different layers. It also shows that surrounding the MIB are the functions that make up any management system; these are present at all layers.

The Implementation Architecture.

The Implementation Architecture concentrates on providing a limited degree of transparency for applications when using common infrastructure components. To this end there is a major division of the architecture into two classes of component: *Management Applications* (MAs) supported by *Common Services* (CSs). In other words the network management functions implemented in the form of MAs are decoupled from CSs which provide a general purpose computing platform tailored to the needs of NM systems. This allows for the sharing of resources and for approaches to the design of applications which encourages their independence from communications, databases and man machine interfaces, see Figure 4.

Management Applications.

Network and customer administration systems split into three broad areas: customer, service and network applications. MAs contain the specific management functionality.

Common Services.

The architecture is mainly concerned with the definition of infrastructure components that are available to many application functions across an open distributed system.

In general there are three types of common components: information in the form of data or knowledge, wide-area communication either OS-NE or OS-OS, and MMI to a range of terminal types (ranging from the customer handset, to full-blown 24-bit colour graphical workstations).

Architecture Components.

The Common Service Manager.

The Common Service Manager (CSM) co-ordinates and provides a mechanism for Inter-Process Communication (IPC) between the elements of the architecture within the same local area network. The CSM provides the mechanism whereby messages can be passed and data exchanged across a number of machines within a management centre. This interface is characterised by the publishing (exporting) of operations or procedures that a CS or MA supports. Published operations and procedures may be used (imported) by other MAs or CSs,
deciding at run-time which handler is appropriate / available and passing on the message or request accordingly.

The CSM is seen to bind all the common service components together, providing a "Application Programming Interface (API)" to the applications designer.

The Access Manager.

The Access Manager is an element of the architecture concerned with the implementation of the security policy of the management system. It manages login and registration of users to the management system. Following validation of authorisation codes via retrieval of users type and preferences, the access manager provides access to selected MAs via an appropriate user interface. The classification of users, and the assignment of access rights occurs at many levels within the architecture. These range from access to particular applications to the functionality offered by handlers and common resources.

The System Monitor.

The system monitor can be seen as an application for the manager of the management system. Its purpose is to monitor, via the CSM, the messages and accesses to data from around the system, and provides a window on the system's operation.

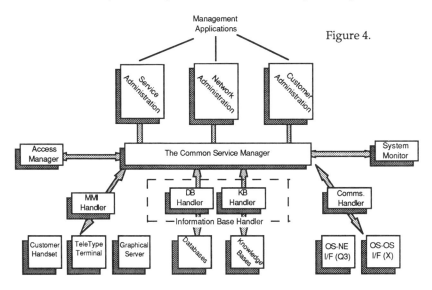

Figure 4.

It has a particular role in the ADVANCE demonstration context where it allows observation of the total system including internal operation.

The MMI Handler.

Conceptually the MMI handler provides a number of specialisations of the presentation of information to the user. The MMI architecture embeds six layers. It uses the Dialogue and Presentation layers in addition to the basic four layers of graphical user interfaces (GUI). The MMI handler selects the most appropriate way of presenting material to each user by a technique known as User Modelling.

This is the subject of a prototype implementation called Application to User Translation Operator (AUTO), which is part of the MMI Architecture as shown in Figure 5.

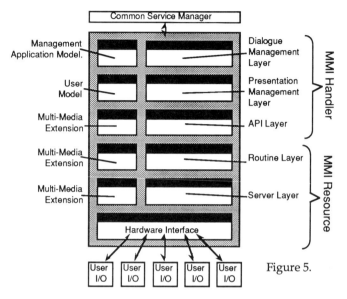

Figure 5.

The Database Handler.

The problems of database placement, database heterogeneity, and local and global schema are addressed by the Database Handler (DBH). The Database Handler provides MAs with three different database views. These are: object, relational and the Generic Message Set (GMS). The GMS is a joint database and knowledge base query language. The three views reflect the different needs of MAs to access managed objects, records, files and other information sources.

In Integrated Network Management Systems it will be important that customer data (such as names, addresses, account details etc..) can be treated in an integrated fashion. To achieve this, substantial re-engineering of databases and cooperation between future and present era data storage and retrieval mechanisms in the NCAS will be essential. All the databases introduced during the new era into the NCAS will be unlikely to use only one type of data model for storage of information. This will result in data being held in a heterogeneous systems distributed over a network.

ADVANCE is addressing data issues by incorporating data handling services into the NCAS architecture by means of a Database Handler module (termed the DBH). The DBH module will provide NCAS applications with a consistent interface to access and manipulate data, independent of actual location or storage format. A number of AIP techniques may be applied in the provision of the DBH for NCAS.

The most prominent approach in the IT world to the handling of data stored in distributed heterogeneous database management systems has been to use the relational model to present a unifying interface to applications and users. A single interface will not be sufficient in the future telecommunications NCAS environments. This is due to the complexity and diversity of applications to be supported. ADVANCE believes that a small number of interfaces will be needed instead. Relational oriented, file oriented and object oriented are good candidates.

By exploring options for the DBH concept in prototypes, ADVANCE has concluded that a mix of data handling services and DBH modules are needed in the NCAS.

The Knowledge-Base Handler.

Access to knowledge bases is co-ordinated via the Knowledge Base Handler (KBH). The interface provided by the KBH is the GMS, (which is also available for the DBH). The KBH offers a mechanism by which a variety of different knowledge sources can be queried. Access to these knowledge sources is obtained independent of how and where it is stored.

The implementation of the KBH utilises a blackboard onto which various MAs and CSs pose queries. The GMS provides a means by which the various part of the query can be handled via different knowledge sources.

The Information-Base Handler.

The Information Base Handler (IBH) supports the functionality of the DBH and KBH. The IBH provides operations that allow a combined way of querying both databases and knowledge bases together. The interface offered is the GMS.

The Generic Message Set.

The Generic Message Set (GMS) provides an open interface across the Management Applications and the Common Services.

The information provision aspects of application processes have to be formulated at different levels of abstraction. Many of these levels are implicit and not extremely useful. As an example, a knowledge base system contains raw data and a structure which transforms the raw data into a knowledge representation model of the system. The knowledge base system will also contain an inference mechanism which provides the intelligence for reasoning about the model.

The power of the proposed GMS is the ability to support access to different levels of abstraction. These information provision levels have been layered as follows:

- Task level.
- Inference level.
- Structure level.
- Object level.
- Data level.

The *Task* layer provides the highest level access point in to a management process. For example a Management Application is able to request the validation of a customer using the Task layer message - "Validate Subscriber 002".

The *Inference* level provides access to the internal inference mechanisms of the management processes. At this level the internal workings of the process are made public to external processes. These internal mechanisms can be examined and questioned by other applications.

At the *Structure* access level the model of the system can be examined and manipulated. These models may contain entities and relationships which interconnect the objects of the model.

At the *Object* layer the objects/entities within the system can be examined and manipulated. Applications will be able to inspect the attribute values of the internal objects, change these values and also create and delete instances of the objects.

The *Data* level provides access to pure data within a system. These data could be internal to a system or may be kept in a database.

The layered approach simplifies the overall structure of the GMS and provide a degree of modularity. The different layers are independent and will be developed separately. To date the design of three layers, Task, Object and Data, are almost complete and their syntax is available. The Inference and Structure layer are currently under definition. A detailed structure for the GMS has been defined. The work on GMS is an evolving activity and will mature as more information processing requirements are identified throughout the project.

The Communication Handler.

The Communication Handler (CH) makes communication services available to applications. There are three main types of communication possible: inter-management centre communication (related to OS-OS communication in M.30 Q3), inter-TMN communication (X) and TMN to network element communication (related to OS-NE communication where the NE supports Q3). All these types of communication are via an OSI 7 layer stack using application layer services such as FTAM, ROSE, ASCE, X.400, X.500 etc... The main type of communication supported by the CH is access via a Q3 interface.

From an integrated management system perspective, the ability for management entities to intercommunicate freely is essential for the provision of integrated management. Without integrated management the full potential of the network resources available will not be realised, leading to inefficient communications. However management systems will evolve gradually, with the integration of existing management systems and with the addition of new management systems. Consequently it is unlikely that all subsystems will communicate in a common manner. Furthermore, there will exist management domains, wherein separate integrated management centres will have jurisdiction over a particular network, for example on a national basis. In this situation the exchange of management data will require compatibility of communications services between the management systems. Therefore a situation will exist whereby management entities will wish to freely intercommunicate, and yet do so in a manner and by a means which will vary immensely from entity to entity. Moreover, this variety will be exacerbated by the fact that the management entities will have been communicating via different network types.

ADVANCE is attempting to address this issue, and will suggest the use of an interface at an equivalent "eighth" OSI layer. This will enable NCAS management entities to intercommunicate freely. The concept put forward and explored is that of incorporating transparent communications facilities into the NCAS architecture by means of a Communications Handler (termed a CH). The CH module is intended to provide NCAS applications with a consistent interface to access and utilise communications services independent of the type of information being transmitted, the end-to-end communications services available, and the intervening network infrastructure type. A number of AIP techniques may be employed in the provision of a CH for NCAS.

Status of the Architecture.

The architecture described above has been prototyped and at present supports a number of applications, including applications that manage service mobility and provisioning, invoice generation and scheduling, and configuration and planning tools for a GSM network.

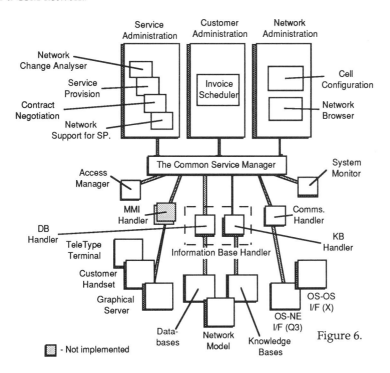

Figure 6.

The project is at present mid way through its 5 year programme, and has just released the first version prototype. The architecture will be developed over the next two years, ie. till project end in 1992. It is hoped that this evolutionary approach, continually assessing the requirements via prototyping, will provide a stable platform for the development of the next generation of management systems.

Figure 6 shows the current state of the architecture's implementation.

Acknowledgements.

The authors would like to acknowledge the help of many individuals during the preparation of this paper: George Williamson, of British Telecom for his part in the reviewing the ADVANCE Prototype Framework document, the members of the ADVANCE System Design Team under which the documentation for the architecture was developed, and the TCG2 team (the systems architecture team of RACE Project\GUIDELINE) who helped form the reference view of the TMN.

References.

1 CCITT Blue Book M.30 General Principles for a Telecommunications Management Network,1988.

2 Willets K., Online, A Total Architecture for Network Management, 1988

3 ANSI, Principles of Functions, Architectures and Protocols for Interfaces between Operations Systems and Network Elements, ANSI T1M1.5/88-LB-04, 1988.

4 Brown, Senior, Williamson, Schepler -"An Architecture for the Management of a Multi-Service Network" - 1990 International Switching Symposium (ISS 90).

A network management architecture for SONET-based multi-service networks

Shaygan Kheradpir, Willis Stinson and Gunilla Sundstrom

Network Architecture and Services Laboratory
GTE Laboratories Incorporated, 40 Sylvan Road, Waltham, MA 02254

Abstract

SONET-based networks promise to transport a variety of new services in the public network flexibly and efficiently. However, Network Management (NM) methodologies developed for Plain Old Telephone Service (POTS) networks are not sufficient for managing upcoming SONET-Based Multi-Service (SBMS) networks. Without a Network Management Architecture (NMA) to organize new NM tools and techniques, the promise of SBMS networks will not be fully realized. This paper motivates and outlines the design principles of an NMA necessary for effective and efficient operation of SBMS networks. These principles are realized in the NMA using three basic types of Operations System: NM applications, NM utilities and an integrated graphical support system. In this operations environment, human network managers cooperate with automated NM applications to "optimize" the SBMS network performance. Automated applications treat faults and congestion in two phases — recognition and resolution — and in two modes — prevention and reaction. An example illustrates how the architecture can be used to recognize and resolve a failure in a sample SBMS network.

1. INTRODUCTION

In this paper we introduce a Network Management Architecture for SONET-Based Multi-Service (SBMS) networks. In the 1990s, these networks exploit such emerging technologies as SONET transmission, broadband switching systems and intelligent remote units to transport a variety of user services. As such, SBMS networks will consist of new high capacity intelligent network elements embedded within the existing network infrastructure. Although there has been a recent flurry of activity in standardizing Network Management (NM) protocols [1–3], notably absent is a methodology for end-to-end management of services in an SBMS network ([4] is an exception).

A Network Management Architecture (NMA) provides the methodology — that is, the techniques, systems and procedures — required for effective and efficient management of network services and resources. The principles of the proposed NMA apply to today's telephone networks, however we elect to describe the NMA in terms of SBMS networks for two reasons: 1) the advent of such networks permits a more structured approach to network management, and 2) the complexity of such networks requires it.

Traditionally, NM Operations Systems (OSs) have been introduced in piecemeal fashion to support new network technologies and services. This ad hoc approach has resulted in a plethora of poorly organized systems generating large volumes of cryptic, redundant and sometimes inconsistent data. As a result, managing services in today's network operating environment is often a difficult, inefficiently performed task yielding less than satisfactory results. Among the problems faced by today's network managers are: 1) multiple points of human-machine interface, 2) large volumes of raw network data presented in cryptic format, 3) lack of coordination across NM tasks, 4) out of date, redundant and inconsistent data, 5) patched-in OSs with proprietary interfaces, 6) difficult to accommodate changes in network domain, and 7) management of facilities, not services.

Intelligent and high capacity Network Elements (NEs) can provide flexible and efficient service transport, but they require powerful new methods for management. For example, a faulty SONET repeater can disrupt thousands of circuits in a fiber span — triggering alarms not only at the faulty repeater but at "healthy" SONET, switching and other equipment supporting the same traffic flow. Inability to quickly correlate a sudden burst of seemingly unrelated alarms to the root cause leads to delays in reconfiguring the network, thereby causing customer dissatisfaction and revenue loss.

To realize the efficiencies offered by SONET and other emerging network technologies, future network management systems must provide a unified view of network performance across the various components of the local access and interoffice networks. To provide the functionality required for effective and efficient management of SBMS networks, we have identified the following principles for NMA design: 1) single point of human-machine interface, 2) knowledge-based graphical representation of network behavior, 3) unambiguous, coordinated responsibility for NM tasks, 4) common network measurement, testing and database utilities, 5) standard NM protocols, 6) easy adaptation to new management requirements, and 7) end-to-end service management.

The remainder of this paper is organized as follows. Section 2 describes the SBMS network domain in greater depth; Section 3 describes the components and functions of the proposed NMA; Section 4 illustrates how the NMA can be used to manage failures; and Section 5 concludes with a brief summary.

2. NETWORK DOMAIN

2.1. Overview of SONET

SONET [5] is the new standard family of transmission interfaces for telephone company optical networks; it provides synchronous transmission from a basic rate of 51.84 Mb/s (OC-1) up to 2.49 Gb/s (OC-48) in increasing multiples of the basic rate.

SONET networks are managed through the use of 1) a Data Communications Network (DCN) carrying operations messages from OSs to gateway NEs, and 2) Embedded Operations Channels (EOCs) carrying operations messages from NE to NE. The DCN and the EOCs together make up a Telecommunications Management Network (TMN [3]) for SONET. Because the EOCs are defined in the SONET standard, installation of a SONET transport network automatically implies installation of part of the associated TMN.

There are two types of EOC. *Bit* or *signal-oriented* EOCs support NE to NE event notification. For example, when an NE detects a failure, it generates a signal, which may, in turn, cause the generation of other signals. Signals propagate around failures, possibly triggering automatic recovery or generating alarms which an OS can use to sectionalize the failure. *Message-oriented* EOCs carry OS-generated or OS-bound messages from NE to NE; typical messages are alarms, queries and remote commands. Figure 2.1 shows an example SONET transport network and the concomitant TMN.

Beyond its clear potential for improving operations, telephone operating companies are already deploying SONET for its many other advantages: ability to support multi-vendor equipment, evolution to higher optical bit rates and broadband payloads, synchronous transmission technology and integration of the world's digital transmission hierarchies.

2.2. SONET-Based Multi-Service Networks

We characterize SONET-Based Multi-Service (SBMS) networks in terms of four attributes:

1) Integrated transmission based on the SONET standard
2) Heterogeneous switching (e.g., narrowband circuit, broadband circuit, packet) *or* integrated switching (e.g., Asynchronous Transfer Mode [6])

3) Local access provided through "intelligent" remote units (e.g., service gateways)
4) Services constructed from basic multimedia capabilities

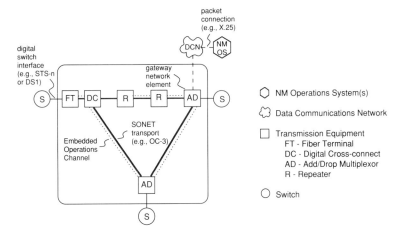

Figure 2.1   Example SONET Transport and Operations Networks

Accordingly, SBMS network elements can be grouped functionally into four planes: *transmission, switching, generic services* and *user services*. Organizing network management information into these planes provides a framework for network monitoring, diagnosis and control. Since the planes are tightly coupled, anomalies can propagate not only horizontally within a plane, but also vertically across different planes. As such, we must be able to manage, concurrently, both specific network elements in a single plane, and entire SBMS networks across multiple planes.

We now describe how an SBMS network could be used to provide a specific multimedia service: Audio Visual Shared Software (AVSS) point-to-point calling. Such a service might be offered between two workstation terminals. An AVSS call provides voice transmission as well as a video window on the workstation screen and a shared computer application enabling both parties to work together on, say, a spreadsheet program.

Figures 2.2 through 2.5 illustrate how such a service may be realized through decomposition into the four functional planes. In the *user services* plane (Figure 2.2), the network is completely transparent to the (CPE-based) user nodes. In the *generic services* plane (included in Figure 2.3), the two endpoint generic service nodes first negotiate with the user nodes for the terms of the service. Next they factor the user service into three generic (or bearer) services. In this case, the voice component of the call maps to a Connection-Oriented Constant Bit-Rate (CO-CBR) generic service; the video component maps to a different instance of the CO-CBR generic service; and the data component maps to a Connectionless Variable Bit-Rate (CL-VBR) generic service.

Whereas the generic service plane provides an end-to-end view of a service, the *switching* plane (included in Figure 2.4) introduces intermediate switching points. In this example, the voice and video traffic are switched along one path, and the data traffic is switched along a different path. Because we do not assume an integrated switching technology, a switching node may contain any of several switch types: packet switches, narrowband circuit switches and/or broadband circuit switches.

Figure 2.2   User Services Plane

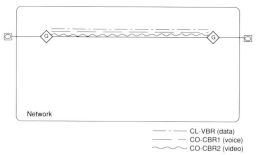

Figure 2.3   Generic Services Plane

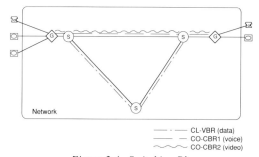

Figure 2.4   Switching Plane

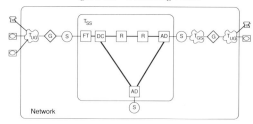

Figure 2.5   Transmission Plane

Finally, the *transmission* plane (included in Figure 2.5) completes the picture of the overall network. The transmission plane consists of different types of transmission networks. Loop networks (e.g., T_{UG}) provide transmission from user nodes to generic service nodes; if necessary, remote access networks (e.g., T_{GS}) provide transmission from remote access nodes to switching nodes; and interoffice networks (e.g., T_{SS}) provide transmission among nodes in the switching plane. SONET's large bandwidth makes it a good candidate for transmission networks T_{GS} and T_{SS} (and possibly T_{UG} networks for large customers). Because SONET is an integrated transmission technology, traffic class is transparent in the transmission plane.

Figure 2.6 illustrates how a simple service, e.g., a POTS call, could be provided across the four planes.

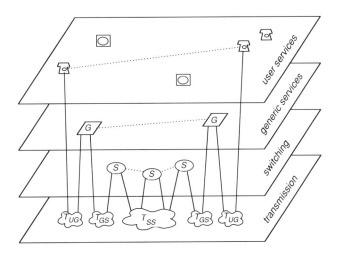

Figure 2.6 Simple Service Broken Down Vertically Across the User, Generic Services, Switching, and Transmission Planes

3. NETWORK MANAGEMENT ARCHITECTURE

A Network Management Architecture (NMA) provides an overall organization for the distribution of NM systems and information flow among them. Figure 3.1 displays an NMA comprising three basic types of Operations Systems (OSs): *NM Applications*, providing automated network management functions, *NM Utilities*, providing data manipulation capabilities shared by the applications, and an *Integrated Graphical Support System*, providing graphical, knowledge-based support for human decision making.

3.1. NM Applications

NM applications comprise systems for automated network management. The three basic types of NM application are 1) fault management systems, responsible for managing network failures, 2) traffic management systems, responsible for managing traffic flows, and 3) a Performance Management system, responsible for coordinating fault and traffic management, in cooperation with the human network manager, to optimize overall network performance.

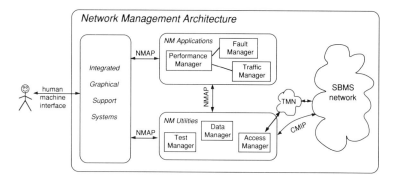

Figure 3.1 Network Management Architecture

Performance Management systems (PMs) are the top level NM applications; they are responsible for monitoring and controlling overall network performance, both within and across network services. PMs coordinate the actions of the lower level, task-oriented applications (e.g., fault and traffic managers) to recognize and resolve network performance problems. PMs must use not only the capabilities of human and automated fault and traffic managers, but also the self-healing capabilities of network elements themselves to ensure that NM resources are allocated effectively and that conflicting resolution paths are not pursued. PMs can operate either autonomously or jointly with a human network manager; all PMs operations can be supervised (and manipulated if necessary) by the human network manager. The nature of cooperation between PMs and the human network manager determines how PMs interact with other applications, i.e., the extent to which queries are actively generated, responses are forwarded, and whether other NM applications are invoked.

Fault and traffic management systems are task-specific applications available for use by the PMs and the human network manager. *Fault Management* systems (FMs) are responsible for managing network failures, and *Traffic Management* systems (TMs), are responsible for managing traffic flows. The PM applies these applications in two phases — recognition and resolution — and two modes — prevention and reaction.

Recognition consists of methods for prediction and detection of network performance degradation. Causes of performance degradation include physical failure, network congestion, security attack, etc. Prediction is the preventive form of recognition, and detection is the reactive form. Although preventive NM is more economical than reactive NM, some kinds of problems are unpredictable — especially certain kinds of failures — so a comprehensive NM methodology must also include mechanisms for detection. Isolation is the identification of the location(s) of a problem following positive detection. An instance of the recognition class of functions is a fault manager which detects and isolates network faults.

Resolution consists of methods for avoidance of, and real-time recovery from, network problems. Resolution actions generally consist of issuing commands which either utilize spare network capacity (e.g., alter communication paths) or confine the effects of problems (e.g., limit network traffic). Examples include downloading switch controls, issuing cross connect commands, switching to backup equipment and exercising admission control. An instance of the resolution class of functions is a traffic manager (such as the Predictive Access-control & Routing Strategy [7]) which computes and implements actions necessary to improve network performance under conditions of congestion.

Table 1 displays typical measurements taken from and controls downloaded to network components in each functional plane. Because the effects of equipment malfunction or congestion can propagate both within and across SBMS planes, human network managers and automated systems can diagnose the cause of a fault by tracking its symptoms within and across the planes. In general, resolving a problem requires the application of some combination of controls. Section 4.3 illustrates an example diagnosis and resolution. In the absence of the network plane abstraction, human network managers and automated systems can be flooded with a large amount of diverse and unstructured information.

Plane	Functions	Network Elements	Measurements	Controls
User Services (e.g. POTS, SMDS, HDTV, AVSS, etc.)	• User Services	• CPE	• Customer reports	---
Generic Services (e.g. CBR-CO/CL, VBR-CO/CL, etc.)	• Map user services to generic services • Service negotiation • Formatting, segmentation, reassembly	• Net. termination points • Access node	• End-to-End: - Blocking - Delay - Throughput - Attempt rates - Holding times - Error count - Alarms/failures	• Admission control • Download software generics
Switching (e.g. packet, narrowband circuit, broadband circuit, etc.)	• Call processing • Switching	• Call processors • Switching elements • Concentrators	• Per NE: - Buffer occupancy - Throughput - Cell loss/delay - Error count - Alarms/failures - Retransmissions	• Local buffer mgt. • Policing traffic • Circuit loop-backs • Routing tables • NE diagnostics • Download software generics
Transmission (e.g. SONET)	• Local access transmission • Interoffice transmission	• Add/Drop MUX • DCS • MUX • Repeater • Lasers/LEDs	• LOS/LOF/LOP • NE failures • Frame occupancy • Code violations • Errored seconds • Prot. switch count	• Path switching • Loop-back • NE diagnostics • Download software generics

Table 1   SBMS Network Planes, Functions, Elements, Measures and Controls

3.2. NM Utilities

NM utilities provide NM applications and the human network manager with generic data manipulation capabilities. The basic utilities are: access manager, data manager and test manager.

The *access manager* utility uses the TMN to provide a uniform communications interface among OSs and NEs. The Common Management Information Protocol (CMIP [8]) is the standard OS to NE application layer protocol for both the DCN and the EOCs; it supports delivery of NE reports, commands, queries and replies. The Network Management Architecture Protocol (NMAP) is the OS to OS application layer protocol; it supports cooperative problem solving among OSs.

The access manager is organized according to planes: it employs a separate Plane Communication Manager (PCM) for communication with NEs in each of the SBMS planes. The PCMs, together with the network database, allow OSs to generate a view of the network appropriate for real-time for diagnosis and control.

The *data manager* utility provides a common interface for all data collection and management functions. Although network elements can provide real-time status directly to NM applications through the Access Manager, applications sometimes require network-wide, historical or other information not stored per network element. Moreover, applications should not be concerned with the mechanics of scheduling, filtering and archiving network data. Data management activities include generating and scheduling NE queries, filtering responses and alarms, forwarding messages to client OSs, serving database requests and maintaining the database.

One way to structure information about NEs is to generate a conceptual view of the managed objects in a network, i.e., their attributes (type, location, etc.) and relationships (connectivity, subcomponents, etc.). This is called an information model [9]; it provides the basis for identifying and addressing managed objects. The information model, instantiated as a database and administered by the data manager, can also be organized functionally according to the four SBMS planes. OSs solve network flow and failure problems using network maps and other information stored in the database. The database can also serve as an archive for network alarms and reports, and as a staging platform for downloading NE software programs (generics).

The *test manager* exercises network functions on behalf of NM applications and the human network manager. This includes generating, scheduling and monitoring tests. Its activities range from element-specific diagnostic testing, to network-wide loopback and service-quality testing. OSs use NMAP to request tests. The test manager interacts with the access manager (using M-ACTION, M-SET and other CMIP messages) and with the data manager (reading configurations and logging test results).

3.3. Integrated Graphical Support System

The introduction of advanced computer technology has changed the role of humans interacting with complex technical systems from that of being primarily manual controllers to that of being decision makers [10, 11]. The Integrated Graphical Support System (IGSS), the human-machine interface for the NMA, supports human decision making through the use of task-oriented, knowledge-based graphical presentations. The human network manager acquires knowledge about the state of the network by extracting information from the graphical displays. The type of knowledge a human network manger can extract is determined by the manager's level of expertise and by the types of information and knowledge represented in the graphical support system.

In order to know *what* information should be displayed, the graphical support system needs models of both the network and NM tasks. In order to know *how* to display information about the network (and the output of the automated applications), the graphical support system needs knowledge about the efficiency of different display formats as they relate to joint human-machine reasoning. Finally, in order to know *when* to display information, the graphical support system needs knowledge about the states of the network, the NM applications, and the human network managers who interact with the system

Figure 3.2 illustrates the architecture of an IGSS capable of generating knowledge as required. This system generates display contents using three sources of knowledge: a model of the human network manager, a model of the current NM tasks, and its own model of the network. The Graphic Support Manager composes display contents in a task-oriented way. For example, displays for situation assessment are generated by the Categorization of States module; displays for choices of actions by the Choice of Actions module; and displays for predictions of future network behavior by the Evaluation of Outcomes module. Composed display contents are then transformed into a graphical presentation. Input from the human network manager is used by the Graphic Support Manager to update all three models. The three models are based on a distinction which has been used successfully in the development of a knowledge-based system that generates task-oriented graphical presentations for conventional power plant control rooms [12,13].

4. EXAMPLE: MANAGING FAILURE USING THE NMA

In this section, we return to the example network and AVSS service. To illustrate the need for the NMA, we introduce a transmission failure, trace its effects, and show how the NMA can be used to diagnose and correct the problem.

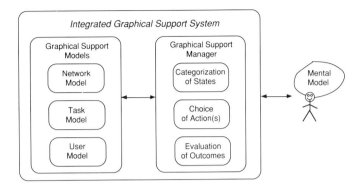

Figure 3.2 Proposed Architecture for the Integrated Graphical Support System

4.1. Failure in the Transmission Plane

Suppose an optical repeater in the transmission plane fails intermittently. The first sign of the problem is that the SONET bit-oriented EOC becomes active with indications of failure; Figure 4.1 traces the flow of resulting bit-oriented signals. If the SONET NEs have been configured to generate alarms as a result of the signals, then the message-oriented EOC and the DCN also become active as the affected NEs send alarms to the transmission PCM.

Which (if any) alarms are received by the transmission PCM depends on two factors. First, the NEs may not have been configured to generate alarms on receipt of the bit-oriented signals. Second, the failure itself may impede the delivery of alarm messages across the message-oriented EOC. The uncertain mapping from failure causes to failure effects makes the jobs of recognition and resolution more difficult.

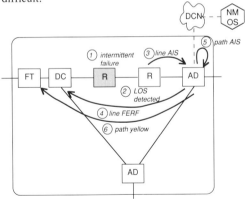

Figure 4.1 Bit-Oriented Signals resulting from Transmission Failure

4.2. Propagation of Failure Effects

We now trace possible effects of the failure on the other network planes. The malfunctioning repeater lies on a path serving AVSS data traffic. As a result, packet switches that depend on the repeater may need to retransmit packets that were lost when offered to the marginal repeater. As the number of retransmissions increases, switch buffer occupancy may also increase, and switch throughput may decrease. If the buffers become badly congested, some packets might be dropped. All of these effects can be revealed through queries sent to switch NEs; if predefined thresholds are exceeded, some will also generate alarms.

In the generic services plane, the CL-VBR generic service may experience loss, increased delay and decreased throughput. Finally, the effect on the user plane (e.g., the AVSS service) is that although the voice and video connections are unimpaired, data delays cause jittery software screen updates. If the congestion is bad enough (and if the shared application program does not exercise flow control) the application software could fail, since the endpoint user nodes may be using inconsistent data. The NM applications have access to most of these effects through the access and data managers. Some effects may be autonomously reported by NEs (e.g., alarms) whereas others can be observed only through explicit query (e.g., performance measures without threshold crossing alerts). Effects in the user plane may not be available to NM applications since user nodes are not part of the network proper. However, if user services are affected before the problem is corrected, user trouble reports will provide a measure of performance in the user plane.

4.3. Application of the NMA

Figure 4.2 traces example network management activity through the NMA. Suppose the failure does not result in explicit alarms, but that NEs still generate performance reports. The reports are passed through the access manager (1) to the data manager (2). The data manager updates the relevant part of the network database, possibly performs plane-specific analysis, and passes the results to the PM (3). The PM periodically forwards status to the fault manager (4) which, on the basis of the performance reports and tests (5) detects and isolates the problem. The fault manager returns the problem ID plus a correlation of the failure effects to the PM (6). The PM can then invoke the traffic manager, requesting problem resolution (7); the traffic manager returns a solution consisting, in this case, of commands to route traffic around the faulty repeater (8). Depending on its configuration, the PM may then forward the problem ID and proposed resolution to the human network manager for ratification (9). If approved, the commands can be forwarded through the Access Manager (10) to the relevant network elements (11), restoring service in the network.

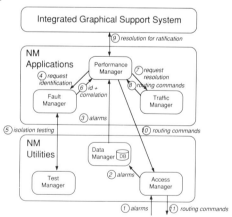

Figure 4.2   Trace of Network Management Activity in the NMA

Figure 4.3 illustrates example fault and traffic managers in more detail. In this example, the PM regularly forwards some subset of network status ($status_1$) to the fault manager. A network state estimator formulates a prediction of network behavior which is compared with actual network behavior (a different subset, $status_2$). If there is a difference, a failure has been detected, and it is passed to a failure identifier. A feedback control loop may be used between the state estimator and the failure identifier to converge on the most probable location/cause of the problem. After the fault manager has generated a probable explanation for the failure, the PM may now invoke the traffic manager to resolve the problem. The traffic manager uses the failure location to narrow its set of available routes before computing a routing strategy (based, for example, on minimizing projected blocking). The traffic manager then passes updated routing tables to the PM for forwarding to the switching PCM.

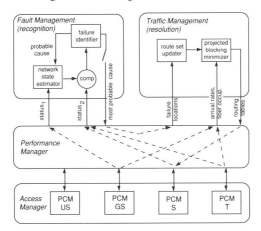

Figure 4.3   Example Task-Specific NM Applications and their use by the Performance Manager

Although this example resolves the transmission failure by sending controls to the switching plane, the PM might, alternately, have resolved the problem through transmission (e.g., cross connect) commands. Note that the NMA works both in the reactive case and the preventive case. In the reactive case, network status is conveyed, in part, through alarms; in the preventive case, the PM schedules queries to look for impairments before they become failures.

5. SUMMARY

This paper has motivated and outlined the design principles of a Network Management Architecture (NMA) necessary for effective and efficient operation of SONET-Based Multi-Service (SBMS) networks. We introduced an NMA comprising three types of network management OSs: NM applications, providing automated performance, fault and traffic management; NM utilities, providing common data manipulation capabilities; and an integrated graphical support system, providing graphical, knowledge-based support for human decision making.

Future work in this area includes prototype development of the basic components of NMA on a laboratory testbed. Of special interest are the development of methodologies for cooperative human-machine fault management and the design of automated NM applications.

6. REFERENCES

1. L. N. Cassel, et al, "Network Management Architectures and Protocols: Problems and Approaches," *IEEE Journal on Selected Areas in Communications*, Vol. 7, No. 7 (September 1989).
2. S. M. Klerer, "The OSI Management Architecture: an Overview," *IEEE Network* (March 1988).
3. V. Sahin, C. G. Omidyar, T. M. Bauman, "Telecommunications Management Network (TMN) Architecture and Interworking Designs," *IEEE Journal on Selected Areas in Communications*, Vol. 6, No. 4 (May 1988).
4. T. W. Callahan, et al, "A Strategy for Broadband ISDN Network Operations," *Proceedings of the IEEE Network Operations and Management Symposium (NOMS)* (1990).
5. Bellcore, "Synchronous Optical Network (SONET) Transport Systems: Common Generic Criteria," TA-TSY-000253.
6. S. Minzer, "Broadband ISDN and Asynchronous Transfer Mode (ATM)", *IEEE Communications Magazine* (September 1989).
7. S. Kheradpir, "PARS: A Predictive Access Control and Routing Strategy for Real-Time Control of Telecommunications Networks," *Network Management and Control*, A. Kershenbaum, ed., pp. 389-413 (1990).
8. ISO/IEC, "Information Processing Systems - Open Systems Interconnection - Management Information Protocol Specification - Part 2: Common Management Information Protocol," ISO/IEC DIS 9596-2 (1988).
9. ANSI T1M1.5, "Telecommunications Management Network (TMN) Modelling Guidelines," APM.PVH.118.001.
10. C. M. Mitchell, R. A. Miller, "Design Strategies for Computer-Based Information Displays in Real-Time Control Systems," *Human Factors*, Vol. 25, pp. 353-369 (1983).
11. C. M. Mitchell, R. A. Miller, "A Discrete Control Model of Operator Function. A Methodology for Information Display Design," *IEEE Transactions on Systems, Man, and Cybernetics*, Vol. 16, pp. 342-357 (1986).
12. G. A. Sundstrom, "Process Tracing of Decision Making: An Approach for Analysis of Human-Machine Interactions in Dynamic Environments," *International Journal of Man-Machine Studies* (in press).
13. G. A. Sundstrom, "User Modelling for Graphical Design in Complex Dynamic Environments: Concepts and Prototype Implementations," Report No. 10, Laboratory for Man-Machine Systems, University of Kassel (GhK), Kassel F.R. Germany (1990).

A Study on an End Customer Controlled Circuit Reconfiguration System for Leased Line Networks

Tetsuya YAMAMURA, Tetsujiro YASUSHI and Nobuo FUJII

NTT Transmission Systems Laboratories
1-2356 Take, Yokosuka-shi, Kanagawa-ken, 238-03, Japan

Abstract

This is a feasibility study on the experimental End Customer Controlled (ECC) circuit reconfiguration management system for leased line networks. We propose a four-layer function architecture to construct the ECC management system and discuss a network modeling method for network management data using an Entity-Relationship diagram. We also describe an operation scenario for circuit reconfiguration management. A prototype system has been designed based on the proposed system architecture, modeling method and operation scenario. This experimental system, implemented in a distributed system environment, manages about 20,000 circuits. Average circuit data retrieve time is 1.5 seconds and elapsed time to create an end-to-end circuit is about 12 seconds.

1. Introduction

End Customer Controlled (ECC) management for leased line networks is expected to provide highly functional network operation services[1] from which customers can obtain network management capabilities such as configuration management, fault management, etc. However, a number of technical problems stand in the way of ECC system implementation.

Firstly, the network Operation System (OpS) must provide an operation interface offering operation management functions controlled by end customers while maintaining operational security. Currently, ISO and CCITT are working on the standardization of the operation interface. Construction of a communication network OpS incorporating the standardized operation interface promises to fulfill the need for an integrated OpS that inter-connects various OpSs and Network Elements (NEs).

Secondly, designing a system architecture that effectively generates the network provider management view and the customer management view concerning various management tasks without sacrificing operational security, is a significant problem. An OpS architecture incorporating a suitable multi-layer structure is necessary to ensure the effective generation of relevant management views.

Finally, it is presumed that any existing communication network, due to its sheer size, is difficult to manage centrally. Therefore, it is important to integrate and distribute communication network management data[2]. Appropriate functional distribution in an OpS architecture would make distributed communication network operation possible. Other basic tasks include devising modeling methods for telecommunication network and data management methods that utilize database systems to facilitate network data management through the use of computers.

To date, several ECC systems have been investigated of which Customer Controlled Reconfiguration (CCR) ones[3],[4] are typical examples. However, these systems do not necessarily solve all of the technical problems involved in providing ECC services. In this paper, ECC system configuration methods that overcome a number of the aforementioned problems are proposed. The feasibility of the configuration, from the viewpoint of network data management, is shown through an ECC circuit reconfiguration management prototype system. First of all, We clarify the ECC circuit reconfiguration management service scenario. Then, we propose an ECC system architecture consisting of four-layer functions and present a network modeling method. After that, we detail the ECC circuit reconfiguration system based on the system architecture, the modeling method and the service scenario. Finally, the prototype system construction and a performance evaluation are discussed.

2. ECC Circuit Reconfiguration System Overview
2.1 Outline

The ECC system offers new services which enable customers to control and administer private networks through a network service providers' operation. Customers can remotely or locally control network configuration as well as collect surveillance data and performance information on their network. The ECC system will have five network administration functions defined as OSI network management capabilities[5]. Among these functions, configuration management is one of the most important.

2.2 Service Scenario

The ECC system offers a network configuration display service. Utilizing displays of their private networks, customers can conduct circuit reconfiguration.

(1) Network configuration display service

The customer's network configuration display request triggers the circuit configuration information display. Display format depends on the Human Machine Interface (HMI) available on the customer's ECC Terminals (ECTs). Service scenarios follow the hypothesis that highly functional ECTs have graphical HMI capability are located at customer offices. Services provided by the ECC system are restricted to the permission levels of the operators. For instance, lowest level operators will only be able to see their own network configurations while special level operators will exist who can see all customers' network configurations. Special level operators will be the network provider's administrators who exercise control over the whole provider's portion of customer's network.

(2) Circuit Reconfiguration Service

Table 1 lists the four categories of circuit Reconfiguration Service. Only two kinds of users are assumed: customers and network provider's operators. In the table, (a), (b) and (c) services are provided using customers' private network resources which consist of permanent use leased lines. On the other hand, temporary network resources are available through (d). The uniform speed circuit reconfiguration, service (a), is described in this paper. Since it is the most basic service of circuit reconfiguration, the other services can be realized through its application.

Below, we present the internal message procedure in the ECC system to accomplish circuit reconfiguration service. Control message flow from an ECT to NEs via the ECC system and response message flow to an ECT after NE control are described:

(i) Reconfiguration request massage flow: The

Table 1 Classification of Circuit Reconfiguration Service

Service Name	Description
(a) Uniform speed circuit reconfiguration	Uniform transmission speed circuits exist in customers' links. Customers can control elemental circuit cross connection and establish circuits.
(b) Multiple speed circuit reconfiguration	Multiple transmission speed circuits exist in customers' links. Customers can establish multiple speed circuits. Circuit speed will not be changed.
(c) Bandwidth control circuit reconfiguration	Elemental circuit speed can be changed by customer control. Customer can establish multiple-speed circuits.
(d) Sharable circuit reconfiguration	Network providers define sharable links and elemental circuits which are temporary available network resources. Customers can get services from (a) to (c).

ECC system receives reconfiguration requests from ECTs and converts them to messages defined for the interface between the ECC system and NEs. Generated messages are sent to NEs;
(ii) NE control: NEs' internal status is altered upon receipt of control messages. NEs' control completion is confirmed;
(iii) Database update: The ECC system's network configuration database is updated;
(iv) Notification: The ECC system receives notification messages from NEs upon control completion and sends a control completion notification message to ECTs.

Circuit reconfiguration services for network provider's operators include customer assignment and circuit provisioning for contracted customers.

3. End Customer Control System Architecture

This chapter describes the proposed four-layer ECC system architecture. ECC functions include the ECC terminal function, ECC management function, network element management function and network element function, as shown in Fig. 1. Each layer's required service capabilities from the view point of circuit reconfiguration management service are clarified as follows:

(1) ECC Terminal Function (ECTF)

ECTs are used for accessing a circuit reconfiguration service. The ECTs' role involves accepting control-commands from customers, displaying results of control activities and managing local information. Required terminal functions will vary according to the kind of a terminal and the level of HMI functions.

(2) ECC Management Function (ECCMF)

ECCMF processes information necessary to establish ECC services as required by customers. It consists of customer management and network operation functions. The former acts to identify and authenticate customers and to confirm access and control privilege for network resources. The later involves command interpretation from ECTs and customers' network resource control.

(3) NE Management Function (NEMF)

NEMF governs access to NE, controls over NE and manages NE data. These functions include access contention control, control-command interpretation and execution as well as network resource management.

(4) NE Function (NEF)

NEF is realized by a telecommunication elemental device or equipment to facilitate communication capability. This function can send alarm information and involves electronic cross connecting functions in order to establish and reconfigure circuits.

Since OpS operates several management systems other than ECC systems, various management views are necessary in order to perform management tasks. Therefore, a network OpS must create and show those management views. Additionally, maintaining operational security is important. We suggest that in the four-layer function architecture the OpS core function be divided into two layers, such as ECCMF and NEMF. This architecture is effective in attaining management services other than circuit reconfiguration.

Generally, ECTs and NEs are geographically distributed. In the four-layer architecture, ECCMF and NEMF will also be distributed for economical and operation work force assignment reasons. Moreover, remarkable progress in technologies has made a distributed processing environment possible. Therefore, the actual target ECC system is a distributed processing system comprised of workstations and mini computers interconnected by the high-speed Local Area Networks (LAN) and Wide Area Network (WAN) shown in Fig. 2, rather than a big center system[2]. The distributed configu-

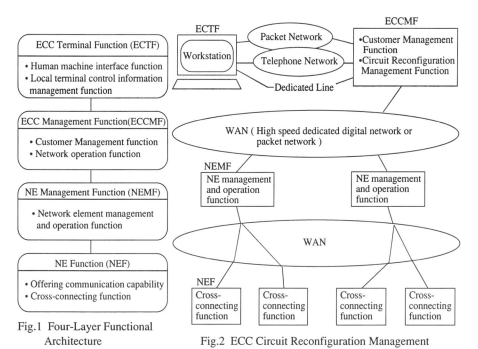

Fig.1 Four-Layer Functional Architecture

Fig.2 ECC Circuit Reconfiguration Management System Overview

ration of the four-layer architecture will accelerate the sharing of information among functions and advance integrated-operation system implementation technologies.

4. Network Modeling

Network modeling and its data representation are vital in network operation system implementation. This chapter describes a network modeling method according to operation objects and their relationships based on the OSI network management methodology currently undergoing international standardization[6],[7]. The Entity-Relationship modeling method[8] can be applied to systematically classify network model elements. Since the network model described here does not precisely conform to international standards, network operation target elements are called "Operation Objects" instead of "Managed Objects."

Telecommunication network configuration management treats communication capability formed over serially connected telecommunication equipment and the relationships between that equipment. The actual connections include logical electronic connections within cross-connect equipment, physical connections between equipment, and so on. Communication capabilities have relationships such as mutual inclusions and serial connections since they are provided through multiplexing and de-multiplexing or connection within transmission equipment.

A process for provisioning communication capabilities to customers must be approached in term of connections between equipment and transmission lines and control of cross-connect equipment. It mainly involves management activities on the various types of connections having relationships such as the mutual inclusion and serial connection of communication capabilities. Management activities and the typical connections they deal with are described below:

1) Equipment and facility management for physical connections between equipment and facilities;
2) Transmission path management for logical communication capabilities consisting of mutual inclusion and serial connections formed by the physical connections;
3) Node and link management for logical communication capabilities provided by nodes (e.g. digital cross-connect equipment) and links, and which consist of mutual inclusion and serial connections formed by the electronic connections established within nodes;
4) Circuit management for logical communication capabilities consisting of mutual inclusion and serial connections actually provided to customers.

A target network configuration example for the ECC circuit reconfiguration system is shown in Fig. 3. Operation objects for managing a dedicated line network have been systematically classified and hence, network modeling has been established based on the network configuration. The result of network modeling using an Entity-Relationship diagram is shown in Fig. 4. Rectangles depict operation objects. Diamonds depict relationships between operation objects.

ECC circuit reconfiguration systems mainly deal with node and link management and circuit management. The operation objects and their attributes required in ECC circuit reconfiguration management are described in Table 2. Operation objects in Table 2 do not contain items related to equipment and facility management and transmission path management. Equipment and facility management and transmission path management are assumed to be conducted in existing network management systems. Operation objects in Table 2 are converted to a data group which is then manipulated to achieve the circuit reconfiguration management function.

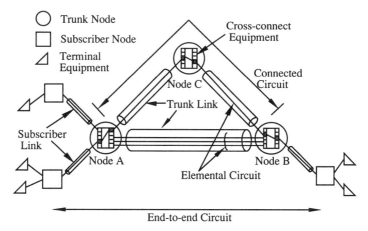

Fig.3 A Network Configuration Example

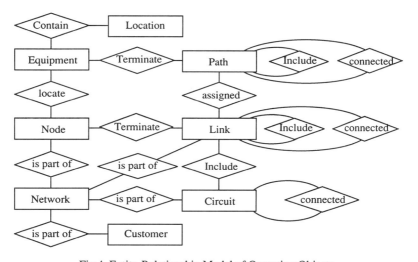

Fig.4 Entity-Relationship Model of Operation Objects

5. Experimental System Design and Performance Evaluation
5.1 Reconfiguration Management Function

Circuit reconfiguration service is defined as follows:

A customer has contracted to use nodes and links which are created between trunk nodes and between trunk nodes and subscriber nodes. The trunk nodes have cross-connect functions. The customer can create and delete end-to-end circuits which are formed by elemental circuits and the cross-connect equipment of trunk nodes.

Circuit reconfiguration service offered

Table 2 Operation objects and Attribute Example

Operation Object	Attribute
Node	node name, node type
Link/Circuit	link/circuit name, terminal nodes name, link/circuit type, transmission speed, status
Network	network name, node name, link/circuit name, customer name
Customer	customer name, customer type, access privilege type
Contain*	component circuit name, end-to-end circuit name
Include*	link name, included circuit name

* They are denoted as relationship objects.

by the ECC system varies according to system users. We suppose two kinds of users, provider's operators and customers. There are four basic functions: retrieve, create, delete and display, that are independent of HMI form. The reconfiguration management function is realized by combinations of these basic functions. Service specifications for a customer and a provider's operator are as follows:

Customer service specification:
(1) network configuration display of customer's private network on ECTs;
(2) circuit reconfiguration: (a) to specify objects for reconfiguration from network resources, and (b) to create circuits using specified network resources and to delete the specified circuits.

Provider's operator service specification:
(1) new registration: to register a new customer and his private network;
(2) circuit configuration display of all customers' private networks on ECTs;
(3) circuit reconfiguration of all customers' private networks.

Both control message sets and data records of database systems should be managed in order to realize circuit reconfiguration. The control message sets are exchanged between ECTF-ECCMF, ECCMF-NEMF and NEMF-NEF. The data records explain network configuration, equipment, facility, customer data and so on.

Uniform speed elemental circuit and end-to-end circuits in contracted links are permitted to exist in uniform speed circuit reconfiguration. An end-to-end circuit consists of serially connected elemental circuits. Uniform speed circuit reconfiguration service is provided after the provisioning of customer's private network. The prototype system described in this chapter is an example of the uniform speed circuit reconfiguration.

5.2 Function Distribution

Figure 5 shows system function distribution of the ECC circuit reconfiguration man-

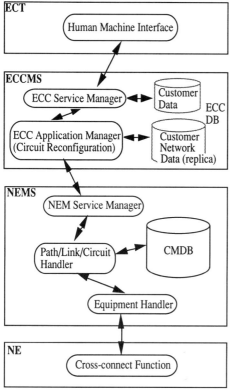

Fig.5 ECC System Function Distribution

agement system and a detailed description of the functions is shown in Table 3. ECTF, ECCMF, NEMF and NEF are implemented in ECT, ECC Management System (ECCMS), NE Management System (NEMS) and NE, respectively. The following is a summary of ECT, ECCMS, NEMS and NE functions implemented with regard to network configuration display and circuit reconfiguration service:

ECT: ECT sends ECCMS a network configuration display service request and a reconfiguration service request, and receives and displays processing results.

ECCMS: ECCMS obtains network configuration data and sends it to ECT when it receives a display request. This display process is completed in ECCMS. We suppose that ECCMS manages ECC DataBase (ECCDB) which contains ECC specific data, for example, customer data, etc. and a database replica of customer's private network configuration data in order to maintain data security and to avoid a drop in performance because of access contention control. ECCMS sends NEMS a reconfiguration request when ECCMS receives a reconfiguration request from ECT. ECCMS receives a reconfiguration acknowledgement from NEMS and sends ECT reconfiguration results.

NEMS: In any display service, NEMS is not accessed by ECCMS since ECCMS manages configuration and customer data. NEMS sends NE a reconfiguration command when NEMS receives a reconfiguration request from ECCMS. NEMS receives a reconfiguration acknowledgement from NE and sends ECCMS reconfiguration results. The concurrent update between the Configuration Management DataBase (CMDB) and its replica should be conducted.

NE: NE is never accessed by NEMS in display service. NE changes circuit configuration based on reconfiguration commands from NEMS and acknowledges reconfiguration results. In this paper, NE is a simple simulator.

It is necessary to set protocol and detailed message sets between ECT-ECCMS, ECCMS-NEMS and NEMS-NE so as to provide ECC service. The message sets can be detailed by clarification of service specifications, necessary data for circuit reconfiguration and circuit reconfiguration processing algorithm.

Table 3 ECC System Main Function Distribution

System Name	System Function	Function Name	Description
ECT	ECTF	•HMI	display configuration, accept control-command, display processing result
ECCMS	ECCMF	•ECC service manager	customer data management, ECT management, security check, access contention management
		•ECC application manager	reconfiguration management, customer network data (replica) management
NEMS	NEMF	•NEM service manager •path/link/circuit handler •equipment handler	access contention management CMDB management control NE
NE	NEF	•cross-connect	change cross-connect, notify control result

5.3 Database Design

(1) Database Allocation

The ECC circuit reconfiguration management system includes two database systems: CMDB managed by NEMS and ECCDB managed by ECCMS. CMDB manages network element/facility (including connection data between equipment). Node/link/circuit management data is located on CMDB since it can be accessed by other management systems. Management data accessed by only ECCMS is located in ECCDB. ECCDB includes a replica of node/link/circuit management data in CMDB.

(2) Schema and Table Design

CMDB manages all of schemas shown in Fig. 4 except customer data. In fact, CMDB has other data concerning transmission equipment, connection relations between equipment and transmission lines. However, such data is not referred to in this paper. Basically, each entity and relationship shown in Fig.4 can correspond to tables, and an entity and relationships can be integrated into one table. Detailed attributes of every entity and relationship are obtained from an analysis of service scenario.

5.4 Protocol Design

A three-level protocol is designed in order to configure the ECC system in a distributed system environment. The levels are classified as follows:

• level 1: protocol between ECC system functions (messages between four-layer functions)
• level 2: protocol between application processes (inter-process communication)
• level 3: protocol between computer systems (communication protocol).

Three levels protocols can be classified into an OSI seven-layer model. The three levels correspond to application/presentation layers, a session layer and transport/network/datalink/physical layers, respectively.

The ECC prototype system is constructed in a UNIX environment via ordinary dedicated or public communication network. The application software is original. In this prototype, protocol design guidelines are that communication protocol is based on international standards or de-facto standards; inter process communication is based on UNIX standards;

and messages between system functions are based on original protocol.

5.5 HMI Design

HMI for communication network operation has been investigated from the view point of reducing human error[9]. We regard the approach to displaying the outcome and the input method as important factors. Therefore, the principle of the direct manipulation method[10], which is expected to decrease human error, was applied to the HMI of communication network operation. The HMI of the ECC prototype system is designed based on the method where operators react to objects represented on a display using a mouse cursor. We utilize high-performance workstations equipped with color bit-mapped displays as ECTs. HMI specifications are as follows:

(1) Customer's private network configuration comprising nodes and links on a map of the Japanese archipelago is displayed on a multi-window environment. Nodes and links are operation objects. Both graphic-base windows and character-base windows are represented on a display.

(2) Another window showing more detailed information is triggered at selection using a mouse on a pull-down menu which emerges by clicking a displayed object, or by pushing a specified button.

(3) Create, delete and update of nodes, links and circuits can be easily performed by direct manipulation of displayed objects.

5.6 Performance Evaluation
(1) Prototype System Overview

The hardware configuration of the ECC prototype system is shown in Fig.6. Workstations are utilized as ECTs and a mini-computer manages CMDB. The workstations and mini-computer are connected through LAN (IEEE 802.3). The ECC prototype system is connected to geographically distributed customers via a publicly accessible packet network or dedicated lines exchanged by a packet switcher (DATAKIT).

(2) Software development environment

General-purpose software packages were utilized as possible. Application program modules were developed on UNIX OS in C language and the X-window system is utilized. INGRES was selected as a database management system. The database access language is SQL.

(3) Performance Evaluation

Customers' private networks which include 6.3Mbit/sec links and 64kbit/sec circuits were supposed. Circuit reconfiguration performance of the prototype system was evaluated. This system manages about 20,000 circuits. Average circuit retrieve time was 1.5 seconds from ECT. Elapsed time to create an end-to-end circuit including appending and updating network management data was about 12 seconds with data integrity checks and database update processes. Figure 7 is a display example. High-performance HMI was realized based on the multi-window environment. Each operation can be performed by direct manipulation of displayed objects and input from a keyboard. Network configuration drawing time (including display map of Japan) was within 1 second. A comfortable operation environment was obtained.

6. Conclusion

The ECC system provides a new network operation service from which customers can manage their own private networks. ECC service is realized by management capabilities of network provider's operation system. The network operation and management method described in this paper enable control over private networks as well as the realization of a future highly-functional integrated network OpS.

There are a number of problems to be solved before an efficient and reliable ECC system for a leased line network can be established. ECC system configuration methods that solve a few of these problems have been investigated and were incorporated into the con-

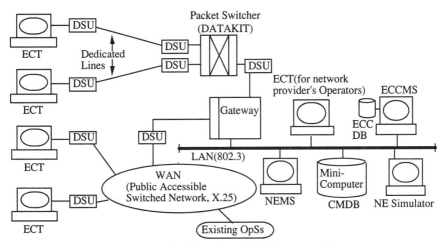

Fig.6 Hardware Configuration of ECC Prototype System

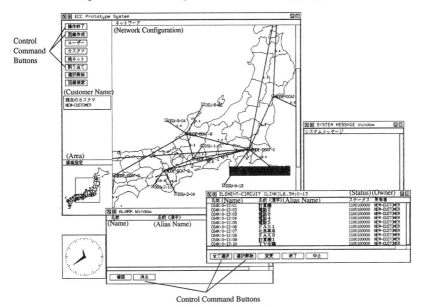

Fig.7 A Display Example

struction of an ECC circuit reconfiguration management prototype system. We have investigated the following items with an aim forward achieving an ECC system:

• Four-layer function architecture;
• Network modeling method using an Entity-Relationship diagram;
• An operation scenario of circuit reconfiguration management service.

The feasibility of ECC system configuration from the viewpoint of network data management is shown through the prototype system's construction. The system performance evaluation reveals that ECC circuit reconfiguration management in a distributed operation environment is feasible as an example of a leased line network configuration which consists of about 20,000 circuits. Average circuit data retrieve time was 1.5 seconds and elapsed time to create an end-to-end circuit including appending and updating network management data was about 12 seconds with data integrity checks. A comfortable network operation environment has been obtained.

Acknowledgment

The authors would like to express their sincere thanks to Dr. Sadakuni Shimada, Dr. Masaki Koyama, Dr. Tetsuya Miki and Kazumitsu Maki of NTT Transmission Systems Laboratories for their continuous encouragement, to Dr. Haruo Yamaguchi and Yasushi Yamanaka of NTT Telecommunication networks Laboratories for technical discussions, and to Kouji Yata, Ikuo Yoda and Yasumi Matsuyuki of NTT Transmission Systems Laboratories for help in prototyping.

References

[1] K.D.Meola and K.K.Verma, "End-Customer Management of Hybrid Networks," NOMS'88, 6.1, 1988.
[2] T.Yasushi, N. Fujii and H. Yamaguchi, "Distributed Configuratuion of End Customer Control Management System," IEEE ICC, 38.4, June 1989.
[3] W.D.Hutchenson and K.K.Snyder,"Control Service Based on Digital Cross-Connect Systems," IEEE J-SAC, vol.SAC-5, no.1, Jan. 1987.
[4] R.K.Berman and R.W.Lawrence,"Customer Network Management and Control Services," ISSLS, 1986.
[5] ISO 7498-4: Management Framework.
[6] ISO DIS 10040: System Management Overview.
[7] CCITT Recommendation M30, "Principle for a Telecommunication Management Network," CCITT AP IX-31-E.
[8] P.P.Chen, "The Entity-Relationship Model Towards A Unified View of DATA," ACM Transactions on Database Systems, vol.1, no.1, 1976.
[9] T.Yamamura, K.Yata, T.Yasushi and H.Yamaguchi, "A Basic Study on Human Error in Communication Network Operation," IEEE GLOBECOM'89, 22.5, Nov. 1989.
[10] B.Shneiderman,"Designing The User Interface," Addison-Wealey Publishing Company, 1987.

V

TELECOMMUNICATIONS NETWORK MANAGEMENT

B
Panel

Can We Really Control Large Distributed Systems?

Panel Chairs: Kenneth J. Lutz
Bell Communications Research
331 Newman Springs Rd.
Red Bank, NJ 07701 USA

Douglas N. Zuckerman
AT&T Bell Laboratories
480 Red Hill Rd.
Middletown, NJ 07748 USA

Abstract

As enterprises expand their scopes in a growing global economy, they increasingly rely on their rapidly evolving computing and telecommunications infrastructures. These infrastructures have developed into large distributed systems, so complex that our best theoreticians and practitioners are challenged to analyze them. We can and have built such systems, but do we know that we can really control them? This panel, composed of the foremost experts in the field, attempts to answer this question from the different perspectives of computer networks, telecommunications networks, and a combination of the two.

PANELIST SUMMARIES

Deborah Estrin, University of Southern California, USA

Large, physically dispersed, tightly coupled, automatically adapting systems are hard to understand, manage, and control. The possible combinations of events are so numerous that such systems are not tractable in a formal manner. We can expect to control our global computer and communication systems in a manner similar to the way human organizations interact by controlling information flows to conform to organization policy. In particular, by placing access controls, routing mechanisms, and accounting feedback at boundaries between administrative domains, undesired system interaction can be identified and contained proportionally to the strength of the enforcement mechanisms applied.

Three types of mechanisms are discussed: access controls, policy based routing and route selection, and accounting and feedback at the boundary points between administrative domains. In summary, resource control mechanisms need to be built into our systems and supporting communication media. At the same time the negative impact on internal, more tightly coupled systems must be minimized.

Ronald Fleming, Bell Communications Research, USA

Network access and control of distributed UNIX-based systems presents significant aspects of managing and controlling large distributed systems. As UNIX based systems have increased in number, many have become part of distributed database systems. For the applications deployed on these systems to perform properly, data communications must be provided between systems. Various networks and data communications methods may be installed to meet this requirement. Effective management and control of these networks and paths can provide the needed inter-system communications availability for the deployed applications.

Mark Mortensen, AT&T Bell Laboratories, USA

The public telecommunications network is an example of a manageable, large distributed system. The network is formed from software applications residing in its distributed switching systems, transmission facilities and network databases. Management and control of this network is distributed across network elements, operations systems, and databases. Network element functionalities incorporate, for example, CCITT Signalling System No. 7 and SONET (Synchronous Optical NETwork) standards. Operations systems are essential to providing either centralized or distributed management of the network, depending on the application. Also, use is made of modern database techniques to handle routings and create new services. Network survivability is essential, and this is achieved through approaches such as alternate routing and self-healing network strategies. Though management of this network historically has been via a largely overlaid set of operations systems, increasingly, these systems are evolving towards hierarchical management models such as CCITT TMN and AT&T's UNMA. Future directions include increased system robustness and flexibility, and multivendor compatibility of the pieceparts. Other trends are: increased OAM&P (operations, administration, maintenance and provisioning) functions in the network (including databases), enhanced OAM&P functions such as network element software download, and changed OAM&P caused by network evolution towards more integration (for example, Service Net-2000 and FiberWorld).

Peter Roden, MIT Project Athena, USA

MIT's Project Athena has provided many insights into whether we can really control large distributed systems. Project Athena is an eight-year experiment in the use of high-powered, networked computer workstations in the MIT education community. Currently in its final year, Project Athena provides academic computing resources to nearly 10,000 users across the MIT campus. About 1,300 workstations in more than 40 clusters are connected to a campus wide network, enabling users to communicate over the system and access their data as well as other resources from file servers and other machines that are network connected. Currently, about 150 courses in more than 20 departments at MIT require students to use Athena. Project Athena is known world-wide for its innovative developments in the areas of system software and applications development for distributed, centrally managed, heterogeneous UNIX environments. Management and control entails customer service and support for both system software and hardware, engineering of new releases and software, and managing quality assurance.

Yechiam Yemini, Columbia University, USA Large distributed systems can *probably* be controlled, but it is going to require a lot of work before the answer is known. We do not yet possess sufficient basic technology knowledge to answer the challenge positively, the problem is likely to become increasingly complex and assume central significance with enterprises (and vendors), and quick fixes via patching of current technologies are not solutions but could become the very problem. To help answer the challenge positively, we must build a significant body of basic and novel technologies and concepts for manageability. We must also invest in basic research and development to develop: architectures and protocols to support coordinated manageability, new database structures to support effective management information base organization, new rule-based system structures and rule-acquisition techniques, new understanding of transient behaviors of complex distributed systems, and visualization techniques to support control of complex scenarios. Solutions through collaboration should be sought: vendors must collaborate to provide users with manageable, multi-vendor systems, and research can help advance collaboration and sharing of basic technologies and provide heterogeneous testbed systems for joint experimentation.

V

TELECOMMUNICATIONS

NETWORK MANAGEMENT

C
Planning and Operations

A NODAL OPERATIONS MANAGER FOR SONET OAM&P

Gary M. Berkowitz[a], Phillip T. Fuhrer[b], Blaine E. Gray, Jr.[a], Alan R. Johnston[a], and Gary L. McElvany[a].

[a]Operations Systems Division, AT&T Bell Laboratories, 480 Red Hill Road, Middletown, New Jersey 07748-4801 USA

[b]Switching Systems Division, AT&T Bell Laboratories, 1200 East Warrenville Road, Naperville, Illinois 60566-7045 USA

1. ABSTRACT

The telecommunications network is evolving to become more dynamic, with Operations, Administration, Maintenance, and Provisioning (OAM&P) functions built-in for rapid response. The coming of Synchronous Optical Network (SONET) network elements (NE) in the telecommunications network creates new demands as well as opportunities for OAM&P. The SONET NE will produce new kinds of data and reports that must be analyzed, filtered, saved, and transmitted to various Operations Systems (OSs). By analyzing and quickly acting on this distributed data, a new opportunity will be created to manage the SONET network dynamically, in near real-time, using intelligence distributed within the NE. A Nodal Operations Manager (NOM) is needed to capitalize on the OAM&P capabilities in the NE as well as to as to faciliate the evolution to the Dynamic Network.

2. DYNAMIC NETWORK OPERATIONS

Dynamic Network is the name given to a network that takes care of itself to the greatest extent possible; one that has operations functions "built in," with only work management, access, and some external databases outside the network.[1] A Dynamic Network is described in terms of three attributes: self-aware, self-adapting, and self-provisioning, which are discussed below.

In the operations environment of the future, NE will undertake more OAM&P intelligence. We view the responsibility of OSs as largely residing in work management and access functions with some information continuing to be stored in stand-alone corporate databases. Work flows, such as customer requests for provisioning or alarm reports from the network, will continue as a fundamental unit for customer-oriented operations planning; the Dynamic Network will speed automation of these work flows.

To achieve the Dynamic Network, many traditional OAM&P functions will migrate into or near the service-providing NE, leaving external OSs with the freedom to provide greater consolidation of information and improved access management capabilities. In addition, a single source of data will be used for database updates and changes, resulting in greater data integrity and simplified operations procedures and flows. The NE will be the primary source of data where possible.

The Dynamic Network will use embedded operations channels for transport of OAM&P traffic, eliminating the need for a separate overlay control network while simultaneously providing additional service-monitoring functionality.

Service nodes will be configured (and re-configured) in real-time under software control, allowing centralization of network control and eliminating the need for manual cross-connects in the network. Reducing technician activity in the network will significantly reduce troubles.

Increasingly, Dynamic Networks have their own information with respect to failures and performance that they can use to provide automatic restoration and thereby create a self-healing network.

2.1 Self-Aware

Self-Awareness is one attribute of Dynamic Network Operations (DNO) and is a characteristic that underlies much of the increased operations power in the Dynamic Network. Self-aware refers to services, inventory, network topology, and performance. Dynamic NE will know which services they are providing, how they are configured, what they are connected to, and how well they are performing.

Dynamic NE will be capable of self-inventory of the equipment resources provided (e.g., CLEI™) and will contain information on the element's location (e.g., CLLI™), the entities connected to it (e.g., CLFI™, pair names) and the services using the element (e.g., CLCI™) equivalent. Network elements will report on the inventory of the channel units in the network elements including type and vintage. Eventually, as the network elements become more intelligent, access to full network topology will be available via the network elements.

The Dynamic Network will also be aware of its performance. It will know the error and traffic status for individual elements, for links, and for the network as a whole, issuing reports when thresholds are exceeded. All this information will be available for use by local network technicians or external OSs. For example, NE will provide performance information in addition to traditional alarms. Network technicians can flexibly assign thresholds and will receive autonomous messages when a threshold has been exceeded. These features simplify acceptance testing during resource provisioning, and will improve detection and analysis of problems during maintenance.

2.2 Self-Adapting

The second characteristic of DNO is self-adapting. This means that the Dynamic Network has enough information and OAM&P functionality both to know that it is being overloaded or is failing, and is able to take corrective action.

With respect to offered load, the network will evolve to route calls using fully distributed flexible non-hierarchical routing techniques, migrating many traditional network congestion management functions into the Dynamic Network. Flexible routing will greatly improve network efficiency and resilience, automatically adapting to changes in calling patterns and automatically routing around node and link failures. Multi-homing of switched services and priority routing by service class will create new services. Manual cross-connects will be eliminated, improving network responsiveness and adaptability. "Private line" services will become more "switch-like." As an example, Digital Cross-connect Systems (DCS) create a "switch-like" environment for the inter-office portion of DS1 and DS3-based services by automating the provisioning process and creating an environment for automated services restoration.

With respect to failures, services will appear to be self-healing from an end-user's perspective. Sparing of modules, links, and equipment, coupled with physical diversity of facilities and services, will dramatically increase network robustness. Each module, link and node will continuously monitor its health to predict and detect service-affecting failures, and to switch automatically to spares or alternate routes to assure service continuity. Self-healing properties of the network will allow many repairs to be done as scheduled rather than on-demand, yielding significantly increased network quality and integrity.

An example of self-adapting for failures is the route diversity functions of fiber carrier systems providing service continuity for links by switching to a hot spare over an alternate path. The alternate path does not need to be "balanced" (i.e., does not need to be of the same length or have the same quantity of repeaters). The network is able to adapt to failure; service becomes self-healing from customers' point of view.

2.3 Self Provisioning

The third characteristic of Dynamic Network Operations is self-provisioning. Self-provisioning means that the network will automatically do resource and service provisioning functions, such as updating available resource lists and assigning resources based on service requests.

As a part of the resource initialization process, network resources will automatically identify themselves to associated entities. Network resources are then added to the network inventory and are ready for service assignment. From a Service Creation perspective, primitives and tools will be provided to assist in the definition of resources (both hardware and software) to provide enhanced network services and features. User-customizable generic features for services and support will replace network element and application specific features.

For service requests, the traditionally centralized service provisioning process will become fully distributed throughout the network. Service requests will be sent directly to the network, and resources will be assigned on a demand basis. The provisioning process will look and feel much like call set-up through the public switched network of today with complete end-customer control of the process, and a real-time network response to requests. Provisioning intervals will be greatly shortened producing a more responsive and adaptive service provisioning process. Billing and any necessary database updates will be done after the provisioning process has provided service to the customer. Examples of self-provisioning private line services as a switched call are the emerging DS1/DS3 dialtone services in which the

DS1/DS3 path is established by station-to-station calling.

3. GENERIC NETWORK ARCHITECTURE

Figure 1 shows the current two-layered OAM&P structure for managing the telecommunications network. In this structure, NE generate data (e.g., alarm and performance data) and centralized OSs analyze the data and do resource assignment and control. For some of the NE, rudimentary control (e.g., protection switching) and data reduction (e.g., performance monitoring analysis) is done in the NE. All control and analysis between elements is centralized within the OSs.

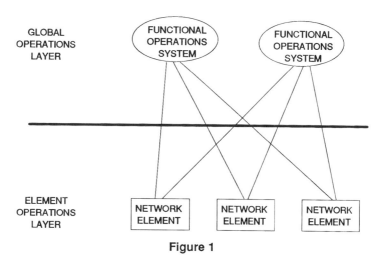

Figure 1

Figure 2 shows the three-layer structure described by Ahrens at NOMS '90 [2] involving a Network Element Operations Layer, a Nodal Operations Layer, and a Global Operations Layer.

The Network Element Operations Layer will do the same functions that are currently provided or contemplated for individual elements (e.g. self-diagnosis, protection switching). The Nodal Operations Layer will contain office-wide data to allow services to be supported from a single access point within each office. The Nodal Operations Layer will therefore assume some of the functions currently done within the OSs that are amenable to distribution within the network. The Global Operations Layer will consist of OSs serving as inter-area managers.

OAM&P STRUCTURE

DYNAMIC TELECOMMUNICATION NETWORKS

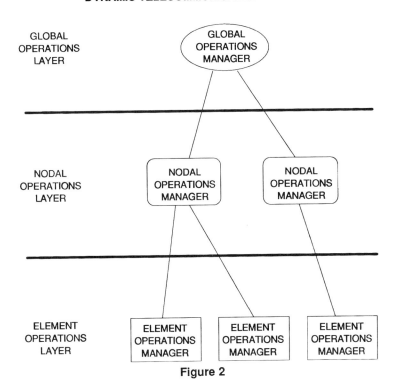

Figure 2

4. NODAL OPERATIONS MANAGER FEATURES

A Nodal Operations Manager (NOM) is needed as the vehicle to evolve to the three-layer operations model. Within a node, the NOM would provide operations access and coordination among the NE. The NOM will utilize the OAM&P functions in the NE plus will provide additional OAM&P capability at the nodal layer. In this way, the self-aware, self-healing, and self-provisioning attributes of a Dynamic Network can be fully realized.

Operations within a node cover several different areas, such as technician interfaces, OS interfaces, logging and processing of fault and performance management information (alarms, performance monitoring, and testing), storage of network configurations and implementation of network reconfiguration plans, and management of NE software and databases.

Technician Interfaces

In the area of technician interfaces, the NOM would serve to present the network technician with a common look and feel operations interface to a variety of network element types by converting the technician interface provided by each NE to a standard technician interface. As NEs move towards using standard technician interfaces, the NOM would continue to serve as a single point of access to multiple NEs.

OS Interfaces

In the area of OS interfaces, the NOM would serve as a lower-layer protocol converter, as a upper-level message translator, and as a mediation device. Protocol conversion, if needed, would match the NE to OS interface to the OS data communications network. Message translation would convert from the NE message set to the various OS message sets, which may be in a variety of formats. Mediation is needed to reduce the volume of messages to a level the OS can handle.

Fault Management - Performance Monitoring and Testing

In the area of collection and processing of fault management information, the NOM would serve as the place where logs would be kept of alarm reports, performance monitoring data, and test results. Simple processing would include routing a report to the appropriate OS, or thresholding the performance monitoring data to only send exception reports to the OSs. Advanced processing could include analyzing several reports to determine trends, initiating tests to confirm fault conditions, and automatically implementing reconfiguration plans as a result of failures detected in the network.

Configuration Management

In the area of network configurations and reconfiguration plans, the NOM would serve as the place for local storage of network maps. These maps would be based on an object-oriented model of the network and would indicate interconnection between units. The NOM could also provide local storage of network maps for reconfiguration of services. As noted above, these maps could be used with fault management information (alarms and test results) to have the network dynamically react to fault conditions.

Software Management

In the area of software and data management, the NOM could be used to store copies of the NE software programs and databases. Having these stored locally would allow for more rapid recovery from fault conditions as well as allow for downloads of new software.

Platform Architecture

The software architecture of the NOM to deliver the functions we have described should be based on a platform that simplifies adding functions as software modules and uses standard messages for communications. This platform approach would allow new features to be rapidly introduced, and would allow each NOM to be customized to the needs of a specific node.

5. APPLICATION TO SONET NETWORKS

The OAM&P features needed for managing SONET networks have have been described in other papers[3], [4]. The SONET operations features have been discussed at length elsewhere[3], [5]. From the perspective of the NOM, we will focus on two features of SONET:

Operations Access
   The presence of the SONET Data Communications Channel (DCC) for interoffice communications and the SONET Local Area Network (LAN) for intra-office communications gives the NOM access to all NEs in the SONET Management Network (SMN) including those in the Fiber Center.

Operations Data
   SONET standards, especially in the performance monitoring (PM) area, require each NE to generate, store and report on considerable quantities of data. In addition, we expect that most SONET NE will have the ability to receive software updates via files downloaded to them over the overhead channels. In both cases, the NOM will be used to help manage the volume of information.

Each of the NOM features be used with the SONET OAM&P features as follows:

OS Interfaces
   It is likely that the initial vendor protocols supported by the SONET NEs are not directly compatible with the protocols supported by the OSs. The NOM is a convenient device for converting between these different protocols and for insulating both NEs and OSs from some of the respective churns. For example, an IEEE 802.3 LAN has been proposed as an intra-office standard for communication between SONET NEs. The NOM could serve to convert between the LAN protocols and the X.25 protocol used by the OSs. At the upper layers of the protocol, the NOM could do translation between the standard Open System Interconnection (OSI) messages on the SONET DCC and existing OS message formats such as Transaction Language 1 (TL1).

Performance Management
   Current standards call for SONET NE to store PM data in 15 minute bins, thereby generating significant amounts of data. Given that one benefit of PM data is to be able to detect degradations in performance, it is necessary to retrieve and store the data even if the counts in a particular bin are well below any alarm threshold. The NOM would be able to be the collection and storage point for this data. The advantages of using a NOM for this function would be to offload some storage requirements from the NEs, and to relieve the OSs from having to process the raw PM data. The OSs would be free to focus on higher-level functions.

Fault Management - Testing

Test access and loopback functions are defined for SONET NEs in TA253[6]. The NOM, with its access to PM data and with its connectivity to the SONET NEs, can function as the access point for requests from technicians or OSs to do tests. Initially, testing will be done for existing services that are carried in SONET payloads. Eventually, we expect that testing will be extended to include new SONET services.

Configuration Management

SONET NEs typically are self-aware at some level, and can report to the NOM on their configurations, hardware and software versions, and equipment utilization. Their support of the SONET overhead also can be used both to retrieve the network maps and to rapidly distribute the changes needed to execute reconfiguration for restoration.

Software and Data Management

The NOM can serve as the access point for upgrading a network of SONET NEs with the latest software generic. With its inventory data on each NE, the NOM will be able to use the DCC to remote download software programs to the NE.

6. CONCLUSION

By utilizing OAM&P functions within a Nodal Operations Manager, we can make use of enhanced features provided by SONET, such as the DCC connecting SONET network elements and extensive performance monitoring data, to provide OAM&P on a dynamic, near real-time, basis. Rather than having to transmit all data to a central OS for processing and waiting for the OS to analyze the data and respond, many of these functions can be distributed to local NOMs, allowing for quicker responses to maintenance, provisioning, and administrative problems. The NOMs will therefore function as the middle layer in an implementation of a SMN.

The NOM will also serve as a single point of access for network technicians, giving a common interface to diverse network elements. It will also serve as the mediating device between OSs and the SONET NEs, including protocol translation until the OSs migrate to standard interfaces (e.g., OSI). We envision the NOM as evolving to support not only SONET NEs, but also the full complement of NEs as described in the Service Net-2000 architecture.

TRADEMARKS

CLCI, CLEI, CLFI, and CLLI are trademarks of Bell Communications Research, Inc.

GLOSSARY OF ACRONYMS

DCC	Data Communications Channel
DS1	Digital Signal 1 (1.544 Mbs)
DS3	Digital Signal 3 (44.736 Mbs)
LAN	Local Area Network
NE	Network Element
NOM	Nodal Operations Manager
OAM&P	Operations, Administration, Maintenance, and Provisioning
OS	Operations System
OSI	Open Systems Interconnection
PM	Performance Monitoring
SMN	SONET Management Network
TL1	Transaction Language 1

REFERENCES

1. G. McElvany, *"The Dynamic Network"*, Western Communications Forum, February 21, 1990.

2. M. Ahrens, L. Dayton, and K. Hanner, *"Toward a Next Generation of Operations"*, IEEE 1990 Network Operations and Management Symposium, February 12-14, 1990.

3. American National Standards Institute, *"SONET OAM&P Draft Standard"*, T1X1.5/89-144R1, October, 1989.

4. CCITT Recommendation M.30, *"Principles for a Telecommunications Management Network"*, 1988 (Blue Book).

5. P. Birkwood, *"SONET Based Fiber Transport Operations and Management"*, IEEE 1990 Network Operations and Management Symposium, February 12-14, 1990.

6. Bellcore, *"SONET Transport Systems: Common Generic Criteria"*, TA-TSY-000253, Issue 5, February, 1990.

A MODEL AND TOOL FOR INTEGRATED NETWORK PLANNING AND MANAGEMENT

Ali Zolfaghari[*,**], Tom Ikuenobe[**] and Stanley Chum[*]

[*]Pacific Bell
2600 Camino Ramon, Room 1S900, San Ramon, California, 94583, USA

[**]Stanford University
Building ERL, Room 202, Stanford California, 94305, USA

ABSTRACT

Advances in network technologies and attendant cost decreases have forced a rethinking of network planning, servicing and management methodologies - three traditionally distinct and separate efforts. To derive maximum benefits offered by these new technologies, an integrated network planning approach is required. In this paper, we describe an integrated feedback model of network planning and management, and present a software package for network optimization that provides an integrated network planning environment.

1. INTRODUCTION

Network planning and engineering (NP&E) consists of providing for required network resources in a cost-effective and timely manner on a continuous basis. The core activities involved are (a) gathering and organization of network state and inventory information, (b) processing this information to come up with network alternatives and (c) analyses of these alternatives and selection of the "optimum" alternative. While it is possible to define many aspects of NP&E, the work described in this paper only deals with issues related to selecting, sizing, deployment and operation of network resources. This aspect of planning is commonly referred to as network dimensioning and optimization.

The two basic variables in network planning and management are capacity and routing (both traffic and circuit), while the objectives are to minimize network costs and maximize utilization of network resources. In the most generic form, network planning and engineering activities attempt to specify a network (topology, routing and capacity, deployment strategy) to satisfy a given set of demands at a minimum-cost by addressing the following questions:
- Where should the network nodes (switches) be located?
- How should the nodes be interconnected?
- How should the traffic be routed?
- What technologies should be used?
- What capacities are needed?
- When are the facilities needed?

The key technologies that have had the most impact on interoffice network planning and engineering in recent times are: Optical transmission systems, which offer cheap and plentiful transmission bandwidth; digital switching and stored program switching, which offer flexible and dynamic network management (traffic and call routing); and digital cross-connects (DCSs), which offered more flexibility in reconfiguring network capacity and topology. These new technologies must coexist with the embedded ones, and a network-wide planning approach is required to derive maximum benefits from their deployment.

This paper is intended to convey an overview of some of our efforts towards developing an easy-to-use integrated network planning and management decision-aid on a workstation. We do not describe, in any detail, the optimization algorithms that we have developed or adapted; these will appear in future publications. Section 2 presents a general discussion of issues and considerations in the integrated network planning and management process. Section 3 discusses guidelines for developing integrated network planning tools. In Section 4, the general structure and different modules of a network planning tool, IPMAT, are discussed. Examples of application of IPMAT in typical network studies based on realistic network models are presented in Section 5.

2. INTEGRATED NETWORK PLANNING AND MANAGEMENT

Traditionally, running a network involves three distinct disciplines: network planning, maintenance (servicing) and operations (network management, including monitoring). Network management entails taking actions in real-time to ensure maximum utilization of installed network resources, commonly via traffic routing control mechanisms. Network servicing involves making routing modifications, capacity assignment and minor facility modifications in response to changing network trends on a weekly or monthly basis. Network planning is an annual activity and results in making major facility and routing modifications. We can therefore consider these to be network planning, each with a different time horizon. A feedback model is shown in Figure 1 to reflect these and the iterative nature of network planning.

Most network planning and engineering problems can be formulated as network optimization problems. Attempting to solve all aspects of NP&E as a single problem leads to large intractable optimization problems. To make the problem more computationally tractable, it is helpful to define smaller problems. In each of these smaller problems, only a small set of parameters are treated as (independent) variables. These problems can be defined such that they (i) closely model practice - in that the smaller problems are often considered at different times and in each network segments, and (ii) can be used to generate an initial solution to the feedback model shown in Figure 1.

The approach taken in this work is to identify and formulate these smaller problems, develop and apply fast (heuristic) algorithms to those problems for which an exact solution cannot be efficiently determined. Our objective is to develop an integrated environment that puts together a number of network optimization algorithms, and where feasible, we employ proven and efficient algorithms. This approach has the potential advantages of
- permitting real-time interactive planning;
- permitting real-time evaluations of "what if" scenarios;
- being applicable to very large networks;
- ease of customization;
- holding down computing costs.

The current problems of interest are:

a) Network Structure and Capacity Optimization: Given the traffic demands, a candidate network and routing, determine a minimum-cost network topology and link capacities to satisfy the demands.

b) Traffic Routing Optimization: Given the network link capacities and traffic demands, determine a traffic routing scheme which maximizes the throughput of the network.

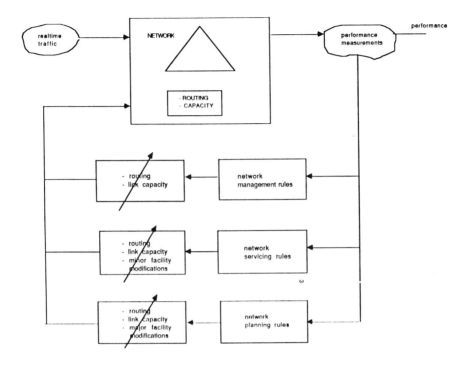

Figure 1: Feedback Model of Network Planning

c) Network Traffic Analysis: Given a network (capacity, traffic demands and routing scheme), determine the traffic performance of the network.
d) Network Expansion Optimization: Given an embedded network and projected demands, generate a minimum-cost augmentation network (network evolution plans).
e) Hub and/or Ring-Network Optimization: Given a set of nodes to be interconnected, and projected demands, specify a minimum-cost hub and/or ring-network (e.g., identify network clusters and gateways, etc.).
f) Network Reconfiguration Optimization: Given existing network facility capacities, determine a logical network connectivity and sizing such that the throughput of the network is maximized.
g) Traffic/Demand Estimation and Projection: Given current traffic measurements and relevant socioeconomic factors, determine projected traffic demands at different times in the future.
h) Basic Network Calculations: A set of procedures for common network calculations, e.g., trunk group blocking probability.

This list is by no means exhaustive; but we restrict ourselves to these problems in this paper.

3. NETWORK PLANNING TOOLS: Desirable Features

A typical network planning system (NPS) consists of a collection of algorithms for solving classes of network optimization problems, bookkeeping routines, network and cost models, and database and user interface routines. Such systems must possess architectures that reflect the changing network planning needs and take advantage of advances in computing technologies; some of the desirable attributes of a NPS are:
- Use of computationally efficient solution algorithms - to accommodate large networks.
- Modular structure, allowing module functions or processes to be easily replaced or modified.
- Open architecture, allowing addition of new modules.
- Consistent data structures for all modules, enabling separation of data from the applications so that all modules can use the same data.
- Integrated functionality, so that the output from one module can be used to drive another module.
- Provide the user with ability to form "macros" employing the basic modules.
- Integrated databases, containing all relevant network state and inventory information. This reduces the amount of effort spent on gathering and organization of data. A significant amount of the work involved in network studies is typically spent on gathering and entering network input data.
- Convenient user interface, such as graphical display of processes and data.

4. IPMAT PACKAGE

We now describe the software package under development. IPMAT (Integrated Planning, and MAnagement Tool) is a network planning and management decision support system being developed along the guidelines mention above. At this stage, the emphasis is on development and adaptation of models and algorithms suitable for large networks and a mix of network technologies. The general architecture of IPMAT is shown in Figure 2. Some of the features of the modules currently in IPMAT are described below; each module addresses the corresponding problem in Section 2.

a) Network Structure and Capacity Optimization (**NSCO**): This module uses a 2-moment traffic model and iterative optimization algorithms described in [2]. Its features:
- Can be used to optimize and design a network, or simply to evaluate the economic feasibility of adding a link to an existing network.
- Can be used to determine the traffic loading on each link of the network for network monitoring purposes.
- Offers flexibility in specifying GOS constraint through the use of traffic weighting factors.
- Handles 1-way and 2-way link optimization.
- Works very well with strongly non-linear cost functions.

b) Traffic Routing Optimization (**TRO**): This module is intended primarily for dynamic (traffic) routing, and generates traffic routes using the algorithm described in [3]. Its features:
- Traffic routing/rerouting to improve network utilization and hence avoid or defer investment in network capacity expansion.
- Investigation of the benefits of introduction of new links into the network
- Investigation of effects of total or partial loss of capacity on some links of the network
- Investigation of network resilience to varying traffic levels .

c) Network Traffic Analysis (**NTA**)
- Network traffic performance analysis (for network monitoring purposes).

• Evaluation of the impact of addition of a link.
• Evaluation of the impact of adding/removing capacity to/from a link.
• Handles various routing strategies.
• Employs a 2-moment traffic model
d) Network Expansion Optimization (**NEO**): This modules combines dynamic programming with iterative improvement optimization techniques. Its features:
   • Can be used to perform capping analysis.
   • Determination of circuit routing.
   • Design of augment network.
   • Evaluation/comparison of circuit routing policies.
e) Generic Functions (**GF**)
   • Computation of internodal distances.
   • Simple tandem-high usage-direct (THD) route calculations (for network design/planning).
   • Traffic analyses such as grade of service calculations, traffic carried by a group of trunks, traffic carried by the last trunk of a group etc. We also plan to include modules to address the other problems identified in Section 2, i.e., Hub and/or Ring-Network Optimization, Network Reconfiguration Optimization and Traffic Demand Estimator. The IPMAT modules generate a large amount of data. The amount of information that is actually presented as output is controlled/specified by the user at run time.

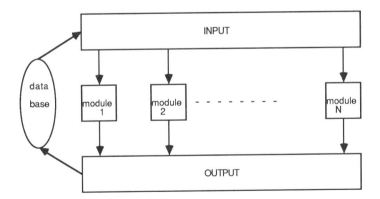

Figure 2: Structure of IPMAT

5. IPMAT APPLICATIONS

The models and algorithms incorporated into the IPMAT package have been tried on networks of varying complexities. The examples used in this paper are based on 3 network models: (a) A 4-node, 6-link network (b) a 9-node, 18-link network and (c) a 62-node, 80-link network. These models and the associated input data reflect actual network topology and traffic patterns. The results discussed here do not represent specific case studies, and are only intended to demonstrate some features and capabilities of IPMAT.

5.1 NSCO (Network Structure and Capacity Optimization)

We use the 4-node all-fiber network as the test network in this example; the relevant input data are
- A candidate network.
- Traffic routes; by default, Minimum distance routing (**MDR**) is used, with the physical distance criterion, to generate paths between node pairs. Alternate routing is allowed.
- Point-to-point traffic matrix (in erlangs).
- Transmission system and switching costs (assumed piece-wise linear).
- Grade of service (GOS) objective - specified as average network point-to-point blocking of 1%.

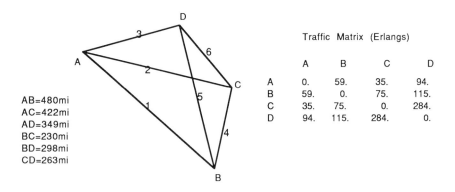

Figure 3: Candidate 4-node Network.

The input candidate network and the resulting optimum network generated by NSCO are shown in Figures 3 and 4 respectively. In general, the cost structure of fiber transmission systems lead to optimum network topologies close to minimum spanning tree (**MST**) topologies. Our assumption of piece-wise linear costs has some influence on the resulting topology; however, this influence decreases with increasing link capacities. The MST of the 4-node network consists of links AD, CD, and BC. This MST network is slightly more expensive than that of Figure 4 due in part to the relatively large traffic volume between nodes B and D. It is important that the candidate network be richly meshed so as to allow for a wide range of alternatives in determining the optimum network.

5.2 TRO (Traffic Routing Optimization)

The test network used here is a 9-node network engineered for a 1% network average blocking under MDR routing scheme. The network is shown in Figure 5 We study the traffic performance of the network under uniform network overload condition and under failure of some network components. It is assumed that there is complete flexibility in choice of routes. At each traffic point we compare the performances using the fixed design routing (MDR), using routes generated by the traffic routing optimization algorithm (**OPT**) and using direct-only routes (**DIR**). MDR, OPT and DIR all use the concept of shortest "distance" between node pairs in path generation. In MDR and DIR, the distance is the physical distance; in OPT, the distance is related to the traffic on links connected the nodes. Also both MDR and OPT allow overflow routing, while DIR does not. Figure 6 shows plots of carried traffic vs offered traffic under the 3 routing schemes - MDR, OPT and DIR.

We also study the effects of loss of link #8 (BH) combined with a 5% uniform increase in traffic over design level. Under MDR, the network blocking is 19%; however this is reduced to 7% using the TRO algorithm to generate alternate paths. Table 1 shows the load and blockage on each link of the network. (In the table -> means a change in value from MDR routing case to the OPT routing case.) The results show improved performance can be realized by a dynamic routing optimization scheme. These improvements are obtained by generating efficient alternate paths and restriction of overflow routing. Other computer runs have given even more remarkable improvement when the traffic variations in the network are nonuniform.

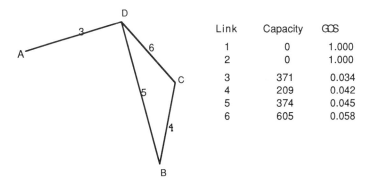

Figure 4: Optimum 4-node Network.

5.3 NEO (Network Expansion Optimization)

The 62-node network is the test network for this trial run. Input data are: Node to node circuit demands for each year of the planning period (10 years); candidate network, i.e., existing network (routes, capacities) and links considered for addition to the network during the planning period; and costs (fiber cable cost per mile, termination equipment costs per module, maintenance costs etc.).

We study the evolution of the existing network towards an all-fiber network. We assume (i) both message and special services demands between a node pair are routed along the same paths (ii) any link capacity expansion is satisfied through the use of fiber technology only and (iii) all non-digital facilities are capped at start of planning period.

TABLE 1
Network Statistics Upon Loss of Link 8: Before and After Routing Optimization. Average Network Blocking goes from 19.42% to 7.30%. (-> means goes to.)

Link #	Capacity	Blocking	Load (Erl.)	Link Utilization
1	272	0.281->0.049	254.9->259.6	0.937->0.955
2	357	0.440->0.059	337.0->345.3	0.944->0.957
3	211	0.563->0.038	176.8->197.5	0.836->0.936
4	140	0.558->0.093	138.9->132.9	0.992->0.949
5	353	0.367->0.087	353.0->344.4	1.000->0.976
6	580	0.226->0.047	575.4->565.3	0.982->0.975
7	430	0.272->0.159	429.5->425.0	0.999->0.983
8	0	1.000->1.000	0.0->0.0	0.000->0.000
9	253	0.056->0.100	241.9->245.7	0.956->0.971
10	435	0.046->0.103	414.1->427.4	0.952->0.982
11	290	0.113->0.100	283.6->282.6	0.976->0.974
12	412	0.132->0.081	397.6->402.8	0.965->0.978
13	146	0.231->0.080	143.2->138.1	0.981->0.946
14	300	0.233->0.025	297.9->281.9	0.993->0.940
15	194	0.317->0.097	194.0->186.7	1.000->0.939
16	122	0.323->0.022	110.4->107.6	0.905->0.982
17	130	0.305->0.078	130.0->122.1	1.000->0.939
18	40	0.617->0.003	39.5->26.1	0.987->0.553

Three main routing policies for additional demands are considered; these are:

A) Fixed routing policy 1: Fixed routing throughout the planning period, utilizing the existing routes. This implies expanding capacities on existing links by augmenting with fiber facilities when needed.

B) Fixed routing policy 2: Fixed routing throughout the planning period, utilizing the routes for some predetermined target network. (Optimized routing for the last year of planning period, not considering embedded routes, is used in the examples below.) This scenario models that in which a target network has been identified and all additional demands are carried by only those links in this network.

C) Dynamic routing policy: Rerouting of additional demand is allowed in each year of the planning period. In this scenario, we attempt to evolve an all-fiber network from the existing network. However, the structure of the resulting fiber network will be strongly influenced by the existing network, except in those cases where the currently satisfied demands are much less than future demands (e.g. very high growth networks).

We also consider two capping options (and assume that all non-digital facilities are capped at start of planning period):
   I) Cap (non-fiber) facilities at start of planning period. This implies satisfying all additional demands with fiber technology.
   II) Cap facilities at exhaust.

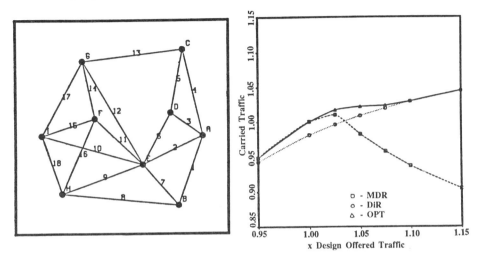

Figure 5: 9-node network   Figure 6: Network Overload Behavior

Circuit rearrangements have not been considered in this study. Such rearrangements generally result in deferring investment in transmission facilities while incurring rearrangement costs.

The routing policies and capping options described above were studied using the NEO module. In the study, we have assumed that, where links currently exist, the active (in-service) capacities are enough to meet the demands in the first year of the planning period. Figure 7a shows the candidate network for expansion. This network consists of the existing network and new links that are to be considered for inclusion in the expanded network. In Figures 8 and 9, we plot the cumulative present worth network expansion costs for each routing policy and capping option. Cumulative network demand is also plotted. Figure 7b shows the links that are expanded during the period.

From the results, if the links are to be capped at exhaust, then the policy of allowing dynamic circuit routing of additional demand in each year of the planning period results in the minimum present worth cost (**pwc**) network. This is an intuitively appealing result since such a routing policy results in better use of installed facilities and deferment of new investments. We also observe that for this capping condition, there is no significant difference in pwc between fixed routing policy 1 and dynamic routing; this reflects the strong influence of the existing routes on the "optimal" dynamic routes chosen for subsequent years. However, if the links are capped at start of planning period, it turns out that minimum pwc network results from a policy of fixed routing that does not consider the embedded routes at start of period. This may be explained by the fact that the network

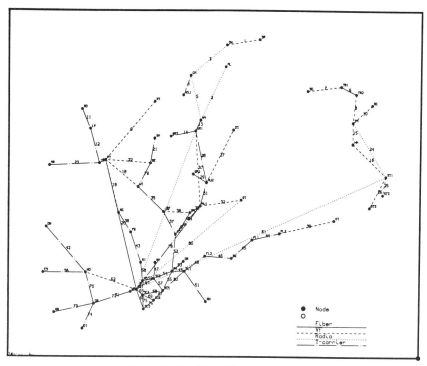

Figure 7a: Candidate 62-node Network

demand grew by about 50% during this period. By utilizing more efficient routing for this additional demand and taking advantage of large transmission bandwidth, this demand is concentrated over a relatively small number of facilities. Thus, fewer transmission facilities are required, hence a substantial savings is realized compared to the policies that grow along the embedded paths.

These preliminary results suggest that choosing the option of expanding along only a "target" network is an appealing long-term policy. Specifically, in our example with capping at start of planning period, it is the optimal policy out of those considered. If rollover is taken into account, then this policy is expected to become more attractive, even for case of capping at exhaust, since there will then be more demand to rapidly fill up the transmission facilities. As explained earlier, rerouting additional demands takes advantage of presence of availability of extra capacity in segments of the network.

In Figures 8 and 9, the concentration of relatively large costs incurred in the first few years is due to assumption of capping all non-digital facilities at start of planning period. About 40% of the links in our test network are non-digital.

Our assumption of back-to-back multiplexing contributes a significant fraction of the termination costs. We note that there is potential for significant reduction in termination costs by incorporating "add/drop" feature, wherein only the demand components that need to be dropped are demultiplexed to the DS1 level.

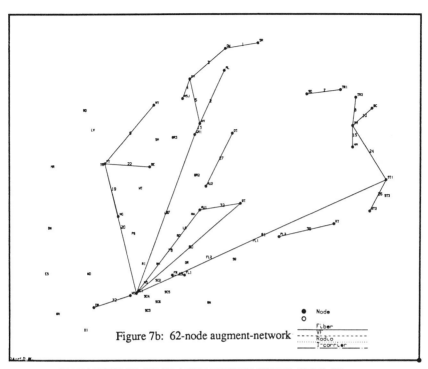

Figure 7b: 62-node augment-network

COMPARISON OF CUMULATIVE PRESENT WORTH COSTS OF
EXPANSION UNDER DIFFERENT CIRCUIT ROUTING POLICIES

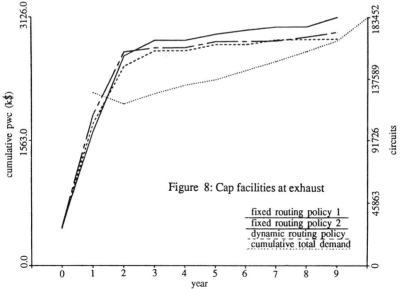

Figure 8: Cap facilities at exhaust

fixed routing policy 1
fixed routing policy 2
dynamic routing policy
cumulative total demand

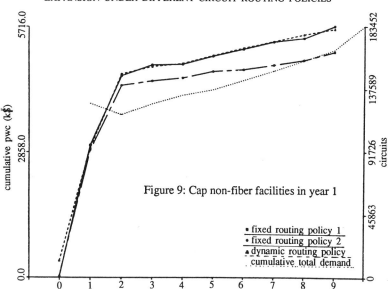

Figure 9: Cap non-fiber facilities in year 1

6. SUMMARY

We have presented and defined a number of problems, often encountered in network planning and engineering, with the aim of studying and developing appropriate models and algorithms for the solutions. In the work covered in this paper we have adapted such models and algorithms and incorporated them into a software package, IPMAT. In its present form, the package addresses problems of present-day networks as well as future networks. Some modules of the package have been applied to network problems of varying complexities to illustrate their features and capabilities. Preliminary results from these studies indicate that the IPMAT package is quite suitable for realistic network planning and engineering tasks. Efforts are under way to improve the optimization algorithms and expand the models in IPMAT.

REFRENCES

[1] Zolfaghari, Ikuenobe, Chum "BIP: Report on Preliminary Network Modeling and Software Development Efforts", Internal Pacific Bell Report, September 1988.
[2] Badr-El-Din, A., *Optimizing Telephone Traffic on Long-Distance Networks*, Ph.D. Thesis, Stanford Univ., CA, (1984).
[3] Ikuenobe, T., et. al., "A Study of Nonhierarchical Networks Employing Alternate Routing", IEEE Comm. Conf. (MONTECH), (Nov. 1987), pp129-132.

ACHIEVEMENT OF PRIVATE VIRTUAL NETWORKS IN A DQDB MAN

A. La Corte, A. Lombardo, S. Palazzo, D. Panno

Istituto di Informatica e Telecomunicazioni,
Facolta' di Ingegneria, Universita' di Catania,
V.le A. Doria, 6 - 95125 Catania, Italy

ABSTRACT
Support of Closed User Groups (CUGs) in the new integrated networks requires appropriate mechanisms to monitor the traffic generated by the CUG users and to enforce the total bandwidth they use within the capacity bounds contractually agreed upon.
This paper introduces a technique of traffic estimation and policing for DQDB Metropolitan Area Networks. The solution proposed is quite compatible with the current draft standard IEEE 802.6.

1. INTRODUCTION

One of the main problems in the management of the new integrated networks derives from the need to mark out private virtual networks for specific groups of users within public networks. Certain businesses, for instance, may need to communicate by means of a private network for reasons of security or availability.
In many cases, however, the installation and running costs of a private network are not economically convenient for users, especially when, for instance, the number of interconnected users is low or when they are spread over a very wide geographical area.
In this scenario it would seem useful to introduce into the new integrated public networks a particular service supporting Closed User Groups (CUGs). This service allows users to form groups, the members of which can communicate with each other but not with users not belonging to the group.
Currently, international standardization organizations have only dealt with the problem as far as ISDNs are concerned [1], while have not yet regarded MANs, which constitute a first step towards the public B-ISDNs of the future. This paper intends to deal with the problem of supplying a CUG service for a MAN complying with the DQDB standard proposed by IEEE 802.6 [2].
The main features which characterize the CUG service are the following:
- access restrictions for the rest of the network

- the portion of the total network capacity used by the members of the CUG.

As far as the first point is concerned, access restrictions can easily be obtained by implementing appropriate address screening of messages in transmission and reception.

The second point, on the other hand, is much more complex. It is quite realistic to suppose that in the contracting stage the network administration and the concern responsible for the CUG will establish a maximum CUG load, according to which accounting will be determined. It is therefore necessary to provide appropriate network control mechanisms which will allow the flow of traffic generated by the CUG to be monitored and enforced within the bandwidth contractually agreed upon.

As for this purpose the CUG is to be considered as a single distributed user, the bandwidth enforcement mechanism also has to be implemented in a distributed fashion. So it is necessary for all the stations in the network to be able to perform local actions taking into account instant by instant the total load generated by the CUG. For this reason, bandwidth enforcement mechanisms like the Leaky Bucket [3] are inappropriate, as they only take the traffic flow generated by a single connection into account and can thus only be applied to individual CUG users.

In previous works, distributed monitoring mechanisms for dual bus networks have been proposed [4][5], but they have not been envisaged for the management of Closed User Groups. In this paper the authors define a mechanism for monitoring in each node in the MAN the traffic generated by the users belonging to a specific closed group distributed over the network. Then they propose two bandwidth enforcement algorithms which take the bandwidth contractually assigned to the CUG into account. The mechanisms introduced are suitable for a DQDB network and compatible with the current proposed standard [2] for the MAC protocol. The current proposed standard specifies the access control and convergence functions required to offer a connectionless MAC data service to support an LLC sublayer in a manner which is consistent with other IEEE 802 LANs. The current standard also specifies additional DQDB functions as a framework for other services, such as isochronous and asynchronous connection-oriented services. These, however, are considered as a subject for further study. For this reason, in the following we will only deal with traffic monitoring and policing for connectionless data services.

Section 2 gives a review of the DQDB architecture. The description is confined to the functions performed by the blocks of the reference architecture which support connnectionless services. In Section 3 the load monitoring mechanism we propose is described. Section 4 introduces the bandwidth enforcement mechanisms for CUG traffic. Finally, table I includes the acronyms which we will use throughout the paper.

2. THE DQDB AND ITS ARCHITECTURE

The architecture of the DQDB metropolitan network is based on a slotted bus topology, as shown in Fig. 1. The two buses support unidirectional communications in opposite directions. Nodes, called Access Units (AU), are connected to both buses and communicate by selecting the proper bus. The head station on each bus generates a frame every 125 µs, which is divided into slots of a fixed and constant length. Each slot contains a control field, the Access Control Field (ACF), which specifies whether access is asynchronous or isochronous, and a segment which constitutes the payload of the slot. In this paper we only refer to asynchronous access to the bus (the Queued Arbitrated mode), controlled by the Distributed Queueing protocol and typically used to provide non-isochronous services.

The operation of the Distributed Queueing Access protocol is based on two control fields in the ACF: the BUSY bit, which indicates whether a slot is empty or not, and the REQUEST field, which is used to indicate when a segment has been queued for access. Each node, by counting the number of requests it receives and the unused slots that pass, can determine the number of segments queued (i.e. in line) ahead of it. This counting operation establishes a single queue across the network for access to each bus.

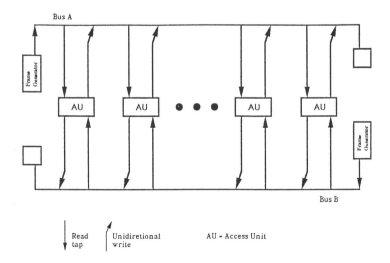

Fig. 1 Topology of the DQDB MAN

Figure 2 shows the functional architecture of a DQDB node, with reference to the QA access mode only. The common Functions block allows the QA Functions block to gain read and write access to the QA slots. The QA Functions block provides an asynchronous data transfer service for QA segment payloads. In transmission, the QA Functions block accepts the QA segment payloads from the convergence block MCF (MAC Convergence Function) and adds the appropriate segment header, including the Virtual Channel Identifier (VCI), to the segment payload to create a QA segment. In reception, according to the VCI value in the QA segment, the QA Functions block delivers the QA segment payload to the corresponding convergence block. In the case of connectionless services a unique VCI default value is defined so that each QA segment payload is delivered to the MCF in all the nodes.

As mentioned previously, the current standard only specifies the convergence function for provision of the Medium Access Control (MAC) service. The MAC service is a connectionless service that supports the transfer of variable length MAC Service Data Units (MSDUs) between peer entities in the LLC sublayer, without the need for the LLC entities to request the establishment of a connection between them. There is no guarantee of delivery of the MSDUs by the MAC service.

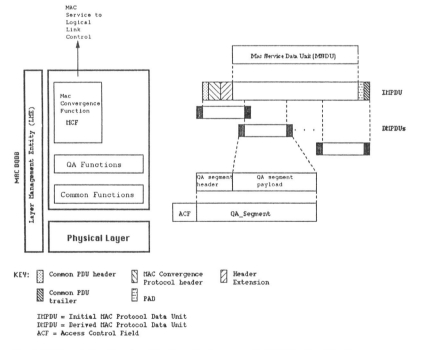

Fig. 2 DQDB node functional architecture for transfer of a MAC Service Data Unit

The MCF block shown in Fig. 2 performs segmentation and reassembly of the MSDUs. In the transmission phase, the MCF forms IMPDUs (Initial MAC Protocol Data Units), adding appropriate fields to the MSDUs, in the format shown in Fig. 3. The Destination Address (DA) and Source Address (SA) are included in the MCP (MAC Convergence Protocol) Header field. In addition, an HE (Header Extension) field is also provided for, the use of which is optional. Its length can range from 0 to 20 octets in multiples of four, and its size is specified by 4 bits (Header Extension Length, HEL) in the MCP Header field. The MCF block takes an IMPDU and divides it into one or more segmentation units, as shown in Fig. 2. Each segmentation unit is 44 octets long. The number and type of segmentation units depend on the length of the IMPDU. The MCF also adds a header and a trailer to each segmentation unit, so as to form a DMPDU (Derived MAC Protocol Data Unit). The DMPDU header specifies whether the segment is of the Beginning Of Message (BOM), Continuation Of Message (COM), End Of Message (EOM) or Single Segment Message (SSM) type. The DMPDU thus obtained constitutes the payload of the QA segment. In the reception phase, the operations performed by the MCF block are the opposite of those defined for the transmission phase, so as to reconstruct the IMPDU and deliver the original MSDU to the LLC sublayer. The MCF block also validates the correctness of each DMPDU received. If the DMPDU is not valid it is discarded. If no errors are found in the transfer of single DMPDUs, the MCF block checks the complete IMPDU again, discarding it if the result is negative.

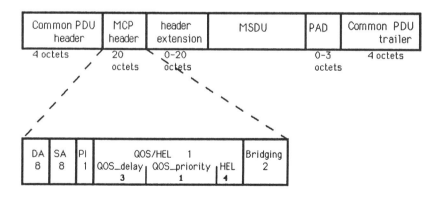

Fig. 3 IMPDU format for connection less data services

3. MONITORING OF CUG TRAFFIC IN A DQDB NETWORK

In order to monitor CUG traffic each station has to be able to:
- identify the CUG to which the transmitting user belongs;
- quantify the intensity of traffic generated by the CUG users.

So appropriate subfields that can contain this information must be defined in accordance with the current standard [2].

If it is supposed that the number of CUGs in the network is no more than 256, it is possible to identify a single CUG through an identifier (called id_cug), expressed by an octet. Let us also assume that whenever a CUG user has to transmit an IMPDU, the corresponding MCF calculates the number of DMPDUs into which it is to be segmented. We indicate this number with DMPDU_No. As the maximum length of an MSDU is 9188 octets, from the segmentation of the IMPDU we can obtain at most 210 DMPDUs. Therefore a single octet is sufficient to indicate the value of DMPDU_No.

In Section 2 we saw that currently the use of the Header Extension (HE) field in the IMPDU is optional. In this paper it is assumed that the Header Extension is only used by the CUG members and that, if present, it is made up of four octets. We use two octets of this field, one for the id_cug and the other for the DMPDU_No. It should be noted that the first segment obtained from the segmentation of the IMPDU (the segment type of which is BOM or SSM) contains the whole HE and therefore all the information necessary for monitoring.

Through this mechanism each node is enabled to know the CUG traffic generated by the upstream stations. It is, however, also necessary for each node to be able to monitor the traffic generated by the downstream stations belonging to the CUG. For this purpose, besides the segments obtained from the division of the IMPDU, it is necessary for each node to send a further signalling segment, which in following we will call SIGNAL_DMPDU, on the opposite transmission bus. This segment is of the SSM type. In addition, in order for it to be recognised as a signalling segment, the Destination Address (DA) is set as equal to the Source Address (SA). The other fields, including the Header Extension, are set as equal to those of the IMPDU to which the SIGNAL_DMPDU refers.

By using the subfields id_cug and DMPDU_No of the Header Extension, contained in the first segment of the IMPDU and in the additional segment SIGNAL_DMPDU, each node can estimate the total load generated by all the users belonging to a specific CUG. When, in fact, the n-th station in the network receives a segment on a bus for which the VCI value is that of a connectionless service, the MCF performs the following operations besides those described in Section 2.

- If the segment is of the BOM type and the HEL value indicates that the length of the Header Extension is equal to four octets, the MCF reads in the HE field, the id_cug and the DMPDU_No values. The latter represents the number of segments that an upstream node belonging to the identified CUG is sending to another user in the group.

- If the segment is of the SSM type, the MCF compares the DA with the SA. If SA is not equal to DA, the MCF performs the operations described in the previous case. If SA is equal to DA, the MCF recognises that the segment is a SIGNAL_DMPDU. The DMPDU_No value read in the HE field represents the CUG traffic which is being transmitted on the other bus by a downstream node.

The amount of traffic detected by the above algorithm is the traffic already queued, which is considered as not being removable from the network.

Let us now see how this monitoring mechanism can be implemented. Let us suppose for the sake of simplicity that in any station users belong to the same CUG. Let us assume that each station is equipped with suitable Load Counters LC_g to count the network traffic generated by the group it belongs to. Two counters are needed - LC_{gA} and LC_{gB} - one for each of the two buses, A and B. Let us consider an interval T. Referring for example to bus A, in the initial instant of this interval, the counter LC_{gA} is set to zero. In the case of a connectionless service, the MCF block of every node, by reading the id_cug in the HE field, can recognise whether the IMPDU segments of BOM and/or SSM type are transmitted by a user belonging to the same CUG. In this case LC_{gA} is increased by the DMPDU_No it finds in the HE field. So, in the interval T, the estimate of the load on bus A generated by users belonging to the CUG being considered is:

$$L_{gA} = n\ LC_{gA} / T + L_{CUG}(T) \quad [bit/s] \quad (1)$$

where n indicates the slot size expressed in bits and $L_{CUG}(T)$ is the load contribution generated by the station in question in the interval T.

This measurement is repeated cyclically. Obviously the same procedure is followed to estimate the load generated by a specific CUG on bus B. For sake of simplicity, below we will use L_g to indicate the load estimated on either of the two buses.

If users in a node belong to different CUGs, it is only necessary to consider two counters for each CUG.

4. BANDWIDTH ENFORCEMENT MECHANISMS FOR CLOSED USER GROUPS

The traffic load generated by a CUG is controlled in accordance with the capacity contractually allocated to the CUG. It is, in fact, necessary to enforce that the members of the CUG do not use more bandwidth than the amount negotiated for each bus, L_{gMx}. At this regard, it should be pointed out that for connectionless services there is no request for establishment of the connection, and the user generates messages (MSDUs) regardless of whether they can be transmitted over the network by the node. Besides, if a DMPDU is lost the receiving station does not activate reassembly of the IMPDU it belongs to. It is therefore appropriate for the bandwidth enforcement mechanism to be applied to each MSDU, and not only to the segments.

A simple policing mechanism would be to impose an upper bound on CUG traffic, equal to the L_{gMx} value which has been established. Whenever a user belonging to a CUG generates an MSDU featuring a DMPDU_No value, the corresponding station calculates what the CUG load on the bus would be if the MSDU were transmitted, as follows:

$$L'_g = L_g + n * DMPDU\_No / T \qquad (2)$$

The comparison between this estimated load and the maximum load for the group, L_{gMx}, allows the management modules to decide whether to accept or discard the new MSDU. If $L'_g < L_{gMx}$, the MSDU is accepted; otherwise it is discarded.

This management scheme ensures that the CUG does not exceed the bandwidth it has been allocated. It must be noted, however, that although this management mechanism is highly efficient at limiting input CUG traffic, it is incapable of improving the total network throughput. In fact, all the MSDUs exceeding the allocated bandwidth are discarded, even if the network load is light and only a few resources are being used.

The limitations, in terms of efficiency, of the policing mechanism described above are due to the rigid division between the above control algorithm and the network congestion control [6][7][8]. For this reason, a bandwidth enforcement mechanism that takes also into account the total network traffic would seem more appropriate.

We calculate the global traffic load in the network as in [5], according to the number of BUSY and REQ bits counted. In fact, the amount of BUSY bits monitored on the transmission bus indicates the load due to the upstream stations, while the number of REQ bits monitored on the reverse bus indicates the load due to the downstream stations. To this purpose we assume each station is equipped with two other Load Counters, LC_A and LC_B, one for each of the two buses A and B (see Fig. 2). In the interval T, counter LC_A is increased by each BUSY bit counted on bus A and each REQ bit counted on bus B. So the estimate of the load on bus A is:

$$L_{tA} = n\ LC_A / T + L'_A(T) \qquad [bit/s] \qquad (3)$$

where n indicates again the slot size expressed in bits and $L'_A(T)$ is the load contribution to bus A generated by the station in question in the interval T.

The same calculation is made for the global load on bus B. For the sake of simplicity, below we will use L_t to indicate the load estimated on either of the two buses.

It should be noted that in monitoring the total amount of traffic in the network, besides the implicit information contained in the flow of packets, which is seen by any node, there is no communication between the stations.

The bandwidth enforcement mechanism works in the following way.

Whenever a user belonging to a group generates an MSDU, the

corresponding station calculates both the CUG load contribution and the total load on the transmission bus, being this MSDU assumed as transmitted:

$$L_t' = L_t + n * DMPDU\_No / T \qquad [bit/s] \qquad (4)$$

The management module decides whether to transmit the new MSDU over the network or to discard it by comparing the estimated loads, calculated as in (2) and (4) respectively, with the maximum loads for the CUG and the whole network.

If $L'_g < L_{gMx}$, the MSDU is automatically accepted. If $L'_g > L_{gMx}$ and $L'_t < L_{tMx}$, the MSDU is still accepted for transmission. In this case a tariff increase for the overload granted to the group may be provided for. Finally, if $L'_g > L_{gMx}$ and $L'_t < L_{tMx}$, the MSDU is discarded.

Let us note that the estimated global load is compared with a maximum load value and not with the total network capacity. In previous papers [9][10][11][12][13], in fact, it has been shown that the grade of performance can be preserved with statistical multiplexing only if the network load is maintained below a certain utilization limit.

In the proposed method, each station performs monitoring and bandwidth enforcement functions independently. It is therefore important to consider the problem of algorithm stability. To this purpose it is preferable for each station to carry out its control operations uniformly in time, rather than have all the measurement periods ending simultaneously. If a large number of active stations is being considered it is sufficient for the station measurement intervals to start at random; in this way no synchronization between the stations is required.

5. CONCLUSIONS

This paper introduces some mechanisms which allow private virtual networks to be created in public metropolitan networks, with particular reference to the ones complying with the standard IEEE 802.6. More precisely, a technique has been defined through which each node in a MAN can estimate the traffic generated by users belonging to a specific Closed User Group (CUG). The protocol mechanisms introduced are quite compatible with the formats and procedures provided for in the current DQDB standard. The authors have also defined and discussed two bandwidth enforcement algorithms. The first enforces the traffic generated by the users of the CUG within the upper bound contractually allocated to the CUG, blocking any load in excess. The second, on the other hand, conditions acceptance of excess load according to the global amount of traffic on the network. The second algorithm thus allows the network to be used more efficiently if the load is light, allowing the CUG to exceed the capacity agreed upon, on possible condition of a possible increase in tariff.

The main value of the mechanisms introduced in the paper is that they are fully distributed and require only a very limit amount of signalling.

Table 1. List of acronyms

ACF	Access Control Field
AU	Access Unit
B-ISDN	Broadband Integrated Service Digital Network
BOM	Beginning Of Message
COCF	Connection-Oriented Convergence Function
COM	Continuation Of Message
CUG	Closed User Group
DA	Destination Address
DMPDU	Derived MAC Protocol Data Unit
DQDB	Distributed Queue Dual Bus
EOM	End Of Message
HE	Header Extension
HEL	Header Extension Length
IMPDU	Initial MAC Protocol Data Unit
LLC	Logical Link Control
MAC	Medium Access Control
MAN	Metropolitan Area Network
MCF	MAC Convergence Function
MCP	MAC Convergence Protocol
MSDU	MAC Service Data Unit
QA	Queued Arbitrated
SA	Source Address
SSM	Single Segment Message
VCI	Virtual Channel Identifier

REFERENCES

[1] CCITT: Draft Recommendation I.255 on "Community of Interest Supplementary Services", Blue Book, 1988.
[2] IEEE: Proposed Standard - Distributed Queue Dual Bus (DQDB) Metropolitan Area Network (MAN), October 1989.
[3] J. Turner: "The Challenge of Multipoint Communication", Proc. 5th TC Seminar on Traffic Engineering for ISDN Design and Planning, Lake Como, May 1987.
[4] J. Limb, "Load-Controlled Scheduling of Traffic on High-Speed Metropolitan Area Networks", IEEE Transactions on Communications, vol.37, No.11, November 1989.
[5] A. Lombardo, S. Palazzo: "An Architecture for a Pure ATM Metropolitan Area Network", Proc. GLOBECOM '89, Dallas, November 1989.
[6] G. Gallassi, G. Rigolio, L. Fratta: "ATM: Bandwidth assignment and bandwidth enforcement policies", Proc. GLOBECOM '89, Dallas, November 1989.
[7] M. Hirano, N. Watanabe: "Characteristics of a Cell Multiplexer for Bursty ATM Traffic", Proc. GLOBECOM '89, Dallas, November 1989.
[8] A. Eckberg, D. Luan, D. Lucantoni: " Meeting the Challenge: Congestion and Flow Control Strategies for Broadband Information Trasport", Proc. GLOBECOM '89, Dallas, November 1989.

[9] F. C. Schoute: "Simple decision rules for acceptance of mixed traffic streams", Proc. ITC 12, Torino, June 1988.
[10] H. Ohnishi, T. Okada, K. Noguchi: "Flow control schemes and delay/loss tradeoff in ATM networks", IEEE J-SAC, vol.6, n.9, December 1988.
[11] G.M. Woodruff, R.G.H. Rogers, P.S. Richards: "A congestion control framework for high-speed integrated packetized transport", Proc. GLOBECOM '88, Fort Lauderdale, November 1988.
[12] J. Filipiak: "Structure of traffic flow in multiservice networks", Proc. GLOBECOM '88, Fort Lauderdale, November 1988.
[13] A. Lombardo, S. Palazzo, D. Panno: "Admission Control over Mixed Traffic in ATM Networks", International Journal of Digital & Analog Cabled Systems, Vol. 3, No. 2, August 1990.

This work has been partially supported by the National Research Council (CNR) in the frame of PFT.

VI

REASONING AND KNOWLEDGE

IN NETWORK MANAGEMENT

Knowledge technologies for evolving networks

Shri K. Goyal

GTE Laboratories Incorporated
40 Sylvan Road
Waltham, MA 02254

Abstract

Expert systems have moved out of the laboratory; a whole new style of automation system design and programming is being introduced in the world of telecommunication systems. The paradigm of rule-based knowledge representation and reasoning is the basic technology involved in this process, and it has been successfully applied to a wide range of "real world" problems: from basic equipment troubleshooting to global network management. Although this paradigm appears better suited to those automated reasoning tasks than any other software technology that came before it, there is some evidence that expert systems may be limited in their abilities to handle anticipated automation requirements of the intelligent network.

The academic and industrial AI communities have been pursuing advanced knowledge technologies, often dubbed second-generation expert systems. In this paper we take a look at some of the characteristics of the future network environment and attempt to evaluate how those knowledge technologies might penetrate even further in automating ever more complex control, problem-solving, and decision-making tasks. We present examples of laboratory technologies in each of three key areas of AI from the standpoint of telecommunication network operation and management: distributed AI, knowledge representation, and machine learning.

1. INTRODUCTION

During the last quarter of the century, and more so in the last decade, worldwide telecommunication network capabilities have rapidly advanced to meet the challenge of the information age. The pace has been further fueled by customer demand for a variety of innovative services that require the support of a high quality reliable voice/data network. Current day network technology enhancements have given rise to many sophisticated network surveillance, control, and decision-support systems [1]. Management and administration of a complex integrated network presents a new challenge for network operators, designers, and technology innovators.

The current telecommunication environment is dynamic and transitional. The US Public Switched Telephone Network (PSTN) is currently operating with a mostly digital switching and inter-office transmission plant. Fiber optics is now the prevalent transmission growth and replacement medium, and is also penetrating the outside plant of the Local Exchange Carriers (LECs). At a different level, SS7 out-of-band signalling is being rapidly introduced, and the hierarchical routing structure of conventional long-distance networks has been replaced by dynamic non-hierarchical routing, resulting in a marked improvement in network availability and robustness at reduced costs. Those major evolutionary steps have created the need for sophisticated Operations Support Systems (OSS) for surveillance, monitoring, testing, and control [2-3].

In this environment, expert systems have moved out of the laboratory and into the PSTN OSS arena. Along with them, a whole new style of automation system design and programming is being introduced in telecommunication systems. The paradigm of rule-based knowledge representation and reasoning has been the basic technology used, and it has been successfully applied to a wide range of "real world" problems, from basic equipment troubleshooting to global network management. Although this paradigm has been a powerful software technology, there is some evidence that expert systems may be limited in their abilities to handle anticipated automation requirements of what is most often referred to as the Advanced Intelligent Network (AIN).

At the most generic level, some of the driving factors in developing intelligent systems to manage the evolving telecommunication network are the following:

- Increasing rate of technology change (causing, among others, a knowledge acquisition and expertise development bottleneck).
- Increased levels of system complexity, causing the decision-making task to be very difficult.
- Inherent distribution of control at some levels and centralization at others.
- Real-time control in a dynamic environment.
- Need for knowledge technologies in the control loop rather than in advisory off-line functions.

The academic and industrial AI communities have been pursuing research in knowledge technologies, often dubbed second-generation expert systems. In this paper, we take a look at some of the factors listed above and attempt to evaluate how knowledge technologies might penetrate even further in automating ever more complex control, reasoning, and decision-making tasks. We also present examples of laboratory technologies in some of the key areas of AI from the standpoint of telecommunication network operation and management.

2. ARTIFICIAL INTELLIGENCE (AI) TECHNOLOGIES

Traditional algorithmic approaches have, in the past, proven satisfactory for managing networks. These approaches alone are not sufficient anymore for the management and control of complex networks [4]. These must be complemented with more powerful heuristic approaches, such as in Artificial Intelligence (AI) systems [5].

AI is, among other things, a set of programming methodologies that focus on the techniques used to solve problems by generating new strategies and plans, and even generate new domain knowledge. Expert systems, a sub-field of AI, provides a new, often successful, way of attacking problems that previously have been considered not solvable by machines. Utilizing human expertise and domain knowledge, the techniques focus on the use of declarative (factual) knowledge and relatively simple rules of inference for putting that knowledge to work on a specific problem.

Research has shown that the basic knowledge about the domain alone was not enough to get the performance exhibited by most experts. The people performing the tasks acquired another kind of knowledge, "experience," that allowed them to concentrate on the most likely causes of a problem and to adapt answers to the specific problems. The people who do this well are regarded as experts. The programs that capture their problem-solving strategies and selectively apply them under specific circumstances are called expert systems.

Most current expert systems utilize rule-based and other simple technologies. Several important technologies, which would potentially provide much-needed power to future AI systems, are under development in various laboratories. Three such key technologies are *Distributed AI* for sharing of expertise and better coordination in problem solving, *Knowledge Representation* for more robust and better utilization of domain experts' knowledge, and *Machine Learning* for providing the self-improving ability to the systems in a dynamic environment. These technologies and their ramifications are further discussed in Section 4.

2.1 Expert systems in support of network operations

A variety of expert systems have been employed in telecommunications as operation support systems (OSS) and for other functions. Many more systems are under development. A recent survey article [6] categorizes the existing systems in three main functional categories: maintenance, provisioning, and network administration. The maintenance systems provide monitoring, troubleshooting, and diagnosis support in keeping the networks operating. Provisioning and planning applications support development of network evolution plans by design, configuration, and execution of these plans. Network administration and management applications help manage the network traffic, and plan and execute a workable strategy when exceptions occur. This application also includes billing, facility assignment, and record keeping. In addition, a miscellaneous application area relates to sales and system configuration support.

Table 1 lists some expert systems that have been tested and/or are in use in telecommunications today [7–16]. Diagnostic expert systems are, by far, the most popular application in telecommunication, spanning public telephone to packet switched networks. Well over a dozen expert systems have been reported to provide monitoring, troubleshooting, and fault diagnosis. Switch maintenance, typically for the older technology switches, is a common application. Perhaps expertise on the older switches tends to disappear as the switch technicians move on and learn the new switches. Also, new switches are quite sophisticated, with much processing power and built-in self-diagnostics to identify and report their own hardware faults. Software faults or bugs are difficult to diagnose internally and by external systems. Therefore, no systems are currently available for the latest generation of switches, despite the clear need in this area.

Table 1
Examples of expert systems in the network management domain

System	Task Performed	AI Method	Type, Status	Environment
Advanced Maintenance Facility (AMF) [7] (BRITISH TELECOM)	Finds faults in TXE4A telephone exchanges	Basically a production system	Diagnostic, field tested	16-bit Micro; UNIX SAGE and LISP
Central Office Maintenance Printout Analysis and Suggestion System [8] (COMPASS) (GTE)	Finds Faults in GTE's No. 2 EAX telephone exchanges	Mix of frames, rules, LISP, and active values	Diagnostic, field tested	Xerox 11xx KEE and INTERLISP-D
Automated Cable Expertise (ACE) [9-10] (AT&T)	Troubleshoots telephone company local loop plant	Forward-chaining rules	Diagnostic, commercial product, over 40 systems deployed	AT&T 3B2; UNIX OPS4 and LISP C
Network Management Expert System (NEMESYS) [11] (AT&T)	Reviews traffic completion data and suggests traffic control changes	Mix of rules, procedures and active rules	Monitor, research prototype	Symbolics KEE
Network Trouble Shooting Ethernet Consultant [12]	Finds problems in DECnet and LANs	Forward-chaining rules with confidence factors	Diagnostic, field tested	VAX EXPERT
Net/Advisor [13] (BB&N)	Monitors real-time network status and suggests actions to take when problems are diagnosed	Back-chaining rules (PROLOG) with LISP interface code	Diagnostic, commercial	Symbolics PROLOG and LISP
DESIGNET [14] (BB&N)	Assists in building a data communications network	Object-oriented programming	Design, research prototype	Symbolics 36xx ZETALISP
MAX (NYNEX)	Analyzes local loop trouble reports and outputs dispatch recommendations	ART rules and mixed paradigm	Diagnostic, deployed in 42 NYNEX locations	SUN Inc. workstation
IAS [15]	Analyzes alarm for Italian packet network	C & OPS-83	Monitoring/diagnostic prototype under evaluation	VAX workstation
ARACHNE (NYNEX)	Supports planning interoffice hierarchical network		Planning/provisioning prototype	SUN communicating with IMS databases
NEC's Network Configurer	Planning for corporate communication and multiservice planning	OPS-83	Planning/provisioning under evaluation	32-bit workstation
NETREX [5] (CALTECH)	Diagnose and repair faults in BANCS network in real-time	Teknowledge S1	Proof of concept prototype	SUN 3/160 and C tools
AT2 [16] (AT&T)	Isolates fault in Special Service Circuits using SARTS remote test system	Rule-based programming OPS-83 (and routines)	Diagnostic, deployed product	OPS-83 and C in UNIX on AT&T 3B2
SSCFI (GTE)	Isolates faults in Special Service circuits using AT2, SARTS, and other test systems	ART-IM rules and schemes, C code, CrossTalk	Diagnostic, prototype under evaluation	ART-IM and C on IBM-PC,DOS

Systems for diagnosing faults in telephone outside plants (ACE [9], MAX), Special Service Circuits (Autotest [16], SSCFI) and specific equipment other than the switches have also been built and several have been tested and deployed. Some of them have had a significant impact on day-to-day operations and company profitability. For example, over forty ACE systems are operational in six RBOCs. MAX is in use in over forty

NYNEX locations. Autotest2 is going through a deployment phase for use in diagnosis of faults in special service circuits by automating testing.

In the provisioning arena, the focus has been on design (e.g., DESIGNET [14]) and system configuration. Only recently is expert system technology being applied to network planning (e.g., ARACHNE).

The most common application in the network administration area is traffic routing or traffic management. In the public network, experts (and, thus, potentially expert systems) monitor traffic data and install switching controls to redirect traffic and relieve congestion in the network. NEMESYS [11] at AT&T Bell Labs is an attempt to automate some of the decision-making abilities of the traffic manager. The routine decisions NEMESYS makes are influenced by network topology and the possible cause of congestion. Several specialized billing applications, such as toll fraud detection systems, are also being developed and evaluated.

Over a period of time, it has been recognized that expert systems should act as background processes to support and complement decision-making activities. Expert systems were originally proposed as stand-alone systems not easily integrated due to their specialized hardware and software platforms and with little intercommunication ability with existing databases. This brought about a lot of resistance from the MIS and operator community. The situation is changing drastically as professional workstations, lately using a UNIX environment, are becoming an acceptable development and deployment environment (to the MIS community) as powerful tools are becoming available for these workstations. Most of these are general-purpose tools. Some efforts to build specialized development tools with knowledge representation suitable for the telecommunication domain have met only limited success, partly due to lack of a formal model of network entities and their interoperability. For a specialized network application tool to be useful, there should be a common structure between the problems to be solved, using a common and complete network knowledge base. Due to the lack of these facilities, network applications have been unable to use specialized, and potentially powerful, knowledge representation mechanisms (e.g., model-based [17], experience-based techniques). These common structures and models are slowly evolving as their need and value is being recognized. The pace of this formalization process is accelerating with the work of various worldwide standardization bodies.

Expert systems also exhibit brittleness in their behavior. Within the region of expertise, they perform at expert level and totally fail beyond these limits. Most systems today are off-line, advisory systems performing a well-defined and confined task. Expansion of the breadth of coverage and, thus, their power and usefulness will be achieved by using advanced technologies. Some of these are discussed in Section 4.

3. EXPERT SYSTEM LIMITATIONS AND THE NEED FOR ENABLING TECHNOLOGIES

It seems that telecommunications has been less successful than other industries in fielding expert systems and reaping benefits from their use. Without detracting from the successes of the systems, we can identify limitations of existing expert system technology, and speculate on why some of these limitations are especially critical for successful use of expert systems in the telecommunications industry.

One of the most significant impediments to fielding expert systems is not strictly technical, but rather reflects the typical organizational structure of the telecommunications industry. Telephone operation organizations tend to have relatively inflexible boundaries between functional groups; this is important for establishing responsibility and for evaluating performance, but it somewhat constrains opportunities for deployment of expert systems or, indeed, of automation systems in general. It is noteworthy that the "successful" expert systems listed above perform largely *within a single organizational* function. The perceived benefit of inter-organizational automation can be much more significant than the benefit accruing to any individual organizations. However, in the present organizational structure, the benefit is often neglected because it is distributed across several organizations.

Despite this organizational limitation, the bridging function is such a compelling opportunity for automation that the payoff to the larger enterprise can sometimes override internal struggles. Unfortunately, an expert system may create boundaries between itself and the rest of the world. All these aspects need some attention, particularly in view of the changing telecommunications environment, if the industry is to reap the benefits of AI.

- One such boundary exists between different expert systems. Significantly lacking from the technology is a standard "knowledge interface" language by which disparate expert systems can communicate. This is the topic of Distributed AI, a technology that is especially critical in telecommunications. Such technology is crucial for an inherently distributed industry where coordination across regions and functions is crucial for any acceptable network performance.
- Another boundary that critically affects the acceptance of any expert system is that which exists between it and the human being whose decisions it is to support. Recent advances in user interface technology (including natural language processing, as well as speech recognition and synthesis technologies) have improved the usability of expert systems for operational personnel. Many more significant developments are on the horizon.
- The boundary between computing environments, especially between AI systems and conventional OSSs, was a serious handicap in the early days of AI for telecommunications. This boundary has essentially vanished in recent years, as platform-independent AI environments have emerged, and as expert system techniques have been recognized as part of the standard embedded-systems toolkit.

The well-known "knowledge-acquisition bottleneck" is another kind of boundary, one that lies between the expert and the expert system. This limitation is experienced across industries, but is manifested in telecommunications in unique ways:

- Knowledge representation schemes developed for other industries are often inappropriate for telephony knowledge. A notable example is the kind of knowledge used by network traffic managers, which is largely experience-based, rather than derivable from precompiled heuristics or well-behaved domain models. Discussed in a later section is research on eliciting and representing this kind of "case-based" knowledge.

- The rate of technological change in telecommunications networks is remarkably high; the expertise needed to operate the network evolves with each introduction of new hardware, signalling systems, routing strategies, or service offerings. Expert systems, as currently fielded, capture only a snapshot of this changing knowledge, and "evolve" only to the extent that they are revised by their maintainers. Machine learning is therefore a clear requirement for future systems in order to keep pace with accelerating change in network technology.

Another feature of network operations, the time-constrained nature of decision-making, has two effects that are especially significant for AI technology:

- The need for real-time response is inherent in many network operations, such as traffic control, for which the timing of an intervention is crucial to its effect on network performance. Real-time expert systems are only beginning to be evaluated, and remain an active research area.
- Because human reaction times are too long for many network operations, especially those in support of future service offerings such as dynamic reallocation of bandwidth, the human role in such tasks must be only supervisory. As automated systems perform more "closed loop" functions, the performance of the network is increasingly dependent on their reliability. For expert system technology, this requires improved verification and validation techniques. These topics have only recently begun receiving attention.

The challenges for expert systems that are posed by network applications will determine the role of this technology in the future of telecommunications. The remainder of this paper will identify some critical technologies and focus on how three of the key technology challenges identified here are being met by emerging AI technology. We will also provide an overview of the research at GTE Laboratories that applies these advanced techniques to solve telecommunication-specific problems.

4. ENABLING KNOWLDGE TECHNOLOGY

4.1 Some critical technologies

The future siblings of the existing systems will be broader in scope, coverage, and more integrated with the control of the systems they monitor. Thus, rather than covering a subset of messages or alarms produced by some piece of equipment, they will cover a majority of meaningful messages from a variety of equipment. Similar systems will be built for pieces of equipment not covered today. Table 2 lists a set of future operations system functions and attempts to match them with a set of enabling technologies. Network managers implement appropriate control even when definite decisions cannot be made in the absence of hard data or under time constraints. Future systems should be able to suggest such controls by using fuzzy logic. The knowledge and accuracy of expert systems will be enhanced by incorporating the knowledge from several experts, leading to the creation of corporate knowledge repositories for each critical expertise area. This corporate resource will be updated and enhanced on a continuous basis through machine learning technologies. All this will free up the brightest staff members to learn newer systems, and this will allow the company to keep pace with changing network technology, while the networks will manage themselves effectively by using integrated automated intelligent systems.

Table 2
Enabling technologies in future operations support systems

Function	Future Operation	Underlying Technology
Testing	Non-vendor-specific monitoring and data analysis systems with expert repair capability.	• Standard interfaces • Data filtering, data correlation • Expert system
DB Maintenance & Software Update	Automatic DB consistency check. Non-interruptive SVR updates.	• Object-oriented programming • Distributed DB's • Advanced software technologies
Network Maintenance	Proactive maintenance of switches and outside plant. Mechanized trouble analysis and dispatch.	• Patterning • Expert system • Operations research
Operator Support	Automated directory/operator assisted functions.	• Speech recognition and synthesis
Provisioning of special and enhanced services.	Automatic provisioning of dynamically reconfigurable circuits through software control. Mechanized design and testing of special circuits. Customer-controlled network configuration for new services.	• Network operating system • Expert system for design, analysis, and testing • Service provisioning using scripts • Enhanced transaction systems • Network security • Machine learning
Traffic Management	Dynamic network reconfiguration by generating alternate routing methods.	• Integration • Operations research • Expert system • Machine learning
Data Collection and Billing	Intelligent billing systems to avoid toll fraud.	• Decision-support system
Planning and Design	Dynamic planning.	• Enhanced planning and database access systems
Performance measurement	Monitoring and support for performance enhancement and tuning.	• Distributed interoperable intelligent system
Network Management	AI-based dynamic network management. Intelligent customer and operator interfaces. Centralized customer contact systems.	• Real-time AI systems • Fuzzy logic • Distributed problem solving • Man-machine interface • System integration • Machine learning

The network domain is a difficult one to manage. It requires a combination of situation assessment, problem solving, planning, and control in near real time, almost on a continuous basis [4–5]. It is a domain for which a formal model does not exist, and expertise, at best, is "spotty" and often uncoordinated. What makes the situation much more difficult is that the domain has been very dynamic with changing architectures and functionality of various network elements as the new generation of innovative services is being integrated to meet the future requirements. Perhaps all these factors are contributing to the slow rate of introduction of expert systems in this domain. Broader and more powerful systems are needed to make a real difference. This requires significant technical developments in key technology areas. Some examples follow.

Networks require a wide variety of coordinated expertise. Activities and operations of a large number and variety of disparate components must be coordinated. This demands cooperation (and negotiations) among individual intelligent systems [18–19]. Issues such as these are being addressed under Distributed AI (DAI) research. At GTE, we are developing a framework for cooperative problem solving and building a Team-CPS testbed for Customer Network Control application [20]. This activity is discussed under Section 4.2.

The dynamic nature of today's evolving network demands utilization of changing expertise. Model-based reasoning and experience-based reasoning support decision making based on a formal problem model or instances of successful experiences, respectively. Such knowledge representation paradigms have proven to be powerful and useful in selected application domains. A testbed for traffic management, called NETTRAC, utilizing Case-Based Reasoning (CBR) is being developed at GTE for network traffic management [21]. This activity is discussed under Section 4.3.

Network configurations and functionality change on a regular basis, and so also does the needed expertise in managing the network. The ability of systems to learn from their successful experiences and mistakes will become an essential requirement as intelligent systems proliferate in the existing telecommunications environment. Major initiatives on machine learning research are being pursued at various universities and industrial organizations. Section 4.4 discusses GTE's activities in the machine learning area.

4.2 Distributed artificial intelligence

The current expert system technology supports developing systems in small, well-constrained domains. In order to tackle larger, more complex problems, several of these autonomous systems have to be integrated in a team and made to work cooperatively. This is the goal of DAI.

A central objective for DAI research lies in the hope of, some day, creating a "society" of expert systems in which a set of selected computer programs can cooperate and share their (local) expertise in much the same way as human experts do, for example, in a meeting room in an ad hoc consultative relationship. A metaphor for this form of cooperative problem solving can be described by a "chef and architect" interaction, each an expert in his domain and with specific viewpoints in designing a kitchen, making decisions and compromises through negotiations [22].

A chef wants a kitchen. He knows all about cooking requirements, recipes, needed tools, and appliances. He can read floor plans and suggest changes to suit his preferences. He can design a convenient kitchen, but he does not know about building codes, architectural issues, and costs associated with a design choice.

An architect knows about the building design constraints, building codes, and architectural requirements for an "aesthetically pleasing" kitchen. He does not know the optimum and detail requirements that suit the cuisine style of the chef.

The joint design process must proceed interactively. The architect proposes an initial layout. The chef fine-tunes the design to meet his specific requirements, and the architect considers and evaluates the proposed changes against his constraints. There may be a need for a negotiating dialogue, for example, to meet the cost requirements.

Among the requirements that stand out in the example are (1) a language for communication between the chef and the architect, (2) a negotiation strategy and (3) a framework to represent the domain expertise of the chef and the architect. It is expected that a better design, for both parties, evolves through a cooperative dialogue as their respective expertise is utilized.

The following section describes the work at GTE Laboratories on cooperative problem solving research. After examining several telecommunications-related application areas, Customer Network Control (CNC) was chosen for the initial experiments. Some key ideas motivated the selection, including the inherent distribution of network control, the occurrence of conflicts among controllers, and the tightly coupled interdependence of problem solvers during crisis management.

Customer networks can include a diverse range of elements: voice/data switches, multiplexers and digital cross-connect systems, the backbone transport network, etc. Some facilities may be customer-owned and some public-network-owned. Currently, the key attributes of CNC are multi-class traffic control, dynamic load balancing and network reconfiguration, and access control of external traffic inbound to the customer networks. A composite example of a private network with extensive customer network control would involve a corporate network that can reconfigure in real time to meet a demand for multipoint video sessions, temporarily offload voice traffic to the public network, reassign incoming 800 traffic to various call answering centers, and balance its use of virtual private network service and dedicated leased public facilities.

Cooperative problem solving for customer network control

TEAM-CPS (Testbed Environment for Autonomous Multiagent Cooperative Problem Solving) is a research testbed that explores the dynamics of cooperative problem solving among dissimilar agents [23]. Interagent cooperation in the testbed is currently demonstrated by having two agents with different views of the network, but with the common purpose of implementing a self-healing network, help each other improve their respective local solutions and jointly solve a facility failure problem.

Two agent types, the private or customer network manager and the public network manager, are defined. An agent in this context is a computer program with autonomous reasoning, problem solving (e.g., an expert system), and communication expertise that can participate in organized joint problem solving with other agents.

TEAM-CPS agents are based on a blackboard inference engine which encodes local expertise, agent control knowledge, and knowledge about the cooperative environment as knowledge sources activated by a common object store. Areas of AI research involved in the design of such agents include automated reasoning, planning, meta-level control of autonomous agents, hypothetical world reasoning, and interagent communications. To support a meaningful cooperative dialogue, agents need a common language, a dialogue control mechanism, knowledge, and goal sharing capabilities, including explicit models of other agents' goals and strategies. We have implemented an approach to interagent cooperation that models shared semantics of action and intent between agents as a goal tree planning, transfer, and integration mechanism. Customer and the public network manager agents cooperate in TEAM-CPS, as illustrated in Figure 1 and discussed in the following scenario.

In the face of a facility failure that affects both the public and the customer network, all the agents begin to plan for restoral. In this initial planning, the customer agents do not know of any resource restrictions in the public network, so they simply request that their lost capacity be restored. Each affected customer agent communicates its request, in the form of a goal tree plan, to the public agent. The public agent then combines these requests, and attempts to fairly distribute the available resources among the requesting agents, denying the remaining, infeasible parts of the customer plans. Again, these replies are communicated in the form of annotations and extensions to the respective customers' goal-tree proposals. Thus, goal trees provide the inter-agent language as well as the internal planning representation used by the agents.

Just as the public agent was able to refine the customers' plans because of its knowledge of the physical limitations of the network, the customer agents can refine the public agent's counter-proposal to conform to their local knowledge of traffic loads and priorities. In addition, they may propose to exercise other control options, such as reconfiguration of existing leased trunks, rerouting of voice traffic to the public network, or re-targeting 800 traffic to bypass the failed facility. When the public agent receives these updated proposals, it makes one last stage of improvement by using both its knowledge of traffic demand on the public network (which could interfere with the customer's off-net or 800 rerouting) and of the physical implementation of existing leased trunks (to create more efficient reconfigurations than could be found by the customer alone).

As this example demonstrates, inter-agent cooperation and negotiation yields a better overall solution, superior to any that could be obtained without the dialogue. This sharing of expertise and interactive exchange of partial solution spaces among the agents is crucial to maintaining coherent control of the customer and public networks, and is thus an essential supporting technology for future networks.

In general, it cannot be assumed that agents have the same reasoning mechanism and are able to communicate using the same representation as for planning and problem solving; e.g., exchanging goal trees like our current agent system. As part of our ongoing activities, we are developing a planner with the capability of planning for partial goal satisfaction and extending TEAM-CPS to handle dissimilar agents which will require more general and explicit inter-agent models and communication language. Some of the other issues relate to the scaling of TEAM-CPS to consider larger systems of agents and the associated problems; role of multiple modes of communication, such as point-to-point, narrowcast, and broadcast; handling of multiple concurrent problems; complexity of dialogues; and termination problems, etc.

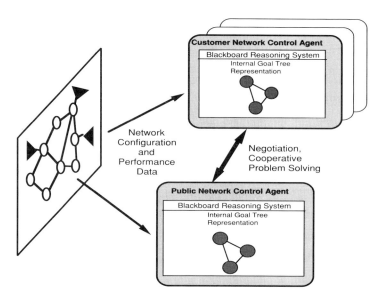

Figure 1. Cooperation in the customer network control domain.

4.3 New knowledge representations

A knowledge representation (KR) is a systematic way of codifying what an expert knows about some domain. Ultimately, a computer must be able to store, process, and utilize this coded knowledge. The main criteria for assessing the power of a representational framework are logical adequacy (ability to express the knowledge we wish to represent), expressive power (a language with well-defined and "complete" syntax and semantics), and notational convenience (agreed-upon conventions which make the information easy to write and read). Many representational schemes have been developed [24–25]. Most expert systems today use production rules or frames for knowledge representations. Rule-based paradigms offer adequate power for general-purpose work but seem to be insufficient for the evolving network requirements.

Model-Based Reasoning (MBR) [17] is particularly well suited to diagnosis and troubleshooting, and can provide an essential KR technology for the network domain in generating a "best guess" solution when all necessary information/data does not exist. MBR works through the interaction of observation (of the behavior of the actual device) and prediction (based on the model). It works on the presumption that if the model is correct, all the discrepancies between the observations and prediction arise from the defect in the device. Model-Based approaches can be practically used when the device or the system can be formally modeled and the problem domain does not contain too many device interactions. It is difficult to model a complete network on a consistent basis because of the wide variety of components it employs for diverse applications and their interactions. Individual devices can be modeled, however, to use this approach in the telecommunications domain. Although it is a serious area of current research, MBR applications to telecommunication systems await more technical developments.

Different representations are appropriate for specific portions of the knowledge that is required in any complex operation. Hybrid representations may increase the robustness of expert systems, thereby overcoming the "brittleness" for which the expert systems are criticized when unable to perform even slightly beyond their explicit domains of expertise.

Case-based reasoning for traffic management

A prototype system, called NETTRAC (NETwork Traffic Routing Assistant using Cases), is being built at GTE Laboratories for traffic management [26]. The system receives network performance data from a (simulated) group of switches under its control, recognizes and interprets abnormal conditions that it observes, and develops plans for installing controls to alleviate network problems. With approval from the user, it installs the controls and monitors [27] their effects, recommending adjustments when needed. Figure 2 shows the functional diagram of NETTRAC.

NETTRAC currently handles the following four broad categories of network traffic problems:

- Trunk group overload (or "isolated-demand overflow")
- Focussed overload
- Partial facility (trunk or switch) failure
- Complete facility failure

In order for NETTRAC to know how to handle a particular type of problem, we provide it with a description of how to recognize, treat, and monitor the problem. These descriptions are the "cases," in the form of past experiences, used by the case-based reasoner. To use these cases for solving network problems, NETTRAC must be able to find a relevant case when confronted with a problem. NETTRAC accomplishes this by indexing its cases according to the crucial features of a situation that together indicate the possible applicability of each case. Final selection of the specific case on which to model its response is done by evaluating the details of a situation to determine which case is most similar and most likely to lead to a favorable outcome.

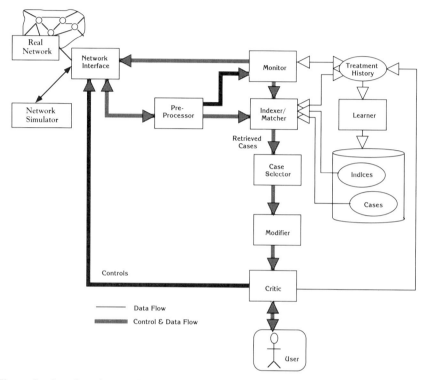

Figure 2. Case-based reasoning (NETTRAC) functional diagram.

Obviously, a large number of cases are required to describe all significant problems in a domain as complex as traffic management, and NETTRAC currently has only a (substantial) subset of the case knowledge it will ultimately need. We have, however, provided the system with a "soft matching" mechanism, the Modifier, that expands the class of problems for which a case is applicable. It does this by defining specific circumstances under which a case may "borrow" attributes from another somewhat similar case. In the example above, a CODE-BLOCK control could be "borrowed" from some other case for any switch types that don't support the GAP control. This avoids the need to explicitly encode every possible combination of features in separate cases.

There are several advantages to be gained from this case-based approach when dealing with complex problems such as network traffic management. Through incorporating a simple learning or case acquisition mechanism, the system can accumulate new cases as it gains experience. This makes CBR especially useful in evolving domains, or where much domain knowledge is missing or difficult to obtain, because the automatically acquired cases can "fill in the gaps" and track gradual changes in the problems of the domain. Also, CBR systems avoid the "brittleness" of other approaches: rather than having to match a precise set of conditions, CBR systems are designed to handle situations that are analogous to, but not exactly like, those of which they have explicit knowledge. This feature is a result of "soft-matching" to determine case applicability, and of adapting the case to apply to a broader range of problems. There are also some efficiencies to be gained through CBR; because the system outlines a complete treatment plan, only one matching and selection cycle may be required for each network problem, compared with dozens or more for a rule-based system. This is the origin of the oft-cited performance advantage of CBR for time-constrained applications.

A significant part of our continuing efforts is the construction of an automatic case-acquisition capability. This feature makes use of the user's expertise as well as the system's own experience to extend its knowledge. Initially, this will involve adding new cases to the system's experience base when the user changes the system's proposed solution. Later, the system may identify additional predictive features as a result of its own experience. As these predictors are identified, they will be used to further index the expanding set of stored cases, thus automatically expanding the experience/case base of the system.

4.4 Machine learning techniques

In order to keep up with the accelerating changes in telecommunication networks, future expert systems will need to be able to adapt themselves, without relying on the intervention of human knowledge engineers. Indeed, it is possible to imagine a rate of change in network technology that will exceed the ability of human organizations to digest the requisite new knowledge. At that point, machine learning will be not only an economic necessity, but a technological imperative for further advancement.

This high degree of pay-off has engendered an explosion of current research activities and a wealth of promising new techniques, encompassing both symbolic methods, and non-symbolic methods such as connectionism and genetic algorithms.

As with the need for multiple knowledge representations, we believe that no single machine learning approach will suffice for complex learning problems. It is much less clear, however, when a particular method is appropriate. Telecommunications applications have been proposed for connectionist networks, for example, from alarm correlation to network design. A pressing need, therefore, is to identify both the appropriate application of emerging techniques and the means by which these techniques can be used together.

One effort to address this issue of integrated learning is ongoing at GTE Laboratories, and is highlighted in the following section.

The integrated learning system [28]

Machine learning is a rapidly growing sub-field of Artificial Intelligence, and a large variety of learning algorithms have been reported in the literature [29-34]. No one algorithm provides a totally satisfactory solution to a wide range of problems. An important issue is how to combine various learning paradigms, how to integrate different reasoning techniques, and how to coordinate distributed problem solvers. An Integrated Learning System (ILS) [28] has been implemented at GTE Laboratories to provide a framework for interpreting and evaluating a variety of learning algorithms distributed across a heterogeneous network of computers and languages. Network Traffic Control has been chosen as the application domain. Figure 3 shows the ILS architecture.

At present, ILS contains three learning paradigms: Inductive (FBI), search-based (MACLEARN), and knowledge-based (NETMAN). ILS also includes a central controller (TLC) which manages control flow and communication between the agents. The agents provide TLC with their own expert advice and critiques on other agents' advice. TLC chooses which suggestion to adopt and performs the appropriate actions. At intervals, the agents can inspect the results of the TLC's actions and use this feedback to learn, improving the value of their future advice. A network simulator (NETSIM) is used to provide a realistic environment to evaluate the learning architecture.

FBI and MACLEARN are written in Symbolics LISP and run on Symbolics Lisp Machines. NETMAN is written in Quintus Prolog and runs on SUN Microsystems workstations. ILS is completely distributed. Communication among instances of the agents and TLC is via TCP/IP streams with a text-based protocol.

Inductive Learning. FBI (Function-Based Induction) is an extension of Quinlan's ID3 [35]. FBI learns decision trees from large numbers of examples. Trees both compress and generalize the experience represented by sets of examples; the trees completely describe the examples. A decision tree can be expanded into a set of rules, one for each leaf of the tree, so that this approach generates classification rulesets from examples.

FBI is also able to discover potentially useful concepts that are combinations of existing functions by examining a tree and finding paths that lead to identical subtrees. The discovered concepts are then used to simplify the existing decision tree and are made available as building blocks in the construction of other decision trees. Often the concepts discovered in this way are useful because they reflect genuine features of the domain. In other cases, the concepts discovered are artifacts caused by random patterns in the example set. FBI can ask other elements of ILS, in particular NETMAN, to assess the utility of a discovered concept.

Search-Based Learning. MACLEARN currently performs best-first search in order to learn useful combinations of operators (called macro-operators or macros) that can be subsequently treated as a single operator. Macro-learning is a form of chunking which can improve search performance by enlarging the set of operators available for the search. The availability of a good set of macros will often drastically reduce the combinatorial explosive nature of a search problem. A key issue in macro learning is to be highly selective in choosing which macros to keep. MACLEARN uses various criteria to perform the filtering process.

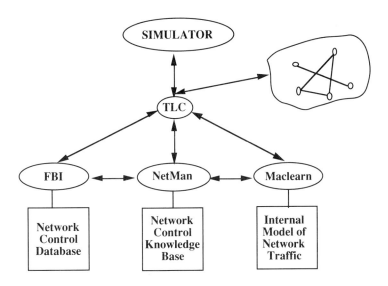

Figure 3. The integrated learning system.

On complex problems, MACLEARN may encounter combinational explosion as the search space of possible operators becomes too large. MACLEARN becomes bogged down in the search and may be unable to find a satisfactory solution. Other agents within ILS can provide assistance to MACLEARN by constraining the search and indicating which part of the search space should be examined.

Knowledge-Based Learning. NETMAN is an example of a knowledge-intensive learning system [34]. Such systems have a large amount of knowledge of the domain, similar to the knowledge base of an expert system. The knowledge need be neither complete nor totally accurate. As a result, NETMAN can make mistakes. (Human experts suffer from the same limitation, of course.)

Ideally, NETMAN would be able to distinguish accurately between situations in which an action will be successful and those in which it fails. Unfortunately, the computation involved, and the stochastic nature of the domain, make this impossible to do precisely. Instead, NETMAN heuristically differentiates the cases by calling on FBI, another component of ILS.

NETMAN learns four major types of information from experience:

- Stored caches: NETMAN stores as a macro the sequence of rule firings that led to advice that worked.

- Support list: The support list indicates how successful or unsuccessful a particular action proved to be. Those that have proved valuable in the past are more likely to be used in the future.
- Possible bugs: When an action fails to have the expected result, NETMAN can classify the cause and severity of this failure. This information is stored and will affect the future use of the action.
- Plans: A plan consists of a sequence of actions that have proved useful, together with the expected effect of each action.

When actions have unexpected results, NETMAN attempts to explain the cause of the unpredicted behavior. This analysis allows NETMAN to discover that sometimes an action will fail due to a bug. NETMAN is able to classify the type of bug and associate it with the action.

The Learning Coordinator. The TLC manages communication between agents and the decision-making process. It first asks the agents to propose a control action. It then asks the agents to critique the proposals of other agents. TLC then chooses among the proposed actions, and executes it. The process is repeated every five minutes when a new set of data and traffic statistics are received from the switches in the simulator.

The success of ILS will be demonstrated when ILS as a whole does better than each individual learning agent. Preliminary results, thus far, have been encouraging. The resulting knowledge/learning of each agent will be utilized more effectively. Efforts are also under way to improve the individual learning algorithms and add other learning paradigms to the ILS.

5. A VIEWPOINT ON THE FUTURE NETWORK

5.1 The ultimate network and its evolution

An obvious way to deal with a complex network is to build automatic controls in the network by building sensory points, reasoning, and control mechanisms—a task much easier said than done! There is still a long way to go before we can take this goal seriously, but limited automatic control functionality is beginning to appear in present systems.

Several generic categories of knowledge engineering applications have emerged in the last several years [6, 36]. Some of the categories of knowledge-based systems relevant to network management, capture, and use expertise in design, planning, interpretation, diagnosis, monitoring, prediction, and control. The intelligent system applications for networks can be mapped into two broad conceptual categories. The first category deals with design and planning issues in establishing and providing network facilities. Flexibility of the network is built into this phase. The second category supports operations, management, administration, and control (OMAC) functions and reflects on the reliability aspect of the network. There are strong pressures to integrate the existing optimizing and algorithmic methods and tools with the emerging intelligence in the network. A variety of stand-alone knowledge-based systems are in field trials or in commercial use in the network. Also, algorithmic design and planning tools such as EFRAP (The Exchange Feeder Route Analysis Program), cash flow, and traffic engineering tools are available today to the network designers. There is a need for building intelligent planning tools and integrating them with currently used databases, so that the same system can be used by different groups, including network planning, network engineering, and implementation planning. Expert system front ends will be providing the consolidated data environments for these groups to make intelligent decisions.

Network management and traffic control systems are currently used in decision support mode to manage networks. As traffic control expert systems become closely integrated with alarm surveillance systems and are tested sufficiently to verify the accuracy of their decisions, they can be trusted to make simple traffic rerouting decisions and to install controls, thus making networks more automatically controlled (autonomous). At the same time, as the systems are extended with more knowledge about operations and alternative controls, they will more closely support human experts in making complex decisions.

5.2 An autonomous network scenario

In an autonomous network, error-free operation is provided by hardware flexibility and intelligent diagnostic functions embedded into the network components and into the network management system. The network management system continuously monitors traffic demand and facility operation, adjusting the normal flow of traffic as needed. This system also has a learning component to permit it to come to anticipate the routine overloads and adjust to the slower shifts (hopefully growth) in traffic demands. One function of this system is to report on the capacity of the network to handle the demands and to warn the planning group when problems seem to be caused by design mismatch rather than temporary demand anomalies. Operators now work with the system to solve long-term operational threats—the system itself handling the small problems.

The autonomous network will also request and direct repair and maintenance functions. The overall system will have a wide variety of sensors and status display systems distributed and integrated in the network. The actions are taken by a collection of processors, with some of the monitoring and analysis functions being performed at switch level, along with some repair and maintenance actions; the rest of the repair actions will take place at regional centers serving a subset of the network (and a single roving repair crew) along with more monitoring and analysis and facility error detection and diagnosis. Regional centers will also be the first line of attack on traffic management functions. For the sake of improved reliability and load sharing, the regions covered by a center will overlap. The centers will be individually responsible for coordinating their actions so that work is not duplicated and that answers to requests from other functions are complete. For the major operational functions, such as diagnosis of network-wide problems and overall coordination of complex problem solutions, there may be one or more central coordination centers. The function of these centers is to monitor the overall performance of the network and the other control centers, and to initiate action when it seems appropriate and is lacking. For this reason, the line between operation and design is less distinct. Operation and design are merged into one process, directing the hardware and software evolution of the network. These centers will also be responsible for interacting with the network owner to provide needed information on network operation and performance. The network owner is now a strategist, orchestrating the decision priorities to meet business objectives.

6. SUMMARY AND CONCLUSIONS

Despite many of the major initiatives in the industry and several high powered developments taking place, it is a fair question to ask, "Are the knowledge systems real?" "Can they perform any 'worthwhile' task with reliability?" Feigenbaum [37] conducted an exhaustive survey of expert systems in day-to-day production use. At the AAAI Conference address in 1988, he reported over 3000 systems supposedly in production use. This number may well have doubled by now. DEC, Dupont, American Express, and IBM reported that their expert systems save ten to hundreds of millions of dollars per year. The advantages of knowledge systems were also established in interviews with the user groups. In many cases, these systems, besides storing the much needed corporate expertise permanently and improving quality and performance, provide up to an order of magnitude productivity increase and 1-2 orders of magnitude speedup of operations. These gains are significant and are being recognized by the corporate management. Some companies have concentrated on building large systems for a complete task. Other companies, like Dupont, built a very large number of small systems to perform minor day-to-day tasks.

In telecommunications, knowledge systems are emerging, but slowly, primarily in diagnosis and maintenance areas. AT&T is using NEMESYS for network management and control. Many more powerful systems are in field trial stage and ready to be launched into day-to-day operations. Integration of these individual systems into a larger system is the natural next step.

Some final questions to ask are, Where might this new technology (AI) lead us? Will there be real benefit by using it? Should there be an autonomous network or a network without human control? Can a communication network control itself effectively, using its own intelligence? These are interesting questions whose answers will emerge through further research. For now, integration of expert systems in network management and operations is paving the way for more reliable, functionally rich, and intelligent networks, and much more is expected as the technology advances.

ACKNOWLEDGMENTS

Much of the work reported in this paper is being conducted in several projects at GTE Laboratories. The author would like to acknowledge the contributions of the staff of these projects. Suggestions and comments of Richard Brandau and Robert Weihmayer have been valuable in the preparation of this paper and are gratefully acknowledged. Larry Bernstein, Pradeep Sen, Rodney Goodman, and John Vittal reviewed the manuscript and gave insightful suggestions. Thanks are due to Wilfredine Chiasson, who shaped the paper in the present form. Finally, the author would like to thank William Griffin for encouragement and support of this work and its publication.

REFERENCES

1. "Expert Systems in Network Management," *IEEE Network Magazine*, Vol. 2, No. 5, September 1988.
2. S. Goyal, "Future of OSSs," Panel, *IEEE Supercom. ICC '90*, Atlanta, GA, April 1990.
3. S. Goyal and L. Kopeikina, "The Evolution of Intelligent Network," ICC'87 Workshop on the Integration of Expert Systems into Network Operations, June 11-12, 1987, Seattle, WA.
4. S. Goyal, R. Weihmayer, and R. Brandau, "Intelligent Systems in the Future Network," Proceedings of *IEEE Networks Operations and Management Symposium*, San Diego, CA, February 11-14, 1990.
5. R.M. Goodman, J. Miller, and P. Smyth, "Real Time Autonomous Expert Systems in Network Management," in *Integrated Network Management*, edited by B. Meandzija and S. Westcott, Elsevier Science Publishers, 1989.
6. S.R. Wright and G.T. Vesonder, "Expert Systems in Telecommunications," in *Expert Systems with Applications*, Vol. 1, pp. 127-136, 1990.
7. M. Thandasseri, "Expert Systems Application for TXE4A Exchanges," *Electrical Communication*, Vol. 60, No. 2, 1986.
8. D. Prerau, A.S. Gunderson, R.E. Reinke, and S. Goyal, "The Compass Expert System: Verification, Technology Transfer, and Expansion," in *Second Conference on Artificial Intelligence Applications*, pp. 597-602, IEEE, Washington, D.C., 1985.
9. P. Zeldin, F. Miller, E. Siegfried, and J. Wright, "Knowledge Based Loop Maintenance: The ACE System," *ICC'86*, pp. 1241-1243, June 1986.
10. G. Vesonder, et al., "ACE: An Expert System for Telephone Cable Maintenance," in *8th International Joint Conference on Artificial Intelligence*, pp. 116-121, AAAI, Menlo Park, CA, 1983.
11. S. Guattery and F. Villarreal, "NEMESYS: An Expert System for Fighting Congestion in the Long Distance Network," *Expert Systems in Government*, IEEE, October 1985.
12. J. Hannan, "Network Solutions Employing Expert Systems," *Phoenix Computers and Communications Conference* (PCCC-87), pp. 543-547, IEEE, Washington, D.C., 1987.
13. L. Mantleman, AI Carves Inroads: "Network Design, Testing, and Management," Da*ta Communications*, pp. 106-123, July 1986.
14. S. Bernstein, "DesignNet: An Intelligent System for Network Design and Modelling," in *International Communications Conference 1987*, edited by D.J. Sassa, New York: IEEE, Seattle, WA.
15. F. Ferrara, F. Giovannini, E. Paschetta, "IAS: An Expert System for Packet-Switched Network Monitoring and Repair Assistance," in *Proceedings of Conference on Artificial Intelligence, Telecommunications, and Computer Systems*, edited by R. Attard, pp. 185-197, Nanterre, France: ECCAI, 1989.
16. J.M. Ackroff, P.T. Surko, and J.R. Wright, "AutoTest-2: An Expert System for Special Services,"in *Proceedings of the Fourth Annual Artificial Intelligence and Advanced Computer Technology Conference*, edited by M. Teitell, pp. 503-508, 1988.

17 R. Davis and W. Hamscher, "Model Based Reasoning, Troubleshooting," AI Labs, MIT, May 1988.
18 E.H. Durfee, V.R. Lesser and D.D. Corkill, "Cooperative Distributed Problem Solving," in *The Handbook of Artificial Intelligence*, edited by A. Barr, P.R. Cohen and E.A. Feigenbaum, Vol. 4, Addison-Wesley, 1989.
19 R. Davis and R.G. Smith, "Negotiation As a Metaphor for Distributed Problem Solving," *Artificial Intelligence*, 20, 1, 1983.
20 R. Weihmayer and R. Brandau, "A Distributed AI Architecture for Customer Network Control," *Proceedings of Globecom'90*, San Diego, CA, December 2-5, 1990.
21 L. Kopeikina, R. Brandau and A. Lemmon, "Case-Based Reasoning for Continuous Control," in *Proceedings Case-Based Reasoning Workshop*, Clearwater Beach, FL (Morgan Kaufmann, Palo Alto, CA), pp. 250-259, 1988.
22 R. Brandau and R. Weihmayer, "Heterogeneous Multi-Agent Cooperative Problem Solving in a Telecommunications Network Management Domain," *AAAI 9th DAI Workshop*, September 12-14, 1989, Orcas Island, Washington, December 1989.
23 R. Weihmayer, R. Brandau, and H.S. Shinn, "Modes of Diversity: Issues in Cooperation Among Dissimilar Agents," *Proceedings of Tenth Distributed AI Workshop, AAAI-sponsored*, October 1990.
24 "*Readings in Knowledge Representation*," edited by R.J. Brachman and H.J. LeVesque, Morgan Kaufmann, 1985.
25 J. Kolodner, R. Simpson, Jr., and K. Sycara-Cyranski, "A Process Model of CASE-Based Reasoning in Problem Solving," in *Proceedings IJCAI-85*, AAAI, Menlo Park, CA, 1985.
26 A.B. Davis, C.V. Lafond, and A.V. Lemmon, "A Proof-of-Concept Network Traffic Management System Using Case-Based Reasoning," December 1989.
27 L. Kopeikina, R. Brandau, and A. Lemmon, "Extending Cases Through Time," Workshop on Case Based Reasoning, *AAAI 1988*, St. Paul MN. Aug 23, 1988.
28 B. Silver, J. Vittal, B. Frawley, G. Iba, and K. Bradford, "ILS: A Framework for Integrating Multiple Heterogeneous Learning Agents," *Proceedings of the Second Generation Expert Systems, 10th International Workshop on Expert Systems and Their Applications*, Avignon 1990.
29 W.J. Frawley, "Using Functions to Encode Domain and Contextual Knowledge in Statistical Induction," in *Knowledge Discovery in Databases (IJCAI-89 Workshop)*, edited by G. Piatetsky-Shapiro and W.J. Frawley, pp. 99-108, International Joint Conference on Artificial Intelligence, 1989.
30 G.A. Iba, "A Heuristic Approach to the Discovery of Macro-Operators," *Machine Learning 3(4)*, pp. 285-317, 1989.
31 R.S. Michalski, J.G. Carbonell, and T.M. Mitchell (editors), *Machine Learning: An Artificial Intelligence Approach*, Vol. 1, Tioga Press, 1983.
32 R.S. Michalski, J.G. Carbonell, and T.M. Mitchell (editors), *Machine Learning: An Artificial Intelligence Approach*, Vol. 2, Morgan Kaufmann, 1986.
33 B. Silver, "Precondition Analysis: Learning Control Information," in *Machine Learning: An Artificial Intelligence Approach*, edited by R.S. Michalski, J.G. Carbonell, and T.M. Mitchell, Vol. 2, pp. 647-670, Morgan Kaufmann, 1986.

34 B. Silver, "NetMan: A Learning Network Traffic Controller," in *Proceedings of Third International Conference on Industrial & Engineering Applications of Artificial Intelligence and Expert Systems,* edited by M. Matthews, pp. 923-931, Association for Computing Machinery, 1990.
35 J.R. Quinlan, "Induction of Decision Trees," *Machine Learning 1,* (1) pp. 81-106, 1986.
36 S. Goyal and R. Worrest, "Expert System Applications to Network Management," in *Expert Systems Applications to Telecommunications,* edited by J. Liebowitz, pp. 3-44, New York: John Wiley & Sons, 1988.
37 E. Feigenbaum, P. McCorduck, and P. Nii, *The Rise of the Expert Company,* TimeBOOKS, 1988.

VI

REASONING AND KNOWLEDGE

IN NETWORK MANAGEMENT

A
Panel

REASONING PARADIGMS FOR INTEGRATED NETWORK MANAGEMENT

Moderator : Prodip SEN, NYNEX Corporation, USA

ABSTRACT

The goal of integrated network management systems is to provide decision tools for network operators and managers, and to automate network management functions, to the extent possible. To this end, we need techniques for the representation of network entities, relationships and behavior, and techniques for reasoning with, and analyzing network event information obtained from the network.

There is a wealth of structural and behavioral knowledge available for computer and communication networks. This includes static information concerning relationships between network entities, as well as knowledge regarding network behavior in the form of analytic and simulation models for network dynamics. On the other hand Artificial Intelligence techniques, in particular Expert Systems techniques, seem to be very promising for reasoning with, and analyzing network event information. These techniques can handle event dependencies, diagnosis procedures, and work with incomplete and possibly conflicting information.

Can we coordinate the use of knowledge encoded in the form of procedural "rules", with that in the form of the behavioral models mentioned above? Are such multiple representation techniques needed, or can one suffice? How should uncertainty about network states and event dependencies be incorporated into the reasoning process? The panel will discuss these and other questions in an attempt to explore reasoning paradigms for integrated network management.

VI

REASONING AND KNOWLEDGE

IN NETWORK MANAGEMENT

B
Knowledge Representation

A multi-agent system for network management.

Ilham RAHALI and Dominique GAITI.

Laboratoire MASI, Université P. et M. Curie, 4 place Jussieu, 75 252 Paris Cedex 05, France.

Abstract.
In this paper, we suggest the use of multi-agent systems for managing heterogeneous networks. The distributed artificial intelligence concepts proposed can represent a departure from the conventional methods used today in communication networks. The main objective of this research is to explore expert systems technology in the domain of network management. We propose a network management system using a set of autonomous agents which cooperate for executing distributed problem solving. The architecture of the system is sufficiently general to allow it to be used for several management domains by substituting different knowledge bases.

1.Introduction.

Previously, with telecommunication networks, there was no need for expert systems, because their architecture was simple and easy to model. They were controlled at one single point and their cost of maintenance was low. However, everything has changed; architectures are now complex, communication links are multiple, flexibility and reliability needs require new approaches in network management. Expert system applications seem to be the right direction to follow. Network management is a really tough task! With second generation network management system, we hope to improve efficiency and reliability, we want a higher degree of network performance with better and more numerous services to offer.

In the communication networks domain, expert system applications are rare and solve only some special problems. The difficulty results from the characteristics of this domain, for example:

- a large amount of knowledge represented by many rules and a lot of information,
- the heterogeneity of the network (multi-manufacturer, multi-data, multi-architecture) where information has different structures and several access techniques (protocols),
- temporal aspects such as real time, statistics, historical and simultaneous arrivals of different problems to solve,
- the distribution through several connected sites which involves a need for interfaces, gateways and correlations (a local problem can have a distant cause).

In view of the above characteristics, the insertion of an expert system in a network management system requires the development of an expert system environment adapted to this application. In this project, we develop a network management system which relies on the

Distributed Artificial Intelligence (D.A.I) paradigm. In this multi-agent system, a set of intelligent entities (agents) cooperate to execute distributed problem solving. Each one is specialized in a network management task and its knowledge is represented according to a proposed model.

2. Approach.

Real time and distribution problems are two fundamental aspects of network management [Rabie 88]. As a matter of fact, time constraints are crucial for telecommunication system features such as response time (from user point of view) and performance evaluation (from network manager point of view). Network applications necessitate very short response times, routing or repairing delays in case of various network faults (e.g. network overload, hardware failure).

The aim is to keep the user unaware of every problem occuring on the network. Consequently, the network manager should be able to detect immediately any network malfunction (fault detection), and distinguish its real causes from the numerous side effects (fault isolation), in order to identify the problem (diagnosis). Then, he should ask for the required knowledge for building a solution to restore the network to an unfaulty behaviour (effective repair) [Sutter 88]. So, the required expert system organization has the difficult task to encompass all aspects of the above operations.

Moreover, a network is characterised by its heterogeneity, its layered structure and the distribution of its resources. Thus, we have to formalize such a network by a system in which knowledge is distributed [Gaïti 89].

Distributed Artificial Intelligence (DAI) allows several processes, called agents, to solve a single problem. Each process uses local power and communicates with remote hosts. In the DAI field, the research aim is to propose a scheme of organization allowing the distribution of the resolution of a shared problem. This organization has to describe what are the mechanisms that can be used by agents to cooperate.

With DAI, problem solving is a cooperation made by a decentralized group. Agents may be small real processes as they can be complex real ones like applications based upon large knowledge bases. This solving process is distributed, meaning that each agent has to share a common information set allowing the entire group to reach the solution. This group of agents is decentralized means that data and control are often logically and physically distributed.

We are using a DAI scheme to allow the management of heterogeneous networks. The reasons for this choice are the following:

❑ DAI allows expert systems connections. Each expert system has its own knowledge base. Several knowledge bases can have a shared part of knowledge. One agent can solve problems whose field is not its own. This feature is well adapted to network management because it needs large amounts of data. This field needs to be shared in several sub fields. These smaller fields should easier to use and model. We have to subdivide each field as long as we don't get a clear and well defined knowledge base.

- ❏ DAI is able to solve problems that are too large for one single host. The only feature you have to consider is not the power of the host but the cost of communication, the speed of the transfer and the reliability in problems that can be only solved in distributed environment.
- ❏ DAI is able to solve the problem of working with several experts at the same time. If the knowledge engineer works with several experts whose answers are not compatible, it does not know what to do. DAI may solve this problem.
- ❏ DAI seems to be the natural solution. For a distributed problem, it is better to find a distributed solution. Management of a large network is a distributed problem.

We consider a distributed network solvers as a set of independent agents. Each agent is able to communicate with other ones. Work has focus on the description of the solving process with sophisticated local means of control. Each solution process has its own knowledge base and its own inference capability to take decisions about what sub-problem to solve and what are the solutions to communicate. Figure 1 shows a system constituted of three networks where each network N is managed by an agent. Three sessions S are opened towards a service center.

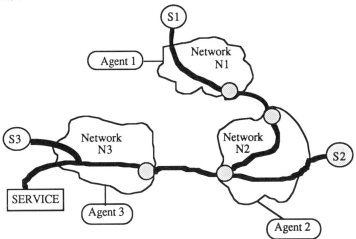

Figure 1 - Scenario.

In such a system, knowledge about the field (network management) is distributed among the solving agent. Each agent has to know what is the sub-field to manage, it gets its information from sub-field sensor or from other agents under its responsibility. Then, it proceeds, analyses and transfers its conclusion to its partners or answers the agent that has previously requested it to solve this specific task. The cooperation has succeeded among the community of network management agents if they own all the knowledge needed to coordinate their acts.

3. Project description.

In this project [Rahali 90], we have proposed a framework based on expert system techniques. The first part of this study focuses on the management information diversity. It describes a knowledge representation model of the management system, which supports appropriate formalisms for each knowledge type (network, services and users). The second part tackles the distribution problem in a network management system. It presents a computational model based on an architecture which employs distributed artificial intelligence principles, and especially multi-agents systems. This dichotomy is possible given that we can separate between computing and data in a knowledge based system.

3.1. Knowledge model.

The use of knowledge based systems (KBS) lets one capture the network management knowledge. One of the main features of a network management system is the large amount of information it handles. Looking at this knowledge domain as one big area would make it impossible to find a single representation formalism covering everything. But if we adopt a top-down approach and try to find some of the main objects in this large domain, we can start looking at these objects one by one.

We have found three objects on a very high level that we think cover a large part of the management system domain and can be fairly representative. These objects are: the **network**, the **services** and the **users** of the management system. For each object, we have proposed knowledge representation formalisms capturing the knowledge properly.

a. Network knowledge modelling.

The network is a concern of most users of the management system. Information about things such as network configuration, topology, status, historical data and components will need to be available to users for them to carry out their tasks.

The process of examining various representation formalisms for network knowledge must begin with an overview of who the users of this knowledge are. The types of users and the tasks they have to perform should influence the choice of formalisms. In particular, the skill levels of the users, the complexity and generality of their tasks have a great significance.

Indeed, a network operator often has a completely different need of knowledge and data than a network planner has. The operator needs fast information on what to do in emergency situations and the planner needs a good overview of the performance during a long period of time.

Thus, each user type has specific needs and therefore his own network model. These different models must be handled by the network knowledge formalisms. In fact, this wide range of needs may force different formalisms to be chosen in order to deal with different models of the network. Even so, different representations must cooperate to ensure consistency of the network knowledge.

The physical view of the network is quite best described in some declarative way that suits the objects involved (e.g. node, switch, link, etc.). For this reason, we have chosen the semantic network formalism to represent this knowledge.

Semantic networks can capture both the topology of the network (i.e. how nodes, switches, lines and trunks are connected together) and describe the objects by using frames. Figure 2 is an example of how to use a semantic network to describe a small part of a network.

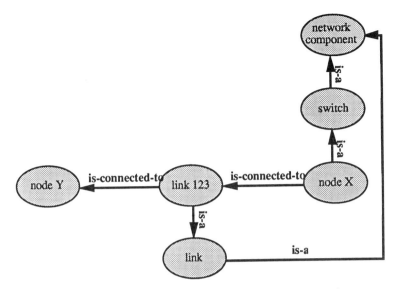

Figure 2 - Semantic network description of a part of a network.

This semantic network approach will probably be able to capture the overall structure of the network. How network components such as switches, nodes or lines are connected are well described in this formalism. Details about how, for example, a switch is built, which components are used, their function and characteristics are more difficult to capture with semantic networks as they were originally defined. One way to overcome this, is to describe the network components using a frame hierarchy. This hierarchy could be a complement or extension to the semantic network.

b. Service knowledge modelling.

The problem of classifying and characterising services is one which is the subject of considerable study in the CCITT and elsewhere. The approaches adopted have all been hierarchical in nature and based on developments of the idea of modelling by attributes. For a particular service, the values taken by the attributes are more or less fixed.

The CCITT Recommendation I.130 introduces a modelling method for the characterization of telecommunication services based on the definition of a serie of attributes and values associated with those attributes. It is possible to distinguish between low layer attributes (basically OSI layers 1-3) and high layer attributes (basically OSI layers 4-7).

Semantic networks are potentially useful here. It is easy to draw a network with a central node representing a service and a serie of arcs labelled with the attributes name linking the service to its attributes values. A hierarchy of services can be set up by introducing "is-a-subset-of" arcs linking a class with its members (see Figure 3).

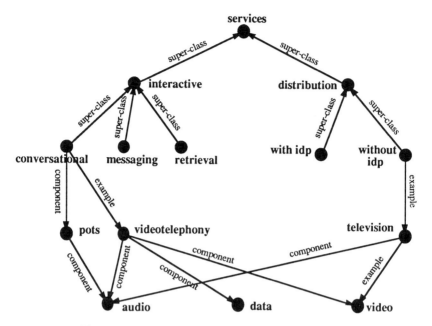

Figure 3 - Semantic network representation of services knowledge.

c. User knowledge modelling.

In order to facilitate the user modelling process, we have decomposed the network management into several functional areas (e.g. accounting, planning, network operational control, etc.). A user of the management system, as maintenance technician or network operator, fulfils one role in his functional area. Each area has its own user roles. One role can be performed by different users of the same area.

Object oriented representation has been chosen to model the users of the management system due to its encapsulation, inheritance and interaction features. Indeed, the model we use relies on hierarchical structure of entities (see Figure 4). The objects are organized into classes (user class, role class, etc.). The general object attributes are the name and the class to which it belongs, the state, the operations (actions an object can execute) and the relation with other objects.

For each role, the object structure is as follow:

```
(object
    name              =
    class             = <role>;
    super-class       = <functional area>;
    sub-class         = <user>;
```

```
        attributes
              activity    =
              input       =
              output      =
              theme       =
              needs       =
              ...
).
```

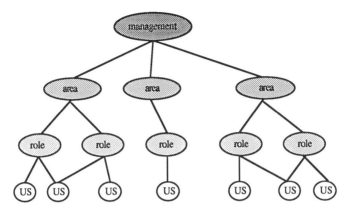

Figure 4 - User classes.

The attributes list is not exhaustive and other attributes can be added. An example of user role can be "network operator", in this case a user who fulfils this role will be described by the following object:

```
(object
        name          = Smith
        class         = network operator;
        super-class   = network operational control;
        attributes
              activity   =    implement procedures, start-up,
                              shutdown and monitor network,
                       collect statistics;
              input      =    network manager;
              output     =    network manager, performance
                              analyst, sales;
              theme      =    procedures, reports;
              needs      =    network and customer information;
              ...
).
```

3.2. Computational model.

In our project, we have designed an intelligent system, using DAI techniques [Lesser 88], [Smith 88], which can solve network management problems. The system proposed can be seen as a set of **agents** that work in a parallel way and interact in an organized way. We can consider that this system is equivalent to an **organization** of problem solvers.

The reasoning in distributed system field can be represented by a collection of agents which constitute a network where domain knowledge is distributed. This distribution often enhances system efficiency and provides a better fault tolerance. Furthermore, agents which are represented by "little" expert systems, are expected to find problem solutions more efficiently.

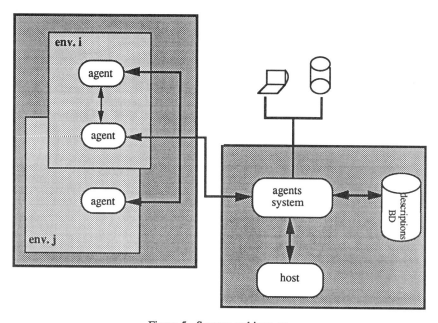

Figure 5 - System architecture.

We have designed a network management system as a set of intelligent and semi-autonomous agents [Gasser 87], it includes the following components (Figure 5):

- ❑ A collection of agents: an agent is the basic computational unit. Inherently, agents are "sociable": they know the other agents of their environment and expect to cooperate with them. Each agent has its own "model of the other agents" which contains information on identity, localisation, goals, skills, etc. of its partners.

- ❑ A set of agents system: these predefined agents provide the command interpretation, a standard user interface, error handler, etc.. These specific agents are used as interactive tools to build and edit behaviours and structures of the other agents.

- ❑ A description data base: the agent descriptions are stored in a data base by agents system group which build, verify new descriptions and build executable agents from descriptions.

The agent constitutes the functional unit of the whole system and is composed by the following elements (Figure 6):

❏ The Module: represents the agent working. It contains a controller and a manager. The controller receives, sends queries and answers, activates and handles the corresponding tasks. The manager provides the access to agent knowledge for acquiring and maintaining it.

❏ The Knowledge: has two types. The first type, knowledge on the agent skill, consists of which management actions the agent can do. This knowledge is procedural one. The second type, knowledge on the agent acquaintances (named environment model), indicates the skill of its partners. This knowledge is declarative one.

❏ The Inference Engine: executes the agent knowledge.

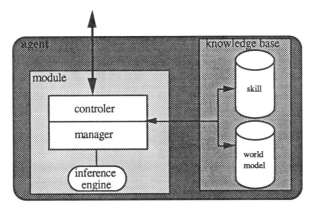

Figure 6 - Agent structure.

The agent objective is to solve any problem for which it has the necessary skill, after queries either from the system manager or from other agents. The agent may have a sufficient domain knowledge to solve the problem alone. If it is not qualified, it asks one or several agents to help him. In this case, the agent distributes tasks on appropriate agents chosen according to his model of acquaintances.

The coordination among the agents is provided by a higher level agent (called manager) that decomposes the problem and distributes the sub-tasks to lower level agents (called workers). These agents can, in turn, decompose the tasks assigned to them into sub-problems and allocate them to specialized agents. We note that the problem decomposition process can be repeated in the agents hierarchy which is created dynamically. We are interested in specializing agents so that we can easily modify them or we can adapt them to the problems they solve. So we obtain an evolutive and flexible system.

We can illustrate the cooperation between agents used to solve a given problem by identifying the following different stages in the case of fault isolation and diagnosis (see Figure 7).

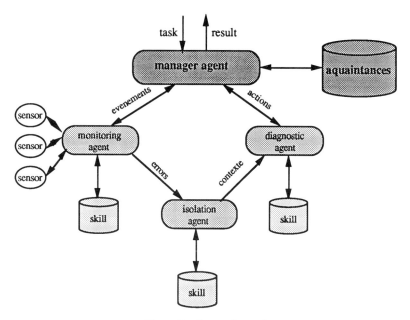

Figure 7 - Errors diagnosis.

The continuous monitoring of the network is done by an automata which carries out tests periodically. The automata compares the current values against pre-set values which correspond to a normal working environment. To isolate a problem, the agent has to be able to recognize the primary symptoms and differentiate between induced failures and the real one. It has also to consider the context, use the past history in a right way and answer as far as possible particular problems. In fact, a failure can mask another one, induce new symptoms without a relationship between them and create a new failure somewhere else. The problem is thus defined and its application area (environment) is sent to a diagnosis agent which determines the actions to take.

4.Conclusion.

The purpose of this paper was to demonstrate the application of distributed artificial intelligence techniques to communication network management. We have designed the management system as a set of agents which cooperate in distributed problem solving. The goal of the project is to manage the campus networks, it is currently being studied at Pierre and Marie Curie University of Paris. A prototype of the management system will be developed in a local software environment using a concurrent programming.

5.References.

[Gaïti 89] D. Gaïti, I. Rahali and J. P. Claudé
"Applying Artificial Intelligence Techniques to the Management of Heterogeneous Networks", Proceedings of *Network Management and Control Workshop*, New York, 1989.

[Gasser 87] L. Gasser, C. Braganza and N. Herman,
"MACE: a flexible testbed for distributed AI research", in *Distributed Artificial Intelligence*, M. Huhns, Pitman, London, 1987.

[Lesser 88] V. R. Lesser and D. D. Corkill,
"Functionally accurate, cooperative distributed systems", in *Readings in Distributed Artificial Intelligence*, A. H. Bond and L. Gasser, Morgan Kaufmann, San Morgan, California, 1988.

[Rabie 88] S. Rabie, A. Rau-Chaplin and T. Shibahara,
"DAD: a real-time expert system for monitoring of data packet networks", *IEEE Network Magazine*, vol. 2(5), 1988.

[Rahali 90] I. Rahali,
"Communication network management: conceptual approach and distributed knowledge based framework", Phd Thesis, University of Paris 6, France, 1990.

[Smith 88] R. G. Smith and R. Davis,
"Frameworks for cooperation in distributed problem solving", in *Readings in Distributed Artificial Intelligence*, A. H. Bond and L. Gasser, Morgan Kaufmann, San Mateo, california, 1988.

[Sutter 88] M. T. Sutter and P. E. Zeldin,
"Designing expert systems for real-time diagnosis of self-correcting networks", *IEEE Network Magazine*, vol. 1988.

[Terplan 87] K. Terplan,
"Communication networks management", Prentice Hall, New Jersey, 1987

[ISO IS 7498-4] / CCITT X700
"Basic reference model, OSI management framework"

Incorporating Non-deterministic Reasoning in Managing Heterogeneous Network Faults

Paul Hong and Prodip Sen

NYNEX Science & Technology
500 Westchester Avenue
White Plains, New York 10604

Abstract
As today's enterprise networks are heterogeneous in nature, a challenging task is to isolate sources of network failures from uncoordinated or incomplete network information, in tackling day-to-day network operational problems. A network uncertainty model using the voting scheme is proposed; and a network belief language is described to represent network uncertainty. To highlight its applications, methods are exhibited on how network failure causes can be isolated in network diagnosis and prognosis. By emphasizing learning from empirical network operational results, issues concerning how such network knowledges can be acquired and refined are then discussed. An example heterogeneous network is used to illustrate the approach.

1. INTRODUCTION

Enterprise networks are heterogeneous in nature and an emerging need for an integrated management system to carry out network management functions is quite eminent [1]. However, one of the challenging tasks faced by today's enterprise network managers is to isolate *primary* sources of network failures from uncoordinated network alarms. It is essential for an enterprise's network management system to be able to parse all reported network measurements, correlate them at the global network level, and translate them into meaningful network trouble information. At the same time, non-reported potential network failures should also be derived in providing a truly pervasive fault management.

OSI-based network management promises a seamless network management scheme in the long run. Migrating toward that goal, an interim solution is necessary to overcome present network management problems where "windowing" into a subnetwork manager via a terminal emulator still dominates the current state of the art, in communicating management information. In addition, there is a critical need to account for the inconsistent and incomplete network information, which inevitably arises in managing a heterogeneously interconnected enterprise network.

In the past, various novel approaches have been used for reasoning in the network management arena [1, 5]. In most cases, network information is assumed to be deterministic. As will be described, the deterministic approach is not sufficient in addressing those heterogeneous network management problems. This paper proposes a reasoning technique which incorporates uncertainty to manage heterogeneous network faults. In this approach, a network proposition is evaluated with a belief metric, which uses a pair of support measures or votes, such that potential *don't know* and *contradicting* conclusions can be expressed and derived.

In this paper, discussions are focused on network events in the context of network fault management functionalities. Translations from network messages to appropriate network events are prerequisites [1]. In section 2, the motivation for adopting uncertainty reasoning is elaborated. In section 3, a high-level modelling guideline is sketched as underlying principles. A network belief model is then described with its associated calculus. Afterwards, the concept of a network belief language is described allowing for human interactions in managing network uncertainty. In section 4, methods are exhibited by isolating a heterogeneous network failure as its applications. Section 5 summarize advantages of our approach, and stresses a systematic view of managing the network knowledges. Discussions are then concluded in Section 6. An example heterogeneous network is used to illustrate our approach.

2. MOTIVATION

Most existing enterprise network management systems are used as passive tools rather than as cohesive intelligent systems; network management tasks remain labor-intensive operations. Human eyes and ears are still commonly used in sensing the status of a network, network uncertainty prevails as a result. The situation is aggravated in a heterogeneous network environment. A simple example enterprise network is shown in Figure 1 (see next page), where a T1 multiplexer subnetwork, a modem subnetwork, an IBM/SNA subnetwork, and an ASCII mini-computer subnetwork, are interconnected together. The T1 and modem subnetworks together form a physical transport subnetwork, upon which the IBM/SNA and ASCII subnetworks provide data communication services via different protocols. Notice that the IBM/SNA subnetwork overlays with the ASCII subnetwork over two 2680 modems to demonstrate the minimal complexity of a prevailing enterprise network. Included in the figure, example network causes and alarms are listed for references throughout the rest of the paper.

Complexities of managing a such heterogeneous network have many folds, some of them can be described as below:

Propagation of subnetwork failures: An enterprise network typically consists of voluminous interconnected network elements, including physical and logical network elements, as well as communication paths as part of its network services. For instance, in the example network, consider an IBM/SNA session between the IBM Host and the SNA Terminal, and assume that the underlying T1 circuit supplies the clock to the underlying modem circuit, then a T1 circuit failure will cause a loss of clock source of the modem circuit, which will eventually cause the failure of the SNA session.

Incompatibility among asynchronous subnetworks: Modern intelligent networks are often built-in with automatic recovery mechanisms, a network element's outage may only be transient, and sometimes it can never be detected without proper logs or further probing. When subnetworks are interconnected without an intelligent control, one subnetwork's transient failure could cause permanent failures in other less resilient subnetworks. Much of these phenomena depends on the timing and coupling mechanisms among subnetworks. For example, a T1 subnetwork trunk failure may, or may not, cause interruptions in the IBM/SNA subnetwork services, depending on how the SNA protocol's recovery timer parameters are set in comparison with to the time required to re-route a T1 circuit by the T1 subnetwork.

Fuzziness due to lost or delayed network information: Unavailability, loss or delay of network messages is common without a robust management protocol. It can significantly reduce the confidence level of a network judgement. Assume the following scenarios are observed from the example network,

(1) C_1 causes C_2[1],
(2) C_1 issues symptomatic network alarms E_1 and E_2,
(3) C_2 issues symptomatic network alarms E_2 and E_5.

Then, consider the question: *What is the real cause if only symptoms E_2 and E_5 are received?* C_2 seems to be the obvious answer. However, if we had discovered the fact that E_1 was issued and lost, C_1 should be the right answer.

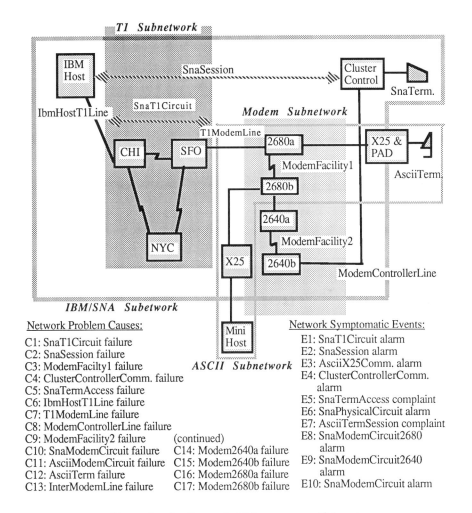

Figure 1. An Example Heterogeneous Network

---

[1] since the T1 circuit supports the SNA session directly

484

All above deficiencies are short-term oriented. From a longer term perspective, an ISO-based management scheme will still have its shortcomings, i.e.,

Abstraction and fusion mechanisms of an integrated management control: ISO-based management standards aim to manage complexities of a large-scale enterprise network in providing *end-to-end* services as required in network provisioning and troubleshooting processes. Consider the situation of troubleshooting the example network when the SNA session fails, a *global-level* test command should be issued to test all subnetwork elements underlying the session, before failed components can be isolated. It requires multiple subnetwork managers coexisting within the management hierarchy to behave cooperatively in completing a such distributed transaction. The ISO management scheme standardizes the concept of *management agent*, which allows a subnetwork manager to manage its network without direct interactions with its managed network objects. As such, the representation of a managed network object will become more abstract as it goes up the management hierarchy. In addition, *artificial* subnetwork objects can be created by a subnetwork manager; and/or artificial subnetwork managers arise as required by standards. For instance, in the example network, an artificial physical subnetwork manager can be created to manage physical communications to support both the SNA and ASCII subnetworks. This subnetwork manager will in turn *interoperate* with the T1 subnetwork manager and the modem subnetwork manager in managing its end-to-end physical subnetwork. All these imposed overheads will introduce uncertainty, due to loss of details and lack of direct control mechanisms.

In short, a present solution is needed for alleviating today's network problems. The theme is to learn a network's behavior from its past by incorporating network uncertainty. In addition, a such solution should also help an enterprise network's management migrate toward an OSI-based scheme in the long run.

3 SOLUTION APPROACH

To deal with the increasing network complexities, most existing approaches have adopted classical logic by designating each network proposition with a definite truth [5]. However, this proves insufficient in dealing with incompleteness of network information. Recently, novel reasoning techniques have been proposed, with a common goal to make a reasoning system able to provide statistically meaningful explanations, such that a conclusion can still be made from a complex situation. To begin with, a generalized theory is presented.

3.1 A General Approach

The true mechanics of network behaviors is typically too complex to be comprehended by a network operator, it forms the main theory of the network. However, as the network is operated for some period of time, the operator can learn and develop his/her knowledge, as part of an auxiliary theory, to augment the main theory. Figure 2 highlights the difference between a main theory and its auxiliary theory in modelling network behaviors, i.e., network causations in particular. The former theory is represented using circles for network causes, and the latter using shaded triangles for corresponding measured network events information. Network causation behaviors, or the so-called *causal network*, are a part of the main theory, which will be modelled by its supporting auxiliary theory.

As illustrated in Figure 2 (see next page), again using the example network, a ModemFacility1 failure causes failures of the SNA modem circuit, the SNA physical circuit, the SNA session and the SNA terminal access, etc. Two special cases deserve special attention: the existence of ($C_5 = E_5$) is called a *direct measurable* relationship, which represents a direct measurement without any uncertainty; and the creation of an *artifact* variable E6, which represents an unmeasurable auxiliary network event, is created in modelling network

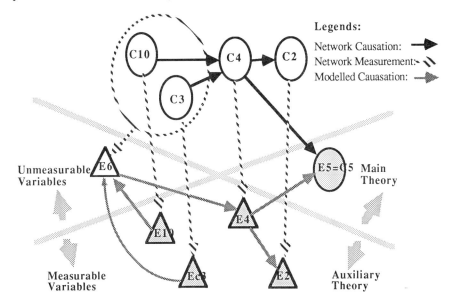

Figure 2. A General Approach Example

causations of the main theory. To effectively characterize the true network complexity, a simplified and plausible auxiliary theory should be built by inputting both of the managed network configuration and empirical network results; and the approach is to incorporate uncertainty into these representations for later network inferences.

To convert empirical network measurement information in building a network model, an *object-oriented* approach is found to be very effective. First, network events are defined as state changes of target network objects. In Figure 3 (see next page), network *evidenced* events represent those information collected from a network management agent or a subnetwork manager. Evidenced event information are typically assigned without any uncertainty, or disbelief. Another class of network events are *hypothetical* events, which are deduced from either an evidenced event or an existing hypothetical event. Hypothetical network events are always deduced with their uncertainty measures; these network events are further subclassified as *causing* as well as *consequent (caused)* network events, in relating to the hypothesis-deducing network event.

The *Pattern matching* procedure among network events is central to how network faults can be managed. The basic rule is stated as follows: when a hypothetical event matches an evidenced event on the event type about the same network object, it postulates a causation relationship between the evidenced event and the hypothesis-deducing event. Also, hypothetical causing events are matched for diagnosis purposes; and hypothetical caused events are matched for prognosis purposes. In addition, a match between a hypothetical causing and a hypothetical caused events can foster a causation relationship, in case of missing correlating events. More details will be clarified as it follows.

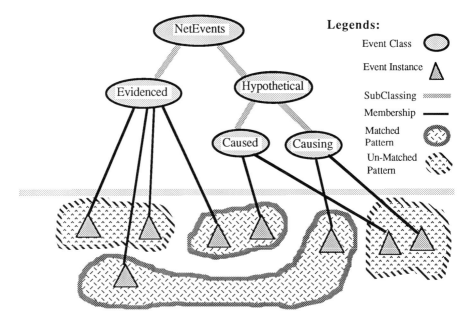

Figure 3. Network Event Classification

3.2 A Network Belief Model

In this section, a simple and intuitive uncertainty model will be reviewed and delineated using Baldwin's *Support Logic Programming* concepts [4]. Inspired by [2, 3], Baldwin introduced a *voting* view of uncertainty measures, or believes, with respect to propositions of network facts or inference rules. The model is reviewed here; more details on how it can be applied to manage heterogeneous network faults will be described later.

An uncertainty measure about a proposition is expressed as a pair of "probabilities", i.e., the probability of positive supports and the probability of negative supports. For example,

Proposition P: A T1 subnetwork trunk failure disrupts the communication between the IBM Host and the Cluster Controller

Belief of P: [By(P)=1/8, Bn(P)=3/4], together denoted as B(P)

where B(P) labels the uncertainty measure of P as a pair of supports: By(P) = 1/8 implies one eighth of tested cases will support the proposition P, and Bn(P) = 3/4 implies that another (1- Bn(P)), or 1/4, portion of the same set of tested cases will support the negation of P. Moreover, the difference between them, i.e., (Bn(P) - By(P) = 5/8), represents the rest portion of tested cases being not certain about the proposition P.

The calculus of these belief measures can now be described. To begin with, the proposition of [**P implies Q**] is denoted as "Q|P"; and **AND, OR** and **NOT** are the usual boolean operators. First, given a proposition P, the boolean NOT operation of P can be derived directly from using C1, which is formulated as follows:

C1: By (NOT P) = 1 - Bn(P)
     Bn (NOT P) = 1 - By(P)

Given independent propositions P and Q, formulae for AND and OR boolean combinations can be expressed by C2 and C3 respectively, as listed below:

C2: By (P AND Q) = By(P) * By(Q)
     Bn (P AND Q) = Bn(P) * Bn(Q)

and,

C3: By (P OR Q) = By(P) + By(Q) - By(P) * By(Q)
     Bn (P OR Q) = Bn(P) + Bn(Q) - Bn(P) * Bn(Q)

By the same token, given a proposition P and an inference rule Q|P, the inferred belief of the proposition Q can be calculated using the following formula C4:

C4: By(Q) = By(Q|P) * By(P)
     Bn(Q) = 1 - By(P) + By(P) * Bn(Q|P)

Finally, a useful rule is to combine two different belief measures about the same proposition. Assume a proposition P is assigned with two belief measures, [By(P), Bn(P)] and [By'(P), Bn'(P)], independently from two different sources, the newly combined belief measure [By"(P), Bn"(P)] about P can be calculated using C5, as described as follows :

C5: By"(P) = [By(P) * Bn'(P) + By'(P) * Bn(P) - By(P) * By'(P)] / C
     Bn"(P) = 1 - [(1 - Bn(P)) * (1 - By'(P)) + (1 - Bn'(P)) * (Bn(P) - By(P))] / C

     where   C = 1 - [By'(P) + By(P)] * [1 - Bn'(P)]

3.3 A Network Belief Language

Using the above-described belief model, diagnosis or prognosis about network states or behavior can thus be evaluated with their belief measures. In a network management control center, a network belief language geared toward human operators can also be designed using the belief model in managing its networks. The ability for a human operator to do the high level reasoning is a critical element for an effective management system. Because numerical belief measures are hard for a human operator to cope with, an English-like interface can expedite a network management process. In Figure 4 (see next page), an example network uncertainty language is suggested based on the distribution of belief measures. Notice that each of the *shaded* uncertainty regions actually includes a wide range of belief measures, which are in a whole uniquely labelled by an English word. Some regions of Figure 4 are explained below:

TRUE :   A fully agreed-upon proposition (characterized by [1,1])
FALSE :   A fully disagreed-upon proposition (characterized by [0,0])
UNKNOWN :   An unknown proposition (characterized by [0,1])
CONTRADICTORY :   A contradictory proposition (characterized by [.5, .5])

The algorithm for mapping from a belief measure to its English word can be generally determined by the network management center. For instance, an additional uncertainty language word *PERHAPS* can be added to represent the region characterized by [.25, .75] in Figure 4, such that further sophistications of reasoning can be fulfilled. On the other hand, whenever necessary, it should also be allowed to translate reversely from an English word back to its characteristic belief measure, so that quantitative inferences can be taken advantages of.

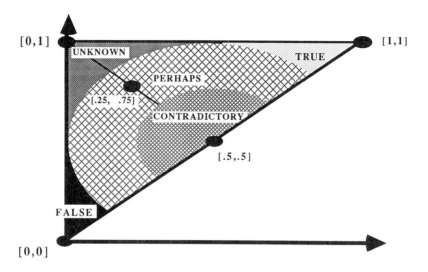

Figure 4. A Belief Language Example

4. NON-DETERMINISTIC NETWORK FAULT MANAGEMENT

In the section, questions regarding how to incorporate the belief model into managing a network will be answered. To start with, network knowledges will be reviewed.

4.1 Network Knowledges

A heterogeneous enterprise network generally consists of elements from various interconnected subnetworks, gateways, and internetwork lines, etc. Operational states of these network objects, or *netobjects*, directly affect provided network services. Important network knowledge are summarized as below:

Structural Knowledge: Most of network complexities come from interconnectivity among netobjects, based on either *compositional* or *functional* considerations. An I/O card residing in a network node is an example of compositional dependency; while a client/server protocol relationship is an example of functional dependency. A graph structure is often used to capture such dependencies among netobjects [6].

Behavioral Knowledge: To manage a network, operational state changes of netobjects, or called *netevent*s, should be evaluated and/or controlled. In particular, knowledges correlating any two netevents are translated into a so-called *causation network*, where consequent and causing netevents are encoded for each netevent. A directed acyclic graph (**DAG**) structure is most suitable to represent a such causation network: each node represents a netevent and each directed edge represents a netevent causation. For a target netevent of DAG, an outgoing edge represents a *consequence* causation relationship; while an incoming edge represents a *causing* causation relationship. A causation DAG network should be *static* in the absence of network reconfigurations.

Empirical Knowledge: When a network is operated, test cases should be accumulated for calculating or modifying network believes. As more test cases can be accrued about a network belief, its belief measure will be gradually converged. To find out those believes of causation rules in a causation DAG network, consider a target netevent, for each *inclusive* outgoing causation, via the Bayes' theorem, a caused belief measure can be independently calculated as the conditional probability of the consequent netevent's existence, given the occurrence of the target netevent. On the other hand, to calculate incoming belief measures regarding a target netevent, the Figure 5 example demonstrates how incoming belief measures can be assigned by listing all potential causes, dictated by the example network, E_1, E_{c6}, E_{c7}, E_{c8}, and E_{c10}, when the target netevent E_6 is observed. Based on E_6 occurrences, (P_i, $i = 0, 1, 6, 7, 8, 10$) should be calculated accordingly as conditional probabilities, in regard to causing the netevent E_6 to occur. For instance, [P_1, P_1+P_0] is the belief measure of the proposition that, given the occurrence of E_6, netevent E_1 causes event E_6. Observe the *exclusiveness* among these causing netevents by assuming that those indirect causation factors had been filtered beforehand.

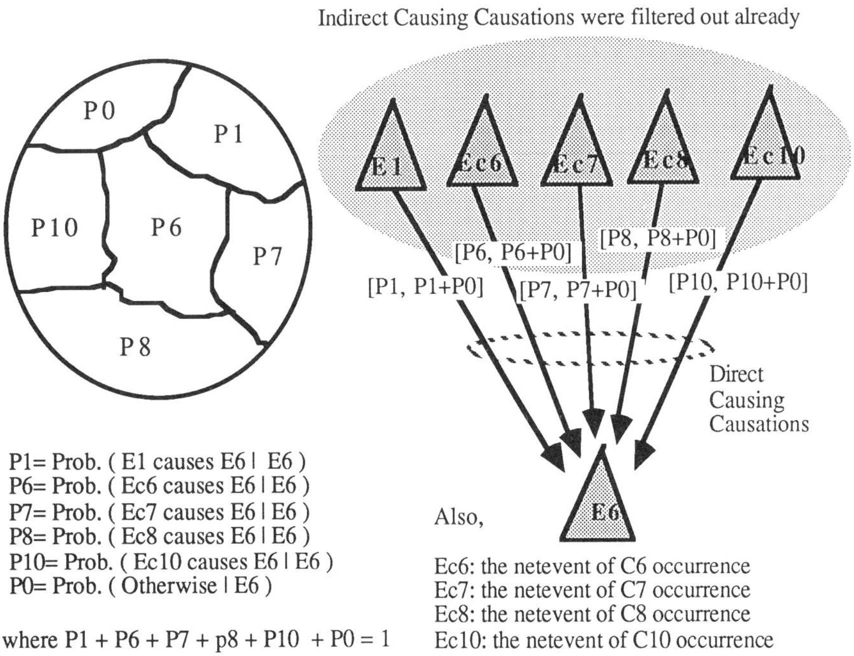

P1= Prob. (E1 causes E6 | E6)
P6= Prob. (Ec6 causes E6 | E6)
P7= Prob. (Ec7 causes E6 | E6)
P8= Prob. (Ec8 causes E6 | E6)
P10= Prob. (Ec10 causes E6 | E6)
P0= Prob. (Otherwise | E6)

where $P_1 + P_6 + P_7 + p_8 + P_{10} + P_0 = 1$

Ec6: the netevent of C6 occurrence
Ec7: the netevent of C7 occurrence
Ec8: the netevent of C8 occurrence
Ec10: the netevent of C10 occurrence

Figure 5. Causing Belief of Network Inferences

4.2 An Example of Network Faults Isolation

When evidenced netevents are observed on various netobjects, one immediate goal is to sort out causal relationships among them. As limited by the scope of the paper, no formal method is formulated; instead, the procedure of isolating netevents causes is exhibited via an example in three major steps:

490

Step 1: Deducing and Matching Netevents: When an evidenced or hypothetical netevent is observed, driven by the DAG knowledge, hypothetical events are thus deduced and attached to their target netobjects accordingly. Each deduced netevent is a potential causing or consequent netevent, which carries the deducing netevent's identifier. In regarding to a single netobject, when an evidenced netevent and a hypothetical netevent do match on the event type, it suggests the proposition that the evidenced netevent either causes, or is the consequence of, the hypothesis-deducing netevent. Figure 6 first shows hypothetical netevents are deduced based on the network connectivity (as decoded as part of the DAG) and can be traced by following the reverse causation directions in Figure 6. It also shows evidenced/hypothetical netevents are matched in providing *correlation*, *diagnostics* and *prognosis*. The combined belief from those of matching netevents will be used as the belief measure of the concluded network causation or netevent. Figure 6

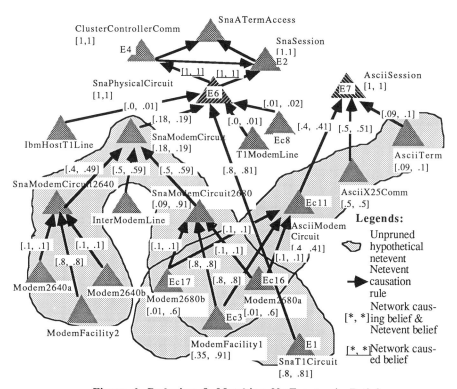

Figure 6. Deducing & Matching NetEvents via Belief

shows how a causation analysis can be conducted after both the SNA physical circuit alarm and the ASCII session alarm netevents are collected. First, by repeatedly applying Baldwin's C4 formula, causing netevents are hypothesized with their belief measures. For instance, using the Figure 5 causing belief distribution, i.e., (P_1=.8, P_{10}=.18, P_6=.0, P_7=.0, P_8=.01 and P_0=.01), the hypothetical SNA Modem circuit failure is calculated with belief [.18, .19], and hypothetical ModemFacility1 failures are calculated with [.07,.98] and [.32,.92] believes from E_{c10} and E_3 respectively. Baldwin's C5 formula are then used to combine derived believes. As such, Modem2680a, Modem2680b, and ModemFacility1 hypothetical failures are matched as potential causes; in particular, the ModemFacility1 failure is calculated with the combined belief [.35, .91], which tells that the ModemFacility1 failure is strongly agreed-upon and is less disagreed-upon than the volatile SNA T1 Circuit, which possesses the [.8, .81] belief, in regard to causing both E_{c10} and E_7 to occur. E_{c3} can thus be chosen for further probing with a high priority.

Step 2: Synthesize Netevent Causations: In correlating netevents, the pattern match process is trying to find a subset pattern within the causation DAG network in rationalizing these evidenced netevents. If simultaneous multiple independent netevents can hardly occur, a tree-like causation relationship can be expected within a cluster of netevents (Note: clusters will be explained later). The root netevent of the tree can be detected as the *primary* cause accounting for the rest netevents of the cluster, which are then classified as *secondary*. Using the same set of netevents in Figure 6, plus additional evidenced netevents E_4 and E_2, a small causation tree can be built in Figure 7, where E_{c3} is detected as the primary cause to be responsible for E_6 and E_7 netevents, and both E_4 and E_2 netevents are detected as consequents of E_6. Nevertheless, there are several difficulties in growing such a causation tree. First the tree grows randomly in fragments in a neither bottom-up nor top-down fashion, as driven by the arriving netevents. Secondly, shaping a tree size is sometimes not straightforward, due to the potential loss/ postponement of received netevents and potential concurrences of independent netevents.

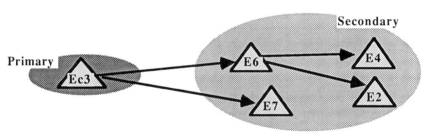

Figure 7. Isolating the Primary Cause

Step 3: Consolidate Hypothetical Netevents: In Step 1, when a hypothetical netevent is deduced with a too-low confidence level, it should be pruned to eliminate unnecessary proliferation of hypothetical netevents, as implied by Figure 6. When netevents are received or deduced, they can be *clustered* based on their timestamps. In other words, two netevents are of the same cluster if and only if they, and/or there exists a chaining of netevents between them, are occurring one-after-another within a certain interval of time. Using a such clustering technique, an *out-of-date* netevent should be retired when no matching can be concluded about it.

It completes our discussions on network faults inferences using the belief model.

5. MANAGING NETWORK KNOWLEDGES

The proposed network event classifications, the belief model, and methods of network inferences described in the previous sections, offer the following advantages:

- *Diagnosis of a network failure with weighted belief:* The backward inferences of causing causations can discover root causes of network failure events, even when they are not reported. Assessed belief measures can be used to guide network testing in

saving tremendous efforts.
- *Prognosis of potential network failures:* The forward inferences of caused causations can detect unreported network failures.
- *Distinguish various types of network uncertainty:* Although uncertainty measures are only complemental network knowledges, they can be incorporated into network inferences and computed in a uniform fashion.
- *Modularity of network knowledges:* The causation rules of the DAG network can be added, modified and deleted on an incremental basis.

As deep knowledges about a network mechanics are complex, a plausible auxiliary theory should be acquired and refined by network experiences; knowledge representation is realized as an integral part of it. At NYNEX, a generalized knowledge-based architecture was conceived for implementation. To ensure the goal to be fulfilled, systematic methods are construed in forming an adaptive learning process, in which empirical data can be converted into expert knowledges in an incremental fashion: first network knowledges are built by an expert, then they will be refined based on analysis of empirical network results.

Of course, a good design of knowledge primitives as well as tools to deduce, input, display, and modify knowledges become imperative. Acquisitions of these network causation knowledges remain a complex task and it begs for efficient software tools to streamline this task of knowledge managements. The process of managing network knowledges should be as important as that of managing network events. The uncertainty model should be used in catalyzing the convergence of network believes. Advantages of using uncertainty measures to control these two processes will hinge upon the successful implementation of the architecture.

From time to time, a network causation relationship depends on too many variables to be effectively modelled and it could be very obscure from empirical network results. Statistical methods should then be applied to test the plausibility of a such network rule or proposition, when a sufficient log of empirical data exists.

6. CONCLUDING REMARKS

In this paper, complexities of managing a heterogeneous enterprise network are examined in the need of incorporating the uncertainty model. A voting scheme, based on Baldwin's *Support Logic Programming* concepts, is proposed to handle fuzziness of management information. Methods on how to apply it to network diagnostics and prognosis are further delineated with examples. At last, a systematic view is presented with emphasis on refining network knowledges from empirical network results, in providing an efficient knowledge management.

9. REFERENCES

[1] J. Goldman, et al, Integrated Fault Management in Interconnected Networks, Integrated Network Management I, IFIP, 1989, pp 333-344
[2] L. Zadeh, A Theory of Approximate Reasoning, Machine Intelligence, Vol. 9, D. Michie and L. Mikulich (Eds), Wiley, New York, 1979, pp 149-194
[3] G. Shafer, A Mathematical Theory of Evidence, Princeton University Press, Princeton, New Jersey, April, 1976
[4] J. Baldwin, Support Logic Programming, International Journal of Intelligent Systems, Vol. I, 1986, pp 73-104
[5] A. Patel, et al, Integrated Network Management and Artificial Intelligence, Integrated Network Management I, IFIP, 1989, pp 647-660
[6] OSI/NM Forum, Object Specification Framework, Issue 1, Forum 003, September, 1989

A Modular Knowledge Base for Local Area Network Diagnosis

Jürgen M. Schröder
Wolfgang Schödl
Institute of Communications Switching and Data Technics (IND)
University of Stuttgart
Seidenstraße 36
7000 Stuttgart 1
Federal Republic of Germany
Telephone: +49-711-121-2488
Telefax: +49-711-121-2477
E-Mail: SCHROEDER@DS0IND5.Bitnet
December 1990

Keywords: Expert System, Diagnostic System, Local Area Network

Local Area Networks (LANs) interconnect individual computing systems to form large distributed systems which are used more and more in Computer Integrated Manufacturing (CIM) and in office automation systems.

Since the proper functioning of the entire system depends highly on the error-free functioning of LAN components, utilities are required to support the user in network trouble shooting. For this a diagnosis system is developed. The kernel of the diagnosis system is an expert system.

LANs for CIM have rather complex structures which are maintained by experts sharing some common knowledge of the entire system. Typically, the knowledge is distributed among all experts involved. Therefore, the solution of problems often calls for cooperation between individual experts.

This approach has been mapped into the construction of the knowledge base of the expert system. The knowledge base has a modular structure.

1. Introduction

In office communication as well as in manufacturing automation the interconnection of computer systems with LANs becomes more and more state of the art. On the one hand, LANs speed up communication between the different divisions; (development, production, store, sale, etc.). On the other hand, the production depends on the reliability of the network. A network failure can stop production and will be costly. Since LANs are the backbone in manufacturing automation systems, the user needs support in trouble shooting.

Networks for CIM consist of many different components, as for example:

- Personal computers (PCs)
- Mainframes
- Workstations
- LAN segments
- Transceiver access points (TAPs)
- Repeaters
- Communication protocols

The different components are normally provided by many different manufactures. Typically, the topology of a network changes daily, as components are added or removed. Therefore, a lot of knowledge is necessary to maintain the network [2]. In most networks the knowledge is distributed between field experts [4], who work together when a failure occurs. This cooperation is coordinated by the system manager. The communication between field experts for problem solving is mapped into a modular knowledge base which is presented in this paper.

2. Diagnosis System

2.1 Architecture

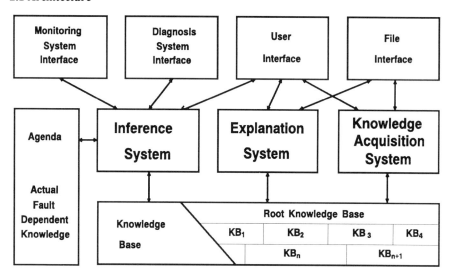

Figure 2.1: Architecture of the Diagnostic Expert System for LANs

The main components of the architecture are the inference system, the knowledge acquisition system and the explanation system. These three components have a connection to the modular knowledge base in which the static knowledge is stored. For user communication, for fault dependent knowledge acquisition with help of a monitoring system, for diagnostic system intercommunication and for file I/O, several interfaces are provided [3, 6].

The major differences between standard expert system architectures and the architecture presented in this paper are the modular knowledge base and the interface to the monitoring system for automatic acquisition of fault dependent knowledge. Figure 2.1 shows the architecture of the diagnostic expert system for LANs.

The inference system works with knowledge stored in the knowledge base and the fault dependent knowledge acquired during the session which is stored in the agenda. Depending on the actual diagnosis state, the inference system uses one of the following problem solving and search algorithms [6,7]:

    Problem solving algorithms:

- Forward chaining
- Backward chaining

    Search algorithms:

- Depth first
- Breadth first
- Refinement
- Tree completion

The knowledge acquisition component is used to acquire the expert knowledge and to store it in the knowledge base. Expert knowledge is mapped into objects and relations between objects. The knowledge typically is described in a description file using a specially defined syntax. The knowledge acquisition system reads the description file and produces the corresponding knowledge base.

The knowledge of the field experts is stored in knowledge base modules. These knowledge base modules can be built independently of each other with the same knowledge acquisition tool. The modules can be used for different networks where the same components are used. Also, the modules are knowledge bases for diagnostic expert systems for the component. Only the so-called root knowledge base, its links and mapping tables to the knowledge base modules are specific to any given network.

Due to the modular architecture of the knowledge base it is easy to build a diagnostic system for a network if modular knowledge bases for the components used are available. The root knowledge base must be built and the links and mapping tables to the modules established. Therefore it is easy to keep track of changes in network configuration.

For fault dependent knowledge acquisition the expert system has an interface to the monitoring system [6,7,8]. With this interface the system is able to monitor and test the network without any user action. This relieves the user from making complicated measurements and setting up LAN monitoring tools. For automatic testing of a component nearly every test program delivered with the component is usable.

At IND a distributed network monitoring system for Ethernet has been developed. This system allows online time-domain reflectometer measurement on Ethernet segments, throughput measurement on Ethernet transceiver cables, connectivity tests (echotests) and protocol stack performance measurements [7, 9]. For this last task distributed measurement systems are controlled via Ethernet by the monitoring system.

For the interaction between diagnostic systems a designated interface is planned. This will be useful for larger networks. Also an interface to the OSI standard network management and Siemens SINEC H1 configuration management is provided through the monitoring interface [5].

2.2 Knowledge Representation

Knowledge in the expert system is represented in the following objects:

- Components
- Final diagnoses
- Diagnoses
- Symptoms
- Symptom test methods
- Explanations
- Knowledge bases

The Components are all of the physical and logical units of the system. Each component can be defective in different manners - it can be totally destroyed or repairable. The diagnoses (the faults the diagnostic system is looking for) are therefore subdivided into final diagnoses and diagnoses. Confirming a final diagnosis means that the inference system stops at this point. No further fault analysis is undertaken.

Symptoms describe the appearance of faults. Each diagnosis has a set of symptoms, so that it is clearly recognisable.

Symptom test methods associate the possibilities to test the appearance of a symptom in the network to the symptoms. The default symptom test method is to ask the user for the appearance of a symptom. A useful knowledge base knows many symptom test methods for automatic LAN inquiry.

Explanations can be text or graphic files with explanations. They can be associated with the objects diagnosis, symptom and knowledge base. They explain, e.g., what the user should do if a diagnosis appears suspect, or how a symptom can be measured.

The object knowledge base is used to create links to other knowledge bases.

There are two ways in which a system can be represented in a knowledge base. The first is to represent the system hierarchically. In a hierarchical representation, a component may have only one predecessor. The number of successors is not restricted. For example, a PC has the sub-components hard-disk, motherboard, graphicboard, power supply etc.. The first component of the tree is called root component. The predecessor component is called the parent of the considered component, a successor is called a child. Figure 2.2 a shows an example structure of a component tree.

This kind of representation is good for the description of local, non-distributed systems. The diagnosis knowledge for e.g. PCs or workstations is described this way.

For network or distributed components description, a heterarchic knowledge description is used. This description model allows as many connections between the objects as necessary. Furthermore, any component may be refined when necessary. The use of refined and unrefined components in parallel is possible. Figure 2.2 b shows the refinement mechanism for a component. In Figure 2.2 b component 2 is most suspicious. For problem solving component 2 needs to be refined. The structure shown in Figure 2.2 c will be used to replace component 2. The problem solving algorithm will use the adjusted model for further searching.

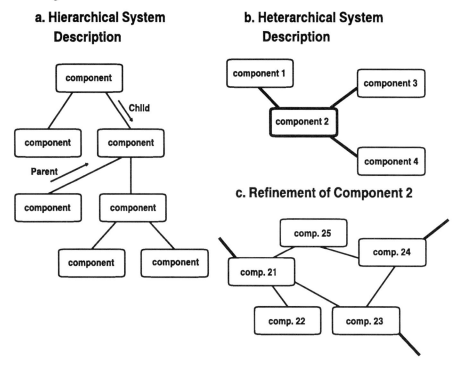

Figure 2.2: Knowledge Modelling Types:
a. Hierarchical system description
b. Heterarchical system description
c. Refinement mechanism

Figure 2.3 shows the modelling of the network depending on the malfunction that occurs, e.g.: A subcomponent within component K2 will communicate with a subcomponent of component K4. Testing components K1 and K3 shows that these components are working properly, so only component K2 and K4 have to be refined. The model used for problem solving consists of the components:

Refinement Layer 1	Refinement Layer 2	Refinement Layer 3
		K21, K22, K23
	K1	
	K3	
		K41, K42, K43, K44, K45

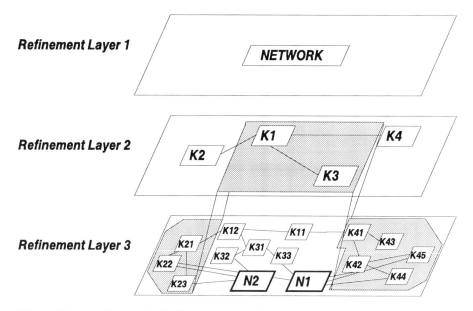

Figure 2.3: Hierarchical Refinement Layers for Dynamic Modelling

In the refinement layer 3, shown in Figure 2.3, new components (N1, N2) and connections appear. With them, new connections between components can be modelled, depending on the possible refinement degree, e.g. communication connections between stations.

Detailed knowledge is represented with the relations between objects. The following relations are usable:

Component	-	Diagnosis
Component	-	Final Diagnosis
Diagnosis	-	Symptom
Symptom	-	Symptom Test Method
Symptom	-	Explanation
Diagnosis	-	Explanation
Component	-	Knowledge Base

Figure 2.4 shows the relationships between objects for the hierarchical knowledge representation. The relationships in a heterarchical knowledge base are shown in figure 2.5.

3. Modular Knowledge Base

3.1 Description

The knowledge base consists of the root knowledge base and a number of subordinate knowledge bases. For each type of network subsystem a knowledge base, treated as a knowledge base module, should be available. A knowledge base can be of the type hierarchical or heterarchical.

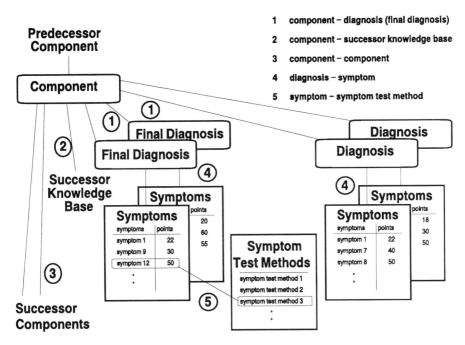

Figure 2.4: Relations between Knowledge Representation Objects in a Hierarchical Knowledge Base.

In the root knowledge base the network is described with all components. This knowledge base is typically a heterarchical knowledge base.

Experts communicate using a so-called agenda. For example, the system manager observes something going wrong in a part of the network. He writes his observations in the agenda and asks a field expert of the suspect network area for help. The field expert will examine the agenda and then start, with its own suppositions, its search for the fault.

Since every knowledge base module is developed independently, a mapping table is used. In this table, symptoms stored in the knowledge base module and symptoms stored in the root knowledge base are associated with each other. That means each symptom that is interesting to another knowledge base module becomes a global symptom. Symptoms exclusively used within the knowledge base module are called local symptoms. This mapping table is built by the knowledge engineer when he links the knowledge base modules to the root knowledge base or to other knowledge base modules. It is not obligatory to build a mapping table. If the mapping table is empty, there are no more suspect diagnoses when a new knowledge base is loaded, so the inference system will start at the root of the diagnosis hierarchy.

As described above, the communication between experts is mapped on the agenda. For that purpose, there is a global and a local agenda for the global and local symptoms, respectively. The inference system uses the current local agenda when it tries to find a problem solution. The global agenda is used when a new knowledge base is loaded. Figure 3.1 shows the loading procedure for a new knowledge base.

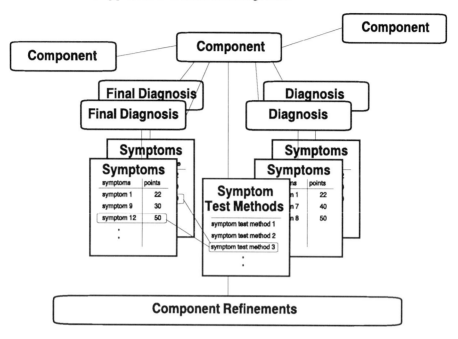

Figure 2.5: Relations between Knowledge Representation Objects in a Heterarchical Knowledge Base.

If a diagnosis has no successor but only a relation link to a new knowledge base module, the inference system first updates the global agenda. For this purpose it uses the mapping table of the current knowledge base. Then it loads the new knowledge base and creates a new local agenda with the information in the global agenda and the mapping table of the new loaded knowledge base. With the symptoms included in the new local agenda, the inference system generates a tentative diagnosis, whereupon it starts the execution of the problem solving algorithm.

3.2. LAN Model

As described above, knowledge can be represented either hierarchically or heterarchically in the knowledge bases. To describe the structure of a LAN, a heterarchically structured knowledge base is useful. The top component is the LAN itself. The following layer refines the LAN into its segments. Each segment can be refined into subsystems describing the segment. New refinements branch out between hardware and software components. For the software description, new relations between components are modelled when a certain level of refinement is reached. This adds the logical part to the LAN model.

When a network fault occurs, the hardware connection between the components will be checked. After that the logical connections can be tested. Logical connections describe the

LAN from another point of view. Protocol types interconnect the stations on which the corresponding communication software is installed.

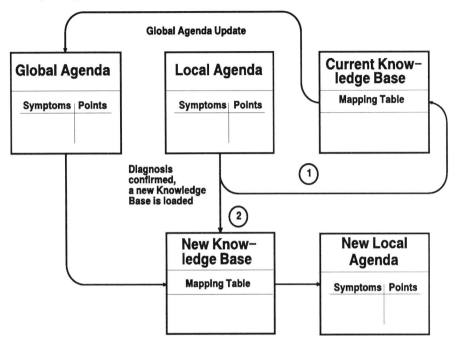

Figure 3.1: Hypothesis Generation for Newly Loaded Knowledge Bases

The root knowledge base refines the whole network only down to the station or protocol system level. For each station or protocol type another knowledge base module will be built and linked to the corresponding station in the root knowledge base.

3.3 Implementation Aspects

A prototype version of the presented diagnosis system on a Personal Computer running under MS-DOS (Registered trademark of MICROSOFT Corporation) and implemented in the programming language C is available since 1989. The database system R:BASE (Registered trademark of MICRORIM Corporation) is used for the knowledge bases in this system.

The MS-DOS operating system was detected as a bottle-neck. The single tasking quality and the support of only 640 kByte memory were the major reasons to look for another operating system. The decision was made to transfer (newly implement!) the diagnosis system to UNIX (Registered trademark of AT&T).

4. Experiences and Outlook

The experience with the system shows the simple adaptation to different LAN configurations. The modular structure of the knowledge base is demonstrated as significant for a diagnosis system for large, complex and distributed systems.

Knowledge bases for the LAN configuration at the IND, DECnet (Digital Equipment Corporation network), Multitronic PC and Siemens PC 16-20 have been created independently of each other. The use of the system shows that it is easy to expand.

A Siemens LAN protocol analyzer [1] and a LAN measurement system developed at the IND is connected to the diagnosis system for automatic testing and network monitoring. Also many manufacturer test programs for different components are used.

It was also recognized that it is not always useful to describe a network only in a hierarchical way, especially for the root knowledge base. This experience led to the construction of the heterarchically organized knowledge base described in this paper.

The new major components implementation under UNIX are almost finished. The inference system and knowledge acquisition system, are available. The work for the explanation system has been started.

Acknowledgement

This work has been supported by SIEMENS AG Department AUT E 51, Erlangen. The authors would like to thank Mr. U. Gemkow, Mr. A. Lederer, Mr. S. Ali, Mr. T. Wagner for fruitful discussions.

References

[1] BATHELT P., PFEIFER K.: "Functions and possible applications of the B5100 LAN protocol tester from Siemens", Proceedings EFOC/LAN 87, P. 392-397, 1987.

[2] BRUSIL P., STOKESBERRY D., DANIEL P.: "Towards a unified theory of managing large networks", IEEE Spectrum, 4/1989, pp. 39-42.

[3] JACKSON P.: Introduction to Expert Systems, Addison Wesley, Reading, 1986

[4] PASQUALE J.: "Using Expert Systems to Manage Distributed Computer Systems", IEEE Network, 9/1988, pp. 22-28

[5] SCHOLLENBERGER W.: "Netzmanagement", Institut für Nachrichtenvermittlung und Datenverarbeitung (IND), Universität Stuttgart, 1989.

[6] SCHRÖDER J. M., GEMKOW U.: "Ein Diagnosesystem für lokale Netze", Kongress der VDI/VDE Gesellschaft Meß- und Automatisierungstechnik,VDI-Berichte Nr.855, 18.-19. September 1990, Baden-Baden, pp. 141-152.

[7] SCHRÖDER J. M.: "Verteiltes Monitoring- und Diagnosesystem für Ethernet", GI und VDI/VDE-GMA Fachtagung Prozeßrechensysteme, Springer-Verlag, 25.-27. Februar 1991, Berlin.

[8] SUTTER M., ZELDIN P. E.: "Designing Expert Systems for Real-Time Diagnosis of Self-Correcting Networks", IEEE Network, 9/1988, pp. 43-52.

[9]   WEIXLER M.: "Distributed Measurement System for Protocols and Applications in ISO 8802/3 LANs", Fourth International Conference on Data Communications Systems and Their Performance, 20.-22. June 1990, Barcelona. Conference Preprints, pp.448-455.

Combining Knowledge-based Techniques and Simulation with Applications to Communications Network Management

Padhraic Smyth*, Joseph Statman*, Gordon Oliver* and Rodney Goodman**

*Communications Systems Research, Jet Propulsion Laboratory 238-420, 4800 Oak Grove Drive, Pasadena, CA 91109, USA

**Department of Electrical Engineering, California Institute of Technology, 116–81, Pasadena, CA 91125, USA

Abstract

In this paper we discuss the idea of combining simulation and knowledge-based techniques for the purposes of designing a communications network management system. While rule-based expert systems have considerable potential in terms of network management applications, the construction of such systems is a complex and expensive task. For a *new* network where no expertise is available, the problem is particularly difficult. We describe here an approach which combines both simulation techniques and the knowledge-based paradigm to solve this problem of designing "expert" systems without an expert. In particular we outline how this approach is being applied to a GPS (Global Positioning Satellite) Packet Radio Datalink network. The main conclusion is that the benefits of building a communications network simulator can be improved significantly by the addition of knowledge-based techniques to the modeller's arsenal of tools.

1. Background and motivation

How does one design an intelligent network management system for a communications network which has not yet been built or fielded? Obviously one would rather not wait several years until operators accumulate sufficient experience that a rule-based expert system can be built using their knowledge. In particular, consider the problem of designing such a network management system for a *new* type of network, perhaps one which exhibits novel characteristics. In this scenario, the performance of the network under *normal* operating conditions may only be partially understood, let alone the problem of trying to characterize the behaviour of the network under *abnormal* conditions (such as traffic overloading, network failures, etc.).

An example of such a network is the proposed GPS (Global Positioning Satellite)

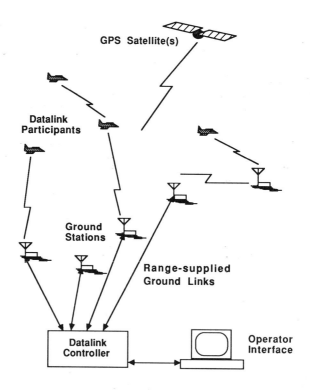

Figure 1. A simplified block diagram of the Datalink network.

Range System Datalink, currently under development for the Range Applications Joint Program Office (RAJPO) [1]. The GPS Datalink network provides two-way digital communications between range participants and a host range command and control center. A simplified block diagram of the network is shown in Figure 1.

Its primary purpose is to support GPS position data collection from range participants to the control center in real-time. The system is implemented as a packet radio network utilising a TDMA participant allocation scheme, with a distributed network routing algorithm. The network supports 200 kbits per second message throughput with data being transmitted in 736-bit packets. Participant rates range from 10 packets per second, down to 1 packet every 10 seconds. Data transmission occurs on an L-band link using 1.6Mhz of bandwidth. The network features both frequency diversity (to combat multipath interference) and spatial diversity (to combat terrain blocking of line of sight reception).

The goal of this paper is not to describe the Datalink system in detail. Rather, we will use it as an example of a complex network whose behaviour is difficult to model in analytical form. The traditional approach to this problem of *characterizing network performance* involves modelling the network in an idealized manner and simulating its behaviour in software. We argue that it is beneficial to go one step further and combine network simulation with knowledge-based techniques. We will show that this approach is advantageous for three important reasons:

1. Knowledge-based techniques can be used to interpret and manage the large volume of data which simulations typically generate, in particular, addressing the problem of correctly classifying the context in which network performance parameters are measured.
2. Machine learning algorithms can identify both performance management rules and fault management rules.
3. The simulator can be used as a test-bed for developing a closed loop network monitor and control system.

2. The benefits of combining simulation and artificial intelligence

The marriage of simulation and knowledge-base technologies is a relatively recent phenomenon [2]. Typically this marriage works in two ways. The first involves using simulation capabilities *within* AI applications, such as a fault diagnosis system — roughly speaking, this type of approach is referred to as *model-based reasoning*. By using a causal model of the system one can in principle reason "from first principles." Given the nature of this paper we will not dwell on this particular approach here, although it obviously has considerable potential in the communications domain [3]. The second approach, and the one which we focus on in this paper, is that of using knowledge-based techniques *within* a more traditional simulation model, i.e., to extend the scope of standard simulation techniques.

To quote Widman and Loparo [2] "simulations generate much more data than they do information." In this paper we will describe the application of knowledge-based approaches to make better use of standard simulation models. The overall objective is the effective use of artificial intelligence as a means of extracting more information from the simulator. In particular, for the Datalink network, we have three primary goals to meet:

1. **Performance characterization**: In addition to answering the basic types of "what-if" questions about the network, we wish evaluate in detail the network performance by studying its transient behaviour, i.e., in a distributed network of this nature we need to evaluate how performance parameters such as bandwidth utilization and burst packet losses are affected by "environmental" factors such as the rate of participant exit and entry to and from the network, terrain blockage, etc. It is essential to transform the raw data into useful information, a function that

cannot be performed manually if we are to obtain statistically reliable amounts of data concerning the events of interest. In addition, since by any definition this abstraction process is knowledge-based (i.e., is more heuristic than algorithmic in nature), it makes sense *not* to implement this function as part of the simulator itself, but rather as an external module.

2. **Fault management and prediction**: We would also like to use the simulator if possible to generate rules for network management, i.e., one would like to be able to give the operator a set of rules of the form "if parameter X deviates by more than y units then event z will probably occur in w minutes from now." This information could either be in the form of text in an operations manual or on-line as part of the system interface. It is conceivable that one could generate these rules manually, by manipulating the simulator until one effectively becomes an expert on how the model works. Note that while the model-designer is in principle an expert regarding its operation, the actual dynamics and temporal characteristics of a model through its state-space can be very difficult to predict, even if the basic rules of behaviour at the component level are relatively simple. Hence, it makes much more sense to take advantage of the wealth of machine learning techniques currently available to *automatically* induce such rules from the data. This marriage of machine learning and simulation appears not to have been tapped, yet the potential gains are significant: real data is usually costly to obtain for learning algorithms and the opportunity to obtain new data is often not available. The limits on the quality of information obtained is in principle limited only by the quality of the simulation model itself.

3. **Real-time network monitor and control**: One might imagine that if one has characterized the performance of the system (and developed prediction rules, etc.), that building a real-time monitor and control system would be quite straightforward. Unfortunately this is not the case. The real-time nature of the problem makes it quite difficult to extrapolate the rule-based expert system technology from the "classic" passive advisory systems such as Mycin [4] to active, real-time control loops. Truly real-time expert systems are difficult to construct and successful field implementations of this approach are relatively rare to date. Hence, for a dynamic system such as the Datalink, where critical events may develop in the time-frame of a few seconds, the implementation of an efficient and robust network controller is quite challenging. The simulator is an ideal test-bed for experimentation. As with machine learning, gaining access to a real system in order to calibrate and test a network control algorithm is often neither practical nor economical. In the case of the Datalink, the system will not be in production until mid-1991 or beyond and the early field systems will be somewhat scaled down versions of the full-fledged version.

Hence, the value of the simulator to the Datalink project extends significantly beyond the traditional benefits of simply answering "what-if" questions and obtaining

performance curves. In the remainder of the paper we will describe in more detail each
of the three components outlined above. First, however, we will give a brief overview of
the overall functionality of the simulator and a description of the proposed architecture
of the real-time network management system.

3. System architecture
3.1 Description of the Datalink simulator

Since the Datalink is a TDMA system, it lends itself to discrete event simulation
at the time-slot boundaries. By choosing this level of granularity we effectively choose
to ignore any sub-slot timing effects which may occur (analysis indicates that timing
and synchronization should not be a factor in network performance). In addition,
with the slot boundary scheme, we choose to simulate at the packet level rather than
at the bit level. This is very much a practical consideration since all of the network
algorithms effectively operate at the packet level (once each participant decodes the
convolutionally encoded packet in hardware). The simulation model has been coded
in C and implemented on a Sun Sparcstation. The most complex portion of the code
involves modelling the central control facility, since this part of the algorithm must
perform all the book-keeping tasks associated with TDMA slot assignment, uplink
message scheduling, broadcast messages, etc.

We initially intended to use the simulator for two primary purposes. The first
purpose was to answer specific detailed "what-if" questions, such as the ability of the
distributed routing algorithm to equally distribute traffic between relayers in accordance with participant duty cycle limits. The process of evaluating these "what-if"
questions is currently under-way. The second primary purpose of the simulator was
to characterize network parameters by running large-scale simulations and collecting
statistical data, e.g., plotting bandwidth utilization as a function of participant entry
or exit activity. However, it quickly became apparent that this latter process was somewhat more involved than simply keeping track of various numerical parameters. In fact,
in order to arrive at sensible results, it would be necessary to tag each measurement
with some type of *context indicator*, since the conclusions which can be drawn tend to
be very sensitive to context. For example, short bursts of packet loss can occur in a
variety of situations, and in particular they depend on the prior history of events immediately preceding the burst. For the simulator itself to track, correlate, and reduce all
this data, seems like an unreasonable approach. After all, the simulator is accurately
reflecting the real system, which itself will generate relatively uncorrelated low-level
information. Hence, it was clear that a knowledge-based real-time "context-classifier"
would considerably enhance the overall information output of the simulation model,
and indeed would be necessary for many of the scenarios of interest. A further requirement on this monitor was that it would need to run in real time (since the simulator

runs roughly in real-time) if it was to be useful.

At this point in the discussion it is instructive to take a closer look at the requirements for a real, field-implemented network monitor and control system for the Datalink, i.e., an operational network management system, the development of which was initially unrelated to the simulation modelling.

3.2 Designing a real-time network management system

As mentioned earlier, the design of a real-time network management system is considerably more difficult than simply implementing a rule-based expert system – rule-based systems are most typically implemented in an *advisory* configuration with no concept of temporal reasoning involved, and, despite claims to the contrary, little work has been done on the more difficult and practical problem of developing truly *autonomous, active* expert systems which can operate in real-time with minimal operator intervention. Based on our previous experience with network management problems [5], we divided the Datalink problem into three parts:

1. **Performance management:** Monitor network performance and adjust network parameters (such as relay-depth and slot assignment rates) in real-time to maximize the probability of correct packet reception from each participant.

2. **Fault management:** Detect when network behaviour deviates from normal and identify appropriate corrective action.

3. **Operator interface:** Implement an intelligent user interface to the network operator in *both* directions, i.e., both interpret and condense incoming data and present it in an informative manner, and filter operator inputs by checking for consistency and providing advisory feedback — note that this advisory component is but a small part of the overall system and is very much context-driven.

In Figure 2 we show a diagrammatic representation of this architecture. Note that this general architecture of a dedicated real-time context classifier, with modular expert sub-systems, is quite generic for real-time monitor and control applications.

4. Bridging the gap: from simulator to real-time system

Initially we intended to develop the real-time monitor and control system using a fielded Datalink system as our testbed. The simulator development was considered a relatively separate activity. However, as should be obvious to the reader at this point, both problems have much in common. Hence, using the simulator as a testbed became the approach of choice. We outline the major steps involved:

4.1 The simulation model:

We built a standard simulation model, using the C language. Given the experi-

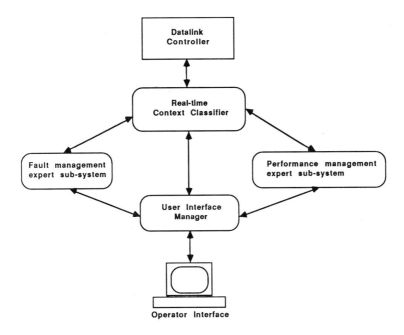

Figure 2. A block diagram of the real-time network manager prototype system.

mental nature of this project, we chose C as the base language for the simulator rather than using a special-purpose simulation package. The two primary reasons for doing this were that first we needed a high degree of flexibility in our software architecture and wanted to maintain control over all portions of the code, and secondly, we wanted to customize the code for speed, so that the simulator would run in as near real-time as possible. This latter requirement is dictated by the fact necessity of real-time simulation if we are to use the simulator as a testbed for developing real-time network management prototypes. The simulation model was implemented in a modular fashion, with the user interface (input syntax, output displays, etc.) being as removed as possible from the actual network model. This approach facilitates both standard manual interaction with the model (via mouse-driven display windows, etc.), and also *automated* interaction, where external programs (such as the knowledge-based techniques to be described below) can interact with the model in a non-supervised manner. In this latter mode, the user-friendly interface modules can be switched off and all simulator output is written directly to a UNIX file. In this way the simulation model serves both as a useful "what-if," user-friendly tool, and as a realistic "low-level" simulation of the actual network.

512

4.2 Knowledge-based simulation analysis:

As we outlined earlier, building a complicated simulation model can result in large volumes of data being generated. In our case we simulate at the packet rate of 330 packets per second in near real-time. Because the routing algorithm is distributed, each of the N participants in the network computes cost and routing information every time a packet is received — these cost tables are computed locally by each participant based on distance, signal quality, and duty-cycle parameters. Typically the number of participants, N, ranges from 5 to 25, although in principle the network can support up to 200 participants. There are a variety of special routing features, such as the use of alternate ground relays, to achieve both spatial and frequency diversity. In addition the Data Link Controller (DLC) operates continuously, aligning downlink packets, managing the TDMA slot assignment table, scheduling uplink packets, etc. Hence, over a typical period of interest of say 30 minutes the simulation model can generate a huge amount of data. The primary characteristics of this data is that the vast majority of it is non-interesting — however occasionally a sequence of noteworthy events will occur.

The questions we are interested in answering are typically of the form "what is the average number of packets lost in succession (burst losses) when network routing changes occur?" In a more abstract sense we are interested in the impact of so-called *environmental* variables on *network* variables, where environmental variables are loosely defined as those parts of the simulation which strictly speaking are not part of the network model itself, i.e., the number of participants, the dynamics of the participants, terrain models, channel models, network entry/exit patterns, etc. The key point to note is that there are a large number of possible states or "contexts" for the network to be in. In addition, when one factors in the time-dependencies which are implicit in the operation of the network (there are unavoidable delays of up to 10 seconds involved with many network state changes), the problem of analyzing the data output from the simulator becomes formidable.

The approach we take is to effectively kill two birds with one stone: we are developing a dedicated "context classifier" as described earlier to operate in real-time on network information. This prototype serves the dual purpose of enhancing simulation analysis, *and* of providing a major component of the real-time network controller. We are developing this context classifier with a higher-level language than C using one of the more recently developed commercial C-based expert system shells. The trend in expert system development tools has been towards more general-purpose, flexible languages, typically C-based object-oriented systems. The era of special-purpose languages, such as Lisp and dedicated hardware for AI applications, has been largely replaced by a more practical, software-engineering attitude in the AI industry. The newer shells typically are quite portable and robust, and most importantly, can be integrated easily with more standard languages. These developments imply that it is now practical to combine large-scale simulation projects with knowledge-based analysis systems using the type of approach we are describing here.

4.3 Learning algorithms for simulation analysis:

Consider that the trajectory of the network through its state-space is somewhat akin to the behaviour of a Markov model. Local regions of the state space are deterministic, reflecting the finite-state machine nature of the underlying network algorithms, while the gross structure of the model is stochastic, reflecting the underlying influence of the environmental variables. In general, it is difficult to characterize the trajectories or dynamics of the model over time since the system is non-linear and quite complex. As mentioned earlier, much of this knowledge of network operations comes only through experience, either by interaction with the simulator, or by direct experience with the real system.

An alternative, and far less costly approach, is the use of automated learning algorithms which exercise the simulator, as a human would, to accumulate expertise over time. Recent work by Smyth and Goodman [6] has described an information-theoretic approach to the problem of finding the most important set of rules from empirical data — in particular their algorithm (the ITRULE algorithm) can find the most informative state transitions in a Markov model, given data samples from the model [7]. The applicability of this style of information theoretic learning to simulation shows great potential, yet these techniques are relatively new and untried. For example, most learning techniques are based on the assumption that the data is initially presented and no further data is available — the idea of an algorithm which could control the inputs to the simulator, and hence generate the type of data it requires to improve its learned knowledge of the system, is quite interesting but relatively unexplored. For the Datalink project we are using the aforementioned ITRULE algorithm for identifying both performance and fault diagnosis rules.

4.4 A network management prototype using the simulator as a test-bed:

As outlined earlier, we can develop a prototype real-time network manager, using the simulator as a test-bed for the development process. The network manager must perform three primary functions: (i) performance management, i.e., choosing network parameters such that performance is optimized, (ii) fault management, both diagnosis, and prediction based on component degradation, and (iii) operator support and user-interface, i.e., the implementation of an intelligent front end which can manage the information flow between the operator and the Datalink central controller. We will not dwell at length on this aspect of the project, except to note that the context classifier described earlier is an integral part of the design, i.e., the events list as output by the classifier is read by a prioritization mechanism which in turn produces a stack of prioritized events for each expert subsystem to deal with. In this manner the expert modules can continuously focus on the most important problems. The development of this part of the system will be quite complex as the actions of each of the expert "agents" must be coordinated such that conflicts do not occur.

5. Conclusions

We have argued in this paper that the combination of simulation modelling with knowledge-based techniques has significant practical advantages. This is not to say that all modelling problems are best pursued in this manner. Nonetheless, it seems quite likely that with the proliferation of large networks, which exhibit complex non-linear characteristics, that this approach may often be worth the extra investment. In particular, we have identified three primary areas where knowledge-based techniques can profitably be used: (i) as an information management aid for improving the simulation analysis of complex models, (ii) in the use of learning algorithms for identifying underlying causal relationships in the model which can only otherwise be discovered by expensive trial and error techniques, and (iii) as a testbed for the development of real-time network management systems.

Acknowledgements

The research described in this paper was performed at the Jet Propulsion Laboratories, California Institute of Technology, for the United States Air Force Systems Command, under a contract with the National Aeronautics and Space Administration.

References

1. M. Birnbaum, R. F. Quick, K. S. Gilhousen, J. Blanda, "Range Applications Joint Program Office GPS Range System Data Link," in *Proceedings of the ION GPS-89 — 2nd International Technical Meeting of the Satellite Division of the Institute of Navigation*, pp.103–108, Colorado Springs, CO, 1989.

2. L. E. Widman and K. A. Loparo, "Artificial Intelligence, Simulation and Modeling: A Critical Survey," in *Artificial Intelligence, Simulation and Modeling*, L. E. Widman, K. A. Loparo, N. R. Nielsen (eds.), John Wiley and Sons, New York, pp.1–44, 1989.

3. R. O. Yudkin, "On Testing Communication Networks," *IEEE Journal on Selected Areas in Communications*, vol.6, no.5, pp.805–812, June 1988.

4. R. Davis and J. J. King, 'The origins of rule-based systems in AI,' in *Rule-based Expert Systems: the MYCIN projects of the Stanford Heuristic Programming Project*, B. G. Buchanan and E. H. Shortliffe, Reading, MA: Addison-Wesley, pp.20–54, 1984.

5. R. M. Goodman, J. W. Miller, P. Smyth and H. Latin, "Real-time Autonomous Expert Systems in Network Management," in *Integrated Network Management*, B. Meandzija and J. Westcott (eds.), Elsevier Science Publishers B.V. : Amsterdam, pp.599–624, 1989.

6. P. Smyth and R. M. Goodman, 'An information-theoretic approach to rule induction from databases,' *IEEE Transactions on Knowledge and Data Engineering*, in press.

7. P. Smyth and R. Goodman, 'Rule induction using information theory,' in *Knowledge Discovery in Databases*, G. Piatetsky-Shapiro and W. Frawley (eds.), MIT Press, in press.

VI

REASONING AND KNOWLEDGE

IN NETWORK MANAGEMENT

C
Applications of Knowledge-Based Systems

A Knowledge-Based System for Fault Localisation in Wide Area Networks

Monica Frontini, Jonathan Griffin and Simon Towers

Hewlett-Packard Laboratories, Filton Road, Stoke Gifford, Bristol, UK

Abstract

This paper describes a knowledge-based system developed for monitoring and fault localisation in a wide area network. The system incorporates some of the network operators' and manager's knowledge on localising faults by using the structure of the network and correlating events.

This paper describes the overall system structure. It focuses on the Event Handler modules describing its submodules and data representation. We also show how this knowledge and its control map onto a blackboard architecture.

1 Introduction

This paper describes a knowledge-based system for monitoring and fault localisation in wide area networks. The first section then describes the networking domain. Following this we discuss the required functionalities of the system. For implementing the system we have chosen to use a blackboard architecture. So, in section 4 we present a general discussion of these architectures, before going on in section 5 to describe the system structure. Section 6 gives details of the demonstration system.

2 The Networking Domain

The demonstration system is based upon HP's internal packet-switched wide area network, HPNet. HPNet is a worldwide network with approximately 50 switching nodes.

The switching technology is based on CCITT X.25 and X.75 packet interfaces. It is comprised of switching nodes connected by backbone links. Users access the network via access links. A network is centrally managed by the network control processor (NCP). The NCP provides administrative, configuration and monitoring functions. Access to the NCP is provided by a Network Operator's Console (NOC). Amongst other things, the NCP displays and logs events generated by the switching nodes [1]. Each switching node reports regularly to the NCP its status, and also notifies it of status changes in devices attached to it (eg backbone or access links). Thus, a single *problem* in the network could result in many messages being received at the NCP (a failed switching node would cause the failure of all of its attached links, and so the nodes at the remote end of each of these would report the links failure) and displayed to the operator on the NOC.

Operators in Bristol and Atlanta alternate the monitoring of HPNet. The network is monitored for 24 hours, 7 days a week. During this time, the operator scans the outstanding events, correlates them with problems and, when required, further investigates the problem or initiates repair when appropriate. Knowldge acquisition interviews and the analysis of test cases show that:

- Events related to a single problem come in bursts;

- Events related to independent problems are often interleaved;

- An action associated to a problem (i.e. a futher investigation or a repair procedure) depends upon several factors such as where (the location) and when (the timezone) the problem occurred; the severity of the problem; the criticality of the component involved.

3 System Functionalities

The system is principally designed to help the network operators and manager in the day-to-day running of the network by providing up-to-date information about the status of equipment in the network, details of problems, and suggested actions to perform when necessary.

The following functions are provided:

- Problem and action definitions.
  The network manager is able to define to the system different types of problems in terms of the events expected to be received, other associated problems, and the states that the problem moves through. The manager is also able to define rules (representing his heuristic knowledge) which confirm or reject state transitions even though not all the expected events have been received.

- Event and problem correlation.
  The system receives "event" messages along the event printer line from the NCP. The system is able to correlate messages which relate to the same problem. For example, when a switching cluster undergoes a "restart" a series of messages will be received relating to the restart and there will also be messages relating to associated problems (backbone and access link problems); there may be 30 or 40 messages received in total. The operator is presented with a one line summary of the problem, but can examine the received events and associated problems if he wishes.

  Thus, the operator does not have to work out for himself which events are related to one another and what the underlying problem is — it is all done by the system.

- Actions.
  The network manager can define the actions to perform for different types of problem or event, for different pieces of equipment and depending on the time of day and

time-zone. It is also defined for each of these actions how long the problem must be open for before the action is to be performed. At the appropriate time then, this information is presented to the operator. The operator can choose to temporarily "hide" this window (for example if he has other more important things to do first), but the action will remain on the "action list" (and thus can be viewed again) until it is "acknowledged".

Thus, the operator does not need to work out for himself what actions to perform for a particular problem — the system works this out, and tells him at the appropriate time.

Note, an action is text presented to the operator — the system can not perform any "tests" itself (such as reseting a port) since it has no access to the NCP or other devices.

- Repeated problems.
  While the occurrence of a problem once may not require any action from the operator (for example a backbone link going down and then up within the space of a few minutes), the repetition of the problem (a backbone link failing five times within an hour) may require operator investigation. The system allows the manager to define thresholds for the number of times a problem is allowed to occur over a period of time before the operator is informed.

- Fail safe window.
  If the system receives any event which it doesn't understand or which refers to a network device which is not in the configuration database, then a window is presented to inform the operator of this.

- Graphic display of the network.

    - The map is automatically drawn from the NCP configuration information.
    - There is a map and configuration editor.
    - Problems with nodes, clusters or backbone links are indicated by highlighting the corresponding icon.
    - It is possible to display just selected regions of the network.
    - For each component, it is possible to obtain its configuration information and the last sets of events and problems related to it.

- Logging facility.
  The system prints to separate log files summaries of problems in the network and actions suggested to the user. This could be modified to generate "trouble tickets".

4 Blackboard Architectures

The first blackboard architecture was developed at Carnegie-Mellon University during the 1970's[2]. Since this time there have been a number of different implementations of blackboard architectures covering a large range of different application areas and incorporating different control strategies. For examples of current work in this area see [4].

The blackboard model is usually described as being comprised of three major components[3] (see figure 1):

The blackboard data structure. At any time during the operation of the system, the current state of problem solving is represented by items stored in a global data area, the *blackboard*. The blackboard is often structured into an "abstraction hierarchy" based upon the data types relevant to the problem. For instance, in the original HEARSAY system [2], the hierarchy was based upon speech elements (phrases, words, sylables, phonemes etc).

Knowledge sources. The domain knowledge required for the application area is partitioned into independent, autonomous modules known as *knowledge sources* (KS's). The KS's contribute to the problem solving by changing the items on the blackboard. There is no direct communication between knowledge sources — all communication and interaction occurs as the result of changes to the blackboard. Hence, the functioning of an individual KS must not depend upon the existence of others. The form of knowledge representation used within each can be chosen to best suit the task of that KS.

Each KS is responsible for knowing the conditions under which it can contribute to a solution. In analogy with production rules, KS's tend to have a *condition* part, and an *action* part. The *condition part* is called by the control KS, and is essentially a request for this KS to examine the blackboard and determine whether it can do anything. If so, it returns a list of *knowledge source activation records* (KSAR's). The *action part* contains the domain knowledge for the application. It is executed following a request to do so from the control KS.

Control. The basic blackboard model does not specify a form of control, but only a general problem-solving behaviour in which the KS's respond "opportunistically" to changes on the blackboard. Clearly, some form of control must be imposed upon this framework in order to result in the KS's being invoked in a coherent manner destined to lead to a solution of the problem at hand. This problem solving strategy may be built into the system, as in HEARSAY [2], or represented explicitly as we have done.

HEARSAY contained a controller, coded in the implementation language, which contained procedural representations of the problem solving strategies to be used by the system. There were two basic strategies: an initial bottom-up synthesis of hypotheses, and then an opportunistic phase during which high-level hypotheses were generated and then broken down in a top-down fashion to compare with low-level data.

In the *Blackboard Control Architecture* of Hayes-Roth [5] there is an explicit separation of control and domain blackboards. Control is explicitly represented as another problem for the system to solve. Thus, in this case, control is viewed as a planning process.

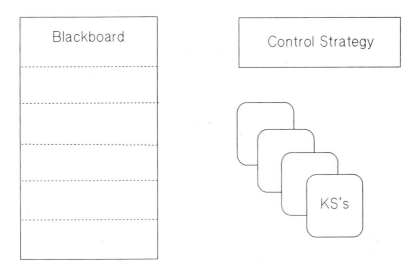

Figure 1: Principal elements of the blackboard model

The term *opportunism* is frequently associated with blackboard systems. Opportunism can be defined as a control strategy which makes best use of the current solution state. For example, if the current strategy is to work top-down, but there is a promising partial solution which is not currently within the focus of this strategy, then the controller could switch strategies to work bottom-up from this partial solution. Some people consider opportunism to be a fundamental part of the blackboard system's definition, while others just consider it to be a class of strategy.

How suitable this architecture was found to be will be discussed later.

5 System Structure

The system manipulates three kinds of input: user data, network events, configuration data. Figure 2 represents the relationship between the fault localisation module, the Event Handler, and other "support" modules.

The Configuration Handler reads in the network configuration as it is in the NCP database and updates the network model. It is initiated on user request and its activation stops the other modules.

The Editor allows the user to modify the topology and status of the network model. The Browser retrieves and displays information on the network status upon user's request. They have been built as part of the User-interface (UI) module. They run without stopping the Event Handler.

The following sub-sections are devoted to the details of the event handler.

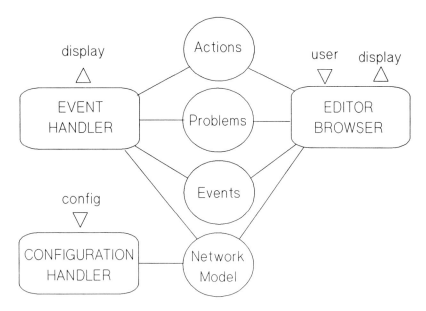

Figure 2: *System Structure*

5.1 Event handler overview

This section provides an overview of the system and describes how the system works. The purpose is to give an understanding of the general framework before moving into the details of KS's and control in the subsequent sections. This framework is independent of the network domain. Therefore, it may be applicable for localising faults in other types of network.

Figure 3 shows the modules (squares) and the data (circles) composing the Event Handler.

The system deals with a continuous in-flow of events.

1. New events are "classified" according to their importance in characterising (proving) hypotheses. (Classifier KS)

2. Classified events are used to "prove" one or more of the existing hypotheses (i.e. event correlation). An hypothesis moves through different stages while its expected events are received. It becomes an open problem when those events which confirm the hypothesis have been received. These are events received at approximately the same time from different network entities reporting their view of the initial problem. Other events represent the time progression of the problem towards its final state. When these temporal events are received the problem is closed. (Prover KS)

3. When a classified event can't be associated to any of the existing hypotheses, a new set of possible hypotheses are identified. This set is "generated" from the predefined

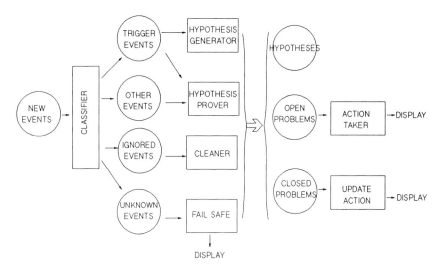

Figure 3: *Event Handler Structure*

set of problem definitions. Associated with each hypothesis are its expected events. (Generator KS)

4. Actions are "taken" upon open problems. The system displays the instructions for the operator within an alarm window. An action can be immediately displayed or delayed. The delay depends on the timezone and the location of the associated problem. (Action Taker KS)

5. Problems and events occurring frequently in an interval may constitute a different problem. Actions are associated to those problems. (Action Updater KS)

6. Old information (events, hypotheses and actions) are regularly "cleaned up". (Cleaner KS)

7. Unknown events and hypotheses are shown to the operator. (Fail Safe KS)

5.2 Classifier KS

This module takes newly received events and classifies them according to the following categories:

- "Must" events: events identified as being able to generate (trigger) at least one hypothesis;
- "May" events: events that are not triggers for any hypothesis;
- "Known" events: events the system doesn't have to deal with;
- "Unknown" events: events the system doesn't know about.

5.3 Hypothesis Generator KS

This module generates a new hypothesis from a "must" event or an open problem — subproblem. The system includes a library of problem definitions. Each problem definition has a list of "triggers", that is, relevent events or sub-problems for this problem. For instance, the problem "backbone link down" (bbl-down) has event type 1048 (bb-link down) and event type 1047 (bb-link up) as triggers. The presence of a trigger event or subproblem identifies a new possible hypothesis. Therefore, if the event or sub-problem that the Generator is considering matches with a trigger then a new hypothesis is generated. This hypothesis is related to the event or sub-problem in consideration and linked to a network object.

5.4 Hypothesis Prover KS

Each hypothesis has a pool of expected events and sub-problems. This module proves existing hypotheses by associating received events and sub-problems with the expected ones. The KS controls the "status" (hypothesis, open or closed) of the problem in relation to which of its expected events it has received and whether any of the manager's heuristic rules have been triggered.

5.5 Action Taker KS

This manipulates actions related to hypotheses or events. The action to perform for a particular problem depends upon, for instance, the location and timezone of the piece of equipment. As for the problem definitions, the system includes a library of action definitions. This KS generates the action for a specific problem or event from the action definition. Some actions are immediately displayed, while others are delayed by the interval specified in the timer slot of the definition. Those actions are "queued" in the waiting queue. When an hypothesis has been "closed" or "rejected" then all its actions are "freed" so that the Cleaner can remove them. If the waiting time expires before this, then a window is generated on the screen and the operator is informed of what actions to perform.

5.6 Action Updater KS

This module allows the user to define thresholds for the number of times problems can reoccur within specified times. If the threshold is exceeded then the operator is informed.

5.7 Cleaner KS

At regular intervals this KS tidies up old material. It removes "closed" or "rejected" hypotheses which are not used by other hypotheses. It deletes "known" events and old events belonging to closed hypotheses. It removes "freed" actions.

5.8 Fail Safe KS

This is another KS running at regular intervals. It looks for "unknown" events, events that have not been used by any hypothesis and events that are related to unknown objects. It rejects hypotheses and open problems which have been around for a long time without beeing closed. It checks for hypotheses which have started to receive their temporal events while still missing a "must" confirmatory event.

5.9 Control

The main control in our architecture is performed by executing a special KS. This privileged KS is called the *control knowledge source*, and to be strictly correct is not a KS since it does not conform to the model outlined in the previous section. This KS performs the following basic cycle of tasks:

1. Choose which (domain) KS to poll.

2. Poll those selected, and receive their KSAR's.

3. Reason amongst the KSAR's to decide upon the best to execute.

4. Execute the KS.

When this application (and one concerned with the design of WAN's) was first considered, it was thought that some form of "planning" would be needed to determine the order of KS execution. However, following the knowledge acquisition it was found that for a particular problem in the network, the human expert followed a fairly fixed sequence of steps. The big issue for control then became dealing with the fact that at any one time there were multiple problems in the network each at a different stage of resolution. Thus, it had to be decided which problem to deal with at any one time — a question of resource allocation.

The Control module specifies the priorities amongst the KS's as follows:

1. the Classifier as the highest priority, therefore new events are immediately classified;

2. the Action Taker as high priority since the "timedout" actions need to be promptly displayed;

3. the Action Updater promptly displays information about repeated problems;

4. the Prover is invoked after received events have been classified;

5. the Generator only works when there are events or sub-problems which have not been used by the Prover;

6. Fail Safe and Cleaner run at regular intervals;

Other rules may be considered as part of the control such as:

- consider first the KS's dealing with inputs (events, hypotheses, problems) from a working day time zone.

- consider first the KS's dealing with inputs which involve more critical equipment.

- consider first the KS's dealing with inputs which involve unreliable equipment.

- consider first the KS's dealing with severe events.

5.10 The Blackboard

The blackboard is segmented into the following entities:

- Network model.
- New events.
- Classified events.
- Hypotheses.
- Open Problems.
- Closed Problems.

6 The Demonstration System

The above system has been implemented on an HP9000/370 workstation running the ART AI environment. The workstation uses a C-interface to connect directly to the event printer line of the NCP.

Figure 4 shows two screens from the operating system. The first shows the network map (the icons are mousable to obtain or alter configuration information), with problem devices highlighted. The top right hand panel shows a summary of the current problems (each of these is mousable to give more information about each problem; in particular its list of correlated event messages and sub-problems). The other panels allow the user to perform such actions as examining the current hypothesised problems, closed problems or actions, controlling the event handler or loading a new network configuration database.

The second screen in Figure 4 gives an example of the type of window that appears when the operator must perform an action.

7 Concluding Remarks

In summary then, we have developed a WAN monitoring and fault localisation expert system. This incorporates the knowledge of experienced network operators and managers in correlating events and relating events to their underlying "problems".

The use of a blackboard architecture proved to be useful; the modular nature of KS's was particularly useful (we could easily add or remove KS's without regard to which other

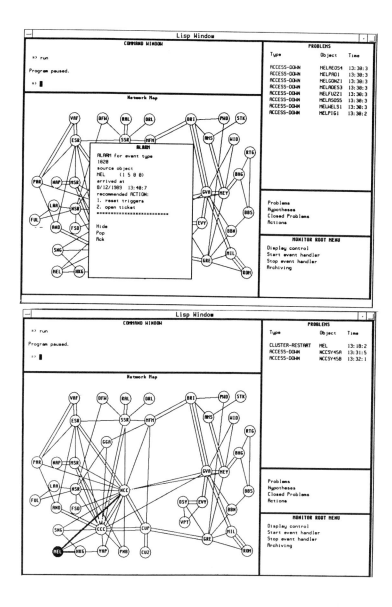

Figure 4: Screen dumps from the working system.

KS's were already present), the use of production rules within each KS was a suitable way of representing the human experts' domain knowledge, the separation of control and domain knowledge allowed us to experiment with different control strategies, the declarative nature of the representation of this strategy makes it easy for the network manager to change it. The opportunistic nature of blackboard control strategies is particularly relevant to event handling applications such as this, and naturally provides the system with the ability to handle multiple problems at the same time.

References

[1] "HP AdvanceNet: HP PPN Consultant Guide", HP Internal Manual 5954-9777 (Sept. 1988).

[2] L.D. Erman, F. Hayes-Roth, V.R. Lesser, R. Reddy, *ACM Computing Surveys,* **12** (1980) pp 213–253.

[3] H.P. Nii, "Blackboard Systems", *AI Magazine,* **7** (1986) pp 38–107.

[4] "Workshop on Blackboard Systems", 11th Int. Joint Conf. on AI (1989), Detroit.

[5] B. Hayes-Roth, "A Blackboard Architecture for Control", *AI Journal,* **26** (1985) pp 251–321.

RAC & ROLE, A Knowledge-Based System for Network Trouble Administration Defined by Rapid-Prototyping

Michael St. Jacques, Delano Stevens, Joseph Sipos and Laura Lau

GTE Data Services, P. O. Box 290152, Temple Terrace, Florida

Abstract

A high level system design for *RAC & ROLE*, a *R*epair *A*nswer *C*enter and *R*epair *O*rder *L*ogging *E*xpert is presented. *RAC & ROLE* is a *knowledge-based* system which assists Repair Answer Center (RAC) personnel in taking, logging and providing updates on trouble reports from business and residence customers who are experiencing problems with their telephone service. The design includes a human interface for display and user input, a knowledge-based system for trouble-shooting problems as well as interfaces to a number of operational systems for other trouble administration and data retrieval functions. Another version includes a voice system to interact with customers and handle a number of problems without human intervention. Key features of the design are flexibility and expandability to handle a growing customer base and changing needs.

Rapid-prototyping is being used as a means of identifying and solidifying requirements for the entire *RAC & ROLE* system. The first implementations of selected features took form in two working prototypes which were written in LEVEL 5 and Microsoft C on an IBM PC/AT. One prototype is a screen based system which prompts RAC clerks with appropriately phrased questions which lead the customer through diagnostics for residential telephone systems aimed at determining the most likely cause of network difficulty. The other prototype employs a computer voice system (TI-Speech) which interacts directly with customers via touch-tone telephones. Using these prototypes as a starting point for requirements development is extremely encouraging indicating the feasibility of this approach for the entire system. As the effort continues, the identification of firm requirements for an internal software product (and possibly a commercial offering) appears likely.

Trademarks

IBM PC/AT[TM] is a trademark of the International Business Machines Corporation (IBM). Level 5[TM] is a trademark of Information Builders, Inc. (IBI). 4TEL[TM] is a trademark of Teradyne Corporation. Microsoft® C and MS-DOS® are registered trademarks of Microsoft Corporation. TI-Speech[TM] is a trademark of Texas Instruments Incorporated. dBASE III Plus[TM] is a trademark of the Ashton Tate Corp. MARK[TM] is a trademark of General Electric Information Services Corp.

Acknowledgments

The authors express sincere appreciation to the following people without whose help this work would not have been possible: **Ms. Michele Neverman** who edited and spoke the voice scripts for the direct customer interface prototype, **Mr. Jim Starks** for directing, promoting and encouraging this work, **Mr. Kent Henrich** for invaluable assistance in the presentations and resolution of issues and to **Mr. Edward Goeske** for providing expertise in problem diagnostics and RAC operation.

1. INTRODUCTION

In the recent years, the need for better methods of identifying software requirements has been recognized by many [Tanik, 1989]. With the advent of Artificial Intelligence (expert and knowledge-based systems in particular), the idea of defining system requirements as a joint or collaborative effort between software provider and consumer was introduced. This joint requirements development effort takes place via an *iterative process* as opposed to the old, sequential method. This iterative approach, sometimes referred to as collaborative development, joint development or most often just called *rapid-prototyping*, must be used in expert systems work because the extent of the knowledge and complexity of the inferencing are unknown ahead of time either by the system developer or by the expert. The same is roughly true of knowledge-based systems where the knowledge is not as rare or as complex as "expert" knowledge. In either case, the system must evolve naturally and be tested often. Rapid-prototyping is evolving and maturing as evidenced by the recent technical literature addressing the topic in all areas of software development [Cielsak et. al. 1989, Duke et. al. 1989, Floyd 1988, Frankfelt 1989, Gupta et. al. 1989, Jordan et. al. 1989, Lewis et. al. 1989, Luqi-1 1989 and Luqi-2 1989].

With rapid-prototyping, more time is spent on requirements definition then with a sequential approach but less time is spent making changes because of missed or incorrect requirements. The process is shown schematically in Figure 1. In essence, a number of prototypes are created which address key issues for the particular system under development. Each prototype is a refinement of the last which fulfills some of the users needs and clarifies key issues. Finally, a complete set of requirements evolves at which time detailed system design (in abbreviated form) begins. Rapid-prototyping helps to avoid the cascading effect of missed or incorrect requirements which are orders of magnitude more costly to correct during code creation than during requirements development. Requirements are better defined because of the so called "Heisenberg Effect" in which the users perceptions of what the system should do (as well as their own roles in interacting with the system) are changed and clarified by their use of the system (i.e. the prototypes). More system is implemented faster and has to be changed less. Phased or incremental introduction with earlier pay off is easier to accomplish. Life cycle issues and the logistics of deploying system enhancements are clearer.

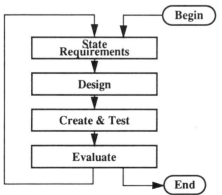

Figure 1. Rapid-Prototyping for Requirements

Recently, a unique rapid-prototyping effort has begun which is closely tied to a busy operations area called the Repair Answer Center (RAC). The RAC acts as a first point of contact for customer problem reports. RACs are strategically located in GTE serving areas across the country and they are an integral part of GTE's continuing effort to maintain and improve customer service and satisfaction. Rapid-prototyping is being employed to define requirements for the entire system (not just the knowledge-based portion) which will involve significant interaction with existing systems and a number of human interface issues. This is part of a continuing effort to improve our software development capabilities.

The focus of the effort is to automate parts of the RAC operation. An initial prototype of a knowledge-based system to assist the report logging and diagnostics functions of the RAC was created. The system was named the Repair Answer Center & Repair Order Logging Expert (*RAC & ROLE*). Two versions of the prototype exist, one for direct customer input (using a computer voice system) and another with a screen oriented approach (i.e. operator input with prompts). In the remainder of this paper, a summary of system features and implementation issues is given.

2. *RAC* OPERATION

When a customer calls the RAC, they can log a trouble report, determine the status of a previously reported problem and sometimes define a more appropriate place to direct their call. Organizationally the RAC is divided into two main groups to handle business and residential customers with a separate phone number provided to each group via automated call distribution splits.

When a trouble call is received at the RAC, the clerk assigned to the call attempts to direct the conversation with a set of key questions in an effort to quickly identify the problem. Today, these questions are printed on a series of five *que cards* which reflect a high level diagnostic search tree. New clerks use these cards extensively, relying on them less heavily as experience is gained. Usually, RAC personnel start with about two weeks of training in residential analysis and trouble reporting after which they work with residential calls for about six to nine months. Then, another two week training period in business trouble analysis and reporting is followed by about six to nine months of handling business customer problems. Generally, after this period, RAC personnel move on to more responsible jobs leaving vacancies which are filled with new people.

Each RAC clerk has a terminal connected to a number of mainframe data base systems as shown in Figure 2. The Service Order Record Computer Entry System (SORCES) is a data base which contains the name, address, type of service and other pertinent information on GTE business and residential customers. Generally, the first question a customer is asked is the phone number they are reporting. The clerk enters this number via the terminal making an inquiry to SORCES which yields appropriate customer information. For example, during the diagnostics it is useful to know if the customer is reporting their own number in trouble (called a first party report) or another number which they were trying to call (referred to as a second party report). There is a third type called a relay report in which an individual other than the owner of the line makes the report, but the possible trouble is identical to first party reports. This might occur if a customer's phone is dead and a neighbor makes the report for them.

For first party (and relay) reports, an attempt is made to determine problem ownership or the likelihood that the customer is experiencing network difficulties as opposed to problems with

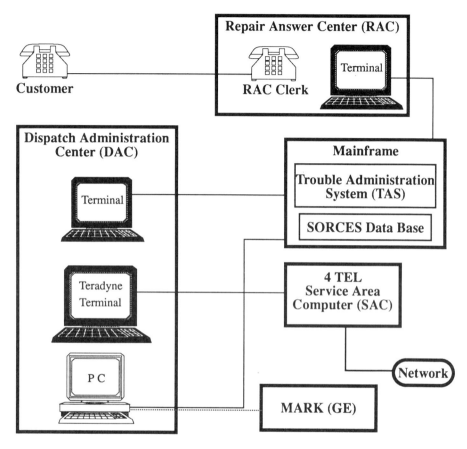

Figure 2. Repair Answer Center (RAC) & Dispatch Administration Center (DAC) Today

their own equipment. This aspect is normally covered sufficiently with one question (such as "Are you having a problem with all of your phones?"). Then the actual problem diagnosis takes place. It proceeds much like a breadth-first search through a group of about eight specific difficulties falling into three main categories in which the customer is unable to call, unable to be called or has transmission problems. Normally the customer simply states the difficulty in a sentence or two with little or no prompting from the RAC clerk. Once the trouble type is identified, further information is gathered to assist in the repair. For example, in the case where a customer complains of a dead phone with no dial tone, it is useful to know what they hear (there being a number of possibilities including a busy signal or static). If the customer does not know the answer to any of the questions, the response is reported as such but the report is still issued. The assumption is always made that the problem exists with the network unless customer responses indicate otherwise.

For second party reports, the problem choices are much more limited due to the second party nature of the report. For example the calling party gets no audible ring or always reaches wrong numbers when calling the number being reported. Again, further information is gathered to abet the repair. A last problem category (other) includes anything which does not fit above. This category includes such things as cable cuts, billing inquiries or problems with customer owned equipment.

After taking the report, the customer is informed of a repair deadline (based on current backlogs and the nature of the problem). There can be a negotiation on the repair time if it is unacceptable to the customer. Finally, the problem information is entered into another data base called the Trouble Administration System (TAS). Various code numbers are entered for typical problem types. Other information is also entered in the appropriate fields on the screen. Most of the problem codes appear on the que cards along with a set of acronyms identifying the gist of the problem. For example, the code CBDT means Can't Break Dial Tone, i.e. the customer takes the telephone off hook, gets dial tone, dials but dial tone continues during the dialing (or touching) action.

Report entry into TAS can be complicated by the pre-existence of other reports on the same phone number. There are four possibilities, permutations of two report types, second party and first party (with the understanding that relay reports are essentially equivalent in this context). Briefly, if the existing report is first party (or relay) it remains until cleared but it can be appended with further information. If the existing report is second party, it can be replaced with a first party (or relay) report or appended by a second party report.

The TAS data base is also accessed by another group which is responsible for dispatching repair personnel. This group resides in the Dispatch Administration Center (DAC) also shown in Figure 2. Reports in TAS entered by the RAC are acted upon via the DAC ultimately resulting in a repair of the problem. When the repair is made, the DAC personnel update TAS reflecting the completion of the work. RAC personnel handle customer inquiries on previously reported problems and can update customers on the progress of the repair based on the input from the DAC. In essence, TAS acts as a communication link between the two groups.

Also available to the DAC personnel are two other systems which provide support for further diagnosis and repair of trouble. One of these is 4TEL which is available from Taradyne Inc. 4TEL accesses the local exchange switching equipment and provides line tests which determine a number of common faults and potential problems on customer lines such as "no ringers." Some isolation capability is also available. 4TEL resides on its own computer system called a Service Area Computer and is currently accessible only through a Taradyne terminal. However, emulation software (and hardware) is available to access 4TEL from a PC.

The other system available to the DAC is called the Mechanized Assignment and Record Keeping (MARK) system which is time-shared from General Electric Information Services Corp. This data base contains information which is useful in isolating and repairing faulty equipment such as facilities inventory data and service path details as well as availability lists for assignment of telephone numbers, originating and miscellaneous equipment. The latter are used in the provision of new service.

The benefits of implementing a knowledge-based system in the RAC are many. They include direct cost savings in training and reductions in the number of reports to be processed. Value added benefits are also possible such as taking and forwarding reports for other agencies. Finally, improved customer perception is likely and the approach may enhance possible RAC center consolidations.

3. PROTOTYPES

Based on initial discussions with operations personnel, two versions of a first *RAC & ROLE* prototype were created, one for direct customer input (using a computer voice system) and another with a screen oriented approach (i.e. operator input with screen prompts). Each is a knowledge-based system using about 120 rules. These were written on an IBM PC/AT (640 Kilobytes of RAM running MS DOS) in LEVEL 5, an inexpensive but very capable expert systems shell from Information Builders, Inc. [IBI 1989] and Microsoft C [Microsoft 1989] from Microsoft Corporation. The PC/AT was selected because of availability and because it is being considered as a possible implementation platform. Voice portions for the direct customer input prototype were created with TI Speech, a voice processing system introduced in 1985 by Texas Instruments. The SORCES and TAS data bases were realistically simulated in dBASE III from Ashton Tate. A simulated link to 4 TEL was also provided. Such a connection is currently not available to the RAC. Both prototypes were designed and created in two weeks by two people full time with another person about half time to create the data bases. This effort included complete user documentation so that our software customers would be able to *play* with the systems as part of their evaluation effort.

Knowledge acquisition for these prototypes was accomplished in one three-hour meeting with an experienced RAC manager. It was possible to achieve this rapid pace because of the nature of the RAC operation, the existence of the diagnostic que cards (which were used extensively in the development of the knowledge bases and inferencing procedures), and the capability of the RAC manager who had a good feel for what we were trying to accomplish and who was extremely enthusiastic about the project. Also, team members had extensive experience in knowledge acquisition techniques from previous projects [Kosieniak et. al. 1988, St. Jacques 1988 and St. Jacques et. al. 1989] thus helping to minimize the time required. Follow-up consisted of a half-day visit to the RAC where we spent time with experienced clerks getting a good feel for the inferencing techniques and the types of problems they encounter.

3.1 Screen Prompted Operator Input Prototype

In this prototype, screen prompts provide questions for the RAC clerk to ask the customer who is on the phone thus guiding the clerk through the diagnostics and generation of the required trouble report. Figure 3 illustrates the functional architecture. Four LEVEL 5 knowledge bases (KBs) were developed (INITIAL, DETPROB, DISPLAY and MODDB) with approximately 120 rules. In addition, four C programs were created. The *goal outline* feature of LEVEL5 was used to create a hierarchical rule structure thus guiding the inferencing. Very simple goal outlines were used in all knowledge bases. As an example, the goal outline for DETPROB (DETermine the PROBlem) is:

1. gather customer information
   1.1 get customer phone number
      1.1.1 chain to appropriate knowledge base
      1.1.2 refer customer to phone repair company
2. end knowledge base gracefully

Since LEVEL5 uses backward chaining, this outline gives procedural goal direction. Goal 1. is attempted first and if it is satisfied, the inference engine tries each of its sub-goals in order. If goal 1. cannot be satisfied then an attempt is made to satisfy goal 2. This process continues

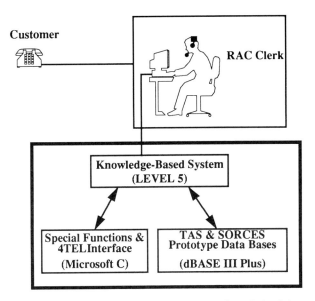

Figure 3. Prompted Operator Input Functional Architecture

until one of highest level goals (in this example 1. or 2.) has been completely satisfied.

Another feature of LEVEL 5 which was used extensively was knowledge base chaining. This feature allows an operational KB to bring a different KB into active memory as needed. Thus a fairly large knowledge area can be broken down into logical segments for design and maintenance. More importantly, it allows access to a large body of knowledge on a PC with only 640 Kilobytes of RAM.

The first knowledge base, INITIAL, gets basic customer background information via SORCES access, determines the type of trouble report (initial, second party or relay) and determines weather the problem is most likely caused by customer equipment or the network. If the problem appears to be with customer equipment, the customer is referred to an appropriate repair company. For network trouble, INITIAL chains to either DETPROB or DISPLAY. If this is a new trouble call or pending trouble is to be updated then DETPROB is used. If no update can be done on a pending report then DISPLAY is used to give the customer status on the existing trouble report.

DETPROB is the major diagnostic KB in the system. Its sole responsibility is to determine the type of network problem and gather information necessary to complete the trouble report. Once the type of network problem is determined, a 4 TEL line test is simulated with a C program. Based on the results of the test (randomly generated) a repair completion time is assigned to the trouble report. A final screen provides the repair clerk with a summary for the customer. Incorporation of the previously unavailable 4 TEL line test is a major enhancement allowing earlier resolution to the problem and greatly improved customer perception.

After the customer is off-line, DETPROB chains to MODDB which updates the TAS pending report database. Automating the TAS transactions is a major improvement because the RAC clerks no longer have to know how to fill out the TAS screens nor do they have to keep track of the detailed procedures of reporting. If this is a first time report then the report information is appended to the database. If there is already an pending report then that report is updated with the new information. Once this is completed MODDB chains to DISPLAY which displays the new or updated TAS pending report. After displaying the pending report MODDB chains to INITIAL and the system is ready for another customer call.

The four C programs were called when appropriate by the various LEVEL 5 KB's. Two of the programs provided system time and date for the report time and date and the promised repair time. Another simulated the connection to 4 TEL and the last provided string concatenation (which LEVEL5 lacks).

3.2 Direct Customer Input (Voice) Prototype

In the second prototype, the human interface exists directly with the customer via the TI-Speech system as shown in Figure 4. The TI Speech system, introduced in 1985, consists of an

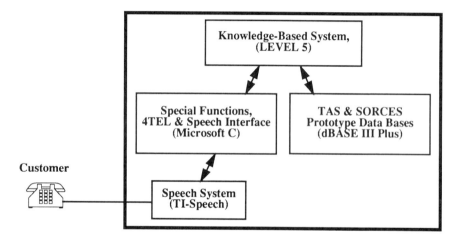

Figure 4. Direct Customer Input Functional Architecture

add-on board for the PC/AT with a standard telephone connection for input/output and software to recognize customer input touch-tone signals (e.g. responses to questions such as 1 for *yes*, 0 for *no*, or a telephone number). Software also supports the creation of scripts which are stored as binary files (digitized speech samples similar to PCM), and the "playing" of scripts when required. The TI-Speech board also answers the phone and disconnects as necessary. Speech input and output routines callable from LEVEL 5 were written in C. Scripts were generated and stored on hard disk in 46 files using 1.5 Megabytes.

The direct customer input prototype used the INITIAL and DETPROB KB's mentioned above. This was possible because the knowledge was essentially the same for both. The human interface differed significantly in the design of the questions. Because of the constraints imposed by touch-tone customer response, many more steps were often required (than the number required in the screen prompted prototype) to reach a conclusion regarding the ultimate source of customer problems.

4. SUMMARY

A number of tests and demonstrations of the *RAC & ROLE* prototypes have been conducted with favorable response. The conceptual design appears to meet the perceived system requirements as defined to date. The flexibility and expandability of current *RAC & ROLE* implementations will allow easy continuation of the rapid-prototyping effort. Requirements are beginning to take shape and we are confident that they will be finalized rapidly as the effort continues. A time frame of a few weeks to iterate a new prototype with complete documentation appears to be acceptable. It is rapid enough to maintain interest and focus of attention yet allows enough time to create something useful. Although we found this to be a reasonable time frame at our experience level, it would probably need some lengthening for those newer to the field.

LEVEL5 was easy to use and allowed rapid creation of the KBs with its English-like rule syntax. The ability to "open" dBASE III Plus files and have access to the database structure made simulating real databases (SORCES and TAS) relatively painless. The capability to call C programs was a major requirement for these prototypes. This capability was provided as part of LEVEL5 (with some assistance from the authors in the area of passing parameters via memory). LEVEL5 debugging capabilities such as examining and changing facts, examining active rules and the ability to see the current line of reasoning are all useful and make KB creation and maintenance straight forward. LEVEL5's *Goal Outline* feature allowed us to structure the KB's in a hierarchical manner therefore making the rules easier to write, understand and maintain. Chaining between KBs was easy and allowed creation of a sizable system (relative to the 640 Kilobyte PC limitation) on a PC/AT.

There are several shortcomings to the tool. Certain string fields in the TAS pending database had to be appended and concatenated (built incrementally). Since this capability was not provided within the shell, a simple C program had to be written performing these types of functions. In addition, LEVEL5's easy question asking facility is limited because only one question can be presented to the user at a time. Finally, external C programs are difficult to debug because when the executable is called from LEVEL 5, the debugger is unavailable.

The TI-Speech board and software used for this project were adequate to show functionality, but they are 1985 technology. For future versions of these prototypes requiring speech capability, an investigation into the current state of this technology will be necessary. Judging from the literature [The Dialogic People, 1990], there is no shortage of information on this popular topic.

As this project evolves, a number of technical issues are surfacing which fall into the three categories of final software functionality, interfaces to existing systems and implementation platform. Without going into extensive detail here, it appears likely that all of these issues are fairly straight forward and resolvable with simple prototypes similar to those previously discussed combined with economic studies. In future prototypes, a number of other features will be demonstrated such as combining direct customer and operator input (i. e. leave direct customer input via switch-out box on request or as necessary), handling large business customers with Centrex, PBXs and leased lines, custom calling features and special services. Also targeted for investigation are inclusion of on-call prompting for after-hours emergencies and inclusion of local serving area demographics.

BIBLIOGRAPHY

Cieslak R., Fawaz A., Sachs S., Varaiya P., Walrand J. and Li A. 1989. "The Programmable Network Prototype System" *Computer*, Vol. 22, No. 5, May, 1989. IEEE Computer Society. pp. 67 - 76.

Authorship shown as *The Dialogic People*, 1990. "So you Wanna Be In Voice Processing?" *Teleconnect Magazine*, Vol. 8, No. 2, February, 1990, Telecom Library, Inc., pp 70-85.

Duke E., Brumbaugh R. and Disbrow J. 1989. "A Rapid Prototyping Facility for Flight Research in Advanced Systems Concepts" *Computer*, Vol. 22, No. 5, May, 1989. IEEE Computer Society. pp. 61 - 66.

Floyd C. 1988. "Outline of a Paradigm Change in Software Engineering", *Software Engineering Notes*, Association of Computing Machinery (ACM), 13(12)25 - 38 April, 1988.

Frankfelt, C. 1989. *Briefing Paper 89-3T: The Diebold Information Technology Scan, 1989: Software Development Technologies - A Management Perspective.* The Diebold Research Program, pp. 5 - 6.

Gupta Rajiv., Cheng W., Gupta Rajesh., Hardonag I. and Breuer M. 1989. "An Object Oriented VLSI CAD Framework, A Case Study in Rapid Prototyping" *Computer*, Vol. 22, No. 5, May, 1989. IEEE Computer Society. pp. 28 - 37.

I. B. I. 1989. *LEVEL 5 Expert System Software, PC Version*, Information Builders, Inc., New York, NY.

Jordan P., Keller K., Tucker R. and Vogel D. 1989. "Software Storming, Combining Rapid Prototyping and Knowledge Engineering" *Computer*, Vol. 22, No. 5, May, 1989. IEEE Computer Society. pp. 39 - 48.

Kosieniak, P., Mathis, V., St. Jacques M. and Stevens D. 1988 "The NETWORK CONTROL ASSISTANT (NCA), a Real-time Prototype Expert System for Network Management." *Proceedings of the First International Conference on Industrial & Engineering Applications of Artificial Intelligence & Expert Systems IEA/AEI-88.* published by the Association for Computing Machinery (ACM). pp. 367-377.

Lewis T., Handloser III F., Bose S. and Yang S. 1989. "Prototypes from Standard User Interface Management Systems" *Computer*, Vol. 22, No. 5, May, 1989. IEEE Computer Society. pp. 51 - 60.

Luqi. 1989. "Rapid Prototyping Languages and Expert Systems" *IEEE Expert*, Vol. 4, No. 2, Summer 1989. IEEE Computer Society. pp. 2 - 5.

Luqi. 1989. "Software Evolution Through Rapid Prototyping" *Computer*, Vol. 22, No. 5, May, 1989. IEEE Computer Society. pp. 13 - 25.

Microsoft Corporation. 1987. *Microsoft C 5.1, Optimizing Compiler for the DOS Operating System, Users Guide.* Microsoft Corporation.

St. Jacques, M. 1988. "An Intelligent Telephone Switch Interface for a Real-time Network Control Expert System." *Artificial Intelligence in Engineering: Robotics and Processes.* Computational Mechanics Publications. Southhampton. pp. 371-385.

St. Jacques M., Stevens D., Mathis, V. and Kosieniak, P. 1989. "*GENESIS*, a Real-Time Expert System for Network Control", *Proceedings of the Network Management and Control Workshop.* Published by the IEEE Communications Society. pp. 156-167.

Tanik, M. 1989. "Rapid Prototyping in Software Development" *Computer*, Vol. 22, No. 5, May, 1989. IEEE Computer Society. pp. 9 - 10.

Texas Instruments. 1985. *TI-Speech, Library.* Texas Instruments Incorporated.

Automated Knowledge Acquisition from Network Management Databases

R. M. Goodman[a] and H. Latin[b]

[a]Department of Electrical Eng., California Institute of Technology, Pasadena, CA 91125

[b]Systems Technology, Pacific Bell, Concord, CA 94520

Abstract

In this paper we describe a technique for automatically learning rules from network management databases. Our motivation for this is to alleviate the knowledge acquisition bottleneck inherent in developing expert systems for integrated network management. We outline our ITRULE rule induction algorithm, show how useful rules can be extracted from trouble ticket and alarms databases, and show how these rules can be automatically loaded into a standard expert system shell, thus virtually instantly producing a prototype expert system.

1. INTRODUCTION

As part of an on-going collaborative project (NETREX) between Caltech and Pacific Bell aimed at producing real-time expert system modules [1], we have been faced with the problem of developing rules via the traditional techniques of knowledge acquisition. This is a very time consuming process in terms of human resources, particularly expert availability. We have therefore investigated various automated knowledge acquisition techniques aimed at speeding up this process. In particular we have been concerned with the automated induction of rules from network management databases. These databases include trouble ticket databases, alarms databases, and topology databases. This area of learning from examples is referred to as machine learning, and a number of statistical and neural network algorithms exist that enable rules or correlations between data to be learned. In this paper we outline our approach to automated rule induction via our own algorithm ITRULE (Information Theoretic Rule Induction). The ITRULE algorithm possesses a number of significant advantages over other algorithms in that the rules that are generated are ranked in order of informational priority or utility. It is thus an easy matter to directly load the rules into a standard expert system shell (such as NEXPERT), utilize an inferencing scheme based on these rule priorities, and have a working expert system performing inference in a matter of minutes. We have implemented the ITRULE suite of programs on a number of platforms (Sun, Mac, PC), and linked these into a number of expert system shells (NEXPERT, KES). This approach means that the expert system developer can "instantly" generate and run a tentative expert system with little domain expertise. This "bootstrap" expert system can then be used to refine the rules in conjunction with the domain expert in a fraction of the time of traditional "cold" question and answer knowledge acquisition techniques.

2. RULE INDUCTION AND UTILITY MEASURES

The motivation for defining rule utilities arises from the desire to provide a more quantitative and rigorous theory for both rule induction (acquisition and learning of rules) and rule-based inference (reasoning using the learned rules).

Let us consider the problem of induction first. The desire to transform such databases into a condensed set of rules demands as a prerequisite the ability to rank rules in some manner. Hence we arrive at the need for a clearly defined rule-preference measure or a rule utility.

Previous work in this area has focussed on either qualitative rule modeling (e.g., Michalski [2]) or else has artificially restricted the nature of the solution to that of a classification or decision tree, usually via the ID3 algorithm or its variants (e.g., Quinlan [3]), or the CART algorithm [4].

It is worth emphasizing that our approach is fundamentally different from these classification and tree-oriented approaches. The problem we are trying to solve is that of *generalized rule induction*. In fact classification is a special case of our approach. In essence, generalized rule induction involves the induction of rules relating *all* attributes and is not just restricted to symptom-class rules. In this manner a model using such a rule set can in principle reason towards any prescribed goal or attribute from any given initial conditions. We feel that this level of generality is a requirement in many domains where the system must exhibit common-sense reasoning. Classification-oriented rule sets (including tree structured rule sets as a special case) do not possess this generality. In domains where the inputs (observable data) and outputs (the goal or class) do not change, a predetermined hierarchy of rules is an appropriate solution, and indeed decision trees have proved remarkably efficient for many such problems [5].

However in some domains, such as network management, all the evidence or attribute variables may not always be observable, i.e., we can only observe a subset of the attributes for any given problem instance. For example, it may not be possible to conduct a particular device test because of the network fault, and we may have to "reason around" the unavailable test. In these circumstances decision trees are severely limited. The ability to deal with such variable inputs is a reasonable requirement of an intelligent reasoning system. In summary, our approach to rule induction can be viewed as a more general and direct method than previous work in this area.

The second motivation for the idea of rule utilities originates from the desire to improve the modeling of rule-based *inference*, in particular the aspect of conflict-resolution or control. Conflict resolution techniques as previously developed in the literature have been primarily qualitative rather than quantitative in nature. While this approach is well motivated it leads to a brittleness and domain-specificity when applied to real problems. By defining a fundamental measure of rule utility, which is in some sense theoretically correct, we can implement a very general and implicit control scheme that automatically incorporates both forward and backward chaining as implemented in most expert system shells.

3. THE ITRULE INFORMATION THEORETIC APPROACH

Consider the problem of quantifying the utility or goodness of a particular rule, where a rule is considered to be of the form

$$\text{If } \mathbf{Y} = y \text{ then } \mathbf{X} = x \text{ with probability } p$$

For our purposes we may treat **Y** and **X** as discrete-valued or categorical attributes. Expressions of the form $\mathbf{Y} = y$ are attribute-value assignment statements, where y is an element of **Y**'s alphabet. Note that $\mathbf{Y} = y$ can be a conjunctive Left Hand Side (LHS), i.e. A = a & B = b. The probability p is simply the conditional probability $p(x|y)$, that is, the probability of the rule right hand (RHS) side given that the rule left hand side is true. While acknowledging that other parameters such as certainty factors or likelihood ratios are often used, we consider the conditional probability as the most basic and well-established rule-belief parameter. Note that we are primarily interested in probabilistic rules, that is, there is an implicit uncertainty in the rule. "Factual rules" where $p = 0$ or 1 are a special case of the theory. This belief parameter p alone is not sufficient to measure a rule's utility. Clearly the utility must be influenced by such factors as the probability $p(\mathbf{Y} = y)$ which reflects the average probability that the left-hand side of the rule will evaluate to true, i.e., that the rule will fire.

Let us adopt an information theoretic approach to this problem and define the information which the *event* y yields about the variable **X**, say $f(\mathbf{X}; y)$. Based on the requirements that $f(\mathbf{X}; y)$ is both non-negative and that its expectation with respect to **Y** equals the Shannon average mutual information $I(\mathbf{X}; \mathbf{Y})$, Blachman [6] showed that the *only* such function is the j-measure, $j(\mathbf{X}; y)$. More recently we have shown that $j(\mathbf{X}; y)$ possesses unique properties as a rule information measure [7]. In general the j-measure can also be interpreted as a special case of the cross-entropy or binary discrimination (Kullback [8]) between the LHS and RHS probability distributions. We further define $J(\mathbf{X}; y)$ as the *average* information content where $J(\mathbf{X}; y) = p(y).j(\mathbf{X}; y)$. $J(\mathbf{X}; y)$ simply weights the instantaneous rule information $j(\mathbf{X}; y)$ by the probability that the left-hand side will occur, i.e., that the rule will be fired. A rule with high information content must be both a good predictor and have a reasonable probability of being fired, i.e., $p(y)$ can not be too small.

The J-measure is then the basis of our rule utility measure in that it defines the average number of bits of information in the Shannon sense that we obtain when the rule is "fired". The final step in developing the general utility measure is to incorporate a cost term $c(y)$. This allows us to incorporate the subjective cost of measuring the LHS variables. The final utility measure used is then $U(xy) = J(\mathbf{X}; y) - c(y)$. This utility measure can then be considered a "goodness" measure relative to the "best" rule.

4. THE ITRULE ALGORITHM

We have developed a suite of algorithms collectively called the ITRULE algorithm [9] which uses the J-measure to derive the most informative set of rules from an input data set. The algorithm takes as input a sample set of features vectors where each of the features is discrete-valued.

As output the algorithm produces a set of K probabilistic rules, ranked in order of decreasing utility. The parameter K may be user-defined or determined via statistical significance tests based on the size of the sample data set available.

For each right-hand side of interest the algorithm begins by first considering first-order general rules, i.e., rules with a single variable on the left-hand side. It then proceeds to search for more informative higher order rules which are specialized versions of the original rule, by adding terms to the left-hand side in a depth-first manner. The search is made highly efficient by using information theoretic criteria to guide the search and constrain the search space. A ranked list is kept so that new rules may be compared with the utility of the rule currently in the Kth position in the list. Rules which exceed this threshold are inserted in the list, those that are not are rejected.

ITRULE uses information theoretic search criteria and small sample statistics to very quickly come up with a candidate set of rules. Furthermore these rules are directly output into the format of the expert system shell in use (NEXPERT, KES in our case) to give an almost instant prototype expert system.

The ITRULE software also has facilities for both manual and automated rule editing, in order to optimize the set of rules to a number of user selected criteria, for example: accuracy, simplicity, lowest cost, self-consistency, etc. In addition, the ITRULE software implements a number of routines for processing continuous attribute data (such as time) into categorical data as required by the algorithm. These routines are manual, information theoretic, statistical, and neural network inspired. Thus for example given a particular time attribute, ITRULE can indicate that it makes sense to categorize the attribute's values into bins of $\leq 5mins, \leq 30mins, and \geq 30mins$.

5. RULE-BASED INFERENCE WITH UTILITY MEASURES

The utility measure associated with each rule is then used in the expert system shell to perform conflict resolution. For example in NEXPERT a priority value is associated with each rule, and after collecting candidate rules which can fire (their LHS's are true) the rule with the highest priority is actually fired. ITRULE associates two utility values with each rule: one for forward chaining based on the j-measure, and one based on the J-measure for backward chaining. In this way ITRULE provides a complete "instant" prototyping system for an expert system shell.

6. EXAMPLES USING NETWORK MANAGEMENT DATABASES

We now describe some of the applications of ITRULE to analyzing network management databases within the Caltech - Pacific Bell collaboration. In Figure 1. we show an ITRULE analysis of a Pacific Bell Network Management trouble ticket database. The database logs trouble reports on a large (25,000 terminals) distributed data network. The number of trouble reports on such a large network run into the hundreds per day. In this case the ITRULE analysis is being used as a data summarizing tool and the rules are being very effectively used for the benefit of humans trying to understand the way in which troubles are closed out, and how long it takes to do so. The top half of figure 1. shows a small portion of the trouble ticket database (the actual historical database is of course huge). The trouble database (and the input format that ITRULE needs) is in the form of a spreadsheet where each line is a trouble record. The attributes (or fields) of the database refer to the equipment ID, the tests that were made in isolating the trouble, who the trouble was referred to, the final close out category, and who fixed the fault. The middle portion of Figure 1. shows the attributes which correspond to the data fields above them. Note that most attributes are categorical, apart from "time - down" which in the raw data is a continuous variable. This attribute was manually categorized by Trouble Center experts who were interested in resolution times of greater or less than one hour. The ITRULE process was directed to produce rules about how long it took to close out troubles, and who actually fixed the problem. The lower part of Figure 1. shows a portion of the ITRULE analysis, giving the 13 most informative rules extracted from the data. The first numeric entry in each rule line is the probability p, the "correctness" of the rule. The last entry is the utility measure (equivalent to the relative J-measure in this case). The rules reveal the fundamental statistics of trouble closeouts in a very human readable way. For example, rule 2 shows that if faults are software based they almost always fixed by the highly skilled testers, alternatively, rule 8 shows that the simpler one terminal call-fail problems are often fixed by the less technically specialized trouble call screeners (usually by resetting the control unit - which is revealed in a rule further down the list). Rule 11 reveals that if there are hardware problems with 4540A terminals then these take over an hour to fix (they involve a referral and a site callout). Note that within the context of this trouble ticket domain rule 1 is an "obvious" rule. When troubles have been referred to the OCS group, then if that group solves the problem they also always fix the problem. The ability of ITRULE to extract these "obvious" rules is of great benefit when bootstrapping a prototype expert system. One of the biggest problems with the conventional technique of interviewing experts is that they have difficulty adjusting themselves to describe the elementary rules of their domain, that is the $2 + 2 = 4$ type rules. These rules are therefore often laborious to extract, however, the ITRULE approach easily finds these rules.

We now show a more complex example in which ITRULE is used to analyze a time varying network alarms database. The objective is to automatically develop rules for a higher level expert system whose output is a (real time varying) prioritized list of the most important network alarms. This alarms list is then presented to the network administrators and gives a real time picture of the most important problems in the network. In order to achieve this we need an expert module that understands the relation between line alarms, so that when presented with the latest alarms evidence it can predict the likelihood of other related alarms occurring, and thus compute a postulated severity measure for each alarm that has occurred. For example, if we can learn that the occurrence of a particular sequence of alarms usually indicates that a serious outage may occur in the near future, thus affecting a large number of users, we can assign a high priority to fixing the detected alarms. The top part of Figure 2. shows a small portion of the raw alarms database. Note the volume of alarms, approximately one every 30 seconds - it is impossible for any human to get an overall picture of the network from such volumes of data, or to understand the relation between alarms. This raw data is then preprocessed into a the spreadsheet form (as in Figure 1.) suitable for ITRULE input. To do this we consider the behavior of each network line over a period of several hours. The attributes of the problem are than the number of alarms, the number of stations with alarms, and the actual alarm types, etc. Each example (row) in the input spreadsheet (not shown) thus refers to the history of a particular line in terms of the type of alarms experienced over the observation period. The lower half of Figure 2. lists the attributes used, and gives an ITRULE analysis of the alarms data, showing a portion of the most important rules. Note that we are not interested in human readability here - just in automatic rule production. The binary value 1 indicates the occurrence of a particular alarm. Note again the "obvious rule", rule 1, which indicates that if a CU fail occurs then a CU restoral will occur soon after.

Figure 3. shows a NEXPERT display of the alarm rules. These rules were generated by ITRULE as in Figure 2. and saved in a NEXPERT loadable file. The alarm expert system shown was generated in a few minutes, and is ready to perform inference. The top screen in Figure 3. shows a portion of the rule network, showing the complex alarm inter-relations. The lower left portion of Figure 3. shows an overview of the several hundred rules generated by ITRULE. The lower right shows a rule being displayed in the NEXPERT rule editor.

7. CONCLUSIONS

The ITRULE process we have outlined provides a valuable tool for automated knowledge and rule induction in the prototyping of expert systems. In particular ITRULE is able to produce meaningful alarm correlations, and could be used to identify higher-level "events" consisting of sequences of alarms. Using ITRULE the knowledge engineer can quickly generate a fully functioning alarms expert system, to significantly improve interaction and knowledge acquisition with the expert. We continue to develop ITRULE applications for real time network management.

8. ACKNOWLEDGMENTS

This work is supported in part by Pacific Bell, and in part by the Army Research Office under Contract No. DAAL03-89-K-0126.

9. REFERENCES

1. R. M. Goodman, J.W. Miller, P. Smyth and H. Latin, 'Real Time Autonomous Expert Systems in Network Management,' *Proceedings of the first IFIP International Symposium on Integrated Network Management*, Boston, May14-17, 1989.
2. R. S. Michalski and R. L. Chilausky, 'Learning by being told and learning from examples', *International Journal of Policy Analysis and Information Systems* 4, pp.125-161, 1980.
3. J. R. Quinlan, 'Induction of decision trees,' *Machine Learning*, vol. 1, 81–106, 1986.
4. L. Breiman, J. H. Friedman, R. A. Olshen and C. J. Stone, *Classification and regression trees*, Belmont, CA: Wadsworth, 1984.
5. R. M. Goodman and P. Smyth, 'Decision tree design from a communication theory standpoint,' *IEEE Trans. Information Theory*, vol 34, no. 5, 979–9 94.
6. N. M. Blachman, 'The amount of information that y gives about X,' *IEEE Transactions on Information Theory*, IT–14 (1), 27–31, 1968.
7. P. Smyth and R. M. Goodman, "The information content of a probabilistic rule," submitted to *IEEE Trans. on Information Theory*, June 1990.
8. S. Kullback,*Information Theory and Statistics*, New York: Wiley, 1959.
9. P. Smyth and R. M. Goodman,'Deriving rules from databases — the ITRULE algorithm,' to be published *IEEE Trans. on Knowledge and Data Engineering*, 1990.

	A	B	C	D	E	F	G	H	I	J	K	L	M	N	O	P	Q	R	S	T
1	TROUBLE TICKET DATABASE																			
2	D0	428	40	4540A	?	CUF	ALLDEV	?	OK	?	?	?	?	?	TERM	CU	RC	CS	CLIENT	under1hr
3	D0	444	C1	TC174	?	CFL	ONE_DEV	?	OK	?	?	?	?	?	TERM	CU	RC	CS	CS	under1hr
4	B0	11	C4	4540A	NO_CO	?	?	OK	NOK	?	?	?	PCO	?	NO TRBL	?	?	?	?	over1hr
5	B0	02D	C1	TC174	NO_CO	CUF	?	?	?	?	?	?	PCO	?	NO TRBL	?	?	?	?	over1hr
6	B0	43E	40	4540B	NO_CO	?	?	?	?	?	?	?	PCO	?	NO TRBL	?	?	?	?	over1hr
7	B0	04D	40	TC174	NDE_DWN	CUF	ALLDEV	?	?	?	?	?	TESTER	?	TELCO	LINE	?	?	?	under1hr
8	B0	57	40	4540A	?	CUF	ALLDEV	?	NOK	NOK	?	?	TESTER	OCS	TELCO	LINE	?	TESTER	PCO	over1hr
9	B0	20C	40	SCC	?	CLF	ALLDEV	?	?	?	?	?	TESTER	?	NO TRBL	?	?	?	?	over1hr
10	B0	46E	C1	4540A	?	CUF	ALLDEV	?	?	?	?	?	OCS	?	TERM	CU	HRDWR	OCS	OCS	over1hr
11	B0	618	C3	4540A	S/R	?	?	?	NOK	?	NOK	TESTER	?	TCU SWR	CU	SFTWR	TESTER	TESTER	under1hr
12	B0	65F	C1	INTRMCU	?	?	?	?	?	?	?	?	OCS	?	TERM	D/S	LOOPBK	OCS	OCS	over1hr
13	B0	65F	40	TC174	?	CUF	ONE_DEV	?	?	?	?	?	OCS	?	DATA SET	D/S	HRDWR	OCS	OCS	over1hr
14	B0	65F	40	TC174	INTRMCU	?	?	?	?	?	?	?	OCS	?	TERM	CU	HRDWR	OCS	OCS	over1hr
15	B0	666	C1	4540A	?	CUF	ALLDEV	NOK	?	NOK	?	?	TESTER	?	DATA SET	D/S	CONFIG	TESTER	CLIENT	under1hr
16	F0	18	C1	4540A	?	CUF	ALLDEV	?	NOK	?	?	?	PCO	?	TELCO	TCXR	HRDWR	PCO	PCO	under1hr
17	D0	58	40	4540A	ALL_DWN	CFL	SOME_DEV	?	NOK	?	OK	?	TESTER	?	TCU SWR	LINE	SFTWR	TESTER	TESTER	under1hr
18	D0	439	40	3274	TRM_DWN	?	?	?	?	?	?	?	OCS	?	TERM	CU	HRDWR	TESTER	OCS	over1hr
19	F0	04C	40	554612	?	LMF	?	?	?	?	?	?	TESTER	?	TCU SWR	LINE MOD	SFTWR	TESTER	TESTER	under1hr
20	F0	51	40	4540A	?	CUF	ALLDEV	?	?	?	?	?	TESTER	?	TERM	CU	HRDWR	OCS	OCS	over1hr
21	F0	O6A	40	4540A	ALL_DWN	?	?	NOK	OK	OK	?	?	TESTER	?	TCU SWR	LINE	SFTWR	TESTER	TESTER	under1hr
22	F0	O6C	40	4540A	ALL_DWN	CUF	ALLDEV	?	NOK	NOK	?	?	TESTER	?	NO TRBL	LINE	?	?	?	under1hr
23	F0	O6C	40	4540A	S/R	CUF	ONE_DEV	?	OK	?	NOK	TESTER	?	MUX	MUX	RESET	TESTER	NCG	under1hr
24	F0	266	40	4540A	ALL_DWN	CUF	ALLDEV	?	NOK	?	?	?	TESTER	PCO	TELCO	DSX JMPR	HRDWR	TESTER	PCO	over1hr
25	F0	268	40	4540A	?	CUF	ALLDEV	?	NOK	NOK	?	?	TESTER	OCS	TERM	CU	HRDWR	OCS	OCS	over1hr
26	F0	41E	40	TC174	?	CFL	ALLDEV	?	NOK	?	?	?	TESTER	OCS	TERM	CU/STN	HRDWR	OCS	OCS	over1hr
27	F0	60D	C1	TC174	?	CUF	?	?	NOK	NOK	?	?	TESTER	OCS	TERM	CU/STN	HRDWR	OCS	OCS	over1hr
28	D0	428	40	4540A	?	CUF	ALLDEV	?	OK	?	?	?	?	?	TERM	CU	RC	CS	CLIENT	under1hr
29	D0	444	C1	TC174	?	CFL	ONE_DEV	?	OK	?	?	?	?	?	TERM	CU	RC	CS	CS	under1hr
30	B0	11	C4	4540A	NO_CO	?	?	OK	NOK	?	?	?	PCO	?	NO TRBL	?	?	?	?	over1hr
31	B0	02D	C1	TC174	NO_CO	CUF	?	?	?	?	?	?	PCO	?	NO TRBL	?	?	?	?	over1hr
32	B0	43E	40	4540B	NO_CO	?	?	?	?	?	?	?	PCO	?	NO TRBL	?	?	?	?	over1hr
33	B0	04D	40	TC174	NDE_DWN	CUF	ALLDEV	?	?	?	?	?	TESTER	?	TELCO	LINE	?	?	?	under1hr
34	B0	57	40	4540A	?	CUF	ALLDEV	?	NOK	NOK	?	?	TESTER	OCS	TELCO	LINE	?	TESTER	PCO	over1hr
35	B0	20C	40	SCC	?	CLF	ALLDEV	?	?	?	?	?	TESTER	?	NO TRBL	?	?	?	?	over1hr

ATTRIBUTES:

```
A  NODE      B  LINE      C  STATION   D  Dev_type     E  SYMPTOM
F  DEV_STAT  G  DEV_no    H  AL_ST     I  RC_CU        J  IL_XIL
K  DL_XDL    L  BCT       M  Refer1    N  Refer2       O  Close_Cat
P  Soln_Dev  Q  Soln_Actn R  Solved_By S  Fixed_By     T  time_down
```

RULES:

Rule Format: IF lhs1 (AND) lhs2.... THEN rhs

Stats: prob(rhs/lhs), j-measure, J-measure, Utility

```
    IF                    AND                        THEN

 1 Solved_By OCS                                    Fixed_By OCS            0.967 0.955 0.379 1000
 2 Soln_Actn SFTWR                                  Fixed_By TESTER         0.912 1.658 0.347  916
 3 Close_Cat TELCO                                  Fixed_By PCO            0.879 1.897 0.285  753
 4 DEV_no ALLDEV         time_down under1hr         Fixed_By CLIENT         0.589 0.863 0.161  425
 5 SYMPTOM NO_CO                                    time_down over1hr       0.923 0.704 0.161  424
 6 Refer1 TESTER         Fixed_By TESTER            time_down under1hr      0.937 0.836 0.143  378
 7 DEV_no ONE_DEV        Refer1 TESTER              Fixed_By NCG            0.690 1.788 0.122  321
 8 DEV_STAT CFL          DEV_no ONE_DEV             Fixed_By CS             0.690 1.788 0.120  317
 9 Solved_By TESTER                                 time_down under1hr      0.763 0.210 0.094  247
10 SYMPTOM ALL_DWN       Soln_Dev LINE              time_down under1hr      0.929 0.629 0.077  204
11 Dev_type 4540A        Close_Cat TELCO            time_down over1hr       0.934 0.558 0.064  167
12 Solved_By CS                                     time_down under1hr      0.900 0.531 0.055  145
13 SYMPTOM NDE_DWN                                  time_down under1hr      0.846 0.309 0.025   65
```

Figure 1. Trouble ticket analysis using Itrule

RAW ALARMS STREAM:

DATE	TIME	NODE	ID	CONDITION
10/28	10:19:49	D2	T061	POL RES 41507
10/28	10:19:40	D0	T049	LINE FAIL 040
10/28	10:19:40	D0	T000	LINE CONDITION ERROR 040 02
10/28	10:19:40	D1	T049	LINE FAIL 270
10/28	10:19:40	D1	T000	LINE CONDITION ERROR 270 02
10/28	10:19:40	D2	T049	LINE FAIL 457
10/28	10:19:40	D2	T000	LINE CONDITION ERROR 457 02
10/28	10:19:34	D0	T013	EBCDIC STATUS *4050* 01914
10/28	10:19:34	D0	T008	CAL FAIL NAK 01914
10/28	10:19:00	D2	T000	BINS 0549,EMBN 1518,MG/M 1554,MRAT 0021
10/28	10:18:00	D3	T000	BINS 0602,EMBN 1548,MG/M 1371,MRAT 0022
10/28	10:17:45	D3	T010	CAL RES 64A00
10/28	10:17:00	D1	T000	BINS 0575,EMBN 1531,MG/M 2088,MRAT 0034
10/28	10:16:45	D0	T049	LINE FAIL 02A
10/28	10:16:44	D0	T052	LINE OUT 02A
10/28	10:16:31	D2	T010	CAL RES 45710
10/28	10:16:19	D2	T010	CAL RES 4570A
10/28	10:15:54	D2	T010	CAL RES 45702
10/28	10:14:18	D3	T011	CU 00 FAILED 64A
10/28	10:14:00	D0	T000	BINS 0522,EMBN 1550,MG/M 2024,MRAT 0030

ATTRIBUTES:

#stats.w/alms	#alms	#unique.alms
CAL.RES	CU.01.FAIL	CU.00.RES
EB.ST:C250-prntr.cvr	EB.ST:4050-dev.nav	LINE.MOD.INOP
AHP.E(FACS).LINK.OK	T060	T015
POL.RES	LINE.OUT	EB.ST:40C2-tx.err
CAL.FAIL.DC	CAL.FAIL.NAK	INTR.NODE.LNK.FAIL
T018	CAL.FAIL.NR	INTR.NODE.LNK.OK
T048	LINE.FAIL	POL.FAIL.DC

RULES:

	IF		AND		THEN		UTILITY
1	CU.01.FAIL	1			CU.00.RES	1	1000
2	CU.00.RES	1			CU.01.FAIL	1	857
3	CAL.FAIL.NAK	0	CAL.FAIL.NR	0	CAL.RES	0	710
4	CAL.FAIL.NAK	0			EB.ST:4050-dev.nav	0	645
5	EB.ST:4050-dev.nav	1			CAL.FAIL.NAK	1	597
6	CAL.RES	0	CU.00.RES	1	CAL.FAIL.NAK	0	570
7	#alms	under10			#stats.w/alms	1	556
8	CAL.RES	0	EB.ST:4050-dev.nav	0	CAL.FAIL.NAK	0	549
9	CAL.RES	0	INTR.NODE.LNK.OK	0	#unique.alms	2	491
10	CAL.RES	0	LINE.MOD.INOP	0	#unique.alms	2	489
11	#stats.w/alms	1	#unique.alms	2	#alms	under10	476
12	CAL.FAIL.NAK	1	CAL.FAIL.NR	0	EB.ST:4050-dev.nav	1	457
13	#alms	under10			#unique.alms	2	455
14	#stats.w/alms	>1	CAL.FAIL.DC	1	#unique.alms	4+	441
15	#stats.w/alms	1	CAL.FAIL.NAK	0	#alms	under10	416
16	#stats.w/alms	1	EB.ST:4050-dev.nav	0	#alms	under10	413
17	#unique.alms	3			CAL.RES	1	413
18	CAL.RES	1	EB.ST:4050-dev.nav	0	CU.01.FAIL	0	413
19	#unique.alms	2			CAL.FAIL.NR	1	410
20	CU.00.RES	1	EB.ST:C250-prntr.cvr	0	#stats.w/alms	1	385
					EB.ST:4050-dev.nav	0	382

Figure 2. Alarms analysis using Itrule

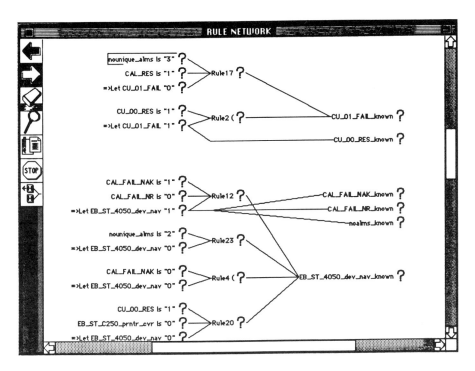

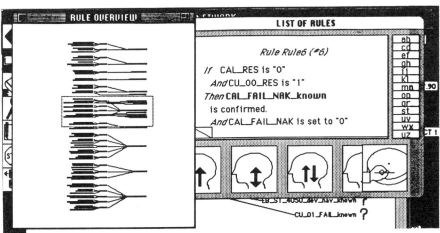

Figure 3. Rules automatically loaded into expert system shell

VII

DISTRIBUTED SYSTEMS MANAGEMENT

Open Distributed Processing and Open Management

P.F. Linington

Computing Laboratory, University of Kent, Canterbury, Kent CT2 7NF, United Kingdom

Abstract

This paper describes the progress to date on the definition of a new set of standards for Open Distributed Processing and examines their relationship to the specification of distributed systems management. It identifies the Reference Model for Open Distributed Processing as a suitable framework for future work on management.

1. Background to ODP

The construction of distributed or interconnected systems in a multi-vendor environment depends on the creation of suitable standards for the required behaviour of the components which make up the systems.

The OSI Reference Model has provided a framework for the standardization of interconnection, but its scope is limited to precisely that - the interconnection of pairs of systems. It has been an essential tool in the creation of a family of basic communications standards, but it does not solve the problem of distribution, largely because of its concentration on peer-to-peer communication rather than description of an arbitrary configuration of objects.

ISO has recognized the standardization of distributed systems as of growing importance, and in 1987 it approved a New Work Item on Open Distributed Processing, aimed at creation of a Reference Model which could integrate a wide range of standards for distributed systems and solve the problem of consistency across such systems, allowing the specification of global properties rather than just behaviour at specific interfaces.

Shortly after the work on ODP started, CCITT also initiated a new study question on a Framework for Distributed Applications (DAF, CCITT QVII/19). Work on this question was carried out in cooperation with ISO from the start, and joint working on the main parts of the Reference Model has now been agreed and is in progress.

2. ODP Scope

Distributed systems take many different forms, and there is no hope for the provision of individual standards tailored to all the different application areas. Indeed, to attempt to create such a spectrum of standards would be misguided, because, as systems evolve, new reasons for integration continue to arise. This process will be greatly simplified if the underlying model used to describe the systems concerned is the same.

The architecture produced needs to cover many different aspects of the operation of a system. It needs to coordinate the functions which make up the system but it must not overly constrain the vendors of distributed system components by specifying irrelevant local detail of their implementations or by specifying unnecessarily features which are internal to them.

To this end, the ODP Reference Model identifies several types of interface at which standardization may be required, and places constraints only at and between these interfaces. They are

a) programmatic interfaces allowing access to a defined function; an example might be the standardization of database languages;

b) man/machine interfaces, as for example in some graphics standards;

c) interconnection standards, like the familiar OSI standards for communication between systems;

d) external physical storage media specifications, to allow information exchange between systems.

For the first time, the ODP Architecture allows the general expression of requirements for consistency between the human interface, the programming interface which supports the application and the OSI protocols which convey information between systems. Specific models in areas such as graphics cover some aspects of this interrelationship, but use of a general architecture will greatly facilitate the combination of standards from a number of different areas.

What is expected of a distributed architecture such as this? Firstly, it must make clear the terminology and underlying concepts which it is to use. This establishment of a descriptive basis is essential if there is to be a consistent interpretation of the resulting standards, but does not, in itself, constrain the types of systems that are to be built. Secondly, the architecture must make specific choices. It must identify features and prescriptions which are to characterize the family of systems eventually built to comply with the architecture. The structure of the Reference Model being produced by ISO reflects this distinction between descriptive and prescriptive requirements.

3. Structure of the Reference Model

The Basic Reference model of Open Distributed Processing will be based on precise concepts and, as far as possible, on the use of formal description techniques for specification of the architecture. It will contain a range of precise definitions and supporting material of a more tutorial nature. However, these two types of material will be separated into normative and non-normative parts, so as to make it clear what obligations the standard implies. There will also be a division between the descriptive parts which introduce basic concepts and specification techniques, and prescriptive parts which state the design choices which characterize the particular architecture chosen for ODP. The model will be published in five parts:

a) Part 1: **Overview**: contains a motivational overview of ODP, giving scoping and explanation of key definitions (with no substantial architectural content), and an enumeration of required areas of standardization expressed in terms of the reference points for conformance identified in Part 3. This part is not normative.

b) Part 2: **Descriptive model**: contains the definition of the concepts and analytical framework and notation for normalized description of (arbitrary) distributed processing systems. This is only to a level of detail sufficient to support Part 3 and to establish requirements for new specification techniques. This part is normative.

c) Part 3: **Prescriptive model**: contains the specification of the required characteristics that qualify distributed processing as open. These are the constraints to which ODP standards must conform, and they are expressed in terms of five viewpoints and the generic functions defined there in. Part 3 uses the descriptive techniques from Part 2. This part is normative.

d) Part 4: **User model**: contains a description of the resulting ODP environment from the users' point of view. This part contains explanatory material on how ODP is intended to be viewed by system engineers designing distributed applications to be run in the ODP environment described in Part 3. This part is not normative.

e) Part 5: **Architectural semantics**: relates the concepts established in part 2 to the standard formal description techniques, so that there is a uniform interpretation of the specifications produced, even when several different techniques are in use together. This part is normative.

4. Approach to modelling

The general approach taken is object-oriented. Systems are modelled in terms of the interactions of a set of component objects at identified interfaces. To aid the identification of common functions, the objects are organized into a family of classes related by subclass to superclass relationships. This general approach is used repeatedly in the various viewpoints described in the next section, but the kinds of objects identified vary from the concrete component of a communications subsystem to the abstract entities in an information or enterprise model.

The architectural concepts needed for the description of ODP systems are constructed from the more basic object-oriented concepts. The architectural concepts include specific kinds of structuring, such as logical organization into coordinated groups or domains, identification of obligations (using the concept of a contract) and temporal organization via shared contexts and liaisons. Specific concepts are also developed to deal with particular aspects such as security or distributed management.

Part 2 of the Reference Model will be sent for ballot from the next international meeting in May 1991.

5. Viewpoints

Rather than attempt to deal with the full complexity of a distributed system, we consider the system from different viewpoints, each of which is chosen to reflect one set of design concerns. Each viewpoint represents a different abstraction of the original distributed system, without the need to create one large model describing it.

It then becomes necessary to consider the framework which organizes these viewpoints. This framework allows verification of both the completeness of the various descriptions and the consistency between them. Depending on the level of detail, the resulting description will also allow the operational characteristics of the described system to be studied.

Informally, therefore, a viewpoint leads to a representation of the system with emphasis on a specific concern. More formally, the resulting representation is an abstraction of the system; that is, a description which recognizes some distinctions (those relevant to the concern) and ignores others (those not relevant to the concern).

The current work on ODP recognizes five viewpoints. These are as follows:
a) the enterprise viewpoint, which is concerned with business policies, management policies and human user roles with respect to the systems and the environment with which they interact; the use of the word enterprise here does not imply a limitation to a single organization; the model constructed may well describe the constraints placed on the interaction of a number of distinct organizations;
b) the information viewpoint, which is concerned with information modelling, providing a consistent common view covering information sources and sinks and the information flows between them;
c) the computational viewpoint, which is concerned with the algorithms and data structures which provide the distributed system function;
d) the engineering viewpoint, which is concerned with the distribution mechanisms and the provision of the various transparencies needed to support distribution;
e) the technology viewpoint, which is concerned with the detail of the components and links from which the distributed system is constructed.

6. Transparencies

The specification of distributed systems requires well structured descriptive tools which allow the various information resources involved to be separated clearly, and unpredictable side effects of interaction to be avoided. This has lead to the identification of a series of fundamental transparencies, each representing a system property which may or may not be preserved in a particular distribution mechanism.

The concept of transparency has much wider applicability than simply to the modelling of ODP systems. Transparency results from the normal process of abstraction and is found in many areas outside distribution. Consideration in ODP is limited to the requirements for various forms of distribution transparency.

A transparency mechanism is placed between a user (a user may be any general object in a model) and a system and enables the user to look through and not perceive some parts of the potentially changing behaviour of the system. The transparency mechanism may be provided by the "transparency related objects".

a) **Access transparency**: The property of hiding from a particular user the way access actually takes place.
b) **Location transparency**: The property of hiding from a particular user where the object being accessed is located.
c) **Address transparency**: The property of hiding from a particular user the way an object is actually addressed. This concept is closely related to location transparency and hides differences which may exist between local and global addressing schemes and changes that may be made to addressing schemes.
d) **Migration transparency**: The property of hiding from a particular user that the object being accessed has changed its location. This concept is related to the concept of location transparency.
e) **Configuration transparency**: The property of hiding from a particular user the way in which the object being accessed is constructed.

f) **Replication transparency**: The property of hiding from a particular user the existence of replicas of an object or of the replication strategy used by the system.
g) **Fault transparency**: The property of hiding from a particular user the occurrence of faults in the system. The user is not aware of faults occurring in the system, nor is it aware of any recovery actions taken.
h) **Concurrency transparency**: The property of hiding from a particular user the existence of concurrency in the system. IP i) **Heterogeneity transparency**: The property of hiding from a particular user specified aspects of the technological variety in the way different parts of the system used are

7. Structures and Functions

The next major milestone for ODP is the creation of a first draft for the prescriptive part of the model, which will set the style for the eventual systems produced on the basis of ODP standards. Much of the preparatory work has now been done, and the next step is to reorganize the material into a suitable form for it to be a published standard. Proposals for the organization of this part of the standard were produced in the most recent ODP meeting, and a consolidated draft is expected from the next meeting.

Computational Model

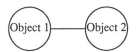

Enginerring Model

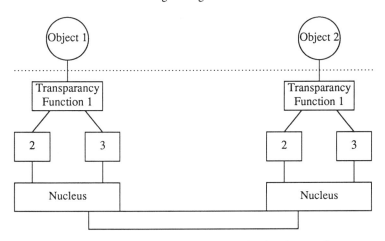

Computational and Engineering views of object interaction

The architecture will include the basic features of the five viewpoint models. For example, the computational model will define the way objects can interact, giving the facilities required from a Remote Procedure Call standard, and will establish the mechanisms by which objects are created and when they can safely be discarded, providing the means by which complex system configurations are built up. These definitions then form the basis for the identification of generic functions needed to support a computational specification, such as the notion of an object factory or a trader of service offers.

In contrast, an engineering model will be much more concerned with how the interactions of the computational model are actually to be realized. Thus it will be concerned with the boundaries between physical systems and the identification of basic infrastructure components, such as a nucleus supporting communication and with the provision of any transparencies assumed in the computational view.

8. Generic Components

Once the architecture is clear, there will be a need for a number of supporting standards giving the specifications of components called for by the set of viewpoint models. Components will be needed to provide transparencies and control object interactions. For example, there may be a number of security related functions, such as authentication units or key managers.

The first generic function to be sufficiently well identified for detailed work to begin is the Trader. It provides the function of maintaining knowledge of the services currently available and matching client requirements with suitable service offers. A service provider wishing to make its services generally available exports the specification and a reference to the service for addition to the trader's records. Any potential client attempts to import the reference to the service by making a request in terms of the specification it requires. This function is central to the flexible configuration and late binding that are expected of resilient distributed systems and will form an essential part of configuration management. It seems likely that the trader will draw on the services of a more basic location service, such as the X.500 directory service, when communication outside the local management domain becomes necessary.

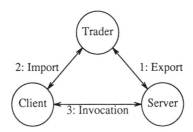

The trader function

The first technical documents on the trader and its services have recently been prepared, and it is expected that approval for a new standard in this area will be given officially during 1991. Work items in other areas will follow rapidly as the architecture matures.

9. Conformance

No architecture is complete without a clear statement of how and when it is to be applied. This raises the question of conformance, and the ODP Reference Model addresses the meaning of conformance as part of its basic set of concepts.

The truth of a statement in an implementation can only be determined by testing and is based on a mapping from terms in the specification to observable aspects of the implementation.

At any specific level of abstraction, a test is a series of observable stimuli and events, performed at prescribed points known as reference points, and only at these points. These reference points are accessible interfaces. A system component for which conformance is claimed is seen as a black box, testable only at its external linkages. Thus, for example, conformance to OSI protocol specifications is not dependent on any internal structure of the system under test.

The RM-ODP identifies certain reference points in the architecture as potentially declarable as conformance points in specifications. That is, as points at which conformance may be tested and which will, therefore, need to be accessible for test. However, the requirement that a particular reference point be considered a conformance point must be stated explicitly in the conformance statement of the specification concerned.

Conformance is a concept which can be applied at any level of abstraction. For example, a very detailed perceptual conformance is expected to a standard defining character fonts, but a much more abstract perceptual conformance applies to screen layout rules.

The more abstract a specification is, the more difficult it is to test. An increasing amount of implementation-specific interpretation is needed to establish that the more abstract propositions about the implementation are in fact true. It is not clear that direct testing of very abstract specifications is possible at reasonable cost using currently available or foreseeable techniques.

There are three major roles in the testing process. These are:

a) the specifier, who constructs a complete specification. To be complete it must contain:

  1) the behaviour of the object being standardized (the declarative statements) and the way this behaviour must be achieved (the imperative statements).

  2) a list of the primitive terms used in the specification when making the imperative and declarative statements.

  3) a conformance statement indicating the conformance points, what implementations must do at them and what information implementors must supply (corresponding to the OSI notions of PICS and PIXIT).

b) the implementor, who constructs an implementation on the basis of the specification. The implementor must provide a statment of a mapping from all the terms used in the specification to things or happenings in the real world. Thus the real interfaces corresponding to the conformance points must be indicated and the representation of signals given. If the specification is abstract, the mapping of its basic terms to the real world may itself be complex. For example, in a computational viewpoint specification, the primitive terms might be a set of interactions between objects. The implementor wishing to confirm to the computational viewpoint specification would have to indicate how the interactions were provided, either by reference to an engineering specification or by providing a detailed description of an unstandardized mechanism (although this course limits the field of application of the implementation to systems in which there is an agreement to use the unstandardized mechanism).

c) the tester, who observes the system under test. Testing involves some shared behaviour between the tester and the system under test. If this behaviour is given a causal labelling, there is a spectrum of testing types from

1) passive testing, in which all behaviour is originated by the system under test and recorded by the tester;

2) active testing, in which all behaviour is originated by the tester.

Normally, the specification of the system under test is in the form of a interface, as is the specification of the tester and test procedures. These interfaces are bound together when testing takes place.

The tester must interpret its observations using the mapping provided by the implementor to yield propositions about the implementation which can then be checked to show that they are also true in the specification. The testing process succeeds if all the checks against the specification succeed. However, it may fail because

a) the specification is logically inconsistent or incomplete, so that the propositions about the implementation cannot be checked (this should not occur);

b) the mapping given by the implementor is logically incomplete, so that the inconsistent or observations cannot be related to terms in the specification; testing is impossible.

c) the observed behaviour cannot be interpreted according to the mapping given by the implementor. The behaviour of the system is not meaningful in terms of the specification, and so the test fails.

d) the behaviour is interpreted to give terms expressed in the specification, but these occur in such a way that they yield propositions which are not true in the specification, and so the test fails.

10. Application of ODP to creation of a Management Model

The main features of the ODP Reference Model have now been introduced. How can they be used to describe the management of distributed systems?

The ODP approach to modelling satisfies a number of the requirements of distributed systems management. It allows for separation of concerns and thereby promotes an orderly and coherent approach to the design process. First, the enterprise viewpoint brings together the various policy issues upon which the design is to be based. It identifies key choices which have to be made, such as whether the management information required from the system is expected to be complete or merely a statistically representative sample. It also establishes the relative costs of various resources and eventualities to give a basis for later engineering choices.

The first stage in the construction of the management model itself is the creation of a single information schema for the resources to be managed and for the management subsystems constructed. It is crucially important that there should be a consistent interpretation of information; a distributed system in which there is a different meaning attached to data by the system gathering readings from that applied by the system performing analysis of it is bound to exhibit some anomalous behaviour.

It is not, of course, necessary for all components in the distributed system to handle either all possible data items, or all the nuances of any single item, but the individual views must be at least derived consistently from some underlying scheme. The basic tools of inheritance refinement and the creation of a class lattice allow the different requirements to be correlated. The identification of the need for such an information view makes it less likely that errors of interpretation will be made.

If there is a clear view of the information being handled, attention can be turned to the other major aspect of the design, the division into objects and the specification of their behaviour. At first, it is enough to consider objects as functional elements which interact with one another. It is neither desirable nor necessary to relate these to the physical boundaries of the system, since there may well be a need for the actual placement of components to change as the system evolves. The emphasis is, therefore, on the identification of coherent objects and interfaces between them, and on the statement of the behaviour of these objects in terms of events observable at their interfaces and correlation between the events at different interfaces. These specifications form a computational description, based on a standardized description of the invocation of one object by another.

At present, the primary emphasis is on the event as the primitive element of behaviour, but this is not the only approach to be taken. There may also be a need for behavioural description in terms of continuous activities, such as the continuing operation of an audio or video channel stream in a multimedia network. Both fall within the scope of the ODP Reference Model.

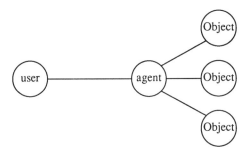

A Computational Model for OSI Management

Once a computational model has been established, we know the configuration of the system components, what information flows take place and what significant events the distributed activity is marked by. It is then necessary to say how the individual steps in the computational viewpoint are to be achieved. The creation of an engineering viewpoint description now turns away from the over all structure of the distributed application and concentrates on how the computational building blocks like invocation and location are achieved.

The engineering model deals with the provision of mechanism to support and interpret the computational steps. It is here that a required levels of transparency must be provided, releaving the computational modeller of concern with the many difficult problems of distribution. Doing this job completely, however, is expensive and demanding, and in some cases undesirable. Thus for an accounting application, location transparency is a definite advantage, but for a management application, knowledge of the location at which control is exercised is of the essence. However, the detailed way in which the information flows from that location, or the positioning of the various pieces of function involved in exercising the control are not of concern.

Transparency is therefore required to a degree and in specific areas which are characteristic of the problem to be solved. The statement of the desired transparencies (or rather the acceptable or desired departures from transparency) are an important part of the computational model; what remains is an engineering problem, and the solutions form the required engineering descriptions.

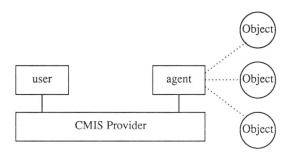

An Engineering Model for OSI Management

Finally, there remains the problem of identifying the most appropriate building blocks for the engineering solution from the available toolkit of standard or proprietary components. The technology viewpoint is basically the catalogue and data book used to select the most appropriate components. It is concerned with the suitability, compatibility and performance of related families of solutions.

11. Conclusions

This paper has outlined the approach to the description and standardization of distributed systems being followed in the work on Open Distributed Processing. It provides a framework within which a growing family of ODP standards will be positioned and an approach to the design and construction of such systems which will allow the construction of coherent but flexible distributed systems.

The management of distributed systems is one particular aspect of distributed systems design. As such it presents particular challenges in its requirements for flexibility, but it is still basically re-using the same techniques and methods found in any distributed system, but is applying them to the control of the infrastructure of the system itself. It is therefore not surprising that the Reference Model for Open Distributed Processing can be applied to the description of distributed systems management, or that it simplifies the design and analysis of such systems.

There will, in future, be an increasingly fertile exchange of ideas and requirements within the standards activity between those concerned with an architectural framework for distributed systems and those with particular interest in their management.

VII

DISTRIBUTED SYSTEMS

MANAGEMENT

A
Conceptual Aspects

Using RPC for Distributed Systems Management

Lorenzo Aguilar

Hewlett-Packard, IAG[1]
1906 Pruneridge Ave 46G
Cupertino, CA 95014
USA

Abstract

This paper proposes to exploit RPC in order to unify the three primordial forms of IPC in distributed systems management: invocation of management operations, distribution and servicing of event reports, and access to management information. It is argued that RPC can provide management applications a unified access to managed resources, to management information, and to distributed, management, and presentation services. The unification bridges different programming models exposed by the three types of services for: invocation, naming, exceptions, persistency, and extensibility. It is shown that RPC can support an elegant realization of the managed object paradigm, scoping of operations over large resource groups, and consolidation of management information. Advanced features can support access control and low overhead implementations.

1. Introduction

Distributed computing environments involve clients and servers that will use remote procedure calls (RPC) as the primary access to distributed services and information. Since management applications can remotely invoke the services implemented by managed resources, RPC can become the primary means for accessing both distributed and management services, and operational and management information. Such unified access is a major step towards leveraging distributed services for distributed systems management, and towards managing the services in a distributed system.

---

1. The views expressed in this paper are those of the author, and they should not be construed nor interpreted as representing an official position, or product commitment from the Hewlett-Packard Co.

This paper argues that RPC provide a uniform access to operational, management, and presentation service interfaces, through which applications can monitor and control resources, retrieve management information, and interact with users. A uniform access to the three types of services alleviates many problems faced by management applications writers, who have to deal with different programming models exposed by the three types of services for invocation, naming, exceptions, persistency, and extensibility. The differences can be bridged through an RPC programming paradigm that presents a uniform model to applications and that is transparently mapped to the diverse models of mechanisms implementing the services. Such a paradigm is sustained by the RPC: semantics, presentation services, run time support, and stub generation facilities.

These programming model differences can be exemplified in an environment consisting of an operating system and services provided by the Open Software Foundation (OSF) technology offerings. The OSF/1 OS is the operating system, the OSF Distributed Computing Environment (DCE) provides the distributed services, the OSF Motif tool kit provides the user interface basis, and the managed resources support interfaces conforming to the Open Systems Interconnection (OSI) Structure of Management Information (SMI[2]). In this environment, management applications must be insulated from dissimilarities between the invocation, naming, registration, extensibility, persistence, and exception models of OSF/1, DCE, Motif, and OSI SMI, such as:

- Invocation of DCE Concert Multithread Architecture threads, SMI actions, or Motif widgets
- Names of SMI managed object instances, or Motif widgets
- Extensibility of SMI managed object classes, or Motif widgets[3]
- Reception and handling of OSI SMI notifications, or Motif input events

The DCE RPC is a key service for bridging the differences above, and this paper examines how the bridging can be accomplished in client-server computing environments in general.[4] The examinations is from the standpoint of critical issues for distributed systems management, namely: application programmatic interface (API), operations scoping, access to management information, event reporting and servicing, interoperation with standard management protocols, access control, and low-overhead implementations. Relevant issues requiring further investigation are also noted, and the paper ends by presenting conclusions.

Remote procedure calls provide programming-language level IPC based on the request-response paradigm that is pervasive in client-server computing environments. The request-response IPC is typically supported over datagram protocols, but is also supported over

---

2. ISO SMI conformance has been mandated by the OSF Request for Technology for a Distributed Management Environment [22].
3. The extensibility models are based on strict and non-strict class inheritance respectively.
4. Note that a practical realization of a client-server environment with available technology will not erase all the model differences. Moreover, there will be management applications that need to be aware of some differences, and the use of RPC should not remove that need.

connection transports. RPC presentation services hide from applications differences in data representations among client and server machines, and marshall (linearize) data structures—as required for network transmission and for dereferencing pointers across address spaces. Most RPC presentation services are supported through client and server stubs, which typically are generated from descriptions of server interfaces and associated parameter and reply data. Additional features integrated into some RPCs include: multicasting, group requests and responses, request and response forwarding, encryption, and authentication. Readers who want a comprehensive introduction to RPC or their use in distributed computing are referred to [7][16][27].

2. Application Programmatic Interface

The emerging RPC de-facto standards for client-server computing can support a high-level programmatic interface that realizes the standards managed object approach. Such an interface can expose to applications a metaphor of invocation of remote services, such as the actions in an OSI SMI managed object. The metaphor is an elegant and powerful realization of the managed object abstraction, and it insulates applications from underlying management services platforms. The suitability of the remote management service metaphor has been recognized in significant standards work, such as the ongoing definition of an API for systems management within the 1003.7 Posix work [14]. The realization of the managed object approach over RPC merges object-oriented design and client-server computing, both of which have been identified as the key design technologies for managing networks during the 90's [25].

Prototype systems in universities and Industry have demonstrated implementations of a remote method invocation metaphor over RPCs, for example [24]. In addition, RPC does not imply a peer-to-peer IPC, and thus its semantics could include direct interactions between a management application (client) and a managed object (server).

A systems management API should expose a consolidated view of management operations and information, and it should expose a level of abstraction higher than a small-grained programmatic interface exposed by individual managed resources. The principal API must expose access to distributed and management services through which resources and management information are accessed. The resource programmatic interface must be accessible to applications that want to directly monitor and control individual resources, as well as be accessible to the distributed and management services through which management operations are scoped over large resource populations, and through which management information is retrieved and fused. The differences in abstraction level and scoping between the two types of programmatic interfaces have been acknowledge by vendors [29] and by emerging Industry standards [22]. The programmatic interface exposed to applications has been acknowledged to be critical to management applications development and evolution [22], and it strongly impacts, portability and extensibility.

The RPC presentation services enable application generality, portability and evolvability by hiding differences across SMIs and management protocols. Applications for multi vendor environments will have to manage resources whose management interfaces conform to the Internet Engineering Task Force (IETF) SMI, as well as resources conforming to the OSI SMI, and also will evolve, as resources manageability conformance transitions. These SMIs exhibit substantial differences in the templates for definition of managed objects, naming of the managed objects, derivation of new categories of objects, and event reporting.[5] Moreover, the OSI and IETF SMIs exhibit dependencies on the CMIP and SNMP management protocols. Applications coded to such SMIs lose generality and have to include code for handling low-level protocol-data-unit details exhibited by the protocols programmatic interfaces.

RPC's separation of presentation services from transport permits independence of the programmatic interface from the communications transport. On the one hand, a particular SMI can be supported with client and server stubs that can map management interface operations semantics to RPC semantics, and that implement an invocation model suitable to a particular programming language. On the other hand, the transport part of the RPC can maintain an SMI and language neutral request-response paradigm enhanced with multicasting, and asynchronous invocation.[6]

Customarily, both client and server stubs are generated by a single compiler and thus interoperability is ensured. In multi vendor environments, a client has to interoperate with a server independently implemented. Agreements will be required for a consistent mapping between the SMI semantics and the protocol data units at both ends, less interoperability be lost. The dependency of the current SMIs on the management protocols ensures interoperability, without a need for further agreements.

From the above, it follows that an RPC programmatic interface insulates applications from both an SMI and management protocol. The RPC transport service convey client data and server responses without any interpretation of the data semantics. The data transparency is an step towards supporting invocation of management interfaces independently of the underlying transport protocols, which the author believes is a desirable goal.

3. Operations Scoping

Integrated management of a distributed system is facilitated by operations scoping and management information consolidation, which bode well with the RPC request-response communications paradigm and both of which are key enablers towards attaining Manager-of-Manager control hierarchies. Distributed computing environments involve aggregations of resources such as resources configured as services, or resources under the authority of an administrative domain. It is required to monitor and control such aggregations as a unit because the number and diversity of the resources preclude an step-and-repeat approach. Accordingly, there should be management services that let a management application mul-

---

5. The SMI model differences occur within the management services, and are in addition to model differences throughout the three types of services identified in the Introduction.
6. It has been experimentally demonstrated that OSI SMI managed object services can be invoked from a C++ application environment [3]. Also, it has been argued that SmallTalk methods can be invoked from a C++ application environment [24].

ticast an operation over all resources in an aggregation, and that deliver to the application a consolidated reply of the operation execution. Applications will request operations over aggregations and over classes of resources and will not know how many replies to expect, nor for how long to wait for them because of possible resource unavailability or unreachability.

Operation scoping and collation of replies are two problems for which the RPC paradigm has advantages over point-to-point, connection communications protocols, like CMIP, and over unreliable-datagram protocols, like SNMP. Some RPCs support multicasting to group addresses [8], whereas others have been integrated with location services and could rely on directory facilities for resolving multicast identifiers to addresses [12] Accordingly, there is a need for underlying mechanisms supporting the resolution of multicast identifiers to individual addresses, such as the Internet Protocol's Host Group Multicasting [10]. Experimental validation is required in order to assess if the mechanisms meet the needs of management operations addressed to unanticipated multicast groups that may have large and rapidly changing memberships.

Further, RPC client stubs can include support for decomposing a single procedural invocation to multiple invocations over several resources, possibly accessing directory services in order to resolve an invocation into multiple addresses. Client stubs could also collate replies from multiple resources because they fan an operation invocation out to the individual resources, and they can time out outstanding responses based on default criteria or client request parameters. Similarly, the RPC server stubs can include support for fanning out and collating replies to/from multiple managed resources in the same node and forwarding a consolidated reply to the client stub. Stubs map management operations semantics to RPC semantics when collating the replies to a multicast operation addressed to resources, whose number, reachability and availability are unknown at application compilation time, and some of which may not reply.

Client and server stubs can also be generated for interfacing applications to management protocols. Accordingly, stub support for multicasting is not exclusive of remote procedure IPC. Generation of stubs for CMIP interfacing was part of the work reported in [3]. Support of multicast and asynchronous remote operations semantics is discussed in [28]. Programming language support of RPC multicast is discussed in [9]. For a thorough discussion of multicasting in distributed systems see [2].

4. Access to Management Information Repository

RPC could present to applications a uniform paradigm for both effecting operations over resources and retrieving persistent state data from a Management Information Repository (MIR). An application writer can see a uniform programmatic interface that supports operation invocation and information retrieval over an indirection abstraction, such as a managed object instance. An RPC client stub can include code that routes invocations to a manager that implements a management operation or to the service through which the MIR is accessed. Point operations—over one resource—are already similar because of the managed object indirection; system-wide operations can be made similar through common scoping over resource instances and over persistent objects recording resource state data in the MIR.

RPC can sustain the retrieval of management information based on powerful queries defined by predicates on managed object attributes and values. Sophisticated retrieval from the MIR does not belong in a management protocol, but in a data base access protocol such as SQL. Also, support for the semantics of the queries belongs outside the RPC, for example in SQL query processors that use the RPC as a request-response service. Nonetheless, an RPC can support both single or scoped operations over managed objects, such as "get" and "set", as well as database queries. MIR access code generation techniques, as those reported in [18], might be applicable to the support of the aforesaid queries. Note that RPC can support a DBMS with an object-oriented retrieval protocol better than it can support a DBMS with an access protocol based on relations, because the former is closer to the remote invocation paradigm.

RPC can support the consolidation of management information, which is required for managing distributed systems. RPC can support information consolidation from diverse information sources, such as from:

- resources, through event report reception and scoping of polling requests
- the Management Information Repository, through services accessing the MIR
- the user interface, through reception of interactive events
- other applications and other management systems,[7] through application data exchanges

5. Event Reporting

RPC can support distribution of event reports from managed resources to interested management applications and services, as well as the invocation of event handlers. Managed resource implementations can generate event reports and send them via RPC to event managers. Alternatively, the events could be notified as remote calls on handlers registered for handling the events [1]. An RPC service can time stamp event reports, and RPC stubs could perform event filtering before delivering to clients or handlers. Last but not least, RPC can be used for registering applications and services interests in receiving event reports, and for registering event handling subroutines to be invoked upon event report reception. The registration information is needed in order to select the parties for distributing an event notification.

Uniform support of event notifications and management operation requests constitutes a significant advance towards uniform monitoring and control in distributed computing. In fact, notifications from resources and requests to resources are the basic IPC modes for resource monitoring and control. Further uniformity could be presented to applications by

---

7. This case covers customer service requests and operator/technician problem tickets, after these are captured through applications.

uniformly handling event reports and signals from the managed resources, the presentation services, and the operating system.

Experimental validation is required in order to assess if a general purpose RPC service can cope with the generation rate and volume of event reports that can occur in large distributed systems. Timely handling of event reports is crucial to effective resource monitoring and to prompt response to failures. For instance, network node failures can generate hundreds of event reports per second, possibly from multiple sources. Experience with general purpose notification facilities, like MIT's Zephyr [11], suggests that RPC can support event report traffic within acceptable delays. Experimental validation could be attained through the porting of a notification facility, such as Zephyr, over one of the Industry standard RPCs, the OSF/DCE RPC [12] or the Sun RPC [26].

6. Interoperation with Standard Protocols

Some resources will not support RPC communications and thus their management requires means for bridging RPC-based communications and communications based on the standard management protocols. Such bridging can be provided either through management platforms that support several communications stacks, or through proxy agents that receive RPC requests and translate them into standard management protocol messages. Interoperation with standard protocols is mandatory since many networked devices and systems will only support standard protocols for a long time, notably SNMP [4].

Management applications could be coded to a systems-management programmatic interface whose program invocations are converted to calls on the service interface of a standard's management protocol and the corresponding protocol data units. Applications attain generality by coding to a managed object abstraction, leaving to underlying support the translation to standard protocols supported by managed resources.

Three cases of translation can be distinguished depending on where the translation or proxy functions are hosted. In a first case, the managed resource supports an RPC server that receives the remote calls and translates to and from a standard protocol supported by the resources' management interface implementation. In a second case, the resource does not support an RPC server and thus a proxy agent receives remote calls and translates to and from a standard protocol supported by the resource that may be accessed remotely. In a third case, an application is coded to an RPC programmatic interface, but neither the application host nor resource support RPC communications and thus the translation has to occur at the application host.

The SunNet Manager^{tm} product is an example of the first case, whereas third party proxy agents for the SunNet Manager constitute an example of the second [19]. Examples of the third case will emerge as applications coded to an RPC interface, such as the SunNet Manager, are ported to management stations without RPC support. The second case will probably be the most common since many of the managed resources are low-end systems or devices with minimal resources, as typically found in SNMP managed devices.

7. Access Control

Authentication, authorization and encryption can be integrated into an RPC in order to control access to management information and management operations, which will be sensitive in most installations. The RPC run time support can access authentication and authorization servers, and can call encryption libraries. The integration follows the client-server computing model, and it constitutes a general solution well documented in the research literature, for example [6], [8]. The OSF DCE RPC will include an authenticated RPC feature based on Kerberos Version 5, and a Privilege Server for authorization checks based on POSIX-conformant access control lists [21].

8. Low Overhead

The low overhead of RPC communications overcomes low-end system limitations and exploits the high-speed, reliable networks on which distributed computing environments will be based. RPC packages like Sun RPC or the Firefly RPC [23] can run on top of datagram protocols, and have been shown to consume limited processing cycles, while attaining round trips under 10 milliseconds. Experimental RPC supports, like the Versatile Message Transaction Protocol (VMTP) [8], have demonstrated that the request-response communications paradigm can leverage low network latencies, high-bandwidths and low error rates, which will be featured by LANs and optical fiber wide area networks.

VMTP's subsettability points to a direction for minimizing the size of RPC client and server implementations by configuring only needed functionality. Minimal RPC configurations are needed for supporting basic systems manageability, like network boot loading or time services, and to support very low end resources, like sensors. Minimal implementations respond to a major requirement on management systems; it should be possible to host them in low-end resources, and they must not generate overhead processing that impacts resource performance.

The remote procedure execution model lends naturally to server implementations capable of servicing multiple requests and providing fast responses, by exploiting lightweight threads [24], [5]. Multiple requests can be serviced by spawning a thread for each, thereby promptly completing short requests, avoiding IPC overhead and multi threading complexity. Fast response times will be mandatory from management services like the Management Information Repository. It is unknown to the author to what extend SNMP agents and managers accessed through CMIP can take advantage of threads.

9. Further Investigation

This section discusses the application of the client-server IPC paradigm of remote procedures to three problem areas, namely the high cost of developing proxy managers, the need for trivial interactive interfaces to new managed resource types, and the embedding of expert systems within a client-server environment. The ideas in this section are in an early stage of development and require substantial investigation. They ideas are discussed only as potential directions for future research.

The automated generation of client and server RPC stubs may enable the realization of an Extensible Proxy Manager (EPM) for simple monitoring and control of resources supporting only proprietary management interfaces. Figure 1 shows how a management application can manage a proprietary resource through standard operations and event reports, which are translated, by the EPM and proxy stubs, into proprietary operations and reports. The EPM must be hosted in a node with protocol stacks for communications both with the requester of the action, and with the proprietary resource. The EPM relies on client and server proxy stubs that perform the translation, in contrast to conventional stubs that do not translate requests between client and server.

The generation of client and server proxy stubs requires an RPC stub compiler capable of processing interface definitions for a family of proprietary resources. Accordingly, in order to manage a family of resources, a compiler would have to be extended with parsing and code generation capabilities specific to the management architecture for such a family. Experimental validation is required in order to assess the feasibility of such compiler extensions. It is presumed that such extensions represent a smaller cost than the development of individual proxies for each resource in the family.

The EPM could provide a partial solution to the expense of developing proxies for managing resources that support only proprietary management interfaces. It is likely that only a reduced set of actions can be translated through proxy stubs, yet the set may provide sufficient functionality for basic management. Multi vendor distributed systems will include proprietary resources that have to be managed in conjunction with standards conformant resources. Administrators may be willing to settle for limited manageability, in exchange for economies in proxy development.

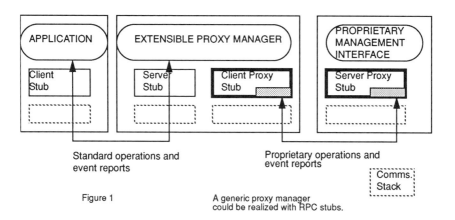

Figure 1   A generic proxy manager could be realized with RPC stubs.

It may be feasible to generate automatically a basic interactive interface from the definition of a new management interface, thereby providing interactive user access to the interface's operations. When a new resource type is introduced into an environment, its management interface could be processed in order to generate linkable client and server RPC stubs through which applications can present and render some of the operations. Applications would effectively be extended with interactive support for a new resource type. Candidate operations are those that can be requested through simple menus or dialog boxes, such as cycling power, or retrieving operational parameter values. Whereas the operations' requests suitable for stub support are simple, it is conjectured that the automated support of trivial interactive requests covers a good portion of the needs for day-to-day systems operation. The two major roadblocks to be overcome are the lack of presentation information in current management interface definitions, and the lack of standards for rendition of management entities. After overcoming them, techniques like constraint-based graphical systems [17] could enable interactive interface generation.

The generation of basic interactive interfaces could make it possible to manage a new resource type without having to modify a management application. An extensible presentation platform server, such as NeWS,^{tm} [13] could be downloaded with a presentation interface directly from a resource that can host a presentation platform client—for example a resource manager. A control console would host the platform's server that can interpret new presentation procedures, downloaded PostScript code for example. A resource management interface would be processed in order to generate the interactive interface, which would be passed to the client after the resource has been configured and its management implementation has been installed. The generation and downloading of the interactive interface can occur without stopping neither the control console or the presentation platform.

Expert system facilities can be embedded within a client-server environment using RPC. Management applications can invoke as servers the expert-system facilities, such as an inference engine. Conversely, the facilities can use RPC in order to access the user interface or the Management Information Repository. Remote procedures can realize a client-server programmatic interface between the application, user interface, and MIR and an inference engine, which thus becomes one more service available to applications [15]. Expert system technology is becoming pervasive in applications like fault isolation and diagnosis, and it is andvantageous to make the technology uniformly accessible under the client-server computing paradigm.

10. Conclusions

The goal of managing distributed computing environments as integrated systems generates needs that call for remote procedure call IPC, which will be the dominant IPC for distributed computing. This paper has proposed the exploitation of RPC for distributed systems management in order to present management applications an unified access to managed resources, management information, and management services. Advanced remote operation features of RPC can be leveraged in order to realize and enhance the standard's managed object approach. Remote calls effectively hide from applications model differences for: invocation, naming, exceptions, persistency and extensibility. Standard management protocols advantages over RPC were noted in some cases, and it is likely that other caveats will emerge, as the application of RPC to systems manageability progresses. Further investigation is required to asses the feasibility of RPC stub generation for low cost and fast development of proxy managers, and of trivial interactive interfaces.

Acknowledgments

The author wants to acknowledge the invaluable technical discussions with multiple members of the Internet community, the Hewlett Packard Co., and in particular the IAG Resource Management Lab. Their insights and expertise greatly contributed to conceive and develop many of the ideas herein reported. Thanks are also due to the technical reviewers and referees of earlier versions of the paper; their suggestions have improved the paper's readability and organization.

REFERENCES

[1] L. Aguilar. NCMA, A Management Architecture that Integrates Enterprise Network Assets. In Proceedings of the IFIP International Symposium on Integrated Network Management, May 1989. Published by North Holland.

[2] Editor Ahamad, M. Multicast Communication in Distributed Systems. IEEE Computer Society Press, 1990. IEEE Computer Society Order Number 1970, Los Alamitos, CA.

[3] J. Balfour, P. Brun, P. Hyland, P. Mellor, J. Read, and P. Toft. Gemini: An Environment for Distributed Network Management Applications. Proceedings of the Second IFIP International Symposium on Integrated Network Management. April 1991.

[4] A. Ben-Artzi, A. Chandna, and U. Warrier. Mind Boggling Complexity. IEEE Network, pages pp 35—43, July 1990.

[5] B. N. Bershad and T. E. Anderson. et. al. Lightweight Remote Procedure Calls. ACM Transactions on Computer Systems ,37—55 ,8(1), February 1990.

[6] A. D. Birrell. Secure Communication Using Remote Procedure Calls. ACM Transactions on Computer Systems, 3(1):1—14, February 1985.

[7] A. D. Birrell and B. J. Nelson. Implementing Remote Procedure Calls. ACM Transactions on Computer Systems, 2(1):39—59, February 1984.

[8] D. Cheriton. VMTP: Versatile Message Transaction Protocol. Version 0.7 available from David Cheriton, Computer Science Dept., Stanford University, February 1988.

[9] E C. Cooper. Programming Lanaguage Support for Multicast Communication in Distributed Systems, Proceedings of the International Conference on Distributed Computing Systems.1990, pp 450-457.

[10] S. E. Deering, Host extensions for IP multicasting. Internet RFC 1112 1989 August; 17. Obsoletes RFC 988, and RFC 1054

[11] A. DellaFera and M. W. Eichin and R. S. French and D. C. Jedlinsky and J. T. Kohl and et. al. The Zephyr Notification Service. ,USENIX Winter Conference Proceedings. February 9-12 1988, Dallas Texas.

[12] Hewlett-Packard Co and Digital Equipment Corp. Network Computing Architecture, Version 2.0.

Hewlett-Packard and Digital Equipment confidential, July 1990.

[13] J. Gosling, D. S. H. Rosenthal, and M. J. Arden. The NeWS Book. Springer-Verlag, 1989.

[14]] IEEE Technical Committee on Operating Systems. System Administration Interface for Computer Operating System Environments. P1003.7/Draft 2. 22 November 1989

[15] F. Lazzeroni and P. Stokely. Embedding Expert Systems. Copyright 1989 Coherent Tought Inc., December 1989.

[16] C. Malamud. RPCs Help Programs Go Places. Data Communications, 19(8):61—70, June 1990.

[17] B. A. Myers and D. et. al. Giuse. The Garnet Toolkit Reference Manuals: Support for Highly-Interactive, Graphical User Interfaces in Lisp. Technical report, Carnegie Mellon U., School of Computer Science, Pittsburgh, PA, March 1990. CMU-CS-90-117.

[18] S. Nakai and Y. Kiriha and Y. Ihara and S. Hasegawa. A Development Environment for OSI Systems Management. Proceedings of the Second International Symposium on Integrated Network Management. IFIP and IEEE CNOM. Published by Elsevier Sciences Publishers. April 1991.

[19] Netlabs. Netlabs CMOT Proxy Agent for SunNet Manager. Technical product description. Netlabs, Los Angeles CA, 1989.

[20] O. Editor Newkerk. Management Services Application Programmatic Interface, Draft. Document Version 1.4, Internet Draft, April 1990.

[21] Open Software Foundation. OSF Distributed Computing Environment Rationale. Open Software Foundation, 11 Cambridge Center, Cambridge, MA, 1990.

[22] Open Software Foundation. Request for Technology, Distributed Management Environment. Open Software Foundation, 11 Cambridge Center, Cambridge, MA, 1990.

[23] M. D. Schroeder and M. Burrows. Performance of Firefly RPC. ACM Transactions on Computer Systems, 8(1):1—17, February 1990.

[24] N. Shacham, J. J. Garcia-Luna-Aceves, and L. Aguilar et. al. Study on Distributed Processing System Topology for Battle Management/C3 Systems. Section 8.3 newblock Technical report, SRI International, Menlo Park California USA, 1987. Final Report prepared for USAF-ASFC, Rome Air Development Center; Contract: F30602-85-C-0186, SRI Project 1041.

[25] K. Shah. Managing Networks of the "90s". Data Communications, pages 127—142, December 1989.

[26] Sun Microsystems Inc. Network Programming. Sun Microsystems, Inc., 2550 Garcia Av.; Mountain View, CA 94043, May 1988.

[27] B. H. Tay and A. L. Ananda. A Survey of Remote Procedure Calls. National U. of Singapore, Department of Information Systems and Computer Science, Singapore, 1990.

[28] Walker, F. E. and Floyd, R. and Neves, P. Asynchronous Remote Operation Execution in Distributed Systems, Proceedings of the International Conference on Distributed Computing Systems. 1990, pp 253-259.

[29] U. Warrier and C. Sunshine. A Platform for Heterogeneous Interconnection Network Management. IEEE Journal on Selected Areas in Communications, 8(1), January 1990.

Dealing With Scale In An Enterprise Management Director

Colin Strutt

Digital Equipment Corporation, 550 King Street (LKG2-2/T2), Littleton, Massachusetts 01460, USA

Abstract
As customer networks expand, we need to be able to deal with the additional complexity caused by the magnitude of the network, in addition to addressing the problems relating to the diversity of components that need to be managed. The approaches to managing a LAN comprised of twenty five PCs and to managing a world-wide enterprise of hundreds of thousands of devices seem to be different.

In this paper we will demonstrate consistent approaches to handling the problems associated with scale in managing an enterprise as defined by the Enterprise Management Architecture (EMA) Director, DECmcc.

1 Introduction

In a traditional network environment, where a few systems were joined together to form a network, the management considerations were straightforward — in general, network management could be ignored. As a network expanded, a certain threshold was reached where "suddenly" it became necessary to employ full-time staff to manage the network. Perhaps it was due to the need to have dedicated personnel to handle the problems that arose; perhaps it was necessary to have trained staff, rather than part-timers; perhaps it was a realisation that the network was becoming a more important part of a company's business, and without a functional, reliable network, a company might not be able to stay in business.

But the available tools did not make the job of a network manager easier. Certainly there were a lot of tools. But each tool provided an island of management — providing some specific management function, for some specific set of network components, with a user interface and a database customised to that environment, and hence inconsistent with the plethora of other tools for managing other components. As the network expanded and new technologies were added into the network, so the number of tools grew. All too often, management of such a complex network meant a row of terminals or workstations for each network manager.

In recent years, much has been written, and systems developed, that address the aspects of complexity in terms of the variety of entities in an enterprise, including various network devices of different technologies, the operating systems, applications and data that make up the enterprise, including [1] [2] [3] [4] [5] [6]. Now we are starting to see the emergence of management systems that integrate management of a variety of network hardware and software components, such as the one described in [7].

In this paper we look at a different aspect of management — that of scale. Ideally, we would like to have a Director that is capable of being used to manage the complete range from the smallest network of a few systems to a worldwide, multinational, multicompany enterprise. In addition, we need to do this in such a way that the management styles employed for the smaller networks can be evolved seamlessly as a network grows, in small or large steps, into a large enterprise.

Some of the important considerations when dealing with a growing network revolve around the need to increase the number of staff involved in management. More people implies the need for more coordination between and organisation of those people, to ensure that they work together efficiently. As the geographic extent of the network increases, there are considerations of staff no longer being co-located. As the network grows to span multiple time zones, and eventually becomes worldwide, the network management, which might have started as an 8:00 to 5:00 operation, now becomes 24 hours a day. In a 24 hour a day operation, one must consider changes of management responsibilities between shifts.

All of these problems mean that we have to address aspects of division of responsibility between people, the potential for overlap of some management of some components, and the sharing of information between people to allow them to work effectively. We illustrate how some of these problems have been addressed in the architecture for Digital's EMA Director, and in products such as DECmcc.

In this paper we present different techniques for handling the problems brought about by addressing scale in the management of an enterprise. First we see how to use multiple directors to divide the management problem; next we see how to share management modules between those directors and then we examine how multiple directors can be employed to provide co-operative management. Finally, we investigate in detail how management domains can be used to good effect when dealing with larger enterprises with multiple managers.

2 Using Multiple Directors

Of course, the simplest solution to dealing with magnitude is the divide and conquer approach. Where there are too many components for one person to manage from one workstation, additional people, each with their own workstation, can each manage their own set of devices independently. This yields a situation where there are "islands of management", as depicted in Figure 1.

For non-integrated management solutions, this can be an appropriate solution, but note that for an integrated Director, this yields some problems. Assuming, as we must, that these different islands of managed components exist in the same network, either as logical portions of the same LAN or as different geographic sites in a small network, then there is a need to deal with those shared aspects of the network. For the LAN, there is the common media, such as wiring, and devices, such as bridges. For the WAN, there are the intersite connections, where each end of the connection needs to be concerned with its operation.

Thus, even in this simple case, we see the need for managers to communicate, though for this solution this communication would be out of band, e.g., via telephone.

The first requirement for any communication, human or otherwise, between these islands, is that there is a common naming applied to the whole enterprise. Thus we observe that even in the simple case of multiple, independent directors, we need to employ a common naming scheme, so that each manager can refer to a managed object unambiguously wherever they are in the network. For DECmcc, a naming scheme based on DECdns [8] is used to provide a consistent, network-wide namespace for all entities in the network.

We now observe that there is no need to restrict each of the island managers to just those components for which they are responsible. Certainly it might be appropriate to limit the capabilities of one manager on components which are in the purview of another. However, the appropriate use of access control or other authorisation schemes, can provide the appropriate security controls to ensure that, say, Manager 1 can change only entities that she "owns", and can non-destructively monitor entities belonging to other managers, as shown in Figure 2.

If the authorisation scheme is suitably flexible, then it is possible to permit a more general scheme whereby each entity defines the set of operations that a particular manager is permitted to perform. This might vary by time of day, or day of week, so that different managers have different capabilities on different shifts. Thus, while management of the components in one site might be handled by on-site personnel during the working day, during non-working hours some of those components could be managed from a different site.

3 Co-operating Management Modules

In the EMA Director, different types of management modules are used to address different aspects of management; some address the user interface (presentation modules), some deal with interfacing to the appropriate technology being managed (access modules) and some implement value added management functions (functional modules) [2].

EMA Director implementations, such as DECmcc, can be used to implement the islands of management described above, while sharing a common namespace to ensure consistent naming of entities across the network.

For example, due to the limitations of some technologies, such as management of LANs or operating systems, there are restrictions on locating Access Modules in the network. Thus, to manage a LAN Bridge, using an 802.1 based protocol, the Access Module must be located on the same extended LAN as the bridge. In some cases, however, the manager may not be working from a machine connected to that LAN, or indeed, be in the same building or country.

In EMA, the communication between management modules need not occur within one director. It is possible for a Presentation Module to reside on the manager's workstation, the Access Module to reside on a remote LAN close to the entity being managed, and one or more Functional Modules, perhaps executing some compute intensive function, to be executing on a third, large machine. This is shown in Figure 3.

Note that this configuration of modules is with respect to one operation on one entity. The manager may be involved with executing operations on many different entities simultaneously. The configuration of modules involved in each such operation may be different. Thus one DECmcc may be involved in requests from many other DECmcc's at the same time. In general, each DECmcc is configured with an appropriate set of management modules, including those that allow local users access to management capabilities, as well as providing access for remote DECmcc's. Figure 4 depicts how modules from different DECmcc's might be used in carrying out a single operation on an entity.

The DECmcc inter-module interface (based on a procedure call paradigm known as mcc_call) is implemented in such a way as to make the remote aspects transparent to both calling and called management module, and operates in a manner similar to a regular remote procedure call. The "protocol" that is used between DECmcc Kernels is known as the Kernel-to-Kernel protocol.

4 Co-operating Directors

Another aspect that becomes important where multiple managers may potentially affect the state of an entity is coordinating their management operations so as not to interfere with each other.

In the simplest case, the network might be configured to permit only one manager to change a particular entity, and any number of managers to monitor the entity. Either an authorisation scheme or a system built on connectivity constraints might be used to achieve these restrictions. This scheme, while perhaps easy to implement, has drawbacks, not the least of which is being a single point of failure — if the manager, or their system, is unable to communicate with the entity, then the entity is, temporarily, unmanageable.

Where more than one manager has the rights to modify the operation of an entity, there must be some way to coordinate their activities. DECnet, for example, has implemented this scheme for many years. The easiest way to provide coordination is via out of band means; thus human communication is an obvious solution. Where all managers share a common goal, conflicts are less likely to occur. Where conflicts do occur, they can be addressed between those involved, perhaps with the help of a common organisational superior.

However, as more and more automation takes place in the Director, we lose the ability for humans to be involved and thus to be able to coordinate their activities on an ongoing basis. Also, simple management operations initiated by humans typically require single directives on single entities, and these have well defined atomicity properties. As more and more complex management operations are defined (perhaps via Functional Modules), whether automated or not, a single management function may end up being decomposed into a long sequence of individual management operations over a potentially large number of entities. Clearly the possibility for undesirable interactions between managers is therefore much higher.

Thus we need to consider the concept of transactions applied to groups of management operations. Again, a simple approach would be to lock out each of the entities from all but one manager for a sequence of management operations. A more comprehensive approach would involve communication between co-operating directors in order to coordinate their management operations with respect to one or more entities. Such a protocol we refer to as the Director-to-Director protocol, which is distinct from the Kernel-to-Kernel communications described above. A full discussion of transactions, as it applies to network management in EMA, is beyond the scope of this paper. Figure 5 shows the communications between directors for a single entity.

5 Management Domains

In the discussion so far, we have described how we would employ multiple people, using multiple directors, to address some aspects of scale in management. These techniques allow us to distribute management functions to a potentially unlimited number of managers, each responsible for their own sphere of managed objects, even where these sets of objects overlap.

Another, equally important aspect of scale is where there is a large number of managed objects under the purview of a single manager. In this case, we need to define a method that allows us to organise the managed objects in a way that eases the job of the manager.

If we consider a workstation where we graphically represent the components of a network as icons on a network map, as the network size increases we reach the bounds of what might reasonably be displayed within one window. While it is possible to argue that "if it doesn't fit into a single window, it is too much for one person to manage", this does not take into account managers requiring, at different times, a detailed view of some part of the network as well as an aggregate view of a larger portion of the network.

Thus with DECmcc we define the concept of a Domain representing a sphere of interest of some set of managed objects for a manager. (Note that others have used the term "domain" with different meanings — some refer to domains as a group of managed objects for security reasons, others as a group of managed objects for which the same policy applies, and yet others define a domain as a means to constrain the set of operations that may be applied to a group of managed objects. For example, see [9]). In DECmcc, the concept of a domain is only known to the Director — the entities themselves are not aware of domains.

Our first observation is that different users have different reasons for defining a set of entities to be managed together, or to appear in one window of a graphical user interface. We can foresee at least the following reasons for grouping entities:

- entities that are the same type (e.g., all bridges)
- entities that are on the same LAN, same site or same floor of a building
- entities that belong to the same group in an organisation
- entities that form the wide area backbone or provide a particular network technology (such as a PC LAN or TCP/IP subnetwork)
- entities for which a particular management function, or functional area, is to be applied (such as for performance management or fault management)

It is also reasonable to anticipate that some managers may desire to define an arbitrary group of entities, that cannot be categorised according to any deterministic scheme.

Thus, within DECmcc, we choose to permit domain membership to be completely arbitrary, allowing the manager to define the appropriate set of entities according to the management tasks to be performed. In addition, an entity may belong to more than one domain, as shown in Figure 6. While initially the concept of allowing the user to define domain membership might appear to shirk the issue of how the contents of individual domains are defined, we note that this is the most general approach possible, allowing the organisational policies of the network administration to be applied without constraints from the management system. Hence, a manager desiring to group all entities on a LAN can do so, either manually or automatically.

In DECmcc, the creation, deletion and listing of members of a domain is handled by the Domain Functional Module. Because of the nature of the DECmcc interfaces, additional layered Functional Modules can use the services of the Domain FM to implement the desired organisational policies, and even to implement capabilities to populate domains with members automatically, where algorithmically feasible.

Domains, therefore, represent an organisation that is independent of the entities and known only within the director. Each domain needs to be named, and thus in order to be manageable in the Director has a global name known to the naming service. As the domain membership changes infrequently, we can conveniently store the domain membership information in the naming service. This means that the names and composition of domains are available to any user of DECmcc — more importantly, it allows these definitions of groups of entities to be shared amongst managers.

Domains not only represent a method to organise the entities in a network, but they also provide a convenient place with which we can store historical data for entities in a domain. [2] discusses how the EMA Director is constructed to provide a consistent way of handling data over time, in addition to active network data. By storing data from the network against a domain, it provides a convenient way to allow subsequent retrieval of that data, as operations referring to past time are invoked through the same domain representation on an iconic map. Additionally, it permits us to associate the data so recorded with the domain owner and thus account for disk space used. The manager who owns the domain is thus able to manage their own disk space as they see fit.

6 Additional Domain Capabilities

As powerful a concept as domains are, there are still additional capabilities that managers need to ease their task of providing management. With the simple organising principle thus far defined, managers can now use a workstation with multiple windows to represent different views of the network, via different domains. However, as such, this only represents a small step from having multiple directors providing a view to a manager, and otherwise acting independently; here we allow one manager to have multiple but independent views.

As domains are named and manageable, as any other entity, we have the ability to represent them on an iconic map along with other more tangible entities. Fig 7 shows a map representing both manageable components as well as references to other domains. Now, a manager can either select an icon representing a real manageable object and apply a management operation to it, or they may select an icon representing another domain and 'navigate' into that domain. We refer to such domains as Reference Domains. Domains have global names and so the domain to which a manager navigates does not have to be a domain defined by, or owned by, that manager — thus domain definitions may be shared.

Sharing of domain definitions means that one manager, starting from their own view of their portion of the network, may now navigate from domain to domain to manage a part of the network in a completely different part of the world. Of course, the access rights that one manager may have on entities in another part of the network are determined by those entities, and may permit only non-destructive monitoring operations — but even this provides considerable management capabilities, such as the ability to trace a path through a network or to determine the possible causes for performance problems in distributed applications.

Also, as historical data is stored with reference to a domain, appropriately authorised managers can access historical information about entities in other parts of the network, having navigated to the appropriate domain.

Within EMA, we have chosen to distinguish three different types of domains, to serve different purposes. These are the Definitional Domains, Dynamic Reference Domains and Static Reference Domains.

6.1 Definitional Domains

Definitional Domains are domains whose sole purpose is to define a common grouping of entities which may be shared between multiple domains. Such domains are never active, in that they are not represented on an iconic map, and historical data may not be stored against them.

Typical uses for Definitional Domains are to define a set of particular entities that more than one manager may wish to use in their own active domain definitions. Thus, if one such domain represents all terminal servers on one floor of a building, another the terminal servers on another floor, and yet another the workstations on one floor, and so on, different managers can include those domain contents in different groupings to suit the needs of the management tasks they need to perform. As the domain contents are enumerated each time the user opens a window for a domain containing a Definitional Domain, each Definitional Domain need only be kept up to date once, and all uses of that domain will retrieve an up to date copy of the information.

6.2 Dynamic Reference Domains

One typical use for domains is to represent a group of entities as a single icon in a workstation window. For a manager responsible for a large number of entities, it may be more convenient to represent a set of icons as the set of aggregations (domains) with which the manager has to deal. However, particularly in a fault management environment, it is important that notification of unusual conditions, or other conditions that require human attention, be made available to the manager expeditiously.

Where a single domain is represented in a window, it is possible to represent notifications via salience changes to the icons representing the appropriate entities by, for example, changing the colour of the icon. Similarly, with Dynamic Reference Domains notifications which apply to icons in a referenced domain are automatically associated with the icon representing that domain in the referencing domain. This is shown in Figure 8. Thus the manager viewing the Campus domain, on seeing the icon for Building 12 change, can then navigate into the Building 12 domain to get a more accurate indication of the problem; the icon for the Floor 1 domain indicates that we need to look another level deeper. Dynamic Domains may therefore be considered to be organised hierarchically. Clearly it is important that there be no loop in the sequence of references.

Independent of passing up of notifications, Dynamic Reference Domains can also be used to navigate around the network for normal monitoring and control activities.

6.3 Static Reference Domains

Unlike Dynamic Reference Domains, Static Reference Domains are intended solely to provide managers with a navigation capability. As such there is no strict hierarchical relationship between Static Reference Domains, so there are no restrictions on the ways in which domains can refer to other domains via such references. These domain references are particularly useful in following network paths between two systems in different parts of a network.

Where Dynamic Reference Domains can be considered to be organised as a directed graph, Static Reference Domains can be organised as a mesh topology.

7 Conclusion

As the size of networks grows rapidly, the use of isolated and independent management systems no longer matches the needs of organisations to manage their enterprises in a coherent fashion.

We recognise the need for an enterprise-wide Director, such as DECmcc, to be designed to handle the various aspects with scaling in an enterprise: the increasing number of components, the geographic extent, global communications, time zones, larger number of people involved, and the different roles of managers.

With EMA, it is possible to deploy multiple independent copies of a DECmcc Director to manage different parts of the network. It is also possible to allow different management modules on different Directors in a network to interoperate, to allow remote management.

Finally, by defining an extremely flexible scheme for organising manageable objects in an enterprise, known as domains, we can allow each manager to deal easily with an ever larger number and diversity of entities from a single workstation, and yet retain all the power of the integrated director approach of EMA and its implementation in DECmcc.

References

[1] Fehskens, L., An Architectural Strategy for Enterprise Management, in Integrated Network Management I (1989) 41-60.

[2] Strutt, C., and Shurtleff, D., Architecture for an Integrated, Extensible Enterprise Management Director, in Integrated Network Management I (1989) 61-72.

[3] Sylor, M.W., Guidelines for Structuring Manageable Entities, in Integrated Network Management I (1989) 169-183.

[4] Shurtleff, D., and Strutt C., Extensibility of an Enterprise Management Director, in Network Management and Control Workshop (September 1989) 129-141.

[5] La Pelle, N.R., et. al., The Evolution of Network Management Products, Digital Technical Journal (September 1986) 117.

[6] Digital Equipment Corporation, Digital's Enterprise Management Architecture General Description (Spring 1988) Order #AA-PD5JA-TE.

[7] Digital Equipment Corporation, DECmcc Introduction (Fall 90) Order #AA-PD57A-TE.

[8] Martin, S., et. al., Development of the VAX Distributed Name Service, Digital Technical Journal (June 1989) 9.

[9] Sloman, M.S., and Moffett, J.D., Domain Management for Distributed Systems, in Integrated Network Management I (1989) 505-516.

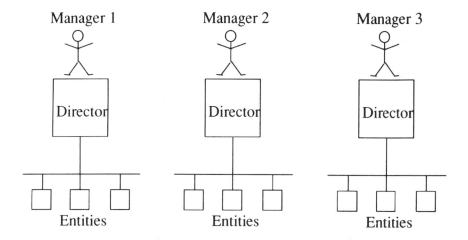

Figure 1. Islands of Management

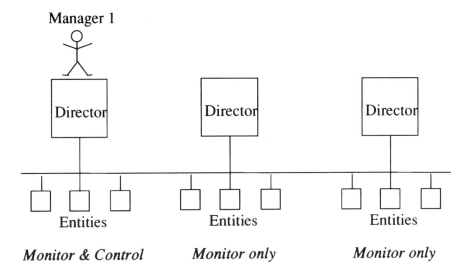

Figure 2. Controlling Management Access Rights

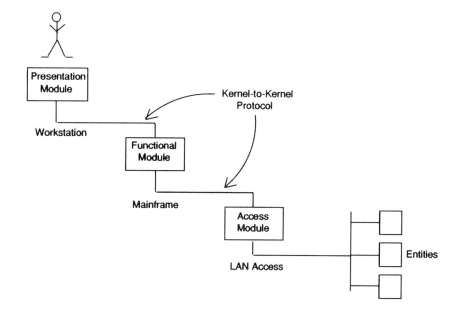

Figure 3. Distributed Management Modules

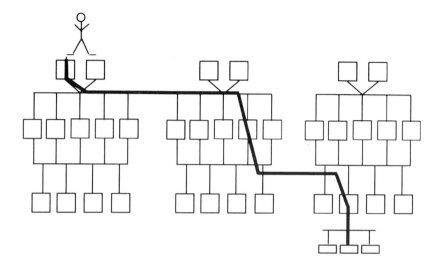

Figure 4. Execution Path Through Multiple MCC's for One Operation

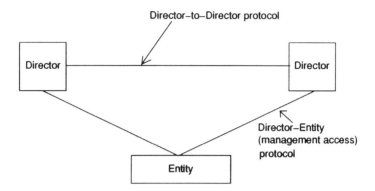

Figure 5. Multiple Directors Coordinating via Director-to-Director Communications

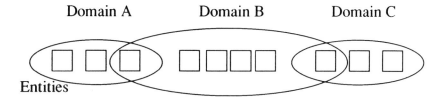

Figure 6. Overlapping Domains of Entities

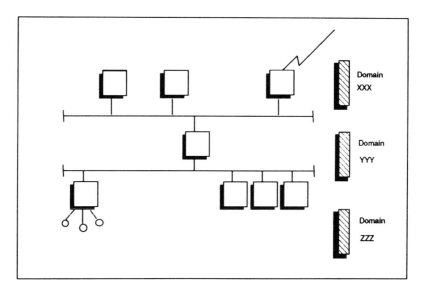

Figure 7. Domain References on a Network Map

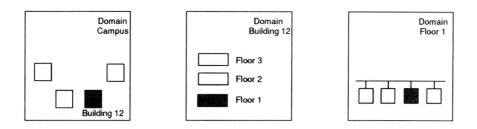

Figure 8. Forwarding Notifications Between Domain Maps

Delegation of Authority

Jonathan D. Moffett & Morris S. Sloman

Imperial College of Science Technology and Medicine
Department of Computing, 180 Queen's Gate, London SW7 2BZ
Email: jdm@uk.ac.ic.doc, mss@uk.ac.ic.doc

Abstract

This paper is concerned with the specification of discretionary access control policy for commercial security and the delegation of access control authority in a way which gives flexibility while retaining management control.

Large distributed processing systems have very large numbers of users and resource objects so that it is impractical to specify access control policy in terms of individual objects or individual users. We need to be able to specify it as relationships between groups of users and groups of objects. The systems typically consist of multiple interconnected networks and span a number of different organisations. Authority cannot be delegated or imposed from one central point, but has to be negotiated between independent managers who wish to cooperate but who may have a very limited trust in each other.

The paper proposes the use of *access rules* to specify, in terms of their domain memberships, what operations a user can perform on a target object. The delegation of authority to allow security administrators to create access rules requires limiting the scope of the users and target objects for whom they can create rules. The paper shows how *role domains* can be used to permit flexible but controlled delegation of authority from an *owner*, via *managers*, to *security administrators*.

1. INTRODUCTION

Large distributed processing systems may contain hundreds of thousands of resources and be used by thousands of users. This implies that it is impractical to specify access control policy in terms of individual objects or individual users. We need to be able to specify it as a relationship between groups of users and groups of objects.

We define *authority* as power which has been legitimately obtained. There are a number of ways in which legitimacy may be obtained, but for our purpose it is obtained in accordance with the currently applicable laws. This paper addresses authority as applied to discretionary access control. In particular it covers the means by which authority can be granted and removed dynamically.

The organisations using and managing distributed processing systems are hierarchical in nature and authority in them is delegated downwards form senior management. In general authority is delegated not to a person but to a position or role within an organisation. Typically the decision is that 'the Payroll Clerk should change the Payroll Master file', and John should have that authority only because he occupies the position of Payroll Clerk.

Distributed systems often consist of multiple interconnected networks and span a number of different organisations. Authority cannot be delegated or imposed from one central point, but has to be negotiated between independent managers who wish to cooperate but who may have a very limited trust in each other.

The model for delegation of authority must reflect the organisational management structure and policy as well as providing mechanisms for the transfer of authority from one agent to another. A resource owner should be able to delegate authority over his resources to another user, within the mandatory constraints of the system. In addition it must be possible to delegate *the authority to delegate,* but limit the scope of this second level delegation. For example, an owner can delegate responsibility to a security administrator to give users, in a defined part of an organisation, access to his objects. Finally, the model should permit users to have personal domains where they are able to control the objects they create.

We make the assumption that there is no inherent right of access of any kind for ordinary users. If a person is not the owner of an object, and has not been given authority, then the computer system should refuse all access.

2 MANAGEMENT CONCEPTS

A detailed discussion of Management of Distributed Systems can be found elsewhere [Sloman 1989]. We concentrate in this paper only on the concepts relating to authority and its delegation.

2.1 Ownership

It is assumed that humans are ultimately responsible for the actions of the system. In many situations they will use automated agents to perform operations within the system, but they will retain responsibility for the actions of the agents, and therefore require the power to control the agents. Thus we will always be able to trace responsibility back to human users, who we call the *owners* of the system (e.g. managing director of a company or board of directors). *Ownership* in this paper is intended to denote a concept as close as possible to normal legal ownership of goods and property. All computer systems in a country such as the UK process resources with identifiable legal owners who have legal powers over them. We regard ownership as the starting point for delegation of authority.

As a very rough approximation, ownership of an object implies the legal power to perform any feasible operation on it, together with responsibility for the operations which are performed. There are of course limitations to this power; for example an owner of personal data may not disclose it except under the terms of data protection legislation, and the owner of petroleum spirit at a bulk distribution plant must ensure that an automated system dispensing it to tanker lorries does so in accordance with certain safety regulations. Restrictions of this kind have to be represented in computer systems as mandatory constraints, and are beyond the scope of this paper.

Many previous authors, e.g. [Lampson 1974], have assumed that the user who creates an object automatically becomes its owner. We do not hold that view, but distinguish between ownership of objects and the delegated power to create them. A data processing clerk who submits a job to create a new version of a file of bank accounts is not the owner of those accounts, but is carrying out the task of creating the file as an agent of the owner of the bank.

For the purpose of discussing authority we assume that we start with an owner, who can then dispose of or share his ownership or delegate a subset of his powers to another person. Provided that this is done legitimately we will describe this other person as having gained authority. In a computer system authority is normally represented as the ability to perform defined operations on objects through specified interfaces.

2.2 Separation of Responsibilities

Separation of responsibilities is an important control concept which is familiar in the context of auditing, e.g. [Waldron 1978]. It is designed to ensure that no-one has excessive authority. [Clark 1987] has pointed out that the concept should be modelled in computer systems. It requires that different aspects of certain transactions should be carried out by different users, so that no one person can carry out the transaction autonomously. An example is the authorisation of payment of suppliers' invoices, where the input of invoices to a computer system must normally be carried out by a different user from the person who can trigger off the actual release of payments. Neither can carry out the other's function, so the payments cannot be made without their cooperative activity. Requirements for separation of responsibility have to be specified at an application level but support for it needs to be provided by the access control system.

This separation of authority manifests itself in the role of a security administrator, who often is responsible for granting access authority in large organisations The security administrator should not be allowed to grant himself access to the resources under his control.

2.3 Management Structures

Each organisation has its own management style, which is reflected by different management structures and policies for delegation of authority. For this paper we have identified four typical roles within an organisation management structure - User, Security Administrator, Manager and Owner. Owners can share Ownership with other users, or can delegate Manager authority. Managers can delegate authority to Security Administrators, who can create access rules which allow ordinary Users to perform operations on target objects. This structure is modelled in more detail in terms of domains and access rules in section 4. Note that it is an application decision whether a user in a management role has authority over himself, e.g. a Security Administrator being able to give himself access rights. We specifically allow for this case to be prevented, but we do not enforce it as part of the model.

3 DOMAINS & ACCESS RULES

3.1 Domains

An approach to the problems of very large distributed systems is provided in [Robinson 1988], which introduced the concept of generic *domains* as a means of managing systems. This implies being able to take a view, not of objects, but of groups of objects. Domains are defined as named groups of objects to which a common policy, such as access control, can be applied. Domains are a fundamental concept in our view of systems. Apart possibly from root domains, all objects are created within domains, and they are the means used to structure systems so as to be manageable. They are described in detail in [Sloman 1989], but are summarised here.

Management requires the ability to perform operations on defined resources. The scope of this may be limited in the range of objects over which it extends and in the operations which can be performed on them. In our model we can simply enumerate the operations but a more powerful means of representation is needed to describe the range of objects covered. Using domains we can achieve both flat and hierarchical grouping, and also refer to user objects indirectly through the roles and positions which they occupy. For example, domain hierarchies can be used to describe the structure of an organisation and the scope of authority of managers over resources and other users. This is illustrated in detail in the example below.

The Domains concept relies on an object oriented approach, which we use throughout. Everything, including access rules themselves, is treated as an encapsulated entity presenting a set of operations as its interface. It is recognised that actual implementations will not take such a uniform approach. In particular access rules are unlikely to be implemented as objects, but will be entries in lists or databases. 'Type' is used in a very broad sense here. Two object instances are said to be of different Type if their external interface or specified behaviour differ in any respect. We do not attempt in this model to address the management of dynamic change of the interface itself.

A *domain* is an object, of type domain, whose purpose is to represent a set of objects which may be resources, workstations, modems, processes, etc, depending on the purpose for which a particular domain is defined. One attribute of a domain is its Object_Set, which is a set of identities of objects which are referred to as *domain members*. A dual attribute of every object is its Domain_Set, the set of identities of domains of which it is a member. Objects are capable of being members of more than one domain at a time. Domains are persistent even if they do not contain any objects - it must be possible to create an empty domain and later include objects in it.

Domains do not encapsulate the objects themselves - managers or external objects may interact directly with an object in a domain. The objects in a domain do not all have to be of the same type, but the domain maintains information on the interfaces to be supported by objects in it. There is a constraint on domain membership that all objects in it must support the minimum operations required.

The following operations are particularly important for an understanding of domains:

Create or destroy an Object - All Create and Destroy operations are viewed as operations on their containing domains. Domain objects can be created or destroyed like any other objects. Creation both creates an instance of an object and adds it to the object set of a domain, while Destroy both removes the object from the domain's object set and destroys the object instance.

Include or Remove Object in a Domain - Include adds the identity an object into the object set of the domain. It does not affect the state of the object or its membership of other domains, except that the object updates its Domain Set. Domain objects can themselves be included in other domains, and are then referred to as *subdomains*. Remove removes an object's identity from the object set of the domain

The possible set relationships between domains are important both in creating access rules and in understanding their effect. Two domains are defined to be *disjoint* if their object sets are disjoint. They *overlap* if there are objects which are members of both domains. A special case of overlapping occurs when the objects in one domain are a *subset* of the objects in another. The representation of hierarchical organisation can be accomplished by creating a domain object, referred to as a *subdomain*, as a member of another domain, see figure 1. We refer to D2 as a *direct subdomain* of D1, and to D4 as an *indirect subdomain* of D1. Also we refer to D1 as a *superdomain* of D2, D3 and D4.

Domain expressions are sets of objects defined by reference to enumerated sets of objects and domains, and the result of applying set operations to simpler domain expressions. The set operations include *set intersection* ($\cap$) and *set difference* (\). An example is D1\D2, meaning all the objects which are members of D1 except those which are members of D2.

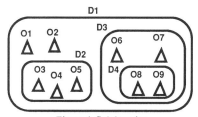

Figure 1 Subdomains

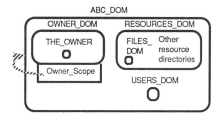

Figure 2 ABC Overall Domain Structure

3.1.1 Example Domain Structure

A commercial organisation, ABC Ltd, with a Research Department which is engaged in a joint venture with an external organisation, DEF Ltd. There is a need to organise access control for ABC itself, and also to allow limited access for members of DEF staff to Research Department files. DEF staff gain access from terminals in their own research laboratory via a communications network. After an audit of the security arrangements at both sites, ABC have taken a policy decision to allow DEF limited security administration rights to control the users who can access ABC. ABC's policy is to use the standard management structure as described in section 2.3, above.

ABC's system consists of a number of computers connected by an company network. The example makes no assumption about what resources are located on what computer, but requires a global identification scheme so that all objects may be referred to uniquely. We show the overall domain structure of ABC in figure 2. The role of OWNER_DOM is explained in section 4.4.

Figures 3 & 4 show the organisations as viewed from ABC.

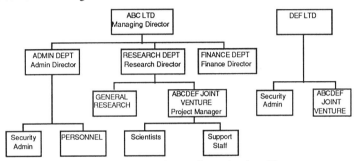

Figure 3   Conventional Organisation Tree

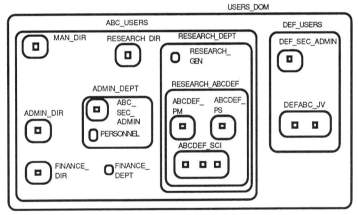

Figure 4   Domain Representation of Organisation

A conventional organisation chart is shown for comparison, and it will be seen that there is a close correspondence between the conventional and domain representations. However, management relationships representing authority in the organisation chart is lost in the domain diagram. We can tell directly from the organisation chart 'who manages whom', e.g. that the Admin Director manages the Security Administrator. We cannot tell that from the domain diagram; it would be perfectly consistent with it for the Security Administrator to manage the Admin Director. Simple domain diagrams tell us only about their membership and not about the authority relations between domains. The extra information is added in access rules and the role domain diagrams as described in section 4.

We show a part of the file structure and its corresponding domain representation in figures 5 & 6 The Personnel and Suppliers Directories contain data which is registered under the Data Protection Act. Unlike the Organisation structure, no information is lost in the translation between the tree structure and the domain diagram.

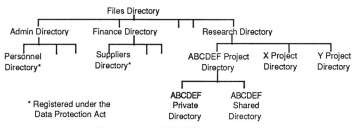

Figure 5 ABC File Structure

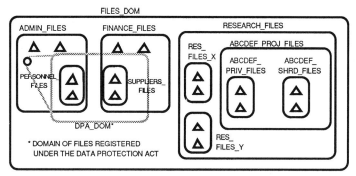

Figure 6 Domain Representation of ABC File Structure

A domain which spans the domains of two departments, such as DPA_DOM in figure 6, may be required for specific purposes. In this case we assume that members of PERSONNEL need to be able to Read the contents of all files containing personal data, so that they can monitor compliance with the Data Protection Act. The domain cannot be set up by one user but requires cooperative activity. ABC_SEC_ADMIN will have to create access rules which allow a suitable member of ADMIN_DEPT to create an empty DPA_DOM domain within ADMIN_DIR and include PERSONNEL_FILES into it, and a suitable member of FINANCE_DEPT to include SUPPLIERS_FILES in it.

3.2 Access Rules

The purpose of access rules is to specify access control policy in terms of what operations a set of users can perform on a set of objects. It therefore enables access control policy to cope with large scale systems. A basic *access rule* specifies a *user domain*, a *target domain* and an *operation set*. Both user and target domains can be the names of domains or more general domain expressions. The operation set defines the names of the operations, defined in the object's interface specification, which the access rule authorises. We make no assumptions that one operation 'implies' another, e.g. Write permission implying Read permission. The attributes of objects are protected by protecting the operations which are used to access them.

The system authorises a request if an access rule exists which *applies* to an *operation request*, which is a triple consisting of (user, target object, operation name). We assume that the system has authenticated the identity of the user. An access rule applies to an operation request if the user in the request is in the user domain of the rule, the target object in the request is in the target domain of the rule and the operation name is in the operation set of the rule. If no access rule exists which applies to the operation request, then authority for the operation does not exist and access is denied.

In addition to an applicable access rule, there has to be compatibility between the interfaces of the user and target objects, to permit the user to bind to the target. This is a configuration issue rather than an access control issue, and is checked by the configuration management system [Twidle 1988].

It is possible that there are overlapping access rules - more than one access rule satisfies the requirement. No conceptual difficulty arises from this, as it indicates that authority has been granted *via* more than one route, but there may be practical difficulties when attempting to remove a user's access rights, which are discussed in [Moffett 1990a].

There is a need to modify access rules dynamically during the life of the system. We model this by treating access rules themselves as objects which are instances of an Access_Rule type. Access permissions are given and removed by creating, destroying and modifying access rule objects. The need to control the granting and removal of permissions is achieved by imposing rules for the creation, destruction and modification of access rule objects. This is discussed in section 4, Delegation of Authority.

3.2.1 Access Rules in ABC

Two example access rules, which we assume to be part of ABC's policy, are as follows:
a) (DEF_USERS, ABCDEF_SHRD_FILES, {Create, Read, Write}) - any member of DEF_USERS can Create, Read and Write files in the ABCDEF_SHRD_FILES domain.
b) (PERSONNEL, DPA_DOM, {Read}) - any member of PERSONNEL can Read files in the DPA_DOM domain.

The access rules do not name the users or objects, and so we cannot tell directly from them whether the access by a particular user to a particular object will be allowed. To interpret the access rule we have to know the domain structure, as shown in figures 4 & 6. The access rules remain valid for members of the domains, whatever objects move in and out of the domains. This achieves our requirement for maximum stability in the face of change.

4 DELEGATION OF AUTHORITY WITH ROLE DOMAINS

When authority is delegated, there is a need to control to whom it is delegated, the resources over which the authority applies, and how the authority can be passed on down a management chain. The existence of an access rule allowing UserA to create an Access_Rule object is clearly not a sufficient control by itself. It provides no constraint on the contents of the Access_Rule object which is created, so UserA would be able to create Access_Rule objects to enable any user to access any objects. We therefore need a means of placing controls on the contents of Access_Rule objects.

A security administrator must be able to grant access authority which he does not possess himself, so the concepts of *having* access and *giving* access must be decoupled. Similarly owners may delegate, to managers, the authority to create security administrators. We therefore need to distinguish between the normal operations which a user can carry out, and operations which give authority.

4.1 Role Domains

The concept of a role domain was developed from the generic domain to define the range and nature of the managerial authority of the members of a domain. A Role Domain object type is a subtype of Domain with additional *scope* attributes, one for each kind of authority defined by management policy. Each scope attribute is a domain expression which identifies the objects over which the domain members have managerial authority. Using Owner as an example, we map authority relationships onto role domains as follows: a user is an Owner of a target object if he is a member of a role domain, and the target object is a member of the role domain's Owner_Scope attribute.

Membership of a role domain gives a user the authority associated with that domain. If a user is a member of more than one role domain he has the authority associated with each domain, e.g. someone may be the manager of more than one department in an organisation. A user can be moved to another role domain if his role in the organisation changes.

In the management structure which we are using here, there are managerial roles of owner, manager and security administrator. Owner and Manager are represented by single scope attributes: Owner_Scope and Manager_Scope. A Security Administrator's authority is represented by two scope attributes: SA_User_Scope and SA_Target_Scope, one to define the users and the other to define the target objects, for which he can make access rules.

4.2 Typical Role Domains

We now discuss our sample management structure using role domains as the means of representation. Normal users of the system have no inherent ability at all to grant access authority, so they are not members of any role domain.

A *Manager* has a defined set of objects (users and/or target objects) over which he has authority. He can define a set of users and/or a set of target objects as the scope of authority of a Security Administrator.

An *Owner* has a defined set of objects (users and/or target objects) over which he has authority. He can give away or share ownership (by 'share' we mean create joint ownership rather than partitioning the objects between two owners), and delegate the Manager authority over these objects, to other users.

We use the term ownership of users to denote responsibility rather than possession. No implication of slavery is intended.

Security Administrator (SA)

A *Security Administrator* can give authority to users to perform operations on objects by creating access rules, by virtue of his membership of a role domain (referred to as an SA Domain). The extent of authority which is delegated to him is limited by the scope attributes of the SA Domain: the users to whom he can grant access authority are limited to those in his SA_User_Scope and the target objects on which he can grant access authority are limited to those in his SA_Target_Scope. He can create domains which contain subsets of the objects defined by the SA_User_Scope and SA_Target_Scope and use these as the User Domain and Target Domain of a new access rule. The system will prevent any attempt to exceed these limits. In our sample management structure we place no limits on the operations which he can put in an access rule.

One of our main aims is to allow a Security Administrator to create access rules for other users while preventing him from giving himself access. This is done by ensuring that the membership of the SA domain and of the SA_User_Scope attribute of the domain do not overlap. Then since the Security Administrator (a member of the SA Domain) cannot himself be a member of its SA_User_Scope attribute he cannot grant access authority to himself. This achieves the desired separation of responsibilities between granting access and having access.

There is of course the need for the Security Administrator himself to have access to some objects, but he cannot create the requisite access rules himself. This need is met by a second security administrator who can allocate access to the first.

Managers

A Manager is defined as a member of a role domain with a non-null Manager_Scope attribute, referred to informally as a Manager Domain. He can create SA Domains, include users in them and alter their SA_User_Scope and SA_Target_Scope attributes. This enables him to appoint Security Administrators and set the scope of their authority.

The limits of the SA_User_Scope and SA_Target_Scope attributes of a role domain are controlled by requiring the invoker of the operation to be a member of a Manager Domain whose Manager_Scope must be a superset of the SA Scope attributes he is altering. Note that we place no restriction on the objects which are members of the new security administrator role domain. We take the view, in this example, that a Manager should be able to decide to appoint any user as a Security Administrator.

Owners

An Owner is defined as a member of a role domain with a non-null Owner_Scope attribute, referred to informally as an Owner Domain. He can create Manager Domains, include users in them and alter their Manager_Scope attribute in order to appoint Managers and set the scope of their authority. He can also share his Ownership. The values of Owner_Scope and Manager_Scope attributes which he alters must be subsets of his own Owner Scope.

We remarked above that the creator of an object is not necessarily the owner of it. This follows automatically from our model; if a user creates a new object in a domain, the ownership of that object, like all others in the domain, remains with the domain's owner. Indeed, the user cannot even read the object after creation unless an access rule also allows him the Read operation.

Ownership is shared if more than one user has Owner authority over an object. At this stage we take a simplistic view of transfer of ownership, by specifying that an Owner should be able to remove Owner authority from another Owner of the same object. This allows both for suicide and for destruction races where the first person to remove the other's ownership wins power. Further constraints on sharing and transfer of ownership will need to be considered.

Users as Domains

A further application of domains is as a means of representing users as persistent objects. We consider user objects to be a specialisation of role domains, which we will call *User Representation Domains* (URDs). URDs will only permit process objects as members. A process is created or included in a URD to represent a human user when he logs on at a terminal or starts a batch job, and deleted or removed when the user logs off or the job finishes. A URD is created as part of the activity of registering a new user on the system. Our view of access rules is, of course, that in general any member of a domain has authority to perform operations which are authorised for the domain itself. Therefore the user process automatically gains all the authority of the URD when it becomes a member, and loses that authority when it is removed from it. This is therefore a method of ensuring that a process has all the authority of a user while, and only while, it is representing him.

Users' Personal Domains

Many systems have an informal concept of a user's personal domain, where the user is automatically allowed authority to give access for other users. A typical general (mandatory) policy is that a user should be allowed to give any other user in the same organisation access to resources in his personal domain. In our terms, a user should have a Security Administrator's authority with an SA_User_Scope of all users in the same organisation and his personal domain as the SA_Target_Scope. We envisage the setting of these scopes being done automatically by the system when a new user is registered.

4.3 Delegation of Authority in ABC

The policy of ABC recognises three kinds of authority: Owner, Manager and Security Administrator. Refer to figures 4 and 6 for these examples.

Owner

The occupant of OWNER_DOM is Owner of all the users and resources known to the ABC system, as shown in figure 2.

Manager

The occupant of OWNER_DOM makes the following Manager Domains; examples are shown in figure 7. MAN_DIR is Manager of RESOURCES_DOM (all resources such as files) and USERS_DOM (all Users). Each Director is Manager of his departmental domains of Users and files directory. In addition ADMIN_DIR is Manager of external domains such as DEF, and is Manager of ABCDEF_SHRD_FILES. Note that both he and RESEARCH_DIR are Manager of these files. They will need to coordinate their activities to avoid conflict.

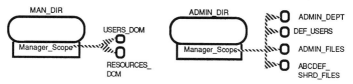

Figure 7   ABC Manager Role Domains

Security Administrator

The occupant of MAN_DIR creates a role domain to make ABC_SEC_ADMIN a Security Administrator, with a user scope of the whole of USERS_DOM (except for ABC_SEC_ADMIN itself), including DEF, and a target scope of the whole of FILES_DOM. So members of ABC_SEC_ADMIN can make access rules for all Users (except themselves) to perform operations on any file in FILES_DOM. For simplicity, there is no way for a Security Administrator to give access for members of ABC_SEC_ADMIN to anything; this would have to be achieved by MAN_DIR creating another Security Administrator.

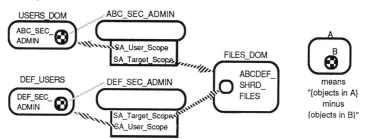

Figure 8   ABC Security Administration Role Domains

The occupant of ADMIN_DIR creates a role domain to make DEF_SEC_ADMIN a Security Administrator, with a user scope of DEF_USERS (less DEF_SEC_ADMIN). The occupant of RESEARCH_DIR then alters the DEF_SEC_ADMIN role domain to give it a target scope of ABCDEF_SHRD_FILES. So members of DEF_SEC_ADMIN can make access rules for members of DEF_USERS (except themselves) to perform operations on any file in ABCDEF_SHRD_FILES directory or its subdirectories. Members of ABC_SEC_ADMIN can make access rules for members of DEF_SEC_ADMIN.

Figure 8 illustrates the security administrators' role domains. It shows clear separation of responsibilities for them. For simplicity we have omitted to show the means by which another user, e.g. the ADMIN_DIR, can make access rules for ABC_SEC_ADMIN.

5   DISCUSSION & CONCLUSIONS

We aimed in our work to model a framework and policies at least as complex as would be met in real commercial conditions. The requirements for it were derived from our experience of large commercial systems. Informal development of the model has taken place throughout in parallel with development of a type-checked specification of it in the Z language [Spivey 1988, 1989]. This has been animated in Prolog and validated with an extended version of the above example.

5.1   Implementation Approach

The use of role domains as a means of representing management authority depends upon the implementation of generic domains and access rules. [Twidle 1988 & Moffett 1990a] describe the implementation approach, which is still under way as part of the DTI/SERC funded Domino project. We here summarise our general approach and show how we intend to build on it.

The lowest level of granularity for access rules will be domains, rather than individual objects, and they will be represented as Access Control Lists (ACLs) held at domain objects. Performance costs will be minimised by limiting access control as far as possible to Bind operations which associate the user and target object. If access control is required at a lower level, e.g. on individual Read and Write operations, this will be done by the checking of capabilities generated at Bind time. An authentication service, probably Kerberos [Steiner 1988] will be used for user authentication.

User objects will be represented by URDs, as discussed above. Thus the only penalty in granularity will be for target objects, which will need to be segregated into separate domains if access control at the level of individual objects is required.

Role domains will be implemented as a subtype of generic domains. No performance optimisation needs to be carried out on the authority-giving operations, as they will be sufficiently infrequent that a high cost is tolerable. The policy relating to these operations will be enforced by the access control system.

5.2   Related Work

Our work builds on that reported in [Moffett 1988], which gives more background on its motivation, and is discussed in more detail in [Moffett 1990b]. Little else has been done on the way in which authority can be transmitted in large organisations. [Lampson 1974] and [Snyder 1981] make explicit assumptions that the authority to give access rights is bound in with the access rights themselves. Stepney and Lord [Stepney 1987] recognise that it is separate, but explicitly regard it as being outside the scope of their model. [Yu 1989] recognises the need for different types of authority in a telecommunications management application, but does not generalise his model.

5.3   Further Work

There are several areas of work on the model which require further investigation:
- Implementation of this model depends upon prior implementation of access rules. This is still at an early stage.
- Authority relations such as we have discussed in this paper are only one case of interaction between people and roles in an organisation. Our analytical approach could usefully be extended to other organisational interactions, particularly peer-to-peer negotiation.
- We used a example management structure in this paper. We are working on extending this work to general management structures of authority.
- Our work is currently targeted at discretionary access control, but at some point it will be necessary to consider how it would integrate with mandatory access control systems.

5.4 Conclusions

We have produced and validated a model which enables the realistic management of discretionary access control in large distributed systems: realistic because it comes to terms with the fact that existing models, using capabilities or access control lists, leave the security administrator with an impossibly large task; realistic, too, because it models the need in real organisations for authority to be passed on in controlled way.

Implementation of the techniques which we have described will enable management control to be applied more effectively to discretionary access control than has previously been possible.

6 ACKNOWLEDGEMENTS

We gratefully acknowledge the support of the DTI/SERC (Grant No. GR/F 35197) for the Domino project. Our colleagues S.C. Cheung, Naranker Dulay, Jeff Kramer, Jeff Magee & Kevin Twidle of Imperial College, John Haberfield & Tony Law of BP and Tony Jeffree of Sema Group have contributed to the concepts described here.

REFERENCES

[Clark 1987] Clark D.C. & Wilson D. R., A Comparison of Commercial and Military Computer, Security Policies, IEEE Security and Privacy Symposium 1987, pp 184-194.

[Lampson 1974] Lampson B.W., Protection, ACM Operating System Review, vol 8 no 1, pp 18-24 (Jan 1974).

[Moffett 1988] Moffett J.D. & Sloman M.S., The Source of Authority for Commercial Access Control, Computer, vol 21 no 2, pp 59-69 (Feb 1988).

[Moffett 1990a] Moffett J.D. Sloman M.S. & Twidle K.P., Specifying Discretionary Access Control Policy for Distributed Systems, Computer Communications, vol 13 no 9, pp 571-580 (November 1990).

[Moffett 1990b] Moffett J.D., Delegation of Authority Using Domain Based Access Rules, PhD Thesis, Dept of Computing, Imperial College, London, September 1990.

[Robinson 1988] Robinson D.C. & Sloman M.S., Domain Based Access Control for Distributed Computing Systems, Software Engineering Journal, vol 3 no 5, pp 161-170 (Sept 1988).

[Sloman 1989] Sloman M.S. & Moffett J.D., Domain Management for Distributed Systems, in Meandzija & Westcott (eds), Proceedings of the IFIP Symposium on Integrated Network Management, Boston, USA, May 1989, North Holland, pp 505-516.

[Snyder 1981] Snyder L., Formal Models of Capability-Based Protection Systems, IEEE Trans on Computers, vol 30 no 3, pp 172-181 (March 1981).

[Spivey 1988] The *fuzz* Manual, J.M. Spivey Computing Science Consultancy, 2 Willow Close, Garsington, Oxford OX9 9AN, UK, 1988

[Spivey 1989] Spivey, J.M., The Z Notation: A Reference Manual, Prentice Hall 1989.

[Steiner 1988] Steiner J.G., Neuman B.C. & Schiller J.I., Kerberos: An Authentication Service for Open Network, Systems, Winter Usenix 1988, Dallas TX. Project Athena, MIT, Cambridge MA 02139, USA, 12 Jan 1988.

[Stepney 1987] Stepney S. & Lord S.P., A Formal Model of Access Control, Software - Practice & Experience, vol 17 no 9 (Sept 1987), pp 575-593.

[Twidle 1988] Twidle K. & Sloman M.S., Domain Based Configuration & Name Management, for Distributed Systems, IEEE Distributed Computing Systems Workshop, Hong Kong Sept 1988.

[Waldron 1978] Waldron R.S., Practical Auditing, HFL (Publishers) Ltd, London, 1978.

[Yu 1989] Yu, Che-Fn, Access Control and Authorization Plan for Customer Control of Network Services, Proceedings of IEEE GLOBECOM '89, Dallas, Texas November 1989, pp 862-869.

VII

DISTRIBUTED SYSTEMS MANAGEMENT

B
Distinguished
Experts Panel

Trends in Integrated Network Management
Convergence or Divergence?

Moderator: *Vint Cerf*, Corporation of National Research Initiatives, USA

Panelists: *Larry Bernstein*, AT&T Bell Laboratories, USA
Paul Brusil, MITRE Corporation, USA
André Danthine, Université de Liège, BELGIUM
Marshall Rose, PSI Inc.,USA
Keith Willetts, OSI Network Management Forum, British Telecom, UK

Computer networking has become an important infrastructure not only for the research, development and academic community but, increasingly in business and in government. As we come to rely on this infrastructure, the need to keep it functioning well and reliably increases. Moreover, as the services become available on a commercial basis, tools for assuring long-term, reliable operation are essential. This panel discussion will explore a number of network management aspects, including but not limited to technology, use of tools for operations and for planning, accounting data collection, management station requirements, dealing with scaling, multiple protocol suite effects and the challenge of Management Information Base development.

VII

DISTRIBUTED SYSTEMS

MANAGEMENT

C
Case Studies

Management in a Heterogeneous Broadband Environment

Jane Hall and Michael Tschichholz

GMD FOKUS, Hardenbergplatz 2, 1000 Berlin 12, Federal Republic of Germany

E-mail: hall@fokus.berlin.gmd.dbp.de and tschichholz@fokus.berlin.gmd.dbp.de

Abstract
Networks have developed from stand-alone systems into internetworks connecting several local and wide area networks. The resulting system is very large and complex, consisting of heterogeneous applications and connected end-systems and spanning several autonomous organisational domains. Managing such a system demands additional tools in order to provide the level of service that users have become used to on their own local network where the distributed system is closed and designed for a specific environment. This paper discusses a management support framework being developed by the BERMAN project within BERKOM, the broadband ISDN project of the German PTT. It describes the BERKOM environment and how management support for distributed applications is being designed for this environment by BERMAN. The management requirements of one BERKOM project and the management support being provided by BERMAN are discussed.

1. THE BERKOM ENVIRONMENT

The introduction of broadband ISDN networks such as BERKOM (BERliner KOMmunikationssystem) will allow high-speed reliable backbone networks to be used to interconnect diverse local and wide area networking capabilities into a very large open internet.[1] Not only can new forms of distributed processing be introduced but various organisations and networks that have previously worked in isolation can now cooperate on a wider scale than before and isolated islands of computing can be integrated into the whole. A variety of heterogeneous and autonomous systems can thus be connected into an open system. The nature of distributed systems will change as it will be possible to use services via the fast internet as speedily and as efficiently as over a LAN regardless of where the services are located, i.e. the performance and reliability characteristics of future LANs and WANs will converge. This can be represented by the idea of "cooperative processing" where the network is used as a local computer is used and where a different kind of access to services is therefore required.[2] In such cooperating open systems, services that were previously available only within closed systems can now be made more widely available, throughout the entire internet if so desired. All kinds of services can be offered to a greater number of users, but there exists no global design to which all components adhere or common basis for coordination between them.

Innovative multi-media[3] applications and a supporting infrastructure based on broadband ISDN are being developed within the BERKOM framework. A wide variety of services is available, and service access is no longer based merely on geographical proximity but also on the most appropriate service. In order to support this, the idea of the marketplace has been adopted where services are offered by various servers and where service users can negotiate on quality-of-service attributes so that the server that best meets the service user's requirements can be selected wherever it is located in the internet. This requires a common view of services regardless of their heterogeneity and could lead to greater competition between servers and to more sophisticated demands being made by users.

The BERKOM environment is thus a distributed service environment comprising a high number of heterogeneous services, resources, users and organisational structures where cooperative agreements can be concluded between autonomous entities in order to pursue specific objectives. The infrastructure being developed makes available a service pool with generally useful services that are intended to support cooperative processing in the BERKOM environment. These services are provided by distributed applications and include services for information storage and retrieval, trading and binding, monitoring and service quality control, authorisation and authentication, domain administration, and various transport services.

2. MANAGEMENT SYSTEM CONTEXT

In a broadband environment with several paradigms for distributed processing, innovative applications and more sophisticated demands on the internet, both from applications and users, distributed processing is developing in a qualitatively different manner from the classical WAN and LAN paradigms and this requires new approaches to management. The management context is much broader and more open and needs to support not only the varying applications already being planned for B-ISDN but also those that have not yet been developed or planned. It is not sufficient to manage the communication alone, there must be an integrated approach so that incompatible services and cooperation between application components can also be managed. The infrastructure supporting B-ISDN services must include management capabilities, and management functionality must be a constituent part of every distributed application.

In an open services environment distributed applications can consist of diverse components cooperating in order to achieve a common goal. The management system must be able to support such cooperation between autonomous entities. If autonomy is to be accepted, then servers should be able to determine how and when their service is offered, and to whom.[4] This is particularly true in an integrated environment where several different organisations are represented, some of whom may be competitors in certain areas and partners in others. They may wish to restrict the use of some services only to those organisations with whom they are currently cooperating and certain services may not be always be available. The management system must be able to manage such distributed applications.

With such an internet it is not feasible to have centralised control, the total network is too large and complex and connects several organisational domains, each with its own management policies and procedures for running its own subnets which must be integrated into the global system. Therefore, one of the issues in managing a very large distributed system is how to ensure cooperation within the internet so that the total system appears as one integrated system and yet at the same time incorporates diverse administrative domains. There must be a smooth interface between the management functions provided for the internet as a whole and those performed within the individual organisational domains. Management policies for the internet must be capable of cooperating with these autonomous areas and must also be valid for the whole internet, i.e. there must be management support for distributed applications spanning different organisational areas.

3. THE BERMAN PROJECT

The BERMAN project is investigating management requirements for distributed processing within the BERKOM environment. The project is adopting an integrated approach to management and is providing support for the management of distributed applications and not just OSI end-to-end communications management. An important design goal is to use standards where applicable and to undertake development in areas not yet standardised. Generic management support tools are being developed that can be used by all components in the distributed system. This is compatible with the idea of an infrastructure where common functions are factored out and offered as a pool of services to all applications.[5] In the same way, the management functions common to all applications are being factored out and offered as a generic set of tools that can be used by any application via a well-defined management interface. Applications thus contain management-related functions and the associated interfaces which allow management operations to be performed on them.

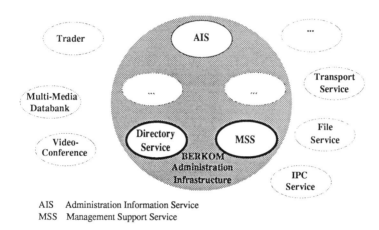

AIS   Administration Information Service
MSS Management Support Service

Figure 1. The BERKOM Administration Infrastructure in the Service Pool

The support being provided is called the BERKOM Administration Infrastructure (BAI) (see Figure 1). The BAI incorporates both directory and management services and its basic functions include: collecting and making available information about services; support for various interactions, such as probes, event notification, manipulating values; domain administration.[6] The advantage of such a support system is that it provides basic services that can be used generally and which do not have to be designed anew by each service designer. It also allows for cooperation between incompatible components which are using the same basic management services.

The BAI is part of the BERKOM service pool and provides services related to administration support for distributed applications. It consists of three services at present, the Management Support Service (MSS), the Directory Service (DS) and the Administration Information Service (AIS), but further services are to be incorporated later as required. The MSS and DS are now being implemented and the AIS is currently being designed. The three component parts are described below.

3.1. The Management Support Service

The Management Support Service (MSS) is based on OSI systems management concepts[7] which have been extended to support distributed applications in the BERKOM environment. It must therefore take into account the autonomy of the components and yet enable them to cooperate within an integrated framework to achieve a common goal. The MSS supports the collection, retrieval and manipulation of management data that is dynamic and liable to change very quickly and so cannot be stored using more permanent information storage tools. It is used for monitoring and controlling the various components of a distributed system and provides generic object types that can be used for application-independent and inter-application management. From this basis further objects can be composed that offer more complex or more specific functionality, for tailored application-specific management for example.

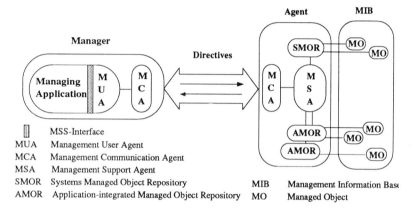

Figure 2. The OSI Management Model and the Basic Management Support System

The MSS is intended to be used by both dedicated managing applications and "normal" applications that contain some managing functionality. The operations offered include functions to establish and control a connection set up for management purposes, and OSI and non-OSI management functions. OSI systems management functions include those for object management, state management, alarm reporting, event control and security alarm reporting. Non-OSI management functions include managing dynamic service attributes (or qualities of service) and one to many management relationships. Management information is held in a management information base consisting of a Systems Object Repository (SMOR) which stores data not directly used for monitoring and control, e.g. configuration data or statistical data, and an Application-integrated Managed Object Repository (AMOR) where highly dynamic data is integrated within the application process being managed so that direct control and monitoring is possible. In addition to the Systems Managed Objects of the SMOR and the Application-integrated Managed Objects of the AMOR, a third kind of MO is allowed for, a Composite MO, which consists of both a Master Part and usually one or more Subaltern Parts. This is a further refinement of the OSI model and better supports the management needs of distributed applications.[8]

Figure 2 depicts the relationship between the OSI management model and the components of the Basic Management Support System (BMSS), which provides the MSS. The operations offered by the service are invoked via Management User Agents (MUAs) and carried out by Management Support Agents. The Management Communication System is hidden from the user and location transparency is achieved as managed objects are accessed and manipulated in the same way wherever they are located in the internet. The MSS thus provides generic tools that can be used by any component via the MUAs. The architecture of the BMSS vis-à-vis Managing Applications and Managed Application Components is shown in Figure 3.

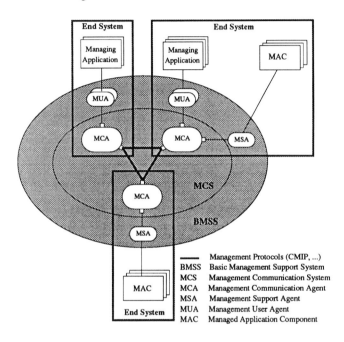

Figure 3. The Basic Management Support System

3.2. The Directory Service

A design aim here, as with the rest of the BAI, is to use standards where possible, and so the BAI is using the standardised X.500 Directory to store semi-permanent information about components.[9] The Directory Service provides global availability for locally stored information. The constraint on the information stored here is that it should be retrieved more often than it is modified and so it is used to store longer-lasting information of global interest, such as addresses, names, information on services, users and other components within the distributed processing system. The operations offered by the Directory Service enable entries to be created, read, modified and deleted. Searches can be undertaken on the information to retrieve entries relevant to a given keyword or combination of keywords. The service thus provides a nameserver functionality as the information stored locally about a component can be accessed using a globally unique name. The Directory Service supports querying by human users as well as by distributed applications.

3.3. The Administration Information Service

The Administration Information Service (AIS) is planned to make available more general operations for administering information and will thus provide transparency regarding the location of the information. Information may be stored in the directory system or in the management system, the user of the AIS does not need to know where. In this way, more complex operations can be made available by the AIS that comprise several operations of one or both of the other two services. For example, if a directory entry is to be created, a corresponding managed object could be created in the management system at the same time using one operation instead of two. Operations concerning groups of objects are also to be provided. The AIS will be particularly useful for domain operations where the same operation has to be invoked on several components, as a group operation will allow the same operation to be invoked on all components in a domain with just one command.

4. THE BERCIM PROJECT AND ITS MANAGEMENT REQUIREMENTS

The administration infrastructure being provided by the BERMAN project is designed to be used by other BERKOM projects to support their management needs. The BERCIM project is one of these projects and this section describes some relevant points of the project and its management requirements. The subsequent section discusses how the BERKOM Administration Infrastructure is being used by BERCIM to meet these requirements.

The BERCIM project is one of several projects investigating specific issues relating to the BERKOM environment. It is concerned with the problems of interdependence and communication between autonomous components and heterogeneous applications and their cooperation within an integrated system.[10] The project is using the distributed CIM environment as an example of independent islands of local processing that are to be interconnected using the fast and reliable data transfer provided by the BERKOM internet. However, the issues have a wider applicability to many other areas where advantage can be taken of high speed networks so that previously autonomous components can cooperate within an integrated system.

As part of the supporting environment a trading service, or trader, is being developed that can help a service user select a server that best meets the user's requirements. The autonomy of servers means that they need not service requests at the quality of service required or indeed at all, and so the idea of a contractual agreement between service requestor and server is introduced. The trader, acting on behalf of the service requestor, negotiates with the server on the service and service qualities that it contracts to supply.

The four aspects of the BERCIM context that interact with the BAI in this example are the domains, the service users, the trader and the servers. These aspects and their requirements on the BAI are described below.

4.1. Domains

The BERCIM environment covers a very large area and it is accepted that the autonomous entities within the internet are not completely open to all users. Partitioning the system by means of domains facilitates the administration and structuring of open systems. A domain is a grouping of components that are characterised by specific criteria or common management policies and can so represent both organisational and operational structures. Domains provide a means of defining "closed areas" in the open environment, thus reducing problems concerned with, for example, autonomy, heterogeneity or decentralisation, as the rules and procedures valid for a component remain under the control of the local organisation.

The autonomous entities are represented by organisational domains which are structured hierarchically. However, a strict hierarchical structure does not allow cooperation to take place between different departments of different firms, for example, and so federal (or cooperative) domains have been introduced that allow cooperative ventures between different subtrees of the hierarchy to be represented. In this way both the autonomous entities and the cooperative agreements can be supported by the architecture. Every component is registered in a hierarchical domain, which represents its organisational context. It can then be registered in federal domains and so request and offer services according to the specific conditions and rules of the federal domain. Each domain has a domain manager which creates and deletes domains and adds and removes components.

Requirements on the BERKOM Administration Infrastructure: The structuring into domains must be represented somewhere in the system so that when users request services it is known which servers are available to them. Also, if operations are to to be invoked on all members of a domain, for example, it must be known who the domain members are. There must therefore be support for defining domains and their relationship to other domains, as well as means for creating and deleting domains and adding and removing domain members.

4.2. Service Users

Service users are those components requesting the use of a service in the open environment. They may be human users or processes, although human users are more likely to request information about services from an information system. When requesting a service, users can specify particular requirements that a server offering this service should meet and can restrict the area within the potentially global internet where the server should be sought to a specific "search area". The user must, therefore, either use the default invocation or know which attributes are defined for the service and the values that can be specified for each attribute. A specific server need not be requested as the most appropriate server can be selected by the trader. However, users can request a service directly from a specific server if they so wish.

Requirements on the BERKOM Administration Infrastructure: Service users must be provided with information about the available services and servers that they can use. They need to know what services are available, the functionality of each service, the service attributes that can be specified and how to use the service. Users who wish to invoke a server directly also need to know how to identify and access the server and possibly the service attribute values of the server, for example, if a particular printer is a PostScript printer. As users are registered in specific domains they also need to know if they have the right to use any particular server if they wish to do this directly.

4.3. The Trader

The trader allows users to state their requirements on a service and it selects the most appropriate server. The task of the trader is therefore to use its knowledge about servers and their attributes to match service requirements with service offers and to establish a binding between requestor and server based on an agreed quality-of-service contract. It has similar functions to that of an intelligent trader currently under discussion in standardisation work, and in ODP terminology it provides a *search service*, a *selection service* and an *invocation service*.[11]

When the trader receives a service request it ensures that it has been correctly formulated, that the service is defined for the domain of the service requestor, and that any quality-of-service requirements are of the type specified for the service. It then selects appropriate servers based on the static service attribute values of the servers offering their service in the domain (or search area, if given) of the service requestor.

If more than one server has static service attribute values meeting the specified requirements the trader checks the current status of their dynamic service attribute values. The service offers are evaluated and ranked according to the suitability of the offer compared with the user requirements. Next comes the binding process which involves negotiating with the selected servers in order to ensure that the service attribute values can be guaranteed and that the server is willing to provide the service to the user. When an agreement has been reached with a server prepared to offer the service with the requested quality-of-service requirements it together with the necessary resources is "reserved" and a communication association is selected and established between the service requestor and the server.

Requirements on the BERKOM Administration Infrastructure: The trader requires information in order to select the optimal server. For services, it needs to know the correct service invocation, for what domains the service is defined, what service attributes are specified for the service and which are dynamic and static. It has to be able to retrieve the static attribute values of servers and obtain their dynamic attribute values on request.

4.4. Servers

Servers in the BERKOM environment offer a specific service that is known within the environment, i.e. the abstract service definition has been defined and registered in the information system and is valid for the domain in which the server is located. Each server can restrict the area in which it offers its service to specific domains within the service definition area and each server has its own values for the service attributes defined for the service. These can be either exactly one value, several values or a range of values.

The servers are autonomous and can make their own decisions as to whether they are willing to cooperate and provide their service. As the servers are heterogeneous they are encapsulated with a uniform interface with which to communicate with the BERKOM infrastructure services. In this interface are the functions required for the server to be part of the BERKOM environment.

Requirements on the BERKOM Administration Infrastructure: Means must be available for providing information about a server in the internet. It must be registered together with the values of its static attributes, and it must be possible to change these values as required. The dynamic attributes defined for a server must be known and the values of an individual server's attributes ascertained at any time. The server interface must therefore interact with the management system so that status information can be requested from the server and event notifications emitted by the server, thus allowing the server's internal activities to be controlled. These management aspects can be thus be handled in a uniform manner and the implementation aspects hidden.

5. BERMAN SUPPORT FOR BERCIM

BERMAN tools are being used to administer the servers and to provide support for the trader (see Figure 4). Information must be provided for the component infrastructure services as well as for the users of the system and made available globally via a uniform interface. How each group interacts with the BERKOM Administration Infrastructure is explained in the following sections.

5.1. Domains

Information on domains is stored by the Directory Service. A directory entry contains the data relevant to an object structured according to a particular schema. An initial schema is defined in the X.500 standard, but it can be extended within Private Directory Management

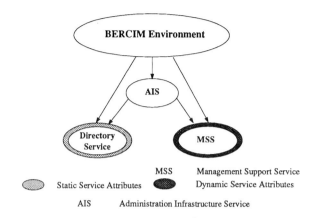

Figure 4. BERMAN Support for BERCIM

Domains by new object classes and attribute types. As the X.500 Directory is based on a hierarchical structure it matches very well with the structure of the organisational (hierarchical) domains. Each node in the Directory Information Tree represents a hierarchical domain and the entries at that node represent the users and servers belonging to the domain. The Directory schema has been extended to include an object class *general domain* to represent the hierarchical domains and from this the object class *federal domain* is derived which enables federal domains to be registered along with their components and a description of the federal domain, e.g. the reason for establishing the domain.

5.2. Service Users

Service users require information on the services available and how to use them. BERMAN has extended the Directory schema to include object classes for service types and for instances of a service, the servers. Use of the directory allows information that has been stored locally but which is of global interest to be available throughout the internet. Service users can thus obtain information on services and servers wherever they are located. The *service description* object class defines the abstract service specification and the semantics of the service, i.e. the functionality of the service and what the service does. The specification contains the service name, the domains for which it is defined, the service attributes of the service, and information on how to use the service.

Users can find the service they require without knowing the name of a particular service by using a keywords option to search for relevant services. They can then read the user information of any service in order to ascertain exactly what the service offers and how to invoke and use it. Users can also find out what service attributes are specified for a service and the values that each service attribute can take (step 1 in Figure 5). This is all the user needs to know in order to use a service as the user can ask the trader to select the most appropriate server offering the service (step 2 in Figure 5).

However, users who wish to select their own server can do so. The object class *service provider* contains information on how a server provides a service and how it can be accessed, including the server's values for the static service attribute values defined for the service. If the server is only available at certain times this information is also stored as well as the domains in which the server can be used.

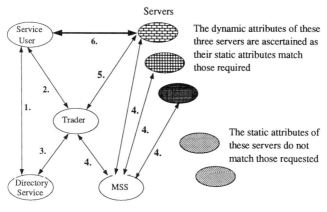

Figure 5. BERCIM Use of the BERKOM Administration Infrastructure

5.3. The Trader

The trader uses the BERKOM Administration Infrastructure to obtain information on the requested service and the servers offering the service. The object class *service description* provides the correct invocation of the service as well as the domains for which the service is defined. The service description also informs the trader about the service attributes defined for the service and which are static and which are dynamic.

The trader invokes a search operation on the Directory Service to obtain those servers providing the service in the domains of the service requestor (or in the search area defined by the requestor) whose static service attribute values match those requested by the user (step 3 in Figure 5). The trader now requests up-to-date values for the dynamic service attributes of these servers via the Management Support Service (step 4 in Figure 5). As the values of these attributes change rapidly, the trader negotiates with the servers until it obtains one that is prepared to provide the service with the requested values (step 5 in Figure 5). How the trader selects a server from the list is the subject of current study.[12] Once a server has been selected the trader links the service requestor with the server (step 6 in Figure 5). Information is stored in the directory about other infrastructure services required by the server including the communication service which must be invoked in order to use the server.

5.4. Servers

The servers are encapsulated with a uniform interface, or BERKOM shell, so that despite their heterogeneity they provide a standard interface to the rest of the environment. This interface incorporates the AMOR which allows the server to become a managed application component vis-à-vis the BMSS and also enables the server to negotiate with the trader about providing its service to the service requestor with the requested quality of service. A managed object template is defined for each service which represents the Management Support Service interface for the service. The managed object template specifies the management information and management operations that can be performed on the service via the Management Support Service. These object types are based on the OSI Structure of Management Information definitions[13] and standardised OSI concepts but have been extended for managing distributed applications in the BERKOM environment. For example, the MSS can be used to ascertain the dynamic service attributes of a server, a concept that does not appear in the OSI management standards. However, the OSI systems management functions[14] defined for the five OSI management functional areas, such as the state management function for configuration management, are also supported.

The template is specified individually for each service and generic operations can be thus combined to provide management support which is tailored to the service. Each server incorporates the management operations defined for its service type in its shell. The shell thus represents the management interface to the server and can, for example, provide information on the server's current state, including its dynamic service attribute values. This dynamic information is obtained on request via the Management Support Service which communicates with the relevant server. A link to the managed object representing the server is contained in the server's directory entry so that a directory search operation will provide its name and location.

6. IMPLEMENTATION

A pilot implementation of the BERKOM Administration Infrastructure is now being realised. The MCS is being based on the Network Management System (NMS) developed by University College London for the Integrated Network Communication Architecture (INCA) Project. BERMAN is extending it for the MCS design so that it conforms to the CMIS/CMIP standards. The MCAs are being implemented using the ISO Development Environment software, the publicly available implementation of the OSI upper layers and of applications which use these. The MUAs provide the interface between user and BMSS and are therefore being implemented in C. The MSAs incorporate the full OSI agent functionality, including scoping, filtering and name resolution and are being developed using C++.

The Directory Service implementation is based on the ISODE Directory System QUIPU also developed within the INCA Project at University College London. As well as extending the X.500 Directory Schema for BERKOM use, an application program interface to the service and an interactive interface suitable for human users are being developed.

7. SUMMARY

The BERMAN project is developing an infrastructure comprising tools that provide administration support to distributed applications in a large heterogeneous environment. The operations provided are generic and intended to be used to build specific management components in a distributed system. The example described has shown how the trader is using the BAI in the BERCIM environment, other scenarios could show how a managing application uses the management operations to exert control. For example, work is planned to investigate how qualities of service can be monitored and controlled in the BERKOM environment using the BAI.

The BERKOM Administration Infrastructure has been designed for a large distributed heterogeneous system. The aim has been to develop a basic support system with clear interfaces so that distributed applications incorporating a managing functionality and distributed applications needing to be managed can all use the BAI. In this way, the management aspects of services in the BERKOM environment can be provided with a homogeneous interface that hides the underlying heterogeneity, and the "building block" approach enables the tools provided to be used and combined as required.

Another aim is to investigate the extent to which current standards can provide the support required. The X.500 Directory is being used for basic retrieval capabilities, but as stated in the DAF Infrastructure document[5] more intelligent trading requires access to information beyond the Directory, such as dynamic management information. The BERMAN project is intending to combine both directory and management functions to provide a useful infrastructure for distributed applications using existing standards where appropriate and extending them where they do not suffice.

References

1. Popescu-Zeletin, R. et. al., "A Global Architecture for Broadband Communication Systems: The BERKOM Approach," in *Proceedings of the Workshop on Future Trends of Distributed Computing Systems in the 1990s, Hong Kong*, pp. 366-373, IEEE, September 1988.
2. Shah, K., "Managing Networks of the '90s," *Data Communications International*, vol. 18, no. 16, pp. 95-106, December 1989.
3. Moeller, E. et. al., "Distributed Processing of Multimedia Information," in *Proceedings 10th ICDCS, Paris*, pp. 588-597, IEEE, May/June 1990.
4. Clark, D.D. and L. Svobodova, "Design of Distributed Systems Supporting Local Autonomy," *COMPCON Spring 1980 Digest Of Papers*, IEEE, February 1980.
5. "DAF: Infrastructure (V5)," DAF 257, Tapiola, Finland, September 1989.
6. "BERMAN Project Deliverable 3," Consisting of *Guide to the Basic Management Support System* and *Guide to the BERKOM Directory*, Berlin, December 1990.
7. *Information Processing Systems - Open Systems Interconnection - Systems Management Overview*, ISO/IEC DIS 10040, June 1990.
8. Dittrich, A., *Composite Managed Objects*. This volume.
9. *The Directory - Overview of Concepts, Models and Services*, CCITT Recommendation X.500, December 1988. Has also appeared as ISO 9594-1.
10. Tschammer, V. et. al., "Support for Cooperative Work in Distributed CIM-Structures," in *Proceedings of the Sixth CIM-Europe Annual Conference, Lisbon, Portugal*, pp. 40-50, Springer-Verlag, London, May 1990.
11. *Working Document on the Trader*, ISO/IEC JTC1/SC21/WG7/N312, October 1990.
12. Wolisz, A. and V. Tschammer, "Service Provider Selection in an Open Services Environment," in *Proceedings of the 2nd Workshop on Future Trends of Distributed Computing Systems in the 1990s, Cairo, Egypt*, IEEE, September 1990.
13. *Information Processing - Open Systems Interconnection - Structure of Management Information*, ISO/IEC DIS 10165, June 1990.
14. *Information Processing - Open Systems Interconnection - Systems Management*, ISO/IEC DIS 10164, June 1990.

Integrated Network Management, II
I. Krishnan & W. Zimmer (Editors)
Elsevier Science Publishers B.V. (North-Holland)
© IFIP, 1991

Gemini: An Environment for Distributed Network Management Applications

J. Read, J. Balfour, P. Brun, P. Hyland, P. Mellor, P. Toft

Hewlett-Packard, Information Architecture Group, 19046 Pruneridge Avenue, Cupertino, CA 95014, and

Hewlett-Packard Laboratories, Filton Road, Stoke Gifford, Bristol BS12 6QZ, England

Abstract

One of the reasons industry has not wholeheartedly embraced distributed systems standards is that it has a large investment in proprietary systems that solve part of the problem. Incorporating standards into these systems can be expensive. Gemini combines a prototype of an object-oriented support platform called Atlas with the OSI network management standards at low cost, so that applications can use a consistent interface to take advantage of both.

1 Overview

Atlas [13], the prototype used in Gemini, is an object-oriented infrastructure developed in HP Labs. One of the uses it was designed for is structuring applications within a machine — an area OSI has just begun to address. Atlas also provides runtime services needed for application development that are not specified by OSI, such as multi-threading and thread synchronization primitives. In terms of standards, Gemini implements the object-oriented network management facilities currently being specified by OSI.

Even though both Atlas and OSI network management are object-oriented, neither of them can access the objects provided by the other. In this paper, both Atlas and OSI network management are considered to be object *worlds*, defined in terms of the policies, protocols, and services that apply within them. For instance, Atlas uses the NCS protocol [14] for inter-object communication, whereas OSI network management uses protocols such as SNMP [10] and CMOT [9]. In addition, OSI network management has user-friendly names, while Atlas has machine-oriented identifiers. The differences between Atlas and OSI network management are described in more detail below.

The combination of Atlas and OSI network management is actually an extension of Atlas to allow applications to interwork with objects in two different object worlds: Atlas and OSI network management. Applications can then use the best of both worlds, because they have access to all of the facilities of each.

The paper discusses Gemini from a technical point of view, concentrating on the facilities provided and how they are implemented rather than on the uses that can be made of them. It begins with a brief description of OSI network management, followed by a description of Atlas. It then describes the facilities provided by Gemini, and possible future work. The final section gives a summary of the main points of the paper.

2 OSI Network Management

Network management is a distributed activity. Management applications need to access remote network components either to get accurate and updated knowledge of the network, or to modify its status.

OSI has been defining a set of standards for managing open systems. They cover the topics listed below:

Model. The network to be managed is seen as a collection of objects on which applications perform operations by sending messages. Operations consist of getting or setting the value of object *attributes*, performing *actions*, and responding to *events* raised by the object [7].

Objects belong to classes defined according to the operations they support. Object instances are structured in a tree based on the containment relationship defined in [7]. Classes are organized in a multiple inheritance hierarchy.

Interface Definition Language. Each managed object class in the OSI environment has a high-level description, called a template, that describes its characteristics. Templates are specified in ASN.1 [1]. The information that appears in the template is defined by a document describing the Structure of Management Information (or SMI) [8].

Protocol. The CMIP protocol [5] is used between the management application and managed objects over OSI stacks. The CMIP protocol has several variations, depending on which communication stack is being used. CMOT is CMIP used over TCP [11], and SNMP is largely CMIP. SNMP was designed by the Internet community to fit in with existing communications stacks and resource location schemes, so it differs from the OSI version. It runs over UDP [12].

Identification. Each class, attribute, action or event is identified through a unique OSI object identifier which is primarily used for registration in an administrative hierarchy. OSI object identifiers are allocated from a global object identifier tree, where each internal node of the tree represents a naming authority. A naming authority may allocate object identifiers, new branches of the tree, or both, as long as uniqueness of object identifiers is preserved. The root of the naming tree is controlled jointly by ISO and the CCITT.

Naming and Location. To access a particular object, an application must know its name, its location, and its class. Thus, rules exist to name and locate objects, and to identify and describe the classes.

Each object is given an X.500 distinguished name, and its location and class are stored in an X.500 directory [2].

Management Services. As part of the OSI layer model, network management is described as an Application Service Element, known as the Common Management Information Service Element (CMISE) [6]. This service element describes the runtime services available from all managed objects.

3 Atlas

Atlas is a prototype of an object management facility that supports large numbers of objects distributed in a heterogeneous wide-area network which supports different machines, different compilers, and different languages.

Atlas is intended to allow the construction of distributed, object-based applications in many application domains, but was initially developed for office information systems. The main features of Atlas are described below:

Model An Atlas object is an encapsulation of state and behaviour. Each object is an instance of a *class* that defines the methods that it implements. How these methods are implemented, and how the state of the object is represented, are hidden from the invoker of the object.

The primary means of communication between objects is method invocation; both synchronous and asynchronous method invocation are supported. An Atlas object is multi-threaded, with a new thread of control created within the object for each incoming invocation.

Atlas provides location transparent method invocation. The only information needed to invoke a method on an object is its Atlas object identifier or *OID*.

Once created, Atlas objects are persistent. Their existence and state do not depend on any given activity in the system. At any time they can be either *active* or *passive*: active objects execute in memory, while passive objects reside on secondary storage. Whether an object is active or passive is transparent to the invoker because Atlas ensures that an object is made active before any method invocation is delivered to it. The implementor of the object provides the mapping between the object's passive and active forms.

Interface Definition Language An Atlas class definition is a description of the method interface provided by a class of objects. It is generally defined using an interface definition language called *MIDL* (Madrigal Interface Definition Language).

Atlas is not restricted to the use of MIDL as the IDL. As long as the IDL can be used to generate the appropriate entries in the *class database* which is consulted before stubs are generated, then it may be used.

Protocol Atlas uses the NCS protocol for inter-object communication. All Atlas method invocations are marshalled into NCS protocol data units. The NCS remote procedure call system is used when method calls are routed to remote machines.

Identification Atlas identification is based on globally unique identifiers. An atlas identifier is a concatenation of a unique global identifier (the machine's IEEE network address) and a unique local identifier (a timestamp). At class compilation time, which is discussed in §4.3, unique identifiers are assigned to each class, and to each method of a class. When an Atlas object is created, it is assigned a globally unique Atlas OID. These OIDs are immutable and identify one and only one object throughout the lifetime of the system. Atlas OIDs are *system-level* identifiers, at a lower level than OSI managed object names.

Naming and Location To access a particular object, an application need only know the object's Atlas OID. When a method is invoked on the object, the Atlas location service determines where the method invocation should be sent. This service is transparent to the programmer.

As yet, Atlas provides no service for mapping meaningful names like "Laser Printer One" into OIDs.

4 Gemini

Gemini does not try to merge object worlds. It instead recognizes that different object worlds exist, and achieves cooperation among them. The environment initially supports Atlas and OSI network management, but the structure can be extended to additional worlds.

Gemini provides a set of facilities which allow both worlds to be accessed in a consistent way. These include a single application programming interface (API) to objects, stub compilers which produce stubs that cooperate to use a single threading mechanism, and a high-level naming scheme. Each of these facilities is described in more detail in the subsections below.

The Gemini style of interworking is shown in Figure 1. There are three types of line in this figure. The dashed lines represent the relationship between client and server, as expressed in the interface definition language (IDL). In this relationship, the services are provided by the server and used by the client. The dotted lines represent the relationship between the IDL and the stubs generated at compile time. The solid lines represent the invocation of methods at runtime.

Note that the figure shows two different IDLs being used to describe objects. This implies that a class is either described in ASN.1 or in MIDL, and that it cannot span the two worlds. That is, Gemini does not provide complete *interworking transparency* between the two worlds. Interworking transparency is defined in [3] as follows: If an object can interact with another object without regard for the environment in which that object resides, then there is interworking transparency.

For example, it is not possible to create a class, such as *modem*, which is known to both worlds. Instead, two different classes, *OSI modem* and *Atlas modem* must be created to describe the set of all modems. If a client is interested in finding all of the modems in the system, it must know how many different classes of modem exist in the system.

Names, however, can span worlds. That is, a named entity can support classes from several different worlds. For example, a printer registered as *myprinter* can support a *managed printer* class in the network management world and a *printer* class in the Atlas world.

The restrictions on interworking transparency are not a serious limitation for network management applications because classes of functionality tend to fall naturally into one

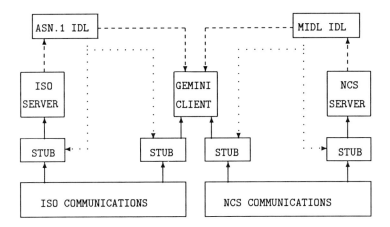

Figure 1: Gemini Interworking

world or the other. Managed object classes are defined to be in the OSI world, and application object classes, such as classes performing I/O to the user's screen, fall naturally into the Atlas world.

4.1 API

Programmers are showing growing interest in object oriented languages such as C++. Moreover, the system to be managed has an object oriented flavour. Therefore, C++ was used to handle managed objects in such a way that they are seen as being local C++ objects. This interface gives the level of location transparency that RPC gives in procedural interfaces.

4.2 OSI Template Compilation

The code for a pair of client and server stubs is specific to a managed object class. It can be generated automatically from the template describing the class. The following sections describe templates in more detail, and then discuss the automatic generation of stubs from the templates.

4.2.1 Templates

Templates are specified in Abstract Syntax Notation One (ASN.1) which is primarily a data typing language. However, it has a *macro* facility that allows new constructs to be defined. These are specified as a set of (essentially BNF) grammar rules describing the syntax of the new construct. Once defined, instances of the new syntax can be used as part of normal ASN.1.

A set of macros have been defined that allow the specification of templates, based on the SMI. An example of a template is shown below. Note that each item in this specification is assigned a unique OSI object identifier with which it can be identified.

```
managed-laser-printer OBJECT
    INHERITS FROM managed-printer
        ATTRIBUTES speed, fonts
        CREATE
        DELETE
        ACTIONS print-test-pattern, status
        EVENTS add-paper
    ::= { hp-nm laser(0) }

speed ATTRIBUTE
    SYNTAX sheets-per-minute REAL
    ACCESS READ
    ::= { hp-nm speed(1) }

fonts ATTRIBUTE
    SYNTAX SEQUENCE OF FontId
    ACCESS READ-WRITE
    ::= { hp-nm fonts(2) }

print-test-pattern ACTION
    PARAM Document
    ERRORS spool-queue-full
    ::= { hp-nm print-test-pattern(3) }

status ACTION
    ACK StatusInfo
    ::= { hp-nm status(4) }

add-paper EVENT
    PARAM sheets-left INTEGER
    ::= { hp-nm add-paper(5) }

spool-queue-full ERROR
    PARAM queue-limit INTEGER
    ::= { hp-nm spool-queue-full(6) }
```

In this example a `managed-laser-printer OBJECT` is being defined. It is itself a subclass of another object, `managed-printer`, from which it will inherit `ATTRIBUTES`, `ACTIONS`, and `EVENTS`. The `print-test-pattern` action in the template above has been shown as a local action for clarity, but in reality, it will almost certainly be inherited. When the `CREATE` (or `DELETE`) keyword is present in the template, the creation (or deletion) of the object can be remotely controlled by management applications. The creation request may specify the initial values of the attributes.

This managed object has two attributes, speed and fonts, which determine the speed of printing and the fonts available. It has two actions, print-test-pattern, which allows the administrator to check that the printer is working, and status, which returns the current status. It can raise a single event, add-paper, asking for paper to be added.

All the items named in the managed object must themselves be defined by other macros. The speed attribute has a SYNTAX showing the type of data stored (in this case a real value), and an ACCESS part showing that this particular attribute is a read-only attribute. The fonts attribute is the same except that this attribute is a read-write attribute.

The actions that an object supports can be defined with the action macro, as in the case of print-test-pattern and status. With this macro a PARAM, or parameter, data type can be specified. This data is provided to the object when it is to perform the action. An ACK, or acknowledgement, data type can also be specified. This is returned from the object after it has performed the action. It can also return an ERROR if something goes wrong. Errors, such as spool-queue-full, can have a data type associated with them.

It is also possible for objects to generate events asynchronously, such as add-paper. This is the equivalent of an action except it is generated by the managed object and performed on all those parties that have expressed an interest in the event.

4.2.2 Client Stub

The ASN.1 compiler compiles the templates into client and server stubs. The client stub represents the server (i.e. the managed object) to the client, and the server stub represents the client to the server. This section describes the mapping from the macros to the C++ stubs.

The client class definitions produced from the above example are shown below:

```
/* CLASS DEFINITION */
class managed_laser_printer_client : managed_printer_client {

    /* attributes */
    MethodErr get_speed(get_speed_out &);
    MethodErr get_fonts(get_fonts_out &);
    MethodErr set_fonts(set_fonts_in &, set_fonts_out &);

    /* methods */
    MethodErr print_test_pattern(print_test_pattern_in &);
    MethodErr status(status_out &);

    /* events */
    EventId register_add_paper(add_paper_handler, void *);
    unregister_add_paper(EventId);

    /* create/delete/lookup */
    create_managed_laser_printer(name &, address &);
    delete_managed_laser_printer();
    lookup_laser_printer(name &);
```

```
    /* construct/destruct */
    managed_laser_printer_client();
    ~managed_laser_printer_client();
}


/* TYPE DEFINITIONS */
typedef get_speed_out ...
typedef get_fonts_out ...
typedef set_fonts_out ...
...
typedef int (*add_paper_handler)(
                int, void *, add_paper_in &, add_paper_out &);
```

The OBJECT macro is mapped to C++ classes. Each of the inherited objects becomes a C++ base class from which this class is derived. Multiple inheritance is supported in C++ 2.0 and in the object templates. The actions of an object are mapped to methods in the C++ class. Each has two arguments representing the input parameter and the output parameter (if these are required). Each method also returns a **status** value in which the errors are encoded.

Two methods are usually generated for each attribute — one to **set** the value and one to get the value. However, these are only generated if there is a corresponding access statement in the template. For example, no **set** method will be generated for a read-only attribute.

The mapping of events is more complex. The client registers interest in an event by calling a method in the client stub and depositing a handler procedure for that event. This handler will be called whenever the event is raised by the server. The client gets back an event handle when it registers its interest, so that it can unregister itself if it does not desire to be notified of further occurrences of that event.

The create and delete keywords (if present) cause the generation of a **create** and **delete** method for the managed object. There is also a **lookup** method that enables the client to bind the client stub to a real server, by supplying the name of that server.

Two methods are generated to control instantiation of the stub object which acts as a proxy for the managed object, a constructor and a destructor. These are the managed_laser_printer_client and ~managed_laser_printer_client methods in the example above. These methods have no effect on the managed object itself.

4.2.3 Server Stub

A similar set of methods are generated for the server stub. These have an extra requirement in that they must be able to map incoming method and OSI object identifiers into real class and method pointers on the local machine. To do this the ASN.1 compiler also produces a set of tables relating these items. These tables are compiled by the C++ compiler and linked into the application.

The tables required are :

method_table{method_id, code_pointer} Each class has a table relating its method identifiers to the actual code for the method.

super_table{superclass_id} Each class has a table listing its superclasses.

class_table{class_id, method_table, super_table} Each application has a table relating the class identifiers of all the classes involved in that application, to the tables for those classes.

4.2.4 ASN.1 Compiler

The template compilation facilities are based on a flexible ASN.1 compiler developed in 1989 by HP Labs [4]. This tool was produced because of the prominence of the ASN.1 language in the standards arena. Uses of the language are becoming more varied as time progresses. In addition, the purpose of ASN.1 is to allow communication between heterogeneous systems. Therefore, it was sensible to develop an ASN.1 compiler suited to such a heterogeneous environment, which produces output in any programming language desired by the user.

The compilers available from other sources are fairly restrictive. This compiler stands out in that it has the following features:

- It supports full 1988 standard ASN.1, including subtypes, import/export, and values. It does not change the syntax of normal ASN.1 in any way.

- It has full macro processing capability. This means that macro definitions are processed on-the-fly, and that instances can then be parsed and processed. Most compilers that handle macros do a simple macro expansion. This compiler also allows semantics to be given to macros. For example, the network management templates are written as a set of macros and annotated to produce C++ stubs.

- It does not use any annotation of the ASN.1 source code (except in macro definitions). This means there is no need to modify the specifications appearing in standards, and that the source code is kept portable, and easy to maintain.

- It can compile recursive module definitions.

- It allows interactive correction of parse errors.

- It gives user definable output. That is, the output can be in any language, and in any form required by the application. The output to be generated is specified by using a set of CORE macros. Several standard sets of CORE macros have been developed, and users with special requirements can write their own or modify the ones supplied.

4.3 Atlas Class Compilation

An Atlas class is described by a *class definition* which plays the same role as an OSI network management template except that the Atlas class definition describes a different set of functionality than the OSI template. Atlas classes are organized in a multiple inheritance hierarchy which also allows a new class to be *generalized* from existing classes

[16]. Generalization is the creation of a superclass from existing classes in the hierarchy. The existing classes effectively inherit the new class. A class is said to *specialize* another class when it inherits it. The class hierarchy provides the ability to treat an instance of a particular class as if it were an instance of one of the inherited classes, which is a powerful mechanism in distributed systems.

The generation of stub procedures from an Atlas class description is a two stage process. The class description is compiled into a set of entries in a *class database*, which are then used to generate stub procedures.

This section briefly describes the Atlas class definition language, class compiler, class database, and stub generation facilities.

4.3.1 Atlas Class Definitions

Atlas classes are generally specified in MIDL, although other IDLs may be used. Atlas provides a compiler to generate entries in the class database from MIDL specifications.

The example below defines the Atlas class for printing a document. It is important to note that there is no overlap in functionality with the managed object template given in §4.2.1.

```
class atlas_laser_printer specializes atlas_printer ;
method [ integer 4 status ] PrintRaster ( charstring image ) ;
...
endclass
```

Atlas class definitions begin with the name of the class being defined, along with the names of any classes that are being specialized or generalized. The methods provided by the class are then defined in terms of their arguments and return values. Arguments and return values may be basic parameter types (integer, charstring, etc.) or composites of the basic types. There is no equivalent of the attributes or events found in OSI network management templates, nor any constraint on the names and forms of the methods.

Atlas class definitions may also contain a description of the *serial form* of an object. The serial form defines the data structure used to transfer the state of an object instance from one implementation of the object's class to another. The transfer of state is part of the process of moving an object.

4.3.2 The Atlas Class Database

The class database contains entries which include the globally unique identifiers associated with classes and methods. Conceptually, there is one global database containing all of the classes ever defined. The class database is distributed so that each Atlas machine contains a part of the database. Once a database entry is created, it can never be changed. Consequently, the problems normally associated with keeping replicated copies of a database consistent are avoided.

4.3.3 Stub Generation

The entries in the class database are used to generate client and server stubs which have the same form as those produced by the OSI network management template compiler.

Parameters are marshalled into the NCS network representation format, and the protocol data unit is sent using the Atlas method calling primitives.

The C++ classes generated as the programming interface to the client stubs have the same form as the C++ classes generated for OSI managed objects, so that objects in either world can be manipulated consistently. Error returns are treated in the same way, and manipulation of a remote object is via a local client stub proxy (a C++ object). An instance of an Atlas class may be referred to by name or by Atlas object identifier. When the instance is identified by name the client binds to the server in the same way as for OSI managed objects. Otherwise, the Atlas location mechanism is used.

4.4 Naming and Location

As stated earlier, Atlas class instances can be accessed either by name or by Atlas object identifier. Managed object instances are always accessed by name. Gemini provides a location facility that is transparent to the application programmer through a utility which binds the appropriate stub to the application program. The choice of stub is based on the name or object identifier used to identify the instance being accessed.

When an Atlas object identifier is given to this utility, an Atlas stub is bound to the application program. When a name is given to the utility, it consults the X.500 directory to determine how to communicate with the desired instance, and to retrieve the information needed to establish communications. This information is recorded and used for all method calls in the class. Applications can communicate with objects that support CMOT, SNMP, or Atlas classes.

The information stored in the directory includes the protocols that can be used to communicate with the object, along with an internet address and a class identifier for CMOT and SNMP objects, and an Atlas OID for Atlas object classes.

Object location and registration facilities are produced as a part of the stubs produced from template compilation.

4.5 Protocols

The network management protocols supported by Gemini are those which run over TCP/IP networks. Consequently, managed objects are accessible through CMOT and SNMP. The protocol supported for Atlas is NCS.

Communication stacks for CMOT, SNMP, and Atlas coexist in Gemini. In addition, the stubs produced from template compilation provide support for association control, request and reply synchronization, and parameter marshalling for each of these protocols.

5 Future Work

The Gemini platform is just beginning to be used to develop network management applications. Consequently, it is too early to give an assessment of how well it has worked in practice. Such an assessment should, however, be done.

Work on Gemini has prompted some thought on interworking transparency, see [3]. Two conditions must be satisfied to give Gemini programmers interworking transparency between the Atlas and OSI worlds. It must first be possible to derive the class definition in Atlas directly from the OSI managed object class template, or vice versa. This can be done in a straightforward manner. Secondly, the client must be able to encode and decode both ASN.1 and MIDL to and from the same set of C++ data structures. This is harder to achieve. Unless the same person writes both of the stub compilers in parallel, it is difficult to generate the same output as an existing compiler in all cases. These two conditions give Gemini programmers interworking transparency, but gateways between worlds would still be needed, for example, to allow an OSI object to access an Atlas object directly. However, as long as the conditions above hold, Gemini-based gateways for direct communication between worlds could be generated automatically.

In addition to this, semantic specifications would be quite useful in templates. This information would, for example, specify when an action could be performed on an object, which would allow the designer of the printer object given earlier in this paper to specify that print was not allowed unless the printer supported all of the appropriate fonts. Work is about to begin on finding methods of adding a suitable level of semantic specification to the IDL.

More work needs to be done in terms of standardization. Two areas for improvement were identified during the design and implementation of Gemini. First, there are two SMI definitions of managed objects, one for SNMP and one for CMIP and CMOT, which made it difficult to provide the application writer with a consistent interface. Second, information about managed objects can be stored either in the management information base (MIB) defined by OSI network management standards, or in the directory information base (DIB) defined as part of X.500. Designers and implementors of managed objects have to choose between these two, potentially redundant, information bases when there is no agreement in the standards community on which information base is more appropriate for a given piece of information.

6 Summary

Gemini is an environment which simplifies the task of writing distributed network management applications. It combines a prototype support platform and an OSI standard at low cost, so that applications can use a consistent API to take advantage of both. The API is object-oriented, gives a great deal of distribution transparency, and allows the application programmer to treat remote objects as though they were local. In addition, Gemini provides coexisting communications stacks, automatic stub generation for both ASN.1 and MIDL, and a high-level naming scheme that can be used to name objects in both worlds.

OSI managed objects are not the only remote objects in an application. The application itself may be composed of many objects which may be distributed throughout a network. This is a powerful capability, limited only because programmers must know which worlds objects are in. The work on Gemini, however, has shown that this restriction can be removed.

References

[1] ISO/OSI. "Specification of Abstract Syntax Notation One (ASN.1)". March 1988.

[2] CCITT. "Directory". Recommendations X.500–X.521. November 1988.

[3] P. V. Mellor. "Compiler Support for Interworking ISO and NCS Environments". Hewlett Packard Technical Report HPL-ISC-TR-90-013. April 1990.

[4] P. V. Mellor, J. Balfour. "A Users Guide to the ISC ASN.1 Compiler". Hewlett Packard Technical Report HPL-ISC-TR-89-053. Nov 1989.

[5] ISO/IEC 9596. "Common Management Information Protocol Specification". April 1989.

[6] ISO/IEC 9595. "Common Management Information Service Definition". April 1989.

[7] ISO/IEC 7498-4. "Information processing systems — Open Systems Interconnection Reference Model — Part 4: Management Framework". April 1989.

[8] ISO/IEC DP 10165. "Information processing systems — Open Systems Interconnection Management Information Services — Structure of Management Information Part 1: Information Model". January 1989.

[9] U. Warrier and L. Besaw. "The Common Management Information Services and Protocol over TCP/IP (CMOT)". RFC 1095. April 1989.

[10] J. D. Case et al. "Simple Network Management Protocol (SNMP)". RFC 1098. April 1989.

[11] DARPA Internet Program Protocol Specification. "Transmission Control Protocol". RFC 793. September 1981.

[12] DARPA Internet Program Protocol Specification. "User Datagram Protocol". RFC 768. September 1981.

[13] Harry Barman, Nigel Derrett, Claus H. Pedersen, Andy Seaborne and Peter Toft. "Atlas Overview". Technical Report, Hewlett-Packard Laboratories, Bristol, England. 1989 — Draft version, to be published.

[14] Apollo Computer Inc., 330 Billerica Road, Chelmsford, Massachusetts. "Network Computing System (NCS) Reference". 1987.

[15] Claus H. Pedersen. "Extending ordinary inheritance schemes to include generalization". OOPSLA 1989 Object-Oriented programming, systems, languages and applications. ACM. September 1989.

[16] Claus H. Pedersen, Harry Barman, Nigel Derrett, Andrew Seaborne and Peter Toft. "The Atlas Class Hierarchy". Technical Report, Hewlett-Packard Laboratories, Bristol, England. 1990.

Design of the Netmate Network Management System

Alexander Dupuy, Soumitra Sengupta, Ouri Wolfson and Yechiam Yemini[*]

Computer Science Department, Columbia University, New York, NY 10027

Abstract

The Network management, analysis, and testing environment (Netmate) project addresses research and experimental issues in distributed network management of large, heterogeneous networks. This paper describes the Netmate architecture, and its model for network management information, which emphasizes the definition of generic network objects and relationships chosen specifically for efficient network management. The model, with its powerful abstraction mechanisms, simplifies the development of Netmate visual, analytic, and auxiliary tools for monitoring, analysis, planning, and testing of complex computer networks.

1. INTRODUCTION

As computer networks become larger (with thousands of elements), more heterogeneous (supporting multiple protocols at many networking layers), and complex (with subtle interactions and inter-relationships between protocols), network management has emerged to be a crucial requirement. For the network to be the information highway for an enterprise, there must exist comprehensive management tools to ensure network integrity and smooth operation.

The Network management, analysis, and testing environment (Netmate) project pursues, as its long term goal, development of fundamental enabling technologies and a comprehensive software environment for network and systems management. We focus on novel technologies and effective tools to support monitoring, interpretation and control of complex dynamic network behaviors. These tools are unified as parts of the Netmate environment (figure 1). Netmate includes software for agents capable of observation/control of networked devices, protocols, and systems (OCP); a Modeler that maintains a management information base (MIB) with a data model of complex networked entities, collects and provides dynamic real-time network data to management

---

[*]*Research supported by DARPA contract #F-29601-87-C-0074 and N.Y. State CAT contract #NYSSTF CAT(89)-5.*

analysis applications (Auxiliary Systems); a network simulation tool for planning and testing of protocols; and a visualization tool that provides human experts with navigation aids to simplify monitoring and control of complex network scenarios. In the Netmate architecture, we focus on experimental studies in order to develop a realistic understanding of the management issues and demonstrate the resulting technology concepts.

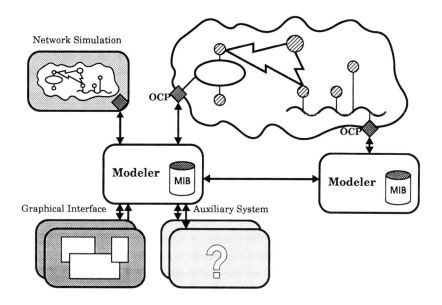

Figure 1: Netmate Architecture

The fundamental issues investigated by Netmate include:
1. Understanding network dynamic behaviors: Why do networks fail? What are the mechanisms by which faults evolve, and how can they be observed and controlled? Modeling and analysis of dynamic (vs. equilibrium) and transient behaviors of networks along different lines of approaches (statistical, neural network-based, inference-based) are expected to develop analytical models to capture and analyze network fault scenarios.

2. Manager-Agent interaction protocol/architecture: How to build manageable networked systems? What language and protocol constructs should be used to specify and delegate management instructions from a manager to an agent? A decentralized management structure supporting a delegation protocol and development of generic agents are necessary for efficient manageability [2].

3. MIB organization: How can we collect/organize/present complex real-time network operational behavior and configuration data? This issue deals with the abstraction modeling of network information that is suitable for analysis and visualization. In this paper, we present the model proposed in Netmate with emphasis on the Netmate Structure for Management Information (SMI) *as determined by the need for efficient network management functions.*

4. Interpretation: How can we correlate/interpret observations of complex network behaviors to diagnose their causes and control their evolution? Due to inherent inconsistencies in the network data, an important requirement in analysis of network behavior is that of incremental diagnosis based on temporal and incomplete observations [10].

5. Visualization: How to provide visual navigation and control of complex scenarios? The visual model of network data is primarily concerned with providing good visual abstraction mechanisms in order to reduce the complexity and the large volume of collected information.

Section 2 presents the Netmate view on the requirements of network data modeling. These requirements decide the SMI for the network model in the Modeler component of Netmate, which is elaborated in Section 3. We discuss other components of Netmate in Section 4. The current status of the Netmate project and conclusions appear in Section 5.

2. MODELING NETWORK INFORMATION

Modeling is necessary for creating a common framework for collection, storage and retrieval of network information. The inherent purpose of these operations is to be able to analyze the information (by human experts or by automated analysis modules) in order to ensure safe and efficient operations of the real network.

The current trend is towards very large and interconnected networks. This implies that the information necessary to manage the network is very large. For example, network entities must be identified at all layers in order to manage network protocols, therefore, the number of such entities is several times larger than the number of physical devices. Furthermore, the management would need to access the dynamic network information repetitively, with the series of values over time available for detailed analysis. Real-life networks are also heterogeneous in all layers. This implies that there are many classes of information with different properties and requiring different treatment. For example, it is common for an enterprise to have both Ethernet and Token-Ring with protocols such as TCP/IP and DECnet on Ethernet and TCP/IP and SNA on Token-Ring operating concurrently.

Thus, the primary function of the modeler is to be a repository of both large volumes of static and dynamic network information about a large class of network entities and protocols. Furthermore, the model must support several levels of abstractions for the collected information in order to allow efficient analysis procedures. The efficiency, size, and heterogeneity requirements dictate that the abstractions be realized as the definition of objects and relationships in the MIB.

A modeler has additional responsibilities regarding coordination of the management entities themselves. For example, an user interface may query the modeler for information, which, as a result, may schedule a query to the real network. When a reply is received from the network, the modeler must correlate that reply with the original query from the user interface. The analysis modules may specify a trigger so that when new information arrives from the network about a specific device, a specific management module function is invoked. There are many crucial issues in the management of the management protocols, which, due to lack of space, are not discussed in this paper.

The objectives and functions of a generic network management system are well described in the current OSI model [5]. Many proprietary management systems, through experience, have also defined the requirements for their individual systems [7, 9]. In order to support a consistent nomenclature, a common MIB is proposed as part of the standards specifications. The two standard data models are the Internet SMI [6] and the OSI SMI [4]. While these two standards provide a common structure for management data for heterogeneous networks, they do not sufficiently address the manageability needs of these networks as discussed in the following section.

3. NETMATE SMI

The central question in constructing the SMI is: What generic objects and generic relationships characterize the domain of network configurations and operations? These objects and relationships should abstract the common behaviors exhibited over a large number of network entities and protocols. Once such comprehensive objects are determined and implemented, it is then possible to construct common logical deductions that are uniformly applicable for different management functions.

For example, a generic property called *status* may be associated for all network entities with a set of possible values including *up* and *down*. A user interface module may then determine appropriate graphical rendering (colored green if *up*, red if *down*) based upon this property regardless of other specific properties of any object. A transition from *up* to *down* state may automatically trigger an analysis module associated with the object. This property, common to all network entities, permits generic inferences to be specified and carried out.

The generic object and relationship definitions should be capable of supporting a rich set of functions which are directly related to efficient network management. Determining such objects and relationships requires experimental dissemination of networking protocols and their relation to the existing, common-place management practices. It is, however, impractical to prescribe a closed set of definitions which do not allow future modifications to the SMI. In fact, extensibility of the SMI is an important requirement to model current network entities, most of which have additional specific properties.

The basic idea of the Internet and ISO SMI standards is to allow a mechanism for the definition and naming of variables containing management information (essentially name-value pairs). Additional structuring is provided in the form of tables of variables which can be defined (although the Internet SMI doesn't support nested tables). The ISO SMI has an object-oriented model, with variables specified as attributes of objects, an inheritance mechanism for defining object classes, as well as a single relationship between objects: containment.

3.1 Network Model

In discussing the Netmate SMI, we will use an example network (figure 2) to examine a number of network management scenarios and how the model would support management operations. The example shows a link layer network of an Ethernet (**E**) with two nodes (**A** and **B**) and a Token-Ring (**T**) with one node (**D**). **E** and **T** are connected by a bridge (**C**), which is also connected to a serial line (**S**). In the physical layer, **E** is divided into two segments (**E'** and **E"**) joined by a repeater **R**. In the application layer, an application client (**D'**) communicates with a server (**B'**). The Ethernet and the nodes connected to it belong to corporation **G**; the Token-Ring and the node connected to it belong to corporation **H**. Using this simple network as an example, we will elaborate a few non-trivial problems associated with network modeling, and present Netmate solutions with qualitative comparisons of the Netmate model with other standard models.

Netmate uses the object-oriented paradigm [3] to represent network objects, data and relationships. The generic objects in this paradigm are *class* and *object*, which is used to represent all entities. The generic relationships are *is-a* and *is-a-kind-of* relationships, and the latter, applicable on the *class* objects, constructs the usual class hierarchy. The class hierarchy is a proper beginning to model networks; however, by itself, it does not succinctly address all generic relationships in the network domain. In the following, we present the our model with careful distinctions and examples.

Netmate SMI (figure 3) currently defines four important object classes: *Layer*, *Node*, *Link*, and *Group*; and five relationships: *is-in-layer*, *is-connected-to*, *is-member-of*, *is-part-of*, and *is-implemented-in-terms-of*. The class *Group*, and relationships *is-part-of* and *is-member-of* have applicability in other domains such as real-time process control.

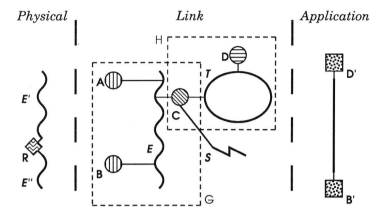

Figure 2: Example Network

Layer. An instance of the *Layer* class represents a network protocol layer with clear functional distinctions and operational boundaries (*e.g.* the ISO Reference Model for networks). It is a generic class because almost all network protocols are implemented in layers. Netmate does not limit itself to any specific layering scheme; instead, it allows layers to be defined for any consistent set of objects that communicate using a common protocol. For example, an Internet network may have TCP, IP, and Link layers, and may coexist with SNA network which has SNA-LLC and SNA-Session layers. The network in the figure 2 is operating at three layers: application, link, and physical.

Node. An instance of the *Node* class, within a specific layer, represents a hardware, firmware, or software element (*e.g.* Ethernet interface in link layer, IP Gateway in IP layer, LU6.2 node in SNA-Session layer) which obeys the protocol rules in that layer. Nodes have the subjective semantics of being endpoints of communication. In figure 2, nodes **A**, **B**, **C** and **D** are in the link layer, **R** is in the physical layer, and **B'** and **D'** are in the application layer.

Link. An instance of the *Link* class, within a specific layer, represents the communication between nodes within the same layer, where the communication is governed by the protocol rules in that layer (*e.g.* Ethernet in link layer, Telnet session in application layer, SNA conversation in SNA-Application layer). **E** and **T** in figure 2 are examples of links. Unlike the Internet and ISO SMIs, links are given equal importance to nodes in Netmate, even when information about links is available only through nodes. This is so because the model distinguishes the *different* properties between nodes and links, and makes the distinction explicit through class definitions. For example, in Token-Ring Source Routing protocol, *max_packet_size* is a property of a connection (*i.e.*, the link), and not of any specific node on that connection.

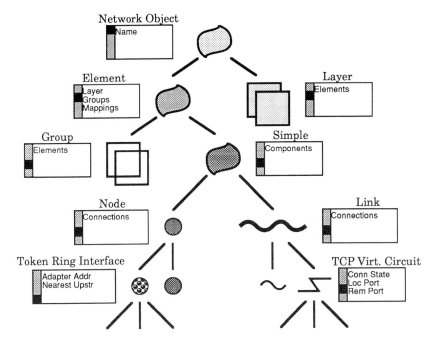

Figure 3: Network Model Class Hierarchy

Group. An instance of the *Group* class, within a specific layer, is a collection of nodes and links (and other groups) which exhibit common operational semantics (*e.g.* Ethernet objects in NY office, Objects on regional T-1 network, All NFS servers). In figure 2, the corporations **G** and **H** form examples of groups. Groups serve many purposes: they may be used for various administrative purposes such as common maintenance contracts, for constructing a set of objects to investigate during a fault analysis period, for constructing a building-by-building network picture on the user interface, etc.

In Internet and ISO SMIs, all network entities are simply objects. The Netmate SMI enhances the classification hierarchy as described above, and then by similarly enhancing the relationships, permits efficient management analysis, as described next.

is-in-layer. A node, a link, or a group may belong to a single layer, which is represented by this relation. Using this relation, a common layer-specific behavior may be prescribed for the nodes and links, and some other behaviors disallowed. For example, it is *a priori* possible to assert that the link layer node **A** in figure 2 may not communicate with the physical layer node **R** using the

Ethernet (link layer) protocols. Such assertions may be assumed by analysis modules, for which the *is-in-layer* relation provides the information which nodes, links, and groups belong to which layer. This relationship is realized in Netmate by the Layer property in *Element* class (which is inherited by *Group*, *Node*, and *Link*), and its inverse, the Elements property in *Layer* class (figure 3).

is-connected-to. A node connects to a link in the same layer. This is a very basic relationship in a network. A node (link) may connect to more than one link (node). The Connections property of *Node* and *Link* classes represent this relation.

The Internet and ISO SMIs do not support direct objects for links, and they do not keep an explicit representation of the *is-connected-to* relationship in their model. Consider a network problem (figure 2) where **B'** is unable to communicate with **D'**. An analysis process may attempt to construct all possible connection paths from **B** to **D**, and test components in each path to determine the problem. The program may discover that these nodes are on different types of links, and then attempt to find a bridge (among all bridges in the network) that may have both nodes in its *forwarding table* property. Using only the information from the nodes, it will have to algorithmically deduce which Ethernet and Token-Ring segments are part of the connection between **B** and **D** to construct the connection paths. This process becomes more complex if the nodes are separated by more than one bridge. Also note that in absence of the connections property, the analysis program has to understand individual nodes' vendor- and protocol-specific properties to construct the paths.

The process is simplified in Netmate SMI. The Connections property in the model is maintained as nodes get connected to the links in the network. Thus, the relation, at the beginning of the analysis, has the information that **E** is connected to **A**, **B** and **C**, and **T** to **C** and **D**. It is then fairly easy to use a graph-topological algorithm to construct the connection path **B-E-C-T-D** before testing any individual component in the network. Also note that with this information, the path searching algorithm need only look at the Connections property in the MIB, and not any protocol-specific properties of the nodes. The same information may be used to create a display of the network as shown in figure 2.

is-member-of. This relation represents the collection of groups, nodes and links into a group, and is realized by the Groups property of the *Element* class and the Elements property of the *Group* class. An element may belong to many groups. Semantically, deletion of a group does not always imply deletion of its members, but only the deletion of the membership relationship, implying the basic independence of element and group objects.

Consider the case when many nodes in corporation **H** are using the application service at **B'**, and **C** is inoperative. When invoked as a result of **D** to **B** communication failure, if the analysis process determines the cause, it need not perform the analysis again for any other node in **H**. This correlation is achieved by grouping all nodes in **H** under a common criterion: the nodes on the Token-Ring. In absence of groups, as in the standard models, the analysis will have to be performed for each such **D-B** pair.

is-part-of. This relation represents the collection of (sub)nodes into a node, or of (sub)links into a link, or of (sub)layers into a layer, and is realized by the Components property of the corresponding classes (figure 3). In contrast with groups, the existence of subnodes are entirely conditional upon the existence of the containing node, *i.e.*, subnodes may not exist independent of the containing node. Furthermore the values of the properties of the containing node are usually aggregations over the values of the properties of its subnodes.

Consider in the example that the communication problem occurs at **C** due to unavailability of memory buffers for the bridging of **E** and **T**. Having detected the problem at **C**, the analysis process needs to examine the traffic behavior on all individual interfaces in **C**, including the one to **S**, since sum-total use of buffers at **C** depends on all three interfaces. Using the Components property, it derives that **C** has three subnode components, each with an independent set of properties contributing to the sum-total behavior at **C**. Also note that if the interface to **S** fails, **C** may continue to bridge between **E** and **T**, but if **C** fails, none of the interfaces may function. The ISO SMI supports this relationship.

is-implemented-in-terms-of. This relation represents the well-understood notion that elements in one layer use the services of elements in other layers, and therefore, are functionally dependent on the well-being of elements in the other layers. The relation is realized by the Mappings property of elements (figure 3).

Assume that in the example network, **E** is found to be faulty. The analysis process, upon querying **A**, determines that there exists a path from **A** to **D**. The problem thus is in the segment that connects **B**, or at **B** itself. To investigate further, the process uses the Mappings property of **E** to identify the physical layer elements (**E"** and **R**) which must be operational for **E** to function properly. The process then checks the hubs to detect cable faults and **R**. An important observation is that this investigation of the physical layer is not warranted until the fault has been isolated to the specific link layer object. In general, the fault diagnosis process may function within a single layer, identifying fault domains using a layer-specific tool, then use the mappings to investigate other layers using perhaps a totally separate set of tools. Due to absence of such a functional-dependency relation in the Internet and ISO SMIs, the process must construct its own dependency information for each individual problem.

In addition to the support of generic object classes and relationships, Netmate SMI is extensible. It allows definition of specific properties for specific layers, nodes, links, and groups, taking advantage of the *is-a* relationship extensions to defined object classes. Indeed, without such definitions, no specific information may be stored about individual network entities. Figure 3 shows the specific classes such as *Token Ring Interface* which has specific properties such as Adapter Address and Nearest Upstream Adapter Address, and the subnode of **C** in figure 2 connecting to **T** is an instance of this class.

A significant number of network queries (model accesses) are likely to access the fundamental relations mentioned above. Thus, if the model fails to provide sufficiently rich set of these fundamental relations, it would mean that applications will incur significant unnecessary cost of constructing these relations. The Internet and ISO SMI models lack such relationships, which are essential for efficient, automated network problem analysis.

4. MODEL DRIVEN TOOLS

Each client of the Modeler services is considered a tool, capable of existing independent of the modeler. This is a design consideration to develop a modular management system. Nevertheless, since the management information ultimately resides in the Modeler, the tools are driven by the need to access and modify that information.

User Interface. The function of the User Interface is to construct a visual model of the network information and provide graphical techniques for navigating the model. It mirrors the abstractions realized in the modeler and converts them into graphical forms. The network layers are depicted as separate views (implemented as windows), and the mapping relation is used to navigate across views. The groups are represented as rectangular boxes containing the elements in the group, and may be iconized to reduce clutter on the screen. The connection relation is depicted as lines between node and link icons. A screen display of an SNA and an Token-Ring network at Columbia-Presbyterian Medical Center generated by the Netmate User Interface appears in figure 4.

Observation and Control Point. OCPs, or proxy agents, are the primary source of network information for the Modeler. As such, their design is critical for issues of heterogeneity and efficiency. One of their most important roles is to perform data reduction on the information which will be stored in the Modeler database. For example, status information which is basically static does not need to be updated in the database unless its value changes; additionally, for variable information, only changes beyond some threshold may need to be reported. Another useful transformation which can be done is to convert polling of devices into alerts, and vice-versa. The OCP-Modeler communication is explained in [2] in detail.

Simulation Tool. In order to support network planning and testing functions, Netmate supports a network simulation tool. An obvious use of a simulator in a network management system is to provide the capability for "what-if" scenarios; for example: take a snapshot of the current network state from the Modeler, load it into the simulator, alter the network configuration, and observe the effects in

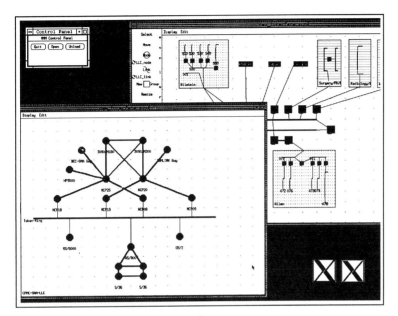

Figure 4: Example of User Interface

the simulated network. From a network protocol research perspective, integration of the simulation with a management system is also useful; real network information such as packet counts and rates, traffic levels and loads become readily available.

Auxiliary Systems. A complete management system should support network-related auxiliary functions, and Netmate maintains information which is useful to many other systems. The network configuration database can easily be used for Inventory Control, identifying specific hardware and software systems in the network. Similarly, if vendor and price information is added, such a comprehensive system may be used by the Purchasing department. Netmate design also includes a Trouble Ticketing system, which is used for keeping track of reported network problems.

5. CONCLUSIONS

Currently, a prototype Netmate system has been developed at Columbia, with a functional Modeler, User Interface, and OCP. The Modeler is implemented using an object-oriented database; the User Interface is an X window system application written in C++. One OCP has been developed, with a generic SNMP-based interface. A Modeler with object-oriented structures based on a relational database and a new OCP, written in C++, which will provide the data transformation and reduction capabilities mentioned above, are under development. The Nest network simulation tool developed at Columbia [1] will be used as the basis for the simulation component.

Netmate addresses management needs of network administrators by supporting a comprehensive set of distributed tools. The object-oriented Netmate data model is well suited to accommodate current as well as future heterogeneous networks and protocols. The model features generic object classes and relationships that are crucial for efficient network management for large and complex networks. The modularity of Netmate system lends toward an elegant and efficient network management solution.

6. REFERENCES

1. A. Dupuy, J. Schwartz, Y. Yemini, D. Bacon. "NEST: a Network Simulation and Prototyping Testbed". *Comm. ACM 33*, 10 (October 1990), 63-74.

2. Y. Yemini, G. Goldszmidt, S. Yemini. How to Build Manageable Systems: The Manager-Agent Delegation (MAD) Model. Proc. of the IFIP TC6/WG 6.6 Second Inter. Symp. on Integrated Network Management, Washington, DC, April, 1991.

3. R. Gupta and E. Horowitz (Ed.) *Object Oriented Databases with Applications to CASE, Networks, and VLSI CAD*. Prentice-Hall, Englewood Cliffs, NJ, 1991.

4. International Organization for Standards. Information Processing Systems - Open Systems Interconnection - Structure of Management Information. International Organization for Standards, 1990.

5. S. M. Klerer. "The OSI management Architecture: an Overview". *IEEE Network 2*, 2 (March 1988), 20-29.

6. M.T. Rose and K. McCloghrie. Structure and identification of management information for TCP/IP-based internets. Network Information Center, SRI International, Menlo Park, CA, May, 1990.

7. D.B. Rose and J.E. Munn. "SNA Network Management Directions". *IBM Syst. J. 27*, 1 (1988), 3-14.

8. S. Sengupta, A. Dupuy, J. Schwartz, Y. Yemini. An Object-Oriented Model for Network Management. In *Object Oriented Databases with Applications to CASE, Networks, and VLSI CAD*, Prentice-Hall, Englewood Cliffs, NJ, 1991.

9. M. Sylor. "Managing Phase V DECnet Networks: the Entity Model". *IEEE Network 2*, 2 (March 1988), 30-36.

10. O. Wolfson, S. Sengupta, Y. Yemini. Active Databases For Communication Network Management. Submitted to SIGMOD 1991.

VIII

USAGE CONTROL

Objective-Driven Monitoring

Subrata Mazumdar
IBM
T.J. Watson Research Center
P.O. Box 704
Yorktown Heights, NY 10598
e-mail: mazum@ibm.com

Aurel A. Lazar
Department of Electrical Engineering
and
Center for Telecommunications Research
Columbia University, New York, NY 10027
e-mail: aurel@ctr.columbia.edu

Abstract

An approach to sensor configuration, installation and activation for real-time monitoring for performance management is presented. An objective-driven measurement strategy for establishing the dynamic and statistical databases of the network is described. Objective driven monitoring allows the activation of sensors for data collection and abstraction based on a set of objectives. The objectives are derived from the quality of service requirements for real-time traffic control and operator submitted queries. The methodology of objective-driven monitoring for selective activation of sensors is implemented as a set of rules in the knowledge base of the monitor.

1. Introduction

Integrated networks consist of many subsystems (switching nodes, multiplexers, links, etc.) that are geographically distributed, carry multiple classes of traffic and have access to different information patterns. Although these subsystems make their own local decisions, they work together for the achievement of the common system wide goal of information transport. The common goal is to guarantee the Quality of Service (QOS) negotiated during the call setup for each of the traffic classes[LAZ90a]. The QOS is specified through a set of performance parameters.

Monitoring of these parameters and of all network resources, such as buffer space, switching and communication bandwidth, and call processing, is required in order to guarantee the QOS[LAZ90a]. A network monitoring system should also be applicable to several representative networks. Therefore, a proposed set of measurement parameters must be network independent [PAW81]. They must be declared in generic terms, such as throughput, time-delay, arrival rate, inter-arrival time, etc. Sensors (measurement points) for these parameters must be made available in all the networks to be monitored. A set of objective criteria or strategies are needed by which sensors can be selectively activated and deactivated among a large number of sensors in a distributed environment. One of the main objectives of the monitoring task is the real-time support of the network control and management system during the decision making process. A consistent view of the network is assumed to be available for monitoring [WU90].

The monitoring of networks can be viewed at different levels of abstraction. Monitoring takes place both at hardware and software level depending upon the hardware and software components that support the information transport. In [BERN83] a network operation center to monitor, control, and manage ARPANET-like packet-switching networks is presented. In [RIT87], [SOH87], [AME82], and [CAS88], network monitoring is done for LANs or interconnected LANs carrying only single class (data) traffic. In the latter work, major emphasis was on the evaluation of usage of *communication resources*. In [LAZ90c], monitoring of an integrated network, called MAGNET II, is carried out by hardware observation units (HOU) connected to network access points. Real-time traffic measurements are reported. The quality of service of traffic classes in the network is evaluated by monitoring the buffer occupancy distribution, the packet time delay distribution, the packet loss, and the gap distribution of consecutively lost packets. In [BRU89], the monitoring of switching resources was considered for managing AT&T's dynamic non-hierarchical routing algorithm for automatic as well as operator oriented control of operations of the network.

Since a network can be considered to be a distributed system, the approaches to monitoring of distributed systems can also applied to monitoring integrated networks. The monitoring of distributed systems can be classified as event-driven monitoring and as database approach to monitoring. Most of the work in event-driven monitoring of distributed systems was done on the application level. Debugging of distributed systems [LEB85], [JOY87], [BAT83] and parallel programming environments [WYB88] are typical examples. Here, major emphasis was given to the performance evaluation of *processing resources*. In [SNO88] a relational approach to monitoring was presented. In the relational approach, monitoring is viewed as an information processing activity and the *historical database*, a class of relational databases that encode time, is considered an appropriate formalization of the information processed by the monitor.

In this paper the steps required to configure, install and activate sensors for monitoring integrated networks are discussed and a knowledge-based approach is presented as a solution to the problem. In order to monitor object behavior, sensors need to be configured and installed in the network. Sensor configuration specifies the characteristics of sensors declared in the knowledge database of the monitor. These characteristics are specified by a set of attributes and a set procedures for operations. Sensor installation involves identification of the measurement points in the network.

The architecture of the knowledge-based objective-driven monitoring system consists of the knowledge database and an inference engine for reasoning on the database [MAZ89]. The inference engine consists, in turn, of two parts: a deductive inference processor and a statistical inference processor. The role of the deductive inference processor is to process the queries about the network behavior and activate sensors in the network. The role of the statistical inference processor is to abstract the information obtained by the sensors. The monitoring system processes queries on system and conceptual level, and sets up sensors to collect information. The system level monitor supports queries only if precise knowledge about the system is available. On the conceptual level, the monitor allows general queries without the precise knowledge of the system architecture of the network.

An objective-driven monitoring scheme is presented to selectively activate and deactivate a subset of sensors among a large number of sensors already installed in the network. Objective-driven monitoring is closely related with the concept of *experimental frame* of [ZIE84] that characterizes modeling objectives by specifying the form of experimentation that is required to obtain answers to the questions of interest. For the class of objective-driven monitoring tasks considered in this paper, the fundamental concepts are derived from the requirements of supporting Quality of Service and of operator submitted queries. The objective-driven monitoring allows us to deal with the problem of complexity in monitoring integrated networks through the concept of an observation frame that we have proposed in [MAZ90a].

In this paper we give an object-oriented definition of sensors in a network and specify a way to configure them. This definition represents an alternative to "a collection of code" given in [SNO88]. Through the specification of object-specific and variable-specific generic sensors, we can define the starting and stopping time for monitoring and also how frequently the samples are to be collected and recorded. Since the various measures for performance analysis are specified through a set of operators, we can easily add a new set of performance measures or selectively activate a subset of measures. Based on our approach, we can select any object, state variable, event or their performance parameter for monitoring.

This paper is organized as follows. Section 2 outlines the architecture of the experimental environment that represents a platform for knowledge-based monitoring of integrated networks. Section 3 describes the key ideas about sensor configuration, installation and query analysis for monitoring. Finally, in section 4, the objective-driven measurement strategy and query based activation of sensors for integrated networks are discussed.

2. The System Architecture of the Monitor

The architecture of the knowledge-based monitoring system was modeled as a real-time system where the monitor asynchronously interacts with the network through an interface [MAZ91]. Thus, the network can be viewed as the environment for the monitor. The interface is all the monitor sees of the network. The characteristics of the interface depend to large extent on the environment. What is and what is not part of the interface depends on the specific requirements of the performance management tasks.

The interface between the network and the monitor consists of a set of state variables. A state variable is persistently present and throughout its existence, it has a value that represents the state, which changes with progression of time. For the task of monitoring the network, the state variables, that represent the interface, get their values from the processes operating in the environment. The semantic information about network objects and the interface are represented by the Entity-Relationship model [CHE76]. A computational model consisting of a set of sample path and performance evaluation operators was defined, to describe various processes that are associated with state variables [MAZ90a], [MAZ91].

Thus, in representing the environment and the interface, the concept of modularity was achieved through the *object* representation. The location and ownership of a state variable was declared through these objects. These objects are responsible for acquisition, manipulation, and dissemination of the information of their state variables. Note that, while implementing the network architecture, one has to explicitly declare a set of state variables that form the interface between the network and the monitor. These variables characterize the observable behavior of the network. The exact specification of the interface depends on the monitor and the specific management tasks, such as performance, fault and configuration management, that the monitor is going to support.

The architecture of the knowledge-based monitoring system (as shown in Figure 1) consists of the knowledge database and an inference engine for reasoning on the database for query processing, sensor activation, and interpretation of data collected by the sensors. The inference engine consists of two blocks: the *deductive inference processor* and the *statistical inference processor*. The role of the deductive inference task is to set up a distributed *observation frame*, *i.e.*, a data space in which a query may be answered.

Figure 2 shows the organization of the knowledge database of the network. The knowledge database is organized as follows. The system level knowledge about the network is represented in the configuration database. The configuration database contains the knowledge about the network entities, such as buffers, source, and servers and their specific instances. Figure 3(a) describes the attributes of those network entities. The dynamic database contains the information about the state and event variables of the objects in the configuration database. Figure 3(c) describes the attributes of the state and event variables. The sensor database contains the generic description of sensor for objects and also any specific class of state and event variables and all of the instances of the sensor object class. The objects in the sensor database represent the specific sampling pattern for data collection and specific sensor instances indicate the activation of the sensors. The sensor database together with configuration database forms the static database. These two databases change much less often than the dynamic database. The dynamic database only exists for those state and event variables which are being measured by activating the sensors in the network. The statistical database is obtained by applying abstraction operators on both state and event variables and provides various performance measures for each state and event variable. Figure 3(b) describes the attributes of the performance parameters.

Typically, a query submitted by the query generator, *i.e.*, a control task or an human operator, requires information about the performance of certain objects in the network. The query is then processed to find out the specific instances of the performance parameters of interest. Based upon these parameters, the deductive inference processor creates a derived object containing the identified performance parameter and their corresponding state and event variables. An associated derived sensor capable of monitoring the derived object is also created, which in turn creates the appropriate sensors in the sensor database for the selected state and event variables. Creation of the sensors for state variables in the knowledge database activates the sensors in the network and data is collected. The statistical inference processor then applies the statistical operators, passed during the query submission or implicit in the performance parameter specification, on the collected information and transmits the processed information to the query processor.

3. Monitoring

As already mentioned in the previous section, the monitor is defined to be a real-time system that maintains an ongoing relationship with its environment, *i.e.*, the network.

The interface between the monitor and the network is defined by a set of state variables. The *interface* is all the monitor sees of the network. Thus, the characteristics of the interface depend to a large extent on the network.

Depending upon the information requirement, a network can be monitored in two ways: *monitoring the change of states* (status monitoring) and *monitoring the event traces*. Status monitoring represents the collection of values of any of the state variables obtained by activating a sensor. The rate at which the information is generated by sensors depends on the speed of operation of the corresponding objects, such as buffers, servers and sources. On the network access level the rate of generation of state information may be equal to the packet arrival and departure rate and at the session layer it may be equal to the rate of arrival of new calls. Event traces are abstractions of state variables over time using the sample path operators of the computation model. An event trace from a state variable is recorded by a sensor as an event variable.

The design of the network monitor can be characterized by the following steps that are in part, adapted from [SNO88]:

- Step 1: Sensor Configuration
  This step involves design of the sensor, *i.e.*, specification of it's attributes and the procedures of operation that handle necessary interaction with the monitor, enabling and disabling the sensors and buffering of monitored data and requested tasks. The sensor attributes specify the starting and stopping times for monitoring, how frequently the monitored events, measures, or resources are to be recorded, and other related performance information to be collected simultaneously;

- Step 2: Sensor Installation
  This step involves identifying the state and event variables that are to be monitored by sensors.

- Step 3: Query Analysis Specification
  This step specifies how to decompose a query, activate sensors and create various dataspaces for abstraction of information.

- Step 4: Execution
  This step is comprised of activating the sensors, generating and abstracting the data collected from the network, transmitting the data from the network to the monitor, and finally presenting the data on a graphics terminal. This step is discussed in section 4.

Even though we have adopted the steps in [SNO88] there are differences between [SNO88] and our approach. In [SNO88] the approach was relational where as our approach is object-oriented. Our definition of sensor as an "object" is alternative to "a collection of code" given in [SNO88]. Since a sensor monitors state variables there exists only one type of sensors that need to be configured for installation in the network. Thus, we need to instrument an object only once in order to obtain various measures, such as, average and variance of buffer occupancy from the variable, buffer state, of a buffer. In our approach the various performance measures are defined as operators for the sensor.

In [SNO88] there is no concept of object specific or variable specific generic sensors. Through the inheritance mechanism of object-oriented approach we can specify object or variable specific generic sensors. In the case of sensor installation, we show how to select the measurement points based on performance management objectives. These measurement points are not based on the measures that we are interested but the actual variables that are responsible for generation of the information and on which the performance measures are to be applied. In the query analysis specification step we show how the data transformation takes place during the collection of the data and how the interpretation is done based on the data. We have clear separation between the raw data that is collected and their abstractions. In this step we have also shown how the information provided by a simple query is used to identify sensors, collect data, detect events and then abstraction and interpretation is done on the collected data.

3.1 Sensor Configuration

A *sensor* is defined to be an object with a set of attributes and a set operators (algorithm or code), implemented either in hardware or software. The sensors installed in the network collect information about the state or event variables of an object and transfer it to the monitor. From an implementation point of view, every state and event variable includes its sensor as a component object. Sensor operations are executed by the set of sample path and statistical operators described in [MAZ90a], [MAZ90b].

Sensors installed in the network to monitor the state variables of an object are termed *primitive sensors*. The primitive sensor corresponds to the object class SENSOR in the sensor database; its attributes are the same as that of SENSOR (as shown in Figure 4(a)). The attribute sensor_code_id specifies the operator to be applied to abstract information from the history of a state variable. The attribute initiated_by indicate the initiator of the query based on which the sensor is activated. Primitive sensors contain the code for sample path and performance evaluation operators.

The attributes of a sensor are defined based on the requirements for both status and event monitoring and they are shown in Figure 4(a). The abstraction operators of a primitive sensor operate on two time levels. The sample path operators abstract events on the same time scale as the state variable. The performance evaluation operators operate on both state and event variables on a time scale based on the interval for statistics collection. The parameters for performance evaluation operators are provided by a set of sensor attributes. These attributes are sample_count, sample_on, sample_off, duration_of_activation, and sampling_interval.

The attribute sample_count of sensor specifies the total number of samples from the state and event variables that are to be collected for statistical inference. The average on a fixed number of samples is computed based on the value of the attribute sample_count and it is the default procedure for evaluating the average over a period. If the value of sample_count is not specified and the attribute sample_on is specified, then the later is used with sampling_interval to compute the total number samples to be collected for statistical inference. The specific values of sample_count or sample_on for monitoring a state variable is determined by the rate at which its values are changing. Their values are also determined by the control algorithm that is managing the object. In order to repeat the statistical inference process, the attribute duration_of_activation of sensor specifies the duration of the monitoring process or the duration of time the sensor remains active. The attribute sample_off of sensor specifies the duration between two consecutive measurement intervals, *i.e.*, sample_on period.

Primitive sensors are activated by sending them a message to. Conversely, a primitive sensor transmits information to the monitor by sending a message. Each message is time stamped with the time of creation of the information sent. If the transmitted message contains the value of a state variable then the time of creation indicates the last sampling time. If the message contains an event indication, then the time of creation indicates the event occurrence time. If the message contains the information about a performance parameter of a state or event variable, then the time of creation indicates when the values of the performance parameters were computed.

Primitive sensors are provided with the capability to queue up multiple requests for monitoring. The messages sent by the monitor to the sensor contain information about the specific sample path and performance evaluation operators to be applied to the collected values. In order to allow multiple users or control algorithms to query the state and event variables, the primitive sensors are provided with the ability of both one-to-one and one-to-many communications.

Two subclasses of primitive sensors, STATUS_SENSOR and EVENT_SENSOR, are defined to monitor the state and the event variables, respectively. The STATUS_SENSOR inherits all the attributes of SENSOR. The relationship type MONITORING-GENERIC-STATE-VAR and MONITORING-STATE-VAR establishes the relationships between the sensor and the subclass of a state variable and the specific instance of the state variable being monitored, respectively. The EVENT_SENSOR is declared as a subclass of the STATUS_SENSOR and thus inherits all of its attributes. The EVENT_SENOSR has an additional attribute, event_operator_id, that defines the operator for extracting the event. Since the behavior of all objects is represented by their state variables, only one type of primitive sensors needed to be configured. Thus, no matter how complex, an object can be monitored as long as its state variables are declared. Therefore, no object specific sensor needs to be configured.

In order to monitor an object, whose behavior is defined by a set of state or event variables, *derived sensors* are defined. The *derived sensors* are an aggregation of a set of primitive and derived sensors. Derived sensors belong to the object class DERIVED_SENSOR, which is a subclass of SENSOR. The DERIVED_SENSOR is obtained based on the primitive sensors associated with the state and event variables of an object to be monitored. An instance of DERIVED_SENSOR created for monitoring an object will contain the corresponding instance of STATUS_SENSOR and/or EVENT_SENSOR. The DERIVED_SENSOR maintains a list of primitive or derived sensors by the relationship attribute DERIVED_FROM. Since the behavior of an object is always expressed by its state and event variables, the sensor that monitors an object is always a member of the subclass DERIVED_SENSOR and is an aggregate object containing the status and event sensors. Thus, in order to monitor the state of a BUFFER, a DERIVED_SENSOR will be created for the BUFFER and it is composed of a STATUS_SENSOR that monitors the state variable representing the BUFFER's state. If an object is an aggregation of a set of objects, the DERIVED_SENSOR for the aggregate object will consist of the DERIVED_SENSOR of the component object.

The DERIVED_SENSOR may be specialized to represent object specific monitoring information such as the sampling pattern. Thus, in order to obtain the behavior of an object NETWORK_STATION, SENSOR_NETWORK_STATION, a subclass of DERIVED_SENSOR, is defined. The relationship between SENSOR_NETWORK_STATION and the corresponding object class, is established by MONITORING-GENERIC-OBJECT and the relationship between the specific instances of the SENSOR_NETWORK_STATION and the specific instance of NETWORK_STATION, which is being monitored, is established by MONITORING-OBJECT-INSTANCE.

Whenever a specific object in the network is to be monitored, an instance of the DERIVED_SENSOR is created in the sensor database. The value of the MONITORING-GENERIC-OBJECT of DERIVED_SENSOR specifies the class name of the object that the derived sensor is monitoring. The instance of the DERIVED_SENSOR and its association with the object in the configuration database is deleted when the sensor is deactivated at the end of the monitoring period. The derived sensors only exist in the sensor database. *Unlike primitive sensors, no counterpart of derived sensor exists in the network*. Both primitive and derived sensor instances are stored in a database called *sensor database*, as shown in Figure 2.

3.2 Sensor Installation

Sensor installation allows the selection of the measurement points in the network, *i.e.*, the state variables of the network objects that define the interface between the monitor and the network. Since network object and state variables can be uniquely identified in the system, the events associated with state variables and their performance parameters can be selected. In the modeling process of the monitor in [MAZ90b] state and event variables were identified based on the performance management requirements. The sensor location was determined by the location of the identified state variables in the network.

Along with sensor configuration, the installation of primitive sensors is the only manual processes associated with our monitoring scheme. Since the specification of state variables is to be decided based on the performance objectives of the control tasks, these two steps will now be required to be carried out during the specification and design step of the network. This is by itself a manual process. Thus, the design of the network monitor has been shifted to the network design phase. This design process results in a robust network design that requires very little tuning during operations. The identification of the sensor location during the specification phase helps to alleviate the reliability and the correctness problem of sensor operations.

3.3 Query Analysis Specification

Query analysis specification derives levels of abstraction of the collected information for performance analysis. It also selects the performance analysis criteria and algorithms for various performance measures. As shown in Figure 5(a), a transaction for a query has three parts: an identification function (I) that selects the state or event variables of a specific object to be monitored, a data transformation function (F) for performance evaluation through statistical inference, and an inference rule (R) to be applied to the abstracted data. The dataspace generated by monitoring a state or an event variable is denoted by the circle F_t in Figure 5(a). It is created after activating the sensor associated with a state or an event variable. The data transformation function (F) (derived from the set of statistical operators) is applied on the dataspace F_t to abstract information from the history of a state or an event variable. Application of F generates a dataspace G_t that consists of only statistical information. If the statistical abstraction is not needed then F is reduced to the identity operator. The inference rule R operates on the dataspace of G_t to further evaluate the statistical information, e.g., for event detection through threshold crossing. If no such operation is needed, then the rule R is reduced to the identity operator.

The deductive inference rule is used to decompose a query into a set of simple queries and then aggregate the information received by servicing the simple queries. For example, in order to find the total average time delay of a call, a set of sensors at the nodes along the route of the call is to be activated to measure the time delay at each node. Once the average time-delay from every node is available, they are aggregated to compute the total average time delay. Therefore, a query for the average time delay of a call will be divided into multiple simple queries and appropriate sensors will be activated to measure the average time delay at every node along the route of the call. Figure 5(b) describes such a scheme, where SENSOR_1 through SENSOR_N measure the time delay at each node along the route of the call. The function $f(D_t^1, \ldots, D_t^N)$ in Figure 5(b) represents the deductive part of the query for data aggregation and it is applied after the data is collected from the appropriate sensors.

We may also want to find out whether certain average throughput-time delay condition, at a buffer of a node, is satisfied. In this case, we need to activate sensors for both the throughput and the average time-delay and send an event indication if the average throughput-time delay condition is not met. In order to do that, the original query will be divided into two simple queries based on the throughput and the time-delay to be computed. Figure 5(b) represents such cases where the original query is divided into multiple simple queries identifying each of the state variables to be measured. $f(D_t^1, \ldots, D_t^N)$ indicates the function that generates an event if the throughput-time-delay condition is not met.

Figure 5(c) represents the case when a state variable is monitored for status monitoring and event reporting or multiple event reporting. $f(D_t^1, \ldots, D_t^N)$ represents any deduction to be done after the data is collected. One such deduction scheme is the correlation between two events generated from the same variable. This scheme can also be used to define higher level events based on the history of event variables.

4. The Objective-Driven Measurement Strategy

In the process of performance management a set of objectives often lead to asking a specific set of questions about the network. The questions of interest could be: does the network support a specified performance or would it be able to provide enhanced performance? They can be answered after appropriate monitoring functions are incorporated into the system and the observed data is processed.

Monitoring as a process determined by objectives is called *objective-driven monitoring*. Objective-driven monitoring is closely related with the concept of *experimental frame* of [ZIE84] that characterizes modeling objectives by specifying the form of experimentation that is required to obtain answers to the questions of interest.

A query to collect information from the network can be submitted to the monitor either by an user from a terminal or by the various knowledge specialists responsible for network control and management. Such agents are called Query Generators (as shown in Figure 1). The submitted query can be of two types: real-time data query and non-real-time data query. The real-time data query represents the query on those objects, whose attributes are updated using the sensors located in the various subsystems of the network. The non-real time data query represents the query on those objects, whose attributes' value do not depend upon the sensory information. The non-real-time queries are handled based on the information available in the knowledge base of the monitor.

The real-time data queries are handled by using the deductive query processing technique, where the inference and retrieval phases of the query have been separated [DUT84], [SNO88]. The schematic of such query processing is shown in Figure 1. Based on the relationships established in the knowledge database between the various objects, the monitor decomposes the requested query into a set of simple queries and analyzes them to determine specific state and event variables that need to be monitored. Once the state and event variable is identified, corresponding sensors in the network are activated. Activated sensors then collect information about the state and event variables and thus update the dynamic database shown in Figure 2. If the query requires statistical abstraction of the collected information, then the statistical inference processor applies the corresponding operators on the state and event variables and updates the statistical database.

As described before, one way to find the total average time delay of a call is to activate sensors to measure the time delay at the nodes along the route of the call and then aggregate the average time-delays. A more elegant solution, however, is to first obtain a *derived object* that contains all the state variables, that exhibit the average time delay of the intermediate nodes along the route of the call. A derived sensor can be attached to this object and finally adding up the time delays leads to the required result. Thus, to answer a query only a restricted data space, called an *observation frame*, is needed. The observation frame contains only the state and event variables, the performance parameters, and the derived sensor and it's components. In the next section a general methodology for an objective-driven measurement strategy is described.

4.1 Deductive Inference

Real-time control algorithms for resource allocation operate based on a set of cost functions and a set of constraints on the behavior of the variables of a system. These system variables could be either describing the state of the system or a statistical abstraction of its state. Thus, control task first leads to monitoring the system variables. In the case of integrated networks, *QOS parameters* define the target operating points and maintaining these QOS parameters near the operating point becomes the control objective of network operations.

Based on the specified QOS parameters, a set of *performance parameters* are identified. The difference between the QOS parameters and the performance parameters is that the latter depend on the systems architecture of the network. From the specification of the QOS parameters, the corresponding performance parameters are derived by the deductive processor based on the knowledge about the system architecture of the network. The latter resides in the configuration database. As an example, the maximum average end-to-end time delay might be a QOS parameter. The average time delay experienced by a call in a given network is the performance parameter associated with it. It is the aggregation of all the average time delays at nodes and links along the route of the call.

Thus, the request for monitoring a QOS parameter is a query consisting of the class name of an object and the corresponding performance parameter. This general query can be made specific by providing values for one or more key attributes of the object class. Based on the submitted query, the deductive inference processor identifies specific objects and the performance parameters that need to be monitored. First, all the objects in the knowledge base, that contain the appropriate performance parameter, are identified. Second, the instances of the performance parameters associated with the selected objects are identified. Third, the instances of state and event variables associated with the selected instances of the performance parameters are identified.

For each of the selected objects a component object is created and relationships are established between the new object and the selected performance parameters and the corresponding state and event variables. In order to monitor the selected objects, a derived sensor is associated with the each of the component objects. The creation of the derived sensor generates the instances STATUS_SENSOR and/or EVENT_SENOR for each of the state and event variables associated with the component object. Creation of the STATUS_SENSOR or EVENT_SENOR activates the corresponding primitive sensors installed in the network. The component object, the associated derived sensor, and the collected information represent a data space called *observation frame*. Thus, the observation frame forms a restricted data space. The answer to queries is obtained by examining, processing, and aggregating monitored information in this space. The statistical inference process takes place only after data has been collected by the sensors.

4.2 An Algorithm for Objective-Driven Monitoring

The operation of the deductive inference processor described above can be formalized into an algorithm consisting of the following steps:

- identify the instances of the object class specified in the query;

- identify the instances of the performance parameter specified in the query associated with the selected objects;

- for each of the selected objects, identify the instances of state and event variables associated with the selected performance parameters;

- create a new object and associate with it the selected performance parameters and the state and event variables of all the selected objects;

- associate a derived sensor with the new object and create sensors that monitor the state and/or event variables;

- activate the sensors in the network;

- apply statistical inference procedures to evaluate the performance parameter

The above steps are in part adopted from the "objectives-driven" methodology for modeling systems of [ZIE84].

4.3 Examples

In the following example, it is shown how to monitor the average throughput of a buffer at a network station of MAGNET II [LAZ90b] network. Let us assume that the query requests the THROUGHPUT of SWITCH_BUFFER with buffer_id 1 at node_no = 0 and st_no = 1.

Based on the rule shown in Figure 6 the specific instance SWITCH_BUFFER_0_1_1 of SWITCH_BUFFER is identified as the object to be monitored. This is accomplished from the specification of the attributes node_no, st_no, and buffer_id, Then, THROUGHPUT_SWITCH_BUFFER_0_1_1, the specific instance of THROUGHPUT for SWITCH_BUFFER_0_1_1 is identified by the procedure **create-object-view-sensor** shown in Figure 7. Based on the relationship PERF-OF-STATE-VAR between the THROUGHPUT and PACKET_OUTOF_BUFFER, the event variable PACKET_OUTOF_BUFFER_SWITCH_BUFFER_0_1_1 is identified. Once the performance parameter and the event variable are identified, an instance of the object class OBJ_VIEW is created using the procedure **create-object-view** (shown in Figure 8). In Figure 8, the line containing the key word **assert** creates the instance of OBJ_VIEW. The new object is considered a weak entity of the SWITCH_BUFFER_0_1_1 and it is uniquely identified by its own class name (OBJ_VIEW), the class name of the object being monitored and key attributes of the latter. The relationship type HAS-OBJ-VIEW-PERF-PARAMETER establishes the association between the new object and the performance parameter THROUGHPUT_SWITCH_BUFFER_0_1_1. Similarly, the relationship type HAS-OBJ-VIEW-STATE-VAR establishes the association between the new object and the state variable PACKET_OUTOF_BUFFER_SWITCH_BUFFER_0_1_1.

Once the instance of OBJ_VIEW is created, a derived sensor is attached to the object by calling the procedure **attach-sensor-to-object-view**. Based on the relationship OBJECT-VIEW-OF-GENERIC-OBJECT, shown in Figure 8, the derived sensor is created as an instance of SENSOR_SWITCH_BUFFER, which is a specialization of DERIVED_SENSOR for SWITCH_BUFFER. SENSOR_SWITCH_BUFFER contains the SWITCH_BUFFER specific sampling information and it is a subclass of DERIVED_SENSOR. If the object class SENSOR_SWITCH_BUFFER does not exist then the derived sensor is created as an instance of object class DERIVED_SENSOR. The relationship MONITORING-OBJECT establishes the association between the instances of OBJ_VIEW and DERIVED_SENSOR. The attributes of SENSOR_SWITCH_BUFFER are shown in Figure 4(b). The existence of the derived sensor implies the creation of an instance of EVENT_SENSOR for the event variable PACKET_OUTOF_BUFFER_SWITCH_BUFFER_0_1_1. It also establishes the relationship MONITORING-EVENT-VAR between the event variable and the instance of the EVENT_SENSOR.

Creation of the sensor causes it to send a message for activation of the primitive sensor associated with PACKET_OUTOF_BUFFER_SWITCH_BUFFER_0_1_1 in the network and for starting the measurement. Once the measurement is completed and the statistical operators are applied, the value of the throughput performance parameter is sent back to the knowledge base.

5. Conclusion

A step by step design procedure of sensor configuration and activation for monitoring network behavior has been presented. The sensor configuration uses the modeling approach for specifying the attributes of the sensors and the procedures for sensor operations.

An objective driven measurement strategy has been presented that selectively activates the sensors needed for collecting the required information. The objectives for monitoring are obtained from the real time control task for resource management or operator submitted queries. The queries are processed by a deductive inference processor that identifies the state variables that are to be monitored. The role of the deductive inference processor is to set up an observation frame, *i.e.*, a data space in which only data relevant to the query is allowed. The answer to queries is obtained by examining, processing, and aggregating monitored information in the data space. The sample path and statistical operators are applied to compute the performance of the network.

Acknowledgements

The research reported here was supported in part by the National Science Foundation under Grant CDR-84-21402 and in part by the New York State Center for Advanced Technology under Project NYSSTF CAT (84)-15 005.

References

[AME82] P.D. Amer, "A Measurement Center for the NBS Local Area Computer Networks," *IEEE Transactions on Computers*, vol. C-31, pp. 723–729, August 1982.

[BAT83] P. Bates and J.C. Wilden, "An Approach to High-Level Debugging of Distributed Systems," *ACM SIGPLAN Notices*, vol. 18, pp. 107–111, August 1983.

[BERN83] Susan L. Bernstein and James G. Herman, "NU: A Network Monitoring, Control, and Management System," in *Proceedings of the IEEE International Conference on Communications '83*, pp. 478–479, 1983.

[BRU89] J.N. Brunken, R. Mager, and R.A. Putzke, "NEMOS - The Network Management System for the AT&T Long Distance Network," in *Proceedings of International Conference on Communications*, pp. 99–114, Boston, MA, 1989.

[CAS88] L.N. Cassel and P.D. Amer, "Management of Distributed Measurement over Interconnected Networks," *IEEE Network*, vol. 2, pp. 50–55, March 1988.

[CHE76] P.P. Chen, "The Entity-Relationship Model - Toward a Unified View of Data," *ACM Transactions on Database Systems*, vol. 1, pp. 9–36, March 1976.

[DUT84] A. Dutta, M.D. Gagle, and A.B. Whinston, "Deductive Query Processing in a Codasyl Data Base," in *IEEE Workshop on Languages for Automation*, pp. 200–208, New Orleans, November 1-3 1984.

[JOY87] J. Joyce, G. Lomow, K. Slind, and B. Unger, "Monitoring Distributed Systems," *ACM Transactions on Computer Systems*, vol. 5, pp. 121–150, May 1987.

[LAZ90a] A.A. Lazar, A. Temple, and R. Gidron, "An Architecture for Integrated Networks that Guarantees Quality of Service," *International Journal of Digital and Analog Communication Systems*, vol. 3, pp. 229–238, April-June 1990.

[LAZ90b] A.A. Lazar, R. Gidron, and A. Temple, "MAGNET II: A Metropolitan Area Network Based on Asynchronous Time Sharing," *IEEE Journal on Selected Areas in Communications*, vol. SAC-8, pp. 1582–1594, November 1990.

[LAZ90c] A.A. Lazar, G. Pacifici, and J.S. White, "Real-Time Traffic Measurements on MAGNET II," *IEEE Journal on Selected Areas in Communications*, vol. 8, pp. 467–483, April 1990.

[LEB85] R.J. LeBlanc and A.D. Robbins, "Event Driven Monitoring of Distributed Programs," in *Proceedings of the 5th International Conference on Distributed Computing Systems*, pp. 515–522, Denver, CO, 1985.

[MAZ89] S. Mazumdar and A.A. Lazar, "Knowledge-Based Monitoring of Integrated Networks," in *Integrated Network Management-I* (B. Meandzija and J. Westcott, editors), pp. 235–243. North-Holland, New York, 1989.

[MAZ90a] S. Mazumdar and A.A. Lazar, "Monitoring of Integrated Networks for Performance Management," in *Proceedings of the IEEE International Conference on Communications/SUPERCOM '90*, Atlanta, GA, 1990.

[MAZ90b] S. Mazumdar, *Knowledge Based Monitoring of Integrated Networks for Performance Management*. PhD thesis, Department of Electrical Engineering, Columbia University, New York, August 1990.

[MAZ91] S. Mazumdar and A.A. Lazar, "Modeling the Enviornment and the Interface for Monitoring and Control of Integrated Networks," in *Submitted for presentation at the IEEE International Conference on Communications ICC'91*, Denver, CO, 1991.

[PAW81] P.F. Pawalita, "Traffic Measurements in Data Networks, Recent Measurement Results, and Some Implications," *IEEE Transactions on Communications*, vol. Com-29, pp. 525–535, April 1981.

[RIT87] Donna Ritter and Marilyn Seale, "A Multi-purpose, Distributed LAN Traffic Monitoring Tool," *IEEE Network*, vol. 1, pp. 32 – 39, July 1987.

[SNO88] R. Snodgrass, "A Relational Approach to Monitoring Complex Systems," *ACM Transactions on Computer Systems*, vol. 6, pp. 157–196, May 1988.

[SOH87] Michael Soha, "A Distributed Approach to LAN Monitoring Using Intelligent High performance Monitors," *IEEE Network*, vol. 1, pp. 13 – 20, July 1987.

[WU90] S.F. Wu and G.E. Kaiser, "Network Management with Consistently Managed Objects," in *Proceedings of the GLOBECOM '90*, pp. 304.7.2 – 304.7.6, San Diego, CA, 1990.

[WYB88] D. Wybranietz and D. Haban, "Monitoring and Performance Measuring Distributed Systems During Operation," in *Proceedings of the 1988 ACM SIGMETRICS Conference on Measurement and Modeling of Computer Systems*, pp. 197–206, Santa Fe, NM, 1988.

[ZIE84] B.P. Ziegler, *Multifacetted Modeling and Discrete Event Simulation*. Academic Press, London, 1984.

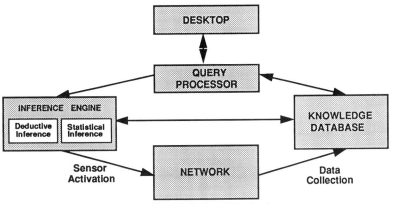

Figure 1. The System Architecture of Knowledge Based Monitoring.

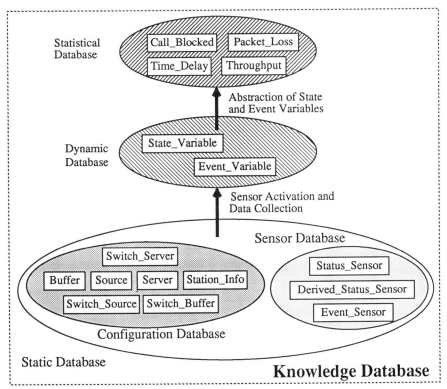

Figure 2. The Organization of the Knowledge Database.

```
(defschema      NETWORK-OBJECT
    (generic_name)              ;;;; Class name of the Object
    (key)                       ;;;; Composite attribute defining the key attribute
    (status)                    ;;;; Indicates the status of object, active or inactive.
)
(defschema      BUFFER
    (IS-A                       NETWORK-OBJECT)    ;;;; Inherits the attribute.
    (generic_name               BUFFER)
    (buffer_id)                                    ;;;; Key attribute of Buffer.
    (buffer_size)                                  ;;;; Size of buffer
)
(defschema      SERVER
    (IS-A                       NETWORK-OBJECT)
    (generic_name               SERVER)
    (server_id)
    (server_rate)
)
(defschema      SOURCE
    (IS-A                       NETWORK-OBJECT)
    (generic_name               SOURCE)
    (source_id)
    (source_rate)
)
```

Figure 3(a). The Attributes of Basic Network Objects.

```
(defschema      PERF-PARAMETER
    (IS-A                       NETWORK-OBJECT)
    (PERF-OF-GENERIC-VARIABLE)
    (PERF-OF-GENERIC-OBJECT)
    (generic_name               PERF-PARAMETER)
    (count)                     ;;; Total number of samples collected for abstraction
    (first_obs_time)            ;;; start time of the measurement interval
    (last_obs_time)             ;;; end time of the measurement interval
    (min_value)                 ;;; minimum value of collected samples
    (max_value)                 ;;; maximum value of collected samples
    (av_value)                  ;;; average of the collected samples
    (var_value)                 ;;; variance of the collected samples
    (time_wtd_av_value)         ;;; time weighted average of samples
    (time_wtd_var_value)        ;;; time weighted variance of samples
    (min_value)                 ;;; minimum value over measurement interval
    (max_value)                 ;;; maximum value over measurement interval
    (av_top_hit)                ;;; average of number of times upper bound is hit
    (av_succ_top_hit)           ;;; average of number of times upper bound is hit successively
    (av_bottom_hit)             ;;; average of number of times lower bound is hit
    (av_succ_bottom_hit)        ;;; average of number of times lower bound is hit successively
)
```

Figure 3(b). Attributes of Performance Parameter.

```
(defschema    STATE_VARIABLE
    (IS-A                     NETWORK-OBJECT)
    (generic_name             STATE_VARIABLE)
    (VAR-OF-GENERIC-OBJECT)   ;;; the class name of object to which it belongs
    (range_of_var)            ;;; list indicating the upper and lower bounds of the variable
    (state_threshold)         ;;; threshold value of state variable

    (value)                   ;;; the value at a sample instant
    (obs_time)                ;;; the time instant of the sample

    ;;; the following attributes represent operators on state variable

    (min_value)               ;;; recorded minimum over a period
    (max_value)               ;;; recorded maximum over a period
    (up_count)                ;;; number of times threshold is crossed from below
    (down_count)              ;;; number of times threshold is crossed from above
    (first_obs_time)          ;;; the starting time of measurement interval
    (last_obs_time)           ;;; the time of observation of last sample
    (top_hit)                 ;;; the rate at which the variable hits the upper bound
    (succ_top_hit)            ;;; the rate at which the variable succesively hits the upper bound
    (bottom_hit)              ;;; the rate at which the variable hits the lower bound
    (succ_bottom_hit)         ;;; the rate at which the variable succesively hits the loweer bound
)


(defschema    EVENT_VARIABLE
    (IS-A                     STATE_VARIABLE)
    (generic_name             EVENT_VARIABLE)
    (event_operator_id)       ;;; the id of the operator to extract the event
)
```

Figure 3(c). The Attributes of State and Event Variables.

```
(defschema    SENSOR
   (IS-A                          NETWORK-OBJECT)
   (generic_name                  SENSOR)
   (sensor_code_id)                           ;;; code for abstraction operator
   (initiated_by)                             ;;; user query or program query
   (duration_of_activation)                   ;;; how long the sensor remains active
   (sample_on_period)                         ;;; length of window open for monitoring
   (sample_off_period)                        ;;; time between two succesive windows
   (sampling_interval)                        ;;; sampling rate
   (sample_count)                             ;;; number of samples to be collected
)

(defschema    STATUS_SENSOR
   (IS-A                          SENSOR)     ;;; inherits sensor's attributes
   (generic_name                  STATUS_SENSOR) ;;; class name
   (MONITORING-GENERIC-STATE-VAR)             ;;; class name of state-variable
   (MONITORING-STATE-VAR)                     ;;; name of specific state-variable
)

(defschema    EVENT_SENSOR
   (IS-A                          STATUS_SENSOR) ;;; inherits sensor's attributes
   (generic_name                  EVENT_SENSOR)  ;;; class name
   (event_operator_id)                        ;;; the operator id for event extraction
   (MONITORING-EVENT-VAR)                     ;;; name of the specific event
)

(defschema    DERIVED_SENSOR
   (IS-A                          SENSOR)     ;;; inherits sensor's attributes
   (generic_name                  DERIVED_SENSOR) ;;; class name
   (DERIVED_FROM)                             ;;; the names of component sensors
   (MONITORING-GENERIC-OBJECT)                ;;; class name of object being monitored
   (MONITORING-OBJECT)                        ;;; name of monitored object
)
(defschema    SENSOR_NETWORK_STATION
   (IS-A                          DERIVED_SENSOR)
   (generic_name                  SENSOR_NETWORK_STATION)
   (key                           (node_no st_no))
   (node_no)
   (st_no)
   (MONITORING-GENERIC-OBJECT     NETWORK_STATION) ;;; class name of object being monitored
)
```

Figure 4(a). The Schema Description of Sensor and Its Subclasses.

```
(defschema    SENSOR_SWITCH_BUFFER_INFO
    (IS-A                     DERIVED_SENSOR)
    (generic_name             SENSOR_SWITCH_BUFFER_INFO)
    (key                      (node_no st_no))
    (node_no)
    (st_no)
    (MONITORING-GENERIC-OBJECT  SWITCH_BUFFER_INFO)
)
(defschema    SENSOR_LINK_BUFFER_INFO
    (IS-A                     DERIVED_SENSOR)
    (generic_name             SENSOR_LINK_BUFFER_INFO) ;
    (key                      (node_no st_no))
    (node_no)
    (st_no)
    (MONITORING-GENERIC-OBJECT  LINK_BUFFER_INFO)
)
(defschema    SENSOR_BUS_SWITCH_FABRIC
    (IS-A                     DERIVED_SENSOR)
    (generic_name             SENSOR_BUS_SWITCH_FABRIC)
    (key                      (node_no st_no))
    (node_no)
    (st_no)
    (MONITORING-GENERIC-OBJECT  BUS_SWITCH_FABRIC)
)

(defschema    SENSOR_SWITCH_BUFFER
    (IS-A                     DERIVED_SENSOR)
    (generic_name             SENSOR_SWITCH_BUFFER)
    (key                      (node_no st_no buffer_id))
    (node_no)
    (st_no)
    (buffer_id)
    (MONITORING-GENERIC-OBJECT  SWITCH_BUFFER)
)

(defschema    SENSOR_LINK_BUFFER
    (IS-A                     DERIVED_SENSOR)
    (generic_name             SENSOR_LINK_BUFFER)
    (key                      (node_no st_no buffer_id))
    (node_no)
    (st_no)
    (buffer_id)
    (MONITORING-GENERIC-OBJECT  LINK_BUFFER)
)
```

Figure 4(b). The Subclasses of Derived Sensors.

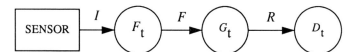

Figure 5(a). Decomposition of a Simple Query about a State Variable.

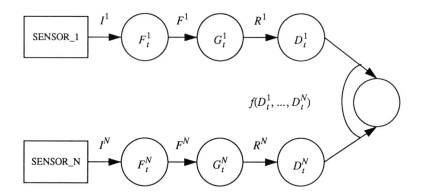

Figure 5(b). Decomposition of a Compound Query about State Variables of Multiple Object.

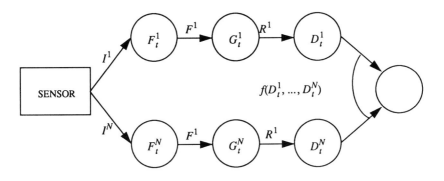

I	:= activates a sensor based on Id to collect data.
F_t	:= raw data acquired by SENSOR.
F	:= data transformation function.
G_t	:= processed or abstracted data after application of F on F_t.
R	:= inference rule *solely* based on G_t.
D_t	:= inferred data after application of R on G_t.

$$f(D_t^1, ..., D_t^N) = \text{inference rule based on } \langle D_t^1, ..., D_t^N \rangle$$

Figure 5(c). Decomposition of a Compound Query about a State Variables.

```
;;; The following rule identifies the specific instances of objects to be
;;;; monitored based on the object class and performance parameter class.
;;;; The variable ?generic_object binds to the name of the object class in Query.
;;;; The variable ?perf-parameter binds the name of performance-parameter.
;;;; The rule selects any objects, that is an instance of the ?generic_object and
;;;; it has a performance parameter that is an instance of ?perf_parameter.
;;;; If the query contains a subset of the key attributes and their values then
;;;; only those objects are identified which match the attribute-value pair of the
;;;; specified query.

(defrule process-query-for-generic-object
   (Query ?generic_object ?perf-parameter ?key_attrib_val_list ?sampling_pattern)
   (schema ?specific_object
                 (INSTANCE-OF          ?generic_object)
                 (HAS-PERF-PARAMETER   ?any_perf_parameter)
   )
   (schema ?any_perf_parameter
                 (INSTANCE-Of          ?perf-parameter)
   )
=>
   (setq ?obj_key_attrib_val_list (get-key-attrib-val-list ?specific_object))
   (if (subset ?key_attrib_val_list ?obj_key_attrib_val_list)
       (create-object-view ?specific_object ?perf-parameter ?sampling_pattern))
)


;;;; The word "defrule" defines a rule.
;;;; The lines above the symbol "=>" are the condition for firing the rule
;;;; The lines below the symbol "=>" are the actions taken if the rule is fired
;;;; The word "schema" defines an object
;;;; The words that start with "?" are defined as variables
;;;; The word "subset" checks if the first argument is subset of the second.
```

Figure 6. The Rule that Selects the Specific Objects Based on the Object Class of a Query.

```
(defaction create-object-view ( (specific_object NETWORK-OBJECT)
                                (perf-param PERF-PARAMETER)) (sampling_pattern)
    (let  (
              (obj_view_object AN_OBJ_VIEW)
              obj_view_object_var perf-param-instance key_names obs_key_names
          )
          (setq    generic_object (get-schema-value specific_object 'generic_name))
          ;;; If the object has no component then
          ;;;;       get the specific performance parameters
          ;;; else
          ;;;; get the specific performance parameter for all component objects.
          (setq    perf-param-instances
              (get-specific-perf-parameters specific_object perf-param))

          ;;; For every performance-parameter instances, get their corresponding
          ;;; state variables.
          (setq obj_view_object_var
              (for param in perf-param-instances join
                      (get-corresponding-svar-of-perf specific_object param)))
          )

          ;;; Set up the key attributes for the instances of OBJ_VIEW
          (setq    key_names (get-key-name-list-object specific_object))
          (setq obs_key_names (cons 'object-view-of-generic-object key_names))

          ;;; Create the instances of OBJ_VIEW in order to setup observation frame
          ;;; with the following attributes and relationships.
          (assert (schema =obj_view_object
                          (INSTANCE-OF                       OBJ_VIEW)
                          (HAS-OBJ-VIEW-PERF-PARAMETER       =perf-param-instances)
                          (HAS-OBJ-VIEW-STATE-VAR            =obj_view_object_var)
                          (key                               =obs_key_names)
                          (OBJECT-VIEW-OF-GENERIC-OBJECT     =generic_object)
          ))
          (attach-sensor-to-object-view obj_view_object sampling_pattern)
    )
)
```

Figure 7. The Procedure that Creates an Observation Frame for Monitoring.

```
(defaction create-object-view-sensor ((object_view OBJ_VIEW)) (sampling_pattern)
    (let    (
                generic_sensor sensor_instance sensor_code eval-list
                (obj_generic (get-schema-value object_view 'generic_name))
                (key_attrib_list (get-schema-value object_view 'key))
            )
        (setq    derived_sensor A_OBJ_SENSOR)
        (setq    sensor_code (form-schema-name "qry" obj_generic))
        ;;; Determine the class name of the derived sensor.
        (setq    generic_sensor (form-schema-name "sensor" obj_generic))
        (setq    sensor_instance 'derived_sensor)
        (if (schemap generic_sensor) (setq sensor_instance generic_sensor))
        ;;; Get the key attributes of the derived_sensor based on the attributes
        ;;; of object_view
        (for key_i in key_attrib_list do
            (assert (schema    =derived_sensor
                            (=key_i     =(get-schema-value object_view key_i))
            ))
        )
        ;;; Create the derived sensor for object_view with following attributes
        (assert (schema    =derived_sensor
                            (INSTANCE-OF            =sensor_instance)
                            (key                    =key_attrib_list)
                            (sensor_code_id         =sensor_code)
                            (status                 created)
                            (MONITORING-OBJECT      =object_view)
        ))
        ;;; Modify the derived_sensor attributes based on the attribute-value-
        ;;; pair list in sampling_pattern, if it is not null.
        (and sampling_pattern
            (setq eval-list (append (list 'schema derived_sensor) sampling_pattern))
            (eval (cons 'modify (list eval-list)))
        )
        derived_sensor
    )
)
```

Figure 8. The Procedure for Creating an Observation Frame.

VIII

USAGE CONTROL

A
Systems Management Functions

Performance Management in an EMA Director

Mohammed Anwaruddin

Digital Equipment Corporation, 500 King Street, Littleton, Massachusetts 01460, USA

Abstract

Digital's Enterprise Management Architecture (EMA) defines the model for building an integrated and extensible management system to manage an enterprise. A management system is called a Director. DECmcc is Digital's EMA Director. Performance management is one of the functional areas that is defined as part of the management function. The Performance Analyzer is a Functional Module (FM) that implements the monitoring aspect of the performance management function in DECmcc. This paper describes a model for Performance management in DECmcc and some aspects of the design of the Performance Analyzer.

1 Introduction

Digital's Enterprise Management Architecture defines a model for building extensible management systems[1]. The architecture is consistent with OSI standards for Network Management. The management system is called a Director. The Enterprise Management Director provides the standard management functions across a potentially unlimited collection of entities [2]. In this paper, we will describe how the Performance Management function is addressed in DECmcc. We begin by reviewing the architecture on which the Management system is based. We then review the Director model and its implementation. Finally, we will review the model adopted for Performance Management and give an overview of a component of the Performance management function. The Enterprise Management Architecture [1], the Director Architecture [2] and the Entity Model [3] are described in detail elsewhere.

2 Background

2.1 Enterprise Management Architecture

The Enterprise Management Architecture (EMA) is made up of

- the Entity model
- the Director model, and
- the Interfaces and the interaction between the two

The Entity model defines a conceptual structure for the management interface to be provided by a manageable entity. Computer systems, LAN bridges, modems, software applications, operating systems, databases are all examples of entities. The model provides a mechanism to group similar objects (entities) into classes. Entity classes are named to identify them. The entities within a class share common properties. These are a set of directives (operations), a set of Attributes (typed values), a set of Events (state transitions) and instance hierarchies of entity classes. A Directive is a request/reply operation from the Director to the entity. Attributes are data that provide management information about the entity. The attributes are partitioned into architecturally defined collections, such as the partition of COUNTERS or the partition of STATISTICS. Each attribute belongs to just one attribute partition. Events are occurrences of normal or abnormal conditions detected by an entity. An entity (called child entity) may be contained in another entity (called parent entity). These containment relationships form an entity class hierarchy which is a tree. The root of this tree is called a Global entity. Figure 1 shows the class structure and entity information.

2.2 The Director

The EMA Director model defines an open, modular platform for managing any type of manageable entity. It defines mechanisms for *access* to entities, *functions* for high-level management, and the *form* of interaction with users.

A Director is a software system that provides the interface between management users and entities that are to be managed by the users. Figure 2 shows the basic components of DECmcc. These are *the Kernel, the Management Modules (MMs)* and *the Management Information Repository (MIR)*.

The *kernel* provides a multithreaded environment for Management Modules to exist and communicate with each other and with management information stored in DECmcc. The *Management Modules* provide services to management users and to each other. There are three kinds of management modules - Presentation Modules (PMs), Functional Modules (FMs) and Access Modules (AMs). Presentation modules provide user interface support. There may be one PM for command line interface, one for windowing interface, etc.. Functional modules provide the management applications. The FMs perform useful management functions on data obtained from entities. Thus there may be one or more FMs for each OSI Management Function. Access modules (AMs) provide access paths to and retrieve management information from various classes of entities. Each AM deals with an entity class using the specific management protocol understood by the entities in the class. They provide a consistent view of the entities that they represent to other MMs and are the key to EMA's ability to support heterogeneous networks with multivendor entities.

The *Management Information Repository* (MIR) is an integrated repository that contains definitions of entity classes, entity instances, entity attributes, and miscellaneous data required by individual management modules.

The Entity attribute data is data that has been collected by polling the entity at user-specified intervals, timestamped and stored in the Repository. This is referred to as Historical data. Function modules, such as the Performance Analyzer, may transform the historical data to provide useful information, such as traffic statistics, that could be used in trend analysis.

2.3 DECmcc is an Integrated and Extensible Director

The Director framework allows the management modules to interact with each other transparently. A client management module does not need to know which management module is providing the requested service. Freedom in choosing a user interface, specifying any function and managing any entity provides an integrated set of management services over an unlimited set of entities.

DECmcc is extensible. As new management modules are added, or when new attributes are supported by an existing entity, the information can be incorporated into the MIR dynamically. This process, known as Enrollment, makes a management module's services and user interface information available to other management modules. (See [4] for a description of the extensible nature of EMA Directors).

3 Performance Management Model

This section describes how Performance Management fits into DECmcc. Performance Management is one of the five Specific Management Functional Areas identified by OSI management [5]. Performance Management allows for the *monitoring* and *evaluation* (analysis) of the performance and *control* of entities. *Monitoring* involves collecting raw attribute data, such as counters and characteristics, over a time interval (data collection) and converting this data into meaningful statistics (data reduction). The statistics could be traffic and error rates, utilization and average message sizes etc.. *Analysis* involves assessing the results of performance measurements for the purpose of fault management and network optimization and planning. *Control* involves adjustment of resources to improve system performance.

One of the questions that is frequently raised is regarding the calculation of statistics - where should the calculation be performed, the Agent or the Manager (Director)? A Director, as described earlier, is the management system that lets users monitor, control and test manageable entities in an enterprise. An Agent is management software that provides a simple remote management interface to the managed object. It provides information about the object's internal state through counters and status messages, and provides a conduit for operations the object supports. Thus, it is the source of observed (or raw) management data. The agent and the managed object form an entity.

There may be convincing arguments for computing some statistics at the agent. In our model we chose to perform the computations in the director rather than in the agent. Some of the reasons that influenced our choice of the model were:

- The requirement that DECmcc be able to manage existing entities. Most of these agents, which predate the entity model, did not perform any data reduction. This meant that we could not depend on the agents producing some simple statistics.

- Limited computational capability at an agent.

- The need for disk storage at every entity in order to provide historical data if statistics were generated by the agent.

- The need to compute statistics that were not anticipated when the agent code was written.

Figure 3 shows the performance management model in DECmcc. Raw data to be reduced is shown as coming from the entity via the access modules. It could also come from a Repository if data was previously recorded there.

In DECmcc, the Functional Modules are organized around the OSI functional areas. However, this organization is not strict. For example the different aspects of performance management are performed by various management modules. A Performance Analyzer Function Module (FM) addresses the monitoring aspect of the Performance Management function. Analysis is carried out by the cooperation of a number of other management modules. For example, an *Alarms* Functional module provides the capability to set thresholds and conditions, that constitute unacceptable performance levels, in the form of *rules* and provide asynchronous reporting when such thresholds are exceeded.

A *Data extract* functional module provides the capability to *export* performance data to a Relational Data base. This historical performance data is used to track network operations over time and do trend analysis. Other modules utilize the performance data to do network tuning and optimization. For example, Performance information, such as traffic statistics, is used in the form of data input to a WANdesigner, a network design application. Figure 4 illustrates the interactions between the Performance Analyzer and some of the other functional modules. These functional modules interact to analyze the results of performance measurements.

Thus, in DECmcc, Management modules cooperate in various combinations to cover the overlap between the OSI functional categories.

3.1 DECmcc Performance Analyzer

The DECmcc Performance Analyzer (PA) is a function module. It forms the basis for performance management. It computes statistics from raw data that is obtained either from DECmcc's repository or from the entities via their access modules. The PA FM knows the formulae to compute statistics. The user specifies the time period over which to compute them. If no time period is specified, a default time period is assumed.

The current version of PA provides statistics for DECnet Phase IV entities, DECnet Phase V entities, TCPIP entities and Bridge Line entities. The statistics provided are simple metrics such as Counts, Averages, Rates, Overhead, Throughput and Utilization.

The following sections will describe how these issues of *Extensibility* and *Flexibility*. were addressed in the design of PA.

3.1.1 Extensibility

The Performance Analyzer computes performance statistics for entities in a multi-vendor enterprise. Statistics are defined by formulae which require observed counts over a time period as well as some entity specific characteristics. For example, the circuit counters *BytesSent* and *DataBlocksSent* observed over some user specified time period *Time* are used to compute the Outbound utilization for a DDCMP Line. The Outbound utilization, expressed as a percent, is defined by the formula

$$\frac{\sum_{\forall circuits} (BytesSent + DataBlocksSent * HeaderSize) * 8}{Speed * Time} * 100 \qquad (1)$$

where the numerator specifies the actual throughput and the denominator the nominal throughput. The variable *Speed* specifies Line speed and is an attribute that is defined as belonging to either the Characteristics or Status attribute partition.

The counters used in the formula are *normalized values* of the observed counts. Counters are simple accumulated counts of the number of times an event occurs since the counters were created (or were zeroed). An event can be the sending or receiving a PDU, an Octet, or it can be the detection of a particular error condition. DECmcc Access Modules sample these counters at user specified periods, but the periods between samples are not precisely controlled because of traffic loads, systems going up/down etc. **Normalization** determines the value of a counter at the end of a specified time period as if the counter had the value zero at the beginning of the period.

Based on the samples collected by the access modules, PA computes normalized differences for requested time periods. The normalized differences are estimates of the difference between the value that the counters had at the end of the requested time period, and the value they had at the beginning. Based on these differences, statistics are computed. Thus, to compute a statistic we require:

- a formula defining the statistic
- a user specified time range over which counts are to be observed
- one or more attribute partitions

Management modules(MMs) in DECmcc act as clients and as servers with respect to other MMs in the reporting chain. For instance, PA is a client of an access module which provides the counters required to compute a statistic and is a server for an Alarms functional module which has requested the statistic in order to monitor performance thresholds.

In DECmcc, a MM's services are requested by a procedure call known as *mcc_call*. The parameters to the call include the entity class and instance, an attribute partition and, optionally, a time interval during which to poll the entity. This information is specified by the user on a management operation such as a SHOW for displaying the value of one or a set of statistics. The information enables the performance analyzer to

- lookup the formulae defining the statistics,
- determine what services are required to compute statistics,
- get raw data and reduce it.

The services required to generate a set of statistics for an entity are a list of SHOW operations to get counters and characteristics (or status) from that entity. These are stored as a list of SHOW directives with place holders for entity instance and time interval. This list of directives with place holders is referred to as *ingredients*. Functions may also be applied to the raw data and are included in the ingredients. For example, the ingredient

```
normalize (show node4 <> line <> all counters)
```

specifies that the function *normalize* must be applied to the Line entity counters before they can be substituted in the formulae. The angle brackets are place holders for entity instances. By substituting the entity instance information and the specified time range in the ingredients, counters and other attributes can be requested from any entity.

By storing the ingredients and the formulae in the Repository, the Performance Analyzer is *data driven* and hence easily extensible. Figure 5 illustrates this.

Figure 6 shows a request to PA for generating DECnet Phase IV line statistics and the associated formulae and ingredients required to compute statistics for entities of that class.

3.1.2 Flexibility

One of the goals of the PA management module is to allow multiple methodologies for generating statistics to coexist. This situation arises when it may be more convenient to provide customized routines to compute statistics for an entity. With DECmcc's dispatching mechanism, it is possible to utilize the customized routines to compute statistics by providing a seperate entry point to the statistics partition for this entity.

Thus, the model allows multiple Performance Analyzers to coexist in the same Director, with the more-specific Performance Analyzer modules taking precedence over the default Performance Analyzer module described above.

4 Summary

This paper has described very briefly how Performance Management fits in DECmcc. It described a model for the Performance Analyzer which forms the core of the Performance Management function. The description emphasized two important aspects of the model namely, extensibility and flexibility. The model fits in well with the architecture for enterprise management specified in EMA.

5 Acknowledgements

The author gratefully acknowledges the work of many individuals associated with the project.

References

1. Fehskens, L., An Architectural Strategy for Enterprise Management, Proceedings of the IFIP Symposium on Integrated Network Management, May 1989.

2. Strutt, Colin and D. Shurtleff. Architecture for an Integrated, Extensible Enterprise Management Director. Proceedings of the IFIP Symposium on Integrated Network Management, May 1989.

3. Sylor, Mark. Managing Phase V DECnet Networks: the Entity Model, IEEE Network, March 1988.

4. Shurtleff, David and Colin Strutt. Extensibility of an Enterprise Management Director, Network Management and Control Workshop, September 1989.

5. ISO/IEC/JTC 1/SC 21 N3313 Performance Management.

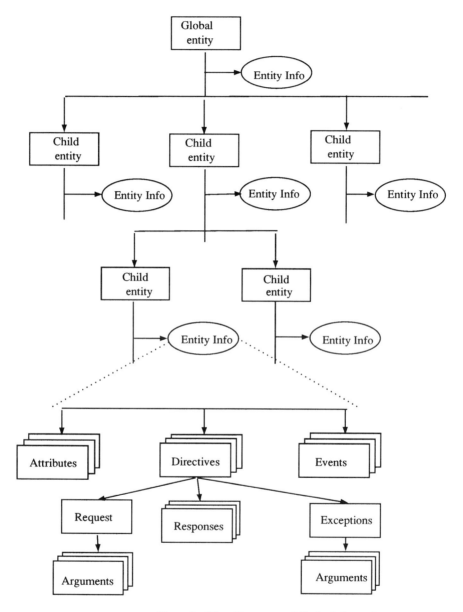

Figure 1: Class Structure and Entity Information

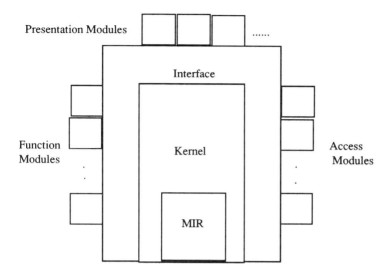

Figure 2: Components of DECmcc

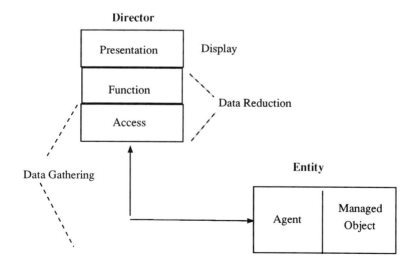

Figure 3: Performance Management Model

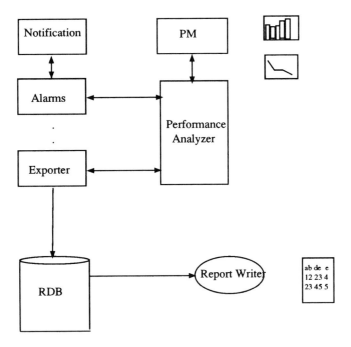

Figure 4: Management module interactions for Performance Management

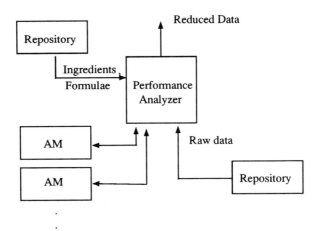

Figure 5: DECmcc Performance Analyzer

REQUEST: SHOW NODE4 BOSTON LINE BNA-0 ALL STATISTICS, -
AT EVERY 1:00:00 UNTIL 17:00:00, FOR DURA 00:45:00

INGREDIENTS:

    sum_normalize(show node4 <> circuit * all counters)

    normalize (show node4 <> line <> all counters)

    show node4 line <> all characteristics

FORMULAE:

    Outbound Block Rate = $\dfrac{BytesSent}{time}$

    .
    .
    .

Figure 6: A Request and associated ingredients

Integrated Network Management, II
I. Krishnan & W. Zimmer (Editors)
Elsevier Science Publishers B.V. (North-Holland)
© IFIP, 1991

Diagnosis of Connectivity Problems in the Internet

Metin Feridun[a]

BBN Systems and Technologies Division, 10 Moulton Street, Cambridge, MA 02138

[a]Present address: IBM Research Division, Zurich Research Laboratory, CH-8803 Rüschlikon, Switzerland

Abstract

Solving problems in the network requires a cycle of detection, diagnosis, and repair. The Connectivity Tool™ (CT) system is designed to diagnose end-to-end connectivity problems in the Internet, where a given site, site A, reports that it cannot reach site B. CT assists the network operator by emulating the steps that an experienced operator would go through to diagnose the connectivity problem. Operating from a third site, CT diagnoses the problem by tracing Internet Protocol (IP) routes in both directions between site A and B to determine where the problem may be located. Although conceptually simple, tracing routes is complicated by the non-uniformity and size of the Internet. CT architecture uses heuristic rules to drive network tools to collect data from the network, and to analyze the results of tool invocations to diagnose the connectivity problem. CT provides a flexible architecture that makes it easy to modify or add heuristic rules or network tools.

1. INTRODUCTION

The Internet has experienced steady growth over the past decade, and this growth progresses as new sites and networks continue to be incorporated at a rapid pace. One result of this expansion has been the division of the Internet into multiple administrative domains, where campus networks (e.g., university or corporate sites) are connected to regional networks (e.g., NEARnet, BARRnet), and these in turn are interconnected through backbone networks such as the NSFnet. In general, although the network operations center (NOC) at each one of these networks is responsible for its network only, it is not unusual for a NOC to accept network problems that cross network boundaries. For example, a local report of a failure to reach a distant site may require investigation of the route to that site. Diagnosis of this type of problem is usually complicated by the non-uniformity and size of the Internet; tools that can assist network operators in solving frequently occurring network problems are in demand.

Solving problems in the network usually requires a cycle of detection, diagnosis, and repair. Detection mechanisms search for evidence of the existence of a problem. Diagnosis mechanisms are tasked to determine the cause of the problem. Finally, repair mechanisms fix the problem.

The Connectivity Tool™ (CT) system is a tool to help network operators diagnose end-to-end connectivity problems in the Internet, when a given site, site A,

---

This work was supported by the Defense Advanced Research Agency (DARPA) under Contract MDA903-89-C-0020. Approved for public release. Distribution unlimited.

reports that it cannot reach another site, site B. Operating at a third site, CT diagnoses the problem by tracing the route between the two sites in both directions to determine where the problem is located. Although conceptually simple, tracing routes is complicated in practice because of the heterogeneous nature of the Internet. Heuristics are needed to handle such issues as deviations in standard protocols, routing policies and device and version-specific problems.

CT contains a set of network tools that can reach into the network to collect and probe for data. These tools include SNMP (Simple Network Management Protocol) pollers, or ping. Using a set of heuristic rules, it integrates these tools to emulate the steps an experienced network operator would go through to diagnose the connectivity problem. CT automates the time-consuming steps, for example the determination of network addresses to poll, and makes available the expertise of an experienced network operator to less experienced network operators, thus reducing the load on the network experts.

In Section 2, we present the algorithm used in CT for diagnosis and explain the reasoning behind it. Section 3 contains a detailed description of the CT architecture, followed by a report on the implementation status.

2. CONNECTIVITY DIAGNOSIS ALGORITHM

2.1 Assumptions

We designed the main diagnosis algorithm for the Connectivity Tool with the following set of constraints and assumptions. First, both the problem-reporting site and unreachable sites are assumed to have an Internet address, and are, in theory, accessible at the Internet Protocol (IP) level. Second, CT looks at IP level routes only, leaving diagnosis of connectivity problems at lower or higher protocol levels for future work. Last, we assume that CT will not necessarily be located at the problem-reporting nor the unreachable site. Therefore, CT needs additional techniques to construct the IP routes between the two sites.

2.2 Algorithm

In a typical diagnosis session, site A reports to the NOC that it cannot invoke or access a service such as telnet at another site B, as shown in Figure 1. There can be many possible causes for the reported problem; a partial list of causes includes:
- Site B not operating ("down")
- Site B operating but the requested service is not available
- The route from site A to site B (or the frequently different return route from site B to site A) is interrupted due to equipment failure, misconfiguration, or other causes
- Congestion of the route from site A to site B (or the return route) is causing loss of packets and consequently deteriorated service.

The objective of the CT diagnosis algorithm is to construct the route between site A and site B in both directions since routes need not be symmetric, with the expectation that examination of the routes will indicate the source of the problem.

The algorithm consists of the following steps:
1. Construct route from NOC to site A.
   This step establishes reachability of site A and helps determine the entry gateway G1 to network A.
2. Construct route from NOC to site B.
   Like step 1, it establishes reachability of site B, and helps determine the entry gateway GN to network B.
3. Force a clockwise loop route from NOC to G1 to GN and back to NOC.
   This route allows NOC to determine the route from site A to site B.
4. Force a counterclockwise loop route from NOC to GN to G1 and back to NOC.
   This route allows NOC to determine the route from site B to site A.

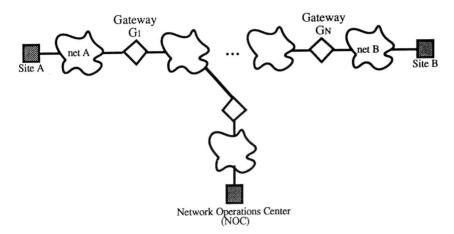

Fig. 1 Connectivity Tool System Diagnosis Scenario

It may not be possible to complete all the above steps for each diagnosis. For example, if B is also unreachable from the NOC, it is unlikely that step 4 will work, as it uses GN, the entry gateway to site B, as its first destination. However, for the same case, step 3 may not provide the complete loop, but will determine the route to site B from A, signalling potential trouble spots.

2.3 Implementation Discussion

The diagnosis algorithm presented in Section 2.2 is conceptually straightforward. Using IP features such as source route option, and exploiting Internet Control Message Protocol (ICMP) messages, this algorithm appears to be easily implementable. However, this turns out not to be the case.

The first problem is that the Internet is heterogeneous in many respects. Protocol implementations do not necessarily follow the standards, and in some cases they are incorrectly implemented. As a result, responses to probes are not guaranteed to return expected results. For example, in response to an IP packet with time-to-live counter equal to zero, not all devices return ICMP "Time Exceeded" message to the source of the packet.

Second, the division of the Internet into backbone, regional, and campus networks implies that each administrative domain can set its own routing policies, and configure and propagate information on devices in its domain in a manner suitable to its mission. Furthermore, a networkwide configuration database that provides information on devices as well as network topology does not exist for the Internet. In the absence of this knowledge, heuristics are needed to discover device information and network topology. For example, we need to "guess" which management protocol, if any, is supported by a device in order to query for its interfaces.

Finally, the level of detail needed to execute each one of the steps and to integrate the results to solve the connectivity problem is not explicitly stated in the diagnosis algorithm. This knowledge is based on what an experienced network manager would use to solve the problem, and has to be emulated in CT using heuristic rules.

3. CONNECTIVITY TOOL ARCHITECTURE
3.1 Architecture Overview
In our design of the CT architecture, we considered the following goals:
- **modularity:** as newer tools, management protocols, or capabilities for connectivity analysis appear, we should be able to incorporate them easily into the architecture.
- **reusability:** there are other network diagnosis problems which are similar to the connectivity problem, and therefore can reuse software modules developed for CT.
- **portability:** the success of tools such as CT depends on how easily they can be disseminated across a variety of platforms.

The Connectivity Tool System architecture is shown in Figure 2. It consists of the following components:
- **network tools** are a set of tools used to directly probe or query the network to collect network information that is used in the diagnostic process. Examples of such tools are icmp_echo or SNMP (Simple Network Management Protocol) poller.
- **network information base** is the object-oriented depository for data collected from the network by network tools, data inferred by the task interpreter, or static configuration data for name or address mapping.
- **diagnostic tasks** are objects that represent the diagnostic functions used in the diagnosis process. They consist of a mixture of procedures and heuristic rules.
- **task interpreter** drives the diagnosis process by activating diagnostic tasks. Depending on the executed diagnosis task, it can invoke network tools or access the network information base. It also drives and is driven by the user interface.
- **user interface** accepts user commands, and displays progress and results of the diagnosis sessions.

We describe these components in more detail in the following sections.

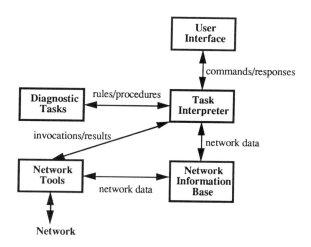

Fig. 2 Connectivity Tool System Architecture

3.2 Network Tools

Network Tools consist of functional modules that can be used to collect information from the network and network devices to facilitate the diagnosis process. These tools use or exploit features of protocols widely available in the Internet, including IP, ICMP, UDP (User Datagram Protocol), SNMP and HMP (Host Monitoring Protocol) [1, 2, 3]. These features are used for such functions as collecting device information, confirming reachability and constructing IP routes. They are invoked by the Task Interpreter module, return results of invocations to the Task Interpreter, and deposit results or network data in the Network Information Base. These tools are described briefly below.

3.2.1 icmp_echo

The icmp_echo sends an "ICMP Echo Request" to check the reachability of an Internet site. If the site responds with an "ICMP Echo Reply" message, the reachability is confirmed. Other responses or the lack of a response are also reported by icmp_echo since they can indicate problems.

3.2.2 rr_route

This tool sends an "ICMP Echo Request" to an Internet site with the IP Record Route option. This option forces each network node on the route to the site (as well as on return) to record its outgoing interface. A maximum of nine address slots is provided, and therefore the usefulness of rr_route is limited because of the large diameter of the Internet.

3.2.3 trace

The trace is based on the "traceroute" program developed by Van Jacobson [4]. Without the IP Loose Source option, it determines the direct path from the local site (e.g., NOC) to a destination site. With the option, longer, indirect routes can be constructed, such as loops originating from the NOC in the CT algorithm.

Trace sends UDP packets to a non-existent UDP port at the destination site. Each probe is sent with an incrementally increasing IP Time-to-Live (TTL) field, starting at 1 for the first probe. Since IP routers are expected to return "ICMP Time Exceeded" when the TTL field in an IP packet is zero, trace uses these ICMP messages to construct the route to the destination, and provides the incoming interface IP address for each traversed node on the route.

3.2.4 SNMP and HMP Tools

A subset of devices in the Internet, mostly IP routers, respond to SNMP or HMP. CT contains a set of tools for both protocols that can query devices. Using these tools, we can collect management information from these devices, such as their interfaces, up/down status, and the like, to help in the diagnosis process.

3.3 Network Information Base

The Network Information Base is an object-oriented depository of information collected by various network tools, or inferred by the task interpreter during the diagnosis process. It is designed to facilitate sharing of information between components of the CT architecture in order to prevent unnecessary queries to the network.

Among the types of information retained in the Network Information Base are:
- device information, such as management protocol-supported, interface and corresponding status information
- result of network tool invocations, such as the result of an icmp_echo invocation
- routes, constructed during the diagnosis process
- inferred facts about devices, such as "device not responding to SNMP or HMP", established by the Task Interpreter.

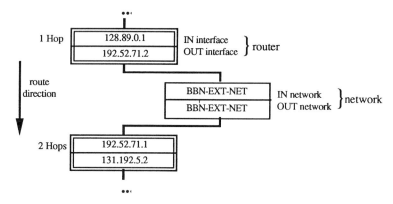

Fig. 3 Route Model

Representation of a route in the Network Information Base uses a model designed to highlight inconsistent or incomplete information returned by network tools. The route model is shown in Figure 3. It consists of a sequence of IP routers and networks. A router is represented by an in and an out interface, corresponding to interfaces used on this route to enter and exit the router. In addition, each router has a hop count, indicating the distance from the start of the route. A network is similarly represented as two halves, i.e., an IN side and an OUT side, to handle inconsistencies arising from responses returned from network tool invocations. It is possible that the "previous" and "next" routers are not neighbors, and that an intervening router is "missing", i.e., not discovered yet. The network representation allows detection of such inconsistencies; for example, in the case of the missing routers, it will have unmatched halves. The hop count fields of adjacent routers can also indicate similar problems.

3.4 Diagnostic Tasks

The procedures and heuristic rules that govern the diagnosis of the connectivity problem are represented by a set of diagnostic task objects. Each task object represents a particular diagnostic task, and includes the following attributes:
- *candidate rules:* these are rules or direct procedure calls to solve the diagnostic task.
- *filter rules:* these rules are used to select among candidate rules those which are most appropriate to the diagnostic task at hand.
- *ranking rules:* these rules rank the filtered rules in the order of expected success.

An example of a diagnostic task is shown in Figure 4. In this example, the diagnostic function is to collect the interfaces of an IP router. There are three candidate rules: *interfaces_already_collected* invokes a subtask, i.e., another task object, that checks whether the interfaces for router object X were already collected by checking the Network Information Base. The other candidate rules, *hmp_interfaces* and *snmp_interfaces,* are direct invocations of network tools.

Three filter rules determine which candidate rules are applicable to a particular invocation of this diagnostic task. These rules are driven by the network management protocol supported by router X; note that if there is no knowledge about the type of X, then all rules (R1, R2, and R3) apply.

Finally, the ranking rules define the order in which the filtered rules (procedures) should be attempted. For this example, the precedence relation is R1 > R3 > R2. R1 has precedence over the others since it encourages reuse of already collected

```
Diagnostic Function: get interfaces for IP router X      ;; X is an IP address
Candidate Rules:
    R1: interfaces_already_collected(X)                  ;; call to a sub-task
    R2: hmp_interfaces(X)                                ;; call to network tool (HMP)
    R3: snmp_interfaces(X)                               ;; call to network tool (SNMP)
Filter Rules:
    F1: if (type_of(X) is HMP) then {R1 or R2}           ;; HMP router
    F2: if (type_of(X) is SNMP) then {R1 or R3}          ;; SNMP router
    F3: if (type_of(X) is UNKNOWN) then {R1 or R2 or R3}
Ranking Rules:
    S1: R1 > R2                                          ;; R1 has precedence over R2
    S2: R1 > R3                                          ;; R1 has precedence over R3
    S3: R3 > R2                                          ;; R3 has precedence over R2
```

Fig. 4 Diagnostic Task Object Example

values. R3 is attempted before R2 since we expect to find more SNMP routers than HMP routers in the Internet.

As implied in the above example, the diagnostic tasks are chained in a graph hierarchy, each path in the graph representing a possible sequence of steps in the diagnosis process. The execution of each diagnostic task either succeeds or fails, based on the success of the execution of its rules and subtasks it may invoke. Execution of each diagnostic task can have a side effect as the results of network tool invocations are recorded in the Network Information Base.

The object representation of tasks provides a flexible representation that simplifies the modification of task hierarchy, as well as the modification of individual diagnostic tasks.

3.5 Task Interpreter

The execution of diagnostic tasks is carried out by the Task Interpreter. Starting with an initial diagnostic task, such as "start diagnosis" task, it searches through an AND/OR graph until either the problem is diagnosed or all possible paths are exhausted. Network tool invocations or data retrieval functions from the Network Information Base represent terminal nodes in the AND/OR graph.

During the execution of a diagnostic task, the task interpreter may invoke subtasks, invoke network tools, check for availability of data, retrieve data from or deposit data into the network information base, or interact with the user through the user interface.

To show the operation of the task interpreter, a subset of diagnostic task objects used in CT is shown in Figure 5; we include only the candidate rules for brevity. These tasks are used in the diagnosis process to label the incoming network address attribute of a network object between two router objects on an IP route, as shown in Figure 6. In the example scenario, we assume knowledge of the incoming interfaces for both routers G1 and G2; in order to find the "in" side of network NET, we need to find the outgoing interface of the "previous" router G1, and determine its network address.

Figure 5 shows eight tasks, T1 through T8. Note that of these tasks, T5, T6, T7, and T8 are terminal tasks, i.e., they cause invocation of network tools or information base access functions, but not another diagnostic task. A branch of the AND/OR graph is shown in Figure 7; we included the values with which the task objects are invoked in Figure 5. In this partial trace of the search, the "in" side of NET will be labeled only if T1 succeeds.

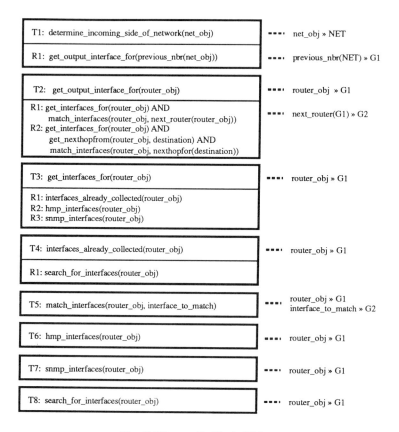

Fig. 5 Diagnostic Task Objects

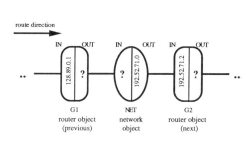

Fig. 6 Diagnosis Scenario

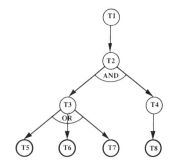

Fig. 7 Partial AND/OR Search Graph

3.6 User Interface

The User Interface module of the CT System allows the network operator to start diagnosis sessions, observe the diagnosis process, and analyze the results. It is driven by the keyboard and the mouse, and is designed to reduce the amount of input by the network operator.

As shown in Figure 8, the user interface consists of the following components:
- *control panel* is the topmost subwindow, and contains the input area, buttons for initiating and quitting diagnosis sessions (Diagnose, Reset, and Quit), and a route display button (Route Menu) to select an already computed route for display. A message line in the control panel indicates the status of the diagnosis process as it proceeds.
- *inference window* is located at the top right of the user interface, under the control panel. It displays the diagnostic task objects executed during the diagnosis process.
- *tool window* is located under the inference window. It displays the results collected by the network tool invocations.
- *route display window* is located at the left of the user interface, under the control panel. It displays routes that are constructed by the diagnosis process. Routers are displayed in the left column, and networks on the right.

Two pop-up windows can be invoked through a menu obtained when a router object in the route display window is mouse clicked. These are:

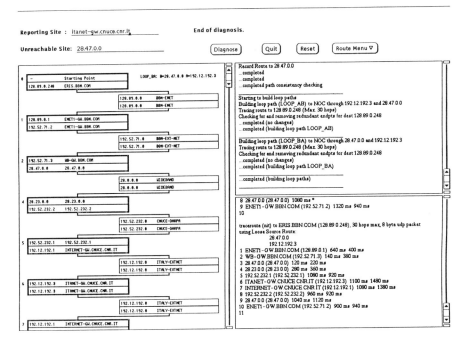

Fig. 8 CT User Interface

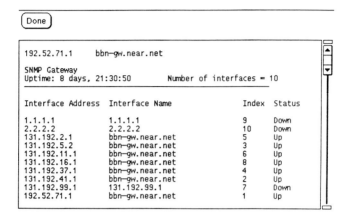

Fig. 9 Gateway Information Pop-up Window

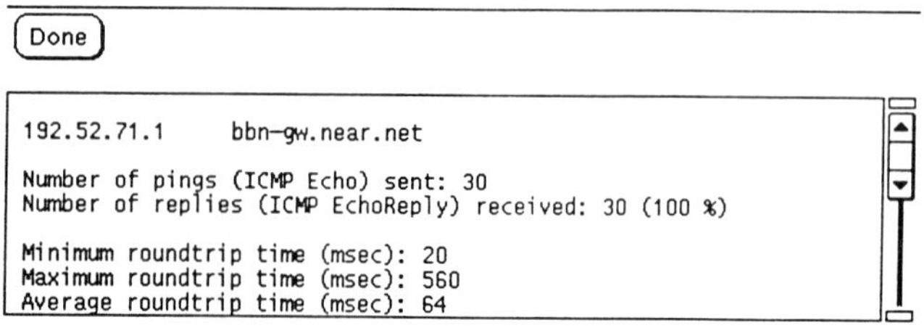

Fig. 10 Ping Status Pop-up Window

- *gateway information window* (Figure 9), which provides information about the router as it appears in the network information base; and
- *ping statistics window* (Figure 10), which displays results of ping (ICMP Echo Request/Reply) round trip statistics to the router.

4. IMPLEMENTATION STATUS

In order to test the ideas and architecture behind the CT system, we implemented a prototype version of the system. The prototype CT runs on Sun UNIX™ workstations, and is implemented using C language. We designed and implemented the network tools described in Section 3.2. For the task interpreter and diagnostic tasks, we chose to implement a simpler version that uses procedure calls and structures to mimic the object-oriented model. This decision has enabled us to test the architecture faster, without deviating much from the intended design. The user interface was implemented using the XView toolkit running under the MIT X Window System distribution (X11R3/R4). The user interface includes all the features described in Section 3.6, and shown in Figures 8 through 10.

The CT System is currently installed at several Internet sites for evaluation purposes.

5. CONCLUSION

We have designed and implemented a diagnosis system to aid network operators in managing networks. Systems such as CT, which integrate network tools with the knowledge of expert network managers, have the potential to reduce network management costs and improve management capabilities.

The architecture of CT provides a flexible basis for adding new capabilities, such as new network tools or diagnostic tasks. At the same time, the architecture is general enough to provide a framework for the development of other, similar diagnosis applications. These applications require the collection of data from the heterogeneous network elements, and the analysis of this data, both using a mixture of procedures and heuristic rules. Examples of such applications are a topology mapper for deriving the logical connectivity for a network, and a performance analyzer for computing the segment and end-to-end delays on a given route.

6. REFERENCES

1. D. Comer, Internetworking with TCP/IP - Principles, Protocols, and Architecture, Prentice Hall, Englewood Cliffs, New Jersey, 1988.
2. J.D. Case, M. Fedor, M.L. Schoffstall, C. Davin, Simple Network Management Protocol (SNMP), Request for Comments 1157, DDN Network Information Center, SRI International, 36 pp., May 1990.
3. R.M. Hinden, Host Monitoring Protocol, Request for Comments 869, DDN Network Information Center, SRI International, 70 pp., December 1983.
4. R. Stine (ed.), FYI on a Network Management Tool Catalog: Tools for Monitoring and Debugging TCP/IP Internets and Interconnected Devices, Request for Comments 1147, DDN Network Information Center, SRI International, 126 pp., April 1990.

Secure Management of SNMP Networks

James M. Galvin[a], Keith McCloghrie[b] and James R. Davin[c]

[a]Trusted Information Systems, Incorporated,[1] 3060 Washington Road, Glenwood, MD 21738; (301) 854-6889; galvin@tis.com

[b]Hughes LAN Systems, Incorporated, 1225 Charleston Road, Mountain View, CA 94043; (415) 966-7934; kzm@hls.com

[c]MIT Laboratory for Computer Science, 545 Technology Square, NE43-507, Cambridge, MA 02139; (617) 253-6020; jrd@ptt.lcs.mit.edu

Abstract

The Internet is a collection of heterogeneous hosts communicating via the TCP/IP suite of protocols. The secure management of this network is of paramount concern. Security in this context refers to three distinct services: authenticating the origin of a management operation, verifying the integrity of the management operation and preventing the disclosure of both the management operation and its response. Recently the Simple Network Management Protocol (SNMP) has been enhanced to provide these security services. The infrastructure to support these services has also been specified.

1 Introduction

The Internet is a collection of heterogeneous hosts communicating via the TCP/IP suite of protocols.[2] The Internet Activities Board (IAB), the coordinating committee of the Internet, recognizes the need for network management in the Internet [2], in spite of the heterogeneity of the communicating hosts and domains. The Simple Network Management Protocol (SNMP) is the current standard protocol to satisfy this need. [1]

---

[1]This work was partially supported by the U. S. Government Defense Advanced Research Projects Agency under contract number F30602-89-C-0125 to Trusted Information Systems, Inc.

[2]The Transmission Control Protocol (TCP) [9] and Internet Protocol (IP) [8] suite comprise a set of protocols that define a variety of network applications, e.g., file transfer and virtual terminals.

The secure management of networks is of paramount concern. Security in this context refers to three distinct services: authenticating the origin of a management operation, verifying the integrity of the management operation and preventing the disclosure of both the management operation and its response.

The SNMP specification identifies an authentication algorithm called trivial authentication. Fortunately, the SNMP specification also defined an authentication service that allows for the specification of alternate authentication algorithms.[3]

Galvin, McCloghrie and Davin have enhanced the SNMP security service to include an abstract service interface. We defined two additional security algorithms according to the interface. [6] The first algorithm provides for the authentication of the origin of a SNMP message and the integrity of the contents of said message. Both security services are based on the application of a suitable Manipulation Detection Code (MDC) algorithm, initialized with a value that is private to the communicating peers, and inclusion of the MDC output with the message to be communicated.

The integrity of the message is guaranteed if the receiver is able to verify the MDC value included with the message. The origin of the message is implicitly known, since the MDC is initialized with a secret, random value, known *a priori* by the communicating peers. To protect against replay attacks, each message includes a timestamp value that is included in the MDC calculation and is evaluated by the receiver.

The evaluation of the timestamp value requires loosely synchronized clocks, but only the clocks of the communicating peers need to be synchronized. Further, the "looseness" of the synchrony is tunable on a per communicating peers basis. This allows a management station responsible for a heterogeneous set of manageable agents to synchronize each according to the variable delays in the communication paths.

The second algorithm enhances the first to provide non-disclosure of the contents of a message. It is based on the application of a suitable cryptographic algorithm to a message authenticated according to the first algorithm. It necessarily depends on a cryptographic key known *a priori* by the communicating peers.

Both algorithms depend on the distribution and management of a secret value to initialize the MDC calculation and of loosely synchronized clock values. The second algorithm also requires the distribution and management of a cryptographic key and its associated parameters. The document describing the two security algorithms assumes the existence

---

[3]In the interests of clarity, it should be noted that the SNMP specification incorrectly uses the term *authentication service* to refer generically to a variety of security services; moreover, the SNMP specification also incorrectly uses the term *authentication protocol* to refer generically to a variety of security mechanisms. Authentication (specifically data origin authentication) is only one of three security services which will be defined at the SNMP service interface. In contrast, this paper adopts a more traditional and less misleading terminology: the terms *authentication service* and *authentication protocol* are applied only in cases pertaining specifically to authentication of data origin and integrity, whereas the terms *security service* and *security protocol* are used to refer more generically to security-related services and mechanisms, including those pertaining to data privacy.

of these values. An additional document proposes two strategies for the distribution and management of these values. [3] The SNMP is used as the transport for the distribution of said values. A third document defines an object suitable for inclusion in the SNMP Management Information Base (MIB) for each of the values. [7]

The remainder of this paper presents the initial analysis upon which the development of the security algorithms. This analysis details the threats, goals, constraints and services defined for the security algorithms. A set of mechanisms were then chosen to be supported by the algorithms. A brief overview of the security algorithms is presented, followed by a brief overview of one of the techniques recommended for administering the communities using the algorithms.

2 Threats

The goals of an SNMP security protocol derive from an account of the threats against which that protocol is to afford protection. Two principal threats are masquerade and modification of information:

Masquerade. The SNMP specification includes a model of access control. Access control necessarily depends on knowledge of the source of a request. The masquerade threat is the danger that management operations which are not authorized for some party, may be attempted by that party by assuming the identity of an administratively distinct party that is authorized for the management operations.

Modification of Information. The SNMP protocol supports a service that allows management stations to alter the value of objects in a managed agent. The modification threat is the danger that some party may alter messages generated by a legitimate party in transit in such a way as to effect unauthorized management operations.

A number of other threats are also identified:

Message Sequencing. The SNMP protocol is based upon connectionless transport services. The message sequencing threat is the danger that messages may be arbitrarily re-ordered, delayed, or replayed to effect unauthorized management operations. This threat may arise either by the work of a malicious attacker or by the natural operation of a subnetwork service.

Disclosure. Distinct from the threat of unauthorized access to objects in a managed agent, the disclosure threat is the danger of passive eavesdropping on the exchanges between managed agents and a management station. Protecting against this threat is mandatory when the SNMP is used to administer private parameters on which its security is based.

Denial of Service. The denial of service threat is the danger that an active eavesdropper will prevent exchanges between managed agents and a management station.

Traffic Analysis. The traffic analysis threat is the danger that a passive eavesdropper will infer information from the exchanges between managed agents and a management station, which may be protected from disclosure, to which the eavesdropper is not authorized.

3 Goals

Based on the foregoing analysis of threats in the network management environment, the principal goals of a SNMP security protocol are enumerated below.

1. The protocol should provide for verification of the community membership of the generater of each received SNMP message.

2. The protocol should provide for verification that each received SNMP message has not been modified during its transmission through the network.

3. The protocol should provide that the apparent time of generation for each received SNMP message is subsequent to that for all previously delivered messages of similar origin.

4. The protocol should provide, when necessary, that the contents of each received SNMP message are protected from disclosure.

A number of other goals can be enumerated, in consideration of the SNMP architecture and future enhancements of the security protocol. For example, in the SNMP, the granularity to which message origin is reckoned is membership in a SNMP community, for, by definition, the members of a community are administratively equivalent parties to the management activity. Further, since communities based on symmetric cryptography necessitate that all members of the community share the community's secrets, and that such communities must have a membership of at least two in order to be useful, the range of policies that can be supported by such communities is accordingly unable to distinguish between the members. Therefore,

- the protocol should not preclude future extension to support single-member communities.

In particular, assymmetric cryptography by definition identifies individual members, since at most one entity should have access to a private key. Therefore,

- the SNMP security mechanisms should, if possible, naturally accommodate future security algorithms based on assymmetric cryptography.

Finally, protection from the remaining two threats identified above need not be supported. In particular, a SNMP security protocol need not attempt to address the broad range of attacks by which service to legitimate parties is denied. Indeed, such denial-of-service attacks are in many cases indistinguishable from the type of network failures with which any viable network management protocol must cope as a matter of course.

In addition, a SNMP security protocol need not attempt to address traffic analysis attacks. It is believed that many traffic patterns are predictable — agents are probably managed on a regular basis by a relatively small set of management stations — and therefore there is no significant advantage afforded by protecting against traffic analysis.

4 Constraints

In addition to the principal goal of supporting secure network management, the design of a SNMP security protocol is also influenced by the following constraints:

1. When requirements of effective management in times of network stress are inconsistent with those of security, the former are preferred.
2. Neither the SNMP nor its underlying security mechanisms should depend upon the ready availability of other network services (e.g., Network Time Protocol (NTP) or secret/key management protocols).
3. A security mechanism should entail no changes to either the SNMP PDU or to the basic SNMP network management philosophy; upwardly compatible changes to the SNMP Message are consistent with the original SNMP specification.

5 Services

The security services necessary to support the goals of the SNMP security protocol are as follows.

Data Origin Authentication. The corroboration that the claimed source of data received is correct.

Data Integrity. The provision of the property that data has not been altered or destroyed in an unauthorized manner.

Data Sequencing. The provision of the property that data is received in the same order in which it was sent.

Confidentiality. The provision of the property that information is not made available or disclosed to unauthorized individuals, entities or processes.

6 Mechanisms

The mechanisms chosen to support each of the security services and the other goals of the SNMP security protocol are as follows.

- In support of data integrity, the use of the MD4 [10] message digest algorithm was chosen. A 128-bit digest is calculated over the appropriate portion of a SNMP message and included as part of the message sent to a receiver.

- In support of data origin authentication and data integrity, the portion of a SNMP message that is digested is first prefixed with a value known only to the members of the community for that message. This value is known *a priori* to each member of the community and is not included in the message whose origin is being validated.

  A message also includes a component that allows the generater of a message to identify itself distinctly from the other members of the community. To the extent the receiver believes that members of the community indicate only their own identity in this component, the precise origin of a message is known.

- To protect against the threat of message reordering, a timestamp value is included in each message generated. A receiver evaluates the timestamp to determine if the message is recent, and it uses the timestamp to determine if the request is ordered relative to other requests it has received. In conjunction with other readily available information (e.g. the request-id), the timestamp also indicates whether or not the message is a replay of a previous message.

- In support of confidentiality, the DES in the Cipher Block Chaining mode of operation is used.

7 Algorithms

[6] not only defines the Abstract Service Interface (ASI) but recasts the SNMP trivial authentication algorithm in terms of this interface. Due to space considerations, neither the ASI nor the recasting of the trivial authentication algorithm is presented here. A brief overview of the SNMP Authentication Algorithm is presented here, and the SNMP Authentication and Privacy Algorithm is described in terms of the former.

7.1 SNMP Authentication Algorithm

This section describes the SNMP authentication algorithm. While the SNMP authentication algorithm affords some assurances with respect to message origin and integrity, it provides no protection against message disclosure. Implementation of the SNMP authentication algorithm entails implementation of the MD4 [10] message digest algorithm.

If the community component of a received SNMP message names a community based upon the SNMP authentication algorithm, then verification of said message is computed according to the following sequence of steps:

1. The secret prefix for the community is obtained, it is concatenated with the message to be authenticated and the MD4 message digest is computed.

2. If the computed message digest is not equal to the digest component of said message, the message is rejected and an exceptional value indicating failed authentication is reported.

3. If the timestamp component of said message does not satisfy an appropriate definition of timeliness with respect to the clock for said community, the message is rejected and an exceptional value indicating failed authentication is reported. A timestamp is considered timely if its value exceeds the value of the most recently received authentic message, and its value does not exceed the sum of the current value of the community clock plus an administratively set value.[4]

4. Otherwise, the message is considered an authenticated message and the management operation is executed.

If the community component of a SNMP message to be generated names a community based upon the SNMP authentication algorithm, then generation of said message is computed according to the following sequence of steps:

1. The secret prefix for the community is obtained, it is concatenated with the message to be authenticly sent and the MD4 message digest is computed.

2. The timestamp component of said message satisfies an appropriate definition of timeliness with respect to the clock for said community. A timestamp is considered timely if its value is precisely the value of the community clock.

3. The SNMP message is constructed with the additional components and sent.

---

[4]The administratively set value is called a *lifetime* in [6]. Its value would be based on the expected transmission delays between a management station and a managed agent, and other local defined values.

7.2 SNMP Authentication and Privacy

This section describes the SNMP authentication and privacy algorithm. This algorithm affords both protection against message disclosure and equal assurances with respect to message origin and integrity. Implementation of the SNMP authentication and privacy algorithm entails implementation both of the MD4 message digest algorithm and the Data Encryption Standard (DES) [4] encryption algorithm in the Cipher Block Chaining (CBC) [5] mode of operation. The privacy afforded by the use of encryption is assumed to justify its additional computational cost. Since the DES encryption algorithm is in the public domain, as are several implementations, it is likely that management stations can easily be upgraded to support privacy. In an agent, however, resource considerations may prevail.

Implementation of the SNMP authentication and privacy algorithm is, to a first-order, a straightforward extension of the SNMP authentication algorithm.

If the community component of a received SNMP message names a community based upon the SNMP authentication and privacy algorithm, then verification of said message is computed according to the following additional step:

- Prior to computing the MD4 message digest, said message must be decrypted according to the DES with the cryptographic key that is secret to the community.

If the community component of a SNMP message to be generated names a community based upon the SNMP authentication and privacy algorithm, then generation of said message is computed according to the following additional step:

- Prior to the SNMP message being sent, it must be encrypted according to the DES with the cryptographic key that is secret to the community.

8 Administration

Two strategies with which to distribute and maintain the shared secrets and common clock required by the new security algorithms are presented in [3]. They are called simply Strategy A and Strategy B. The objects referenced by the strategies should be part of the Management Information Base (MIB). MIB definitions for the objects appear in [7].

Strategy A is appropriate to the administration of a SNMP community for which every participating security service affords a reliable representation of the clock for said community. Strategy B is appropriate to the administration of a SNMP community for which some participating security service may not afford a reliable representation of the clock for said community. A brief description of strategy A is present below.

8.1 Distribution of Shared Secret

For each SNMP community, its secrets enjoy non-volatile, incorruptible, and private representations at each participating security service.

The secrets are initialized by their manual distribution to the security service for each participating SNMP peer. When a managed system is first installed, it is configured with a SNMP community based on a security algorithm (e.g., the SNMP authentication and privacy algorithm defined in [6]) that supports both authentication and privacy of SNMP messages.

Ideally, knowledge of these initial values should be minimized. To this end, the manually distributed values are required to be immediately altered by a responsible management station to be values known only to the software of said management station and the managed systems participating in said community. The values are altered by application of authenticated, private SNMP SET operations. The SET operation must be contained in an authenticated SNMP message to prevent unauthorized modification of the values. The SET operation must be contained in a private SNMP message to prevent disclosure of the secret values.

The secrets for said community are subsequently distributed by successful application of authenticated, private SNMP SET operations to those instances of the MIB objects representing said secret values.

The alteration of the secrets for said community by application of SNMP SET operations also requires the alteration of the clock for said community to a known and recognizably novel value (e.g., zero) by the same SET operations.

By this practice, if the response is not received by the originating SNMP peer, then the alteration may be confirmed by trivially authenticated queries of an instance of the MIB object that represents the clock for said community. A response to a SNMP SET of this kind may not be received owing to

- a failure of the relevant management agent,
- the loss or corruption of the response on the network, or
- the authentication or encryption of the response according to the newly altered secret values.

This latter case assumes the secret values for said community are managed by the community itself. If the secret values for said community were alterable only by a SNMP community other than itself, a response would not be lost due to this latter case.

8.2 Distribution of Common Clock

For each SNMP community, the clock for said community enjoys an incorruptible representation within the security service for each participating SNMP peer.

The epoch associated with each clock is implicit in the local practice by which said clock is manipulated. For example, if, in local practice, said clock is reset to zero simultaneously with any alteration of the secrets for said community, then that clock's epoch is the moment of the most recent alteration of said secrets.

The interval represented by each clock is reckoned in the units attributed in [11] to the TimeTicks defined type, and its duration is limited by the maximal TimeTicks value. In particular, neither clock's value may advance beyond the maximal TimeTicks value.

A clock for said community is distributed by

- authenticated SNMP queries of instances of the MIB object that represents said community clock and
- by successful application of authenticated SNMP SET operations to the MIB object that represents said community clock, simultaneously with the alteration of its corresponding secret values.

By altering a clock simultaneously with its corresponding secret values, said community is not vulnerable to a replay of the SET operation. Thus, any SNMP SET operation that refers to an instance of a clock MIB object, also properly refers to an instance of its corresponding secret MIB object.

Distribution of the clocks by trivially authenticated SNMP queries introduces vulnerability to replay attack whenever such a retrieved clock value might form the basis for subsequent protocol operations. Accordingly, retrieval of the clock for said community in this way is properly used only

- as tentative evaluation of the need for clock (re-) synchronization among community participants or
- as the basis for a subsequent authenticated, private SNMP SET operation that alters both the clock and its corresponding secrets for said community.

Either usage of trivially authenticated clock queries admits vulnerability to attacks that may entail denial of service, but such attacks are readily detected and do not entail unauthorized invocation of management operations.

9 Summary

The secure management of networks is of paramount concern. This proposal integrates the following security services with the Internet standard network management protocol the SNMP:

- origin authentication,
- message integrity,
- message sequencing and
- message confidentiality.

The MDC algorithm chosen to support message integrity is MD4. Origin authentication is implicitly supported by prefixing the data to be integrity protected with a secret. Confidentiality is supported by the application of the CBC mode of operation of the DES to the message to be protected.

Both the algorithms and the administrative techniques documents each have a section addressing security considerations. A few items are worth noting here.

- An authentication community based upon the trivial authentication algorithm affords no security whatever.
- The granularity of the authentication is a function of the number of participants in a community. In particular, to the extent that a SNMP PDU does not explicitly identify the origin of a SNMP message, a SNMP community comprising more than two participants is vulnerable to attacks involving the duplication or redirection of authentic SNMP messages to unintended destinations with unintended effect. SNMP communities that are based on algorithms that authenticate both data origin and integrity and that comprise only a single management station and a single management agent are not subject to this class of attack.
- In general, the administration of a SNMP community exclusively by SNMP messages associated with another SNMP community has a positive effect upon both the security and robustness of the administrative activity. A particularly secure practice is the institution of SNMP communities whose only use is the distribution of the secrets used by other SNMP communities.

References

[1] Jeffrey D. Case, Mark S. Fedor, Martin L. Schoffstall, and James R. Davin. A Simple Network Management Protocol (SNMP). RFC 1157, DDN Network Information Center, SRI International, May 1990. Obsoletes RFC1098.

[2] Vint Cerf. Report of the second Ad Hoc Network Management Review Group. RFC 1109, DDN Network Information Center, SRI International, August 1989.

[3] James R. Davin, James M. Galvin, and Keith McCloghrie. Administration of SNMP Communities. RFC DRAFT, DDN Network Information Center, SRI International, 1990. Submitted.

[4] FIPS Publication 46-1. *Data Encryption Standard*. National Institute of Standards and Technology, U. S. Department of Commerce, Washington, D.C., January 1977. Federal Information Processing Standard (FIPS).

[5] FIPS Publication 81. *DES Modes of Operation*. National Institute of Standards and Technology, U. S. Department of Commerce, Washington, D.C., December 1980. Federal Information Processing Standard (FIPS).

[6] James M. Galvin, Keith McCloghrie, and James R. Davin. Authentication and Privacy in the SNMP. RFC DRAFT, DDN Network Information Center, SRI International, 1990. Submitted.

[7] Keith McCloghrie, James R. Davin, and James M. Galvin. Experimental Definitions of Managed Objects for Administration of SNMP Communities. RFC DRAFT, DDN Network Information Center, SRI International, 1990. Submitted.

[8] Jon B. Postel. Internet Protocol. RFC 791, DDN Network Information Center, SRI International, September 1981. Obsoletes RFC 760.

[9] Jon B. Postel. Transmission Control Protocol. RFC 793, DDN Network Information Center, SRI International, September 1981.

[10] Ronald L. Rivest. The md4 message digest algorithm. RFC 1186, DDN Network Information Center, SRI International, November 1990.

[11] Marshall T. Rose and Keith McCloghrie. Structure and Identification of Management Information for TCP/IP based internets. RFC 1065, DDN Network Information Center, SRI International, August 1988.

VIII

USAGE CONTROL

B
Distinguished Experts Panel
(Same as VII-B)

VIII

USAGE CONTROL

C
Accounting management

Design Considerations for Usage Accounting and Feedback in Internetworks

Deborah Estrin[1] and Lixia Zhang[2]

[1] D. Estrin, Computer Science Department mc0782, University of Southern California,
Los Angeles, CA 90089, estrin@usc.edu.

[2] L. Zhang, Palo Alto Research Center, Xerox Corporation,
3333 Coyote Hill Road, Palo Alto CA 94304, lixia@parc.xerox.com.

Abstract

This paper investigates the design of resource usage feedback mechanisms for packet switched internetworks. By usage we mean the utilization of network resources. By feedback we mean information passed from network service providers to users concerning their usage. After a discussion of the *motivations* for feedback mechanisms, *feedback channels* and *policies* are described. We then outline issues raised by the design of *mechanisms* to realize these policies, including: network service disciplines, accounting granularity, metrics, authentication, and coordination among transit carriers.

Usage-based charging is only one means of feedback. Our purpose is to begin a systematic discussion of the technical issues associated with a range of usage feedback alternatives. Therefore the paper should not be read as a policy statement promoting usage sensitive charging in internets. In fact, one of the goals of the feedback mechanisms explored in this paper is to allow network service providers and users to *avoid* the introduction of usage sensitive charges if they so wish; while still realizing the benefits of statistical resource sharing offered by packet switching and the benefits of efficient resource utilization offered by usage feedback.

Keywords: Network Accounting, Inter-Enterprise Networking.

1 Introduction

This paper concerns resource usage feedback for interconnected, packet-switched, computer communications networks; hereafter referred to as internetworks, or internets. The global internetwork has developed through the interconnection of thousands of commercial and private networks.[3] As the technology matures, the role of commercial service providers is expected to grow, along with the demand for accounting mechanisms. At the same time, increasing connectivity brings with it the need for mechanisms that motivate efficient behavior on the part of the larger and more heterogeneous user population.

In this paper we investigate the design space for resource usage accounting and feedback mechanisms in a large scale, packet-switched internetwork. In particular, unless stated otherwise, most of our discussions below assume a *connectionless* internet that provides datagram services. By usage we mean the utilization of network resources; the primary resource concerned in this paper is network bandwidth for packet forwarding. By feedback we mean information passed from network service providers to users concerning their usage. As we will explore in this paper, this information may take various forms, such as usage reports, billing, or performance penalties. The traditional circuit-switched telephone network provides a possible model for resource usage accounting and feedback. However, many of the mechanisms do not translate directly into a packet switched environment.

Although we do not address issues of cost recovery specifically, charging is one form of feedback and therefore this discussion is of relevance to cost recovery as well. Cost recovery entails additional tasks such as setting prices based on a careful assessment of both fixed and incremental cost factors; further discussion is beyond the scope of this paper.

1.1 Internet model and terminology

Internet technology has developed primarily within private and consortium networks. Commercial carriers have participated mostly through leasing of lines used to connect network nodes within the private networks. More recently there has been increased interest in commercial offerings of datagram delivery services, e.g., Switched Multi-Megabit Data Service (SMDS) [14]. The advent of commercial offerings introduces new incentives, and in some cases a necessity, for resource usage feedback mechanisms; and the accounting necessary to collect information for the feedback channel.

We refer to the different administrative entities and their associated network resources as Administrative Domains (ADs). As described in [10, 13, 12], an AD is a set

---

[3]Commercial networks refer to those that offer services to anyone and for any purpose, so long as they pay the established fees (e.g., AT&T, GTE Telenet, MCI, PSI Inc.). Private networks refer to those that are operated and used by a restricted set (often one) of organizations (and/or for a restricted set of uses) based upon administrative, instead of (or in addition to) monetary, arrangements (e.g., NSFnet, Xerox Corporation's internal network).

of resources (network links, routers, bridges and end systems) under the control of a single administrative authority. In this context, a *stub* AD is one that does not carry transit traffic for other ADs, e.g., private customers/consumers of communications services. That is, all traffic entering a stub AD is destined for end systems within that AD, and all traffic exiting a stub AD originated within the AD. Most campus and corporate networks are examples of stub ADs. *Transit* AD refers to an AD whose primary function is to provide transit services for other ADs. Long haul backbone and regional networks are examples of transit ADs. In addition, some private networks that are connected to more than one transit or stub network offer limited transit services to selected ADs. We note the existence of bypass links, along side the more common hierarchical structure. The term *end user* refers to the human beings who make use of the communication resources via the end-systems that lie within the ADs. These distinctions are relevant to our discussion because we must identify which entities provide the feedback, and likewise to which entities the feedback is provided.

2 Motivations

There may be multiple purposes served by accounting and feedback for resource usage. One goal may be to recover costs. Another may be to motivate individual users to behave more efficiently from the perspective of the shared resources (i.e., the network). In the latter case, feedback signals should be different when the network is lightly loaded than when it is heavily loaded. Although both cost recovery and efficient network usage can be achieved using accounting and feedback, accomplishing one does not necessarily accomplish the other. Moreover, a usage-sensitive charging mechanism in one part of the internet may introduce the need for a feedback scheme in another part (e.g., a transit carrier's charging mechanism may motivate a stub AD to introduce usage sensitive feedback in order to motivate efficient use of the communications budget). This paper focuses on design considerations for usage feedback mechanisms. However, because of the potential interaction and frequent confusion, we begin with a brief discussion of cost recovery.

2.1 Cost Recovery

The most basic goal of cost recovery is to generate revenues that are adequate to pay for physical facilities (links, routers, etc.), operation, maintenance, software development, personnel, etc. The cost model is complicated somewhat by the need to generate enough revenue to fund improvement and expansion.[4]

A more unique problem in the context of data networking is an environment in which additional capacity can be called up on demand (at greater expense than had it

---

[4]This problem has been studied extensively by economists in the areas of telephony and utility company capacity planning and tariffs.

been planned for and installed privately). As traffic load increases, decisions must be made concerning a) whether to dial up additional resources, b) how long and under what conditions to maintain them, c) how to distribute this additional cost among users, d) whether to redistribute existing capacity, and e) at what point to invest in permanent facilities instead.

Cost is recovered by charging users for their network usage. Therefore the charging itself is one means of feedback. As such the charging policies may have a great impact on users' behavior. For example, the most common form of cost recovery today in packet switched networks is a fixed-fee per physical connection, where the fee is often a function of the bandwidth of the leased lines utilized by the connection. Neighboring transit ADs agree upon procedures for carrying each others traffic. The mechanisms for supporting various settlement and allocation procedures among the transit ADs is an intersting issue beyond the scope of this paper; it has been addressed extensively in the case of telephony.

This approach provides no feedback to the end user regarding the actual resource usage and so does little to encourage efficient network usage. The feedback only provides a signal to the organization as to what bandwidth connection to select, or whether to connect at all. In the absence of any other feedback, connected users would have little incentive either to upgrade a poor protocol implementation to the best available one (which may cost both effort and money), or to carefully plan their network usage to avoid congesting the network unnecessarily.

Another concern is the desirability, from a policy perspective, of exposing all users and usage to usage-sensitive billing. It may be preferable in some environments to decouple cost recovery and usage feedback in order to encourage communication among all, or some special subset, of users (e.g. promoting communication among members of the research community). In other words, global efficiency is very hard to measure when one takes into account the externalities (goods and costs) associated with communication. Therefore, it is not appropriate to simply minimize network usage to the exclusion of other factors. For this reason, we discuss alternative feedback models below.

2.2 Feedback

Feedback is needed in any service system to motivate users to make globally-efficient use out of existing resources. From the systems' perspective, when the system is lightly loaded feedback should encourage (or at least not discourage) usage to maximize system throughput. When the system is heavily loaded (i.e., demand approaches or exceeds the finite capacity) feedback should motivate deferable users to delay submitted traffic or expendable users to back off altogether. That is, an ideal feedback system would encourage intelligent usage while preventing the system from being overloaded. In the context of internets, there are two particular types of efficiency that we want to motivate: efficient implementations and efficient end-user behavior. For example, a good transport protocol implementation that eliminates superfluous retransmissions should help reduce the probability of network congestion.

An example of motivating efficient user behavior is feedback that encourages users to shift time-insensitive traffic to off peak hours. The current Internet, for example, may be considered as providing a very crude form of such feedback, e.g, during peak hours the network performance degrades so that performance-sensitive users are forced to shift their work to less crowded times of day. However, users less sensitive to performance degradation, might even have an incentive to transmit more to compensate for the losses caused by congestion. The inefficient users are not penalized adequately by the total queueing delay increase or packet losses that is caused by their action. The current Internet provides a first-come-first-serve (FCFS) datagram service, therefore the increased delay and losses are shared among all users. When demand exceeds capacity, the result is a network that is overly congested during peak hours and consequently provides poor performance to all users. In other words, what is locally efficient behavior for some users results in globally inefficient usage from the perspective of the network resources. It illustrates the tragedy-of-the-commons phenomenon [9].

As a first step in supporting usage feedback, some form of usage accounting mechanism must be implemented in packet switched networks. Next we discuss some of the issues raised in packet accounting.

2.3 Accounting in Packet Switched Internets

The effort required to account for traffic depends upon the network architecture. Circuit switched networks reserve resources for each user call, and therefore feedback and accounting can be performed along with call setup and teardown. Connection-oriented, packet-switched networks maintain state per connection inside the network and successive packets in a connection typically travel via a fixed route (although some architectures allow the connection to switch routes in midstream). If the connection protocol reserves resources then the accounting and feedback needs are analogous to the circuit switched case. If there is no reservation, then connection state and switch function must be augmented with accounting related information and packet-counting, respectively. In a pure datagram network there is neither resource reservation nor per-user state maintained within the network. Packets from the same end-to-end connection (i.e., source-destination, transport level association) are forwarded independently and may travel through different routers.

There is also a further interaction among application types, network architecture, and accounting. In computer communications, the range of application behavior and desired services is much greater than in voice telephony. Human to human voice communication represents a single type of application, and the entire telephone network has been built to optimize the service quality and pricing mechanisms of the application.[5] Moreover, whereas voice traffic is handled relatively efficiently with circuit switching, computer communications are often bursty. The more varied and bursty the traffic patterns, the more important it is to avoid inefficient forms of resource reservation.[6] The diversity of traffic patterns presented by computer communication applications implies that the network should distinguish between classes, or types, of service (e.g. delay sensitivity). When different *types of service* (TOS) are provided, the accounting function will need a more complex mechanism than a simple packet meter. The extra packet processing involved in supporting TOS specific performance guarantees may offer some opportunity for supporting accounting related functions *if* the TOS is implemented on a connection basis. Otherwise if TOS is offered on per-packet basis, additional work would be required to account for usage on a per packet, per TOS basis.

Another difficulty with respect to accounting in a packet switched, computer communications context is the unit of accounting. The units of accounting in packet switching potentially are much smaller than in circuit switching (i.e., a packet instead of a call) so the overhead of accounting could be much higher. This small unit is also problematic for the end user. A user can easily estimate the cost of a telephone call based on the call duration. In the current computer communications environment, however, it would be difficult for a user to predict the network usage implied by his or her application-level actions, if the network accounting is based on the unit of packet. The packet is too low level of an abstraction for the user; and today's applications and transport protocols are not instrumented to translate packet counts into units that are meaningful to the end user.

Despite these difficulties, usage accounting and feedback have some particular benefits in the context of packet switching. Computers served by packet switched internets differ in many ways from human users served by telephone networks. Real time, voice communication exhibits rigid requirements for stable transmission delay and rate. Many computer applications, on the other hand, exhibit "softer", more elastic, service requirements. For example, a circuit switched phone call must have a 3 Khz channel allocated, otherwise the call cannot start. A packetized voice session, however, can tolerate some degree of packet loss and still support intelligible communication.[7] Due to their asynchronous characteristics, certain applications can even accept temporary postponement of services; electronic mail and background file

---

[5] Although today's telephone networks also carry non-voice applications, such as FAX and dialup terminal-to-computer connections (through the use of a modem), voice remains the dominant load in the system. FAX traffic makes efficient use of the communication circuit. Terminal-to-computer connections, however, make relatively inefficient use because of their bursty nature.

[6] However reservation may be necessary whenever the network must guarantee a service and it is possible to implement efficient reservation for bursty traffic; this is the subject of ongoing research.

[7] One could argue that the telephone systems do not exploit the complete market. There is a potential of multi-TOS for circuit-switched voice communication as well. For example, calls can be sorted to interruptable and non-interruptable ones, so that the former can be cut off during peak hours but also receive a lower charging. However, unlike the electrical power market and data communications, the market size and network efficiency gains do not appear to warrant the overhead associated with differentiating between the traffic types.

transfers are such examples. Therefore it is possible to regulate packet traffic by usage feedback, and thereby enable a service provider to offer better service, at lower prices, to all end users. For this reason, mechanisms for usage feedback could benefit both service providers and consumers, if appropriately designed and implemented.

In summary, there are a number of interesting technical issues raised by the question of accounting and feedback in packet switched internets. We discuss motivations and models for usage feedback in Sections 2 and 3. Section 4 outlines several issues associated with the design of supporting mechanisms. Before describing feedback models in Section 3 we address the interaction and distinction between different internet participants.

2.4 Transit Carrier, Stub AD, and End User Goals

Motivations for usage accounting and feedback differ for transit and stub ADs, and for end systems.[8] At the same time, the motivations of the three types of internet entities interact with one another in some predictable ways.

Transit carriers are concerned with cost recovery through collection of user fees or third party subsidies. In a competitive internet environment, cost recovery increases in importance. Carriers compete by offering attractive services at relatively low prices while still covering expenses and expansion. To keep the price low, transit carriers are concerned with efficient usage of their resources. If users behave efficiently then the same service can be provided to the same number of users at lower cost than if if users behave inefficiently.

Stub ADs want to minimize, or at least contain, costs in the presence of whatever feedback scheme transit carriers implement. In a flat rate environment, stub ADs may be concerned with recovering costs of network attachment charges, and/or with promoting efficient use of a limited capacity connection. Where transit carriers introduce usage feedback, some stub ADs may want to pass such signals back to some or all end systems or users in order to encourage their more efficient behavior. In addition, as transit carriers introduce usage sensitive pricing, stub ADs will be increasingly concerned with verifying that their bills are accurate, i.e., they will want to take measures to prevent fraud. Stub ADs will be concerned with developing accurate models of usage in order to anticipate, plan for, and detect anomalies in usage and charging.

End systems and users will similarly want to minimize, or at least contain, costs in the presence of transit and stub AD feedback mechanisms. Some end system administrators may wish to avoid the overhead and inhibited communication that can result from too fine-grain accounting (while still controlling costs), while others will want to propagate feedback signals all the way to the end user. In addition, some users may be considered billable and others not.

---

[8]Of course the motivations for feedback are not identical for all stub ADs, or for all transit ADs. However, there is more commonality among entities of a particular type.

3 Feedback Models

Feedback schemes can be characterized by the *feedback channel* used and the *policies* implemented.

3.1 Feedback Channel

Usage sensitive *charging* implies billing for services, by definition. But feedback to end systems or users regarding resource usage can also be achieved in terms of network signaling, service quality degradation (e.g. delay), or even administrative means; as an alternative, or in addition, to actual monetary feedback. Each of these can be thought of as a different *feedback channel*. The feedback is usually sent to the traffic source, but in some cases may be sent to the destination or some third party. Below we discuss the features and merits of different feedback channels.

Monetary feedback has very explicit impact on user behavior. Individuals and groups have limited budget resources, and therefore are motivated to economize on their usage (i.e., communication expenses). However, explicit, direct impact does not necessarily mean that this channel is always optimal or desirable. Consider the research community as an example. Externalities such as inhibiting communication based on price-elasticity may well be *undesirable* from the perspective of the social good of "research communication, productivity, and technology transfer", for example. From the perspective of global efficiency, individuals may make suboptimal decisions to underinvest in communications. In other words, some individuals will experience all the cost but not all of the benefit of their expenditure, when the benefit of their communication is partially (or largely) to other members of the community. Consequently, if each individual optimizes his or her own behavior based on local costs and benefits, a social optimum may *not* be achieved.

Allocation or quota schemes can act as a proxy for monetary billing. Unlike real money, the quota is not exchangeable for other goods or services. Traffic sources (which may be end systems or stub ADs, depending upon the accounting granularity) are encouraged to behave efficiently because they have a limited resource, their quota. Various quota schemes have been used in computer systems for usage accounting (e.g. MIT Multics). Such allocation schemes do have drawbacks. For example, users may overly constrain their communication early in the quota period and over utilize at the end or vice versa (i.e., a user could flood the network with traffic at the beginning of the period and then starve for the duration).

Performance feedback can take different forms: an explicit message from the network warning of overload condition (e.g. ICMP source-quench [16]), or an implicit increase in delay or packet-loss rate. This type of feedback has no relation to cost recovery. Its function is to influence user behavior (directly, or indirectly through intermediate protocol layers). For example, upon receiving an ICMP source-quench message requesting a slow down in data transmission, users who find the service inadequate may decide to shift to a less congested time of day.

In the absence of other mechanisms, however, applications or users who are insensitive to the performance parameters may not modify their behavior. For example, electronic mail runs in the background and the end user would not notice whether transfer of some message incurred 50% retransmissions. This can lead to a situation in which performance-sensitive users under-utilize the system (because they find it of less value) and performance-insensitive applications over-utilize the system, from a global efficiency perspective. Therefore performance feedback is most effective when TOS support mechanisms are in place, so that performance-sensitive users can be given priority in utilizing network resources.

Administrative feedback, such as monthly usage reports or allocation schemes, may be used alone, or in combination with performance feedback. Administrative feedback can be effective to the extent users are sensitive to administrative (or peer) pressures. Usage levels can be posted or broadcasted at regular intervals; the performance-insensitive users described above might then be discouraged via administrative pressure from overutilizing the resources. The result would be a more attractive network for performance-sensitive users, and relatively little degradation for the performance-insensitive users who could shift usage to uncongested times of day.

Depending on the feedback channel(s) in use, the receiver of the feedback signal can be different. For example, performance feedback will be received directly by the end user. Administrative feedback may target the stub AD, which may then redistribute the signal internally through whatever channel it deems appropriate. Regardless of the channel type, in order for feedback to be most effective end users should be the ultimate receiver of *some* form of the feedback signal. But how the feedback is provided internally, and whether to associate it with internal accounting and billing actions, is the stub AD's decision.

In summary, the granularity of the feedback recipient is tightly coupled to the intended objective. We suggest that the first objective is to carry the collection of users' traffic in an efficient manner, e.g., introduce delay for deferrable traffic such as asynchronous mail when the network is heavily loaded. This may be achieved, at least in part, through relatively coarse-grained feedback. A second objective may be, in some cases, to provide feedback to finer grain traffic sources (human users) in order to alter users' demand, i.e., offered load, most directly.

3.2 Feedback Policies

In this section we describe four alternative usage accounting and feedback policies: flat per-packet fee, TOS based, peak load, and priority based. We are interested in the potential impact each policy may have on the user's behavior, and thus its effectiveness in regulating network usage. These policies typically are described in the context of monetary feedback, i.e., billing. However, schemes can be devised using the other types of feedback channels. Unless otherwise stated, we assume the network serves each datagram on a first-come-first-serve (FCFS) basis. Moreover, we assume that some form of feedback signal is passed to end users, directly or indirectly.

3.2.1 Flat Per-Packet Feedback

To the extent carriers' costs are related to usage, flat per-packet feedback schemes provide a means for distributing costs among users (e.g. SMDS) [14]. Moreover, this approach provides fine grain feedback to the user to promote efficient use of network resources. However, because the feedback is based on a flat per-packet fee, which is independent of current system load and service quality received, it does not encourage users to delay non-time critical usage and may lead to under-utilization when the network is not loaded. The network provides all users with either a best-effort service (e.g., IP) which may be inadequate for real-time applications, or with a guaranteed high-quality service (e.g., SMDS), which may not be cost effective for less demanding applications. The uniform service type provides no incentive (or support) for users to sort their applications into different categories in order to share the resources most efficiently.

3.2.2 TOS Based Feedback

If internets offer different types of services, the accounting should be based on the TOS service quality provided. When the network is fully loaded, however, additional traffic that requires a high quality TOS will have to be rejected in order to guarantee service qualities to the current users. In this case, users can predict the cost for a required level of service. They either get the requested level of service or nothing.

If the accounting in a TOS-guarantee network is independent of the current or expected system load, and the network simply applies a FCFS policy to resource requests, late comers during peak hours will be forced to shift to different usage times. However, the network would prefer to encourage demanding-TOS users when the network is under-utilized, and discourage them when the network is loaded, by having load-sensitive TOS accounting. This can be achieved by one of the load-sensitive accounting policies discussed next.

3.2.3 Peak Load Feedback

Peak load pricing provides different feedback (e.g., charges different rates) depending upon the aggregate demands placed on the system [2]. If there are regular, predictable times of day at which the network will be heavily loaded, then the charge for transmission during those hours can be raised significantly to shift flexible users off the peak. The charge may be in terms of real money, monthly-report-units, or allocated credit-units; corresponding to the different types of feedback channels.

This scheme is most effective when peak periods are predictable so that users can plan and behave accordingly. Network traffic measurements from different sources have shown consistent gross patterns of network busy hours [1, 8]. If traffic patterns are not so predictable, peak load rates could be varied dynamically with network load. In this case, however, traffic sources would not be able to predict their charges accurately, thereby undermining the utility of the feedback channel for budget planning purposes.

Network facilities may be expanded to meet demand on a dynamic basis, i.e., the network provider may dial up additional facilities to meet peak hour demand.[9] There is a symbiotic interaction between peak load policies and dynamic network costs. By setting a higher charge for peak hours, the revenue may be used to cover the extra cost of dialup lines. More generally, if traffic load variations are predictable, the need for dialing up additional capacity can be predicted and the situation can be made to resemble the fixed capacity case.

3.2.4 Priority Based Feedback

An alternative to peak load rates is priority pricing.[10] Under this policy, the network will serve users in the order of their priority levels, and the rate charged for carrying traffic will be computed accordingly. These rates are slow to change and are advertised to all traffic sources. This scheme is more adaptive than peak load schemes because the priority labels provide a basis for the network to delay lower priority traffic in favor of higher priority when loaded.

Priority pricing has been implemented by utility companies and appears quite promising for network resources as well [17]. For example, in electrical power systems, at peak load the service provider may not be able to meet the peak demand from all users. The priority pricing implementation charges less to customers who are willing to have their service curtailed/cut-off when demand rises above capacity. Inflexible users pay more to receive a guaranteed continuous service. The scheme is relatively static because users vary their priority level infrequently and slowly (i.e., they put in a request and expect it to take some time to go into effect).[11]

In the data network context, performance feedback and priority adaptation could be more dynamic. For example, a user first sets a certain priority level; if the experienced delay is too great (or some other quality metric is too low), and if the users' demand is relatively elastic to performance but inelastic to price (or the administrative equivalent), they may increase the priority levels until acceptable performance is achieved. This means that the actual cost of a particular transaction will depend on the network conditions at that time. To the extent network load is predictable, users will distribute their usage more evenly. The net result is more efficient use of network

---

[9]Even a fixed-facility network is not faced with static costs, i.e., capacity planning decisions are made continually regarding installation of additional facilities. But in the dynamic case, dial-up circuits interact directly with real-time performance and monetary feedback channels.

[10]Scott Shenker of Xerox PARC originally proposed this approach for use in datagram internets.

[11]Another analogy for priority pricing is the airline industry, in which you pay more for flexibility (i.e, making reservations with short notice, without restrictions for cancellation, and without restrictions on time of day and day of week) and certainty (standby pays less than reservations). To be efficient the airline scheme also requires some predictability—certain spaces are reserved for the higher-cost, last minute reservations. If predictions are not accurate, the seats will go unused or will be sold at lower standby rates.

resources. However, a concern here is the potential inefficiency of highly dynamic, real-time, tuning of priorities to optimize end-user service and cost.

If there is no accounting system associated with a priority scheme, however, all users have incentive to set high priority on all traffic, and the scheme will not be effective. Consequently, whether through administrative means, or using an actual or proxy (quota system) monetary channel, users' priority setting must be regulated.

In the subsequent section we enumerate the supporting mechanisms required, and the design issues raised, by consideration of usage based feedback in internets, with a particular interest in load-sensitive, TOS feedback.

4 Design Issues

In this section we enumerate several essential choices that must be made in designing usage-based feedback mechanisms for transit and stub ADs, i.e., network service discipline, accounting granularity, feedback frequency, cost metrics, dynamic capacity issue, authentication, and coordination required among transit ADs. Based on the very early state of work in this area, we raise more questions than we answer. Much more work is needed to analyze design choices and tradeoffs in detail.

Network Service Discipline and TOS Implementation: The network service discipline employed influences the feedback signals directly. Thus far we have assumed FCFS packet handling. We must consider the interaction of network service and feedback channels in more detail. An internet may provide multiple levels of resource control through the use of multiple service disciplines. One level may implement a fairness mechanism that simply insulates all users from one another. A second level may provide a resource guarantee to particular users (or user groups).[12] A third level may implement complete TOS support mechanisms to fully exploit the benefit of statistical sharing in packet switching and allow each user to pay the minimal possible while receiving adequate service.

Accounting Granularity: There is a cost tradeoff associated with fine grain accounting. In general, finer granularity offers more accurate control at the expense of greater overhead to the system. Granularity decisions must be made regarding both the unit of traffic and user monitored.

Frequency of feedback: Another dimension of all feedback schemes is the frequency with which the information is collected and returned to the traffic source. Network management protocols can be used to collect aggregated statistics and return them to the traffic sources on a regular, but infrequent basis. In contrast, some feedback mechanisms are based on real time (minimal delay) information akin to congestion and flow control feedback. Feedback rate presents a clear tradeoff between the rapidity of user adaptation and the cost of realizing the scheme itself.

Cost Metrics: Whether the feedback channel is monetary or administrative, there is

---

[12]Suggested by D. Clark, private communication.

the question of cost metrics, i.e., the appropriate measure or metric for network cost recovery and feedback. The simplest metric is a function of the number of packets. If we introduce a TOS and/or priority mechanisms, the metric should be a function of packets, bytes, hops, TOS, and priority.

Dynamic Capacity: If a service provider routinely dials up additional capacity when the load exceeds a certain threshold, the service provider's costs may rise significantly during crowded periods and may need to be reflected back in user charges. In addition, the users will not experience sustained reduced performance at overload because the service provider will compensate by dialing up additional circuits. Therefore both the performance feedback and cost recovery assumptions are different in the dynamic resource-cost case. Since the natural performance feedback that one gets with a fixed capacity system is now absent (or diminished), monetary or administrative channels must be used.

Identifying Collection Points and Billable Entities: Another issue in a multi-transit AD internet (e.g. backbones and regionals are transits) is how transit ADs and end users will be identified, and how they will be authenticated.

Coordination among Transit Carriers: Coordination is required among transit carriers with respect to both billing arrangements (in the case of monetary feedback channels) and TOS. Billing schemes can vary in several dimensions: who is billed and who is paid, what unit of traffic is billed for, and the nature of the payment. Similarly, tos guarantees are useful to the source and destination only if they are supported and coordinated by all the transit ADs along a path.

Additional Stub-AD and End-System Requirements: Stub ADs must manage their connections to transit carriers. Consequently, they face several additional requirements. For example, in order to control their communications budgets, stub ADs must either be able to predict or bound variable costs, or they must be able to recover over-expenditures from end systems. Stub ADs should be able to select the particular type of feedback channel or mechanisms used internally. Stub ADs also need the ability to verify the actual service quantity and quality delivered. End systems require similar cost control and verification capabilities. In addition, end systems require instrumented applications that can assist users in developing communication cost expectations and that can translate low-level usage feedback signals into higher level units that are meaningful to the end user (e.g., cost per electronic mail message distribution instead of per packet).

5 Conclusions

At its best, resource usage accounting and feedback presents an opportunity to promote efficient usage of network resources, and to reduce end-user communication costs by setting charges that reflect the statistical resource sharing possible with packet switching architectures. Design, simulation, and experimental research is needed to develop appropriate technical mechanisms to realize these benefits, and to avoid the many negative behavioral and technical consequences of poorly designed approaches. Moreover, there is much to be learnt from existing economic theory. This paper represents a first attempt to articulate the design space of usage feedback mechanisms.

Before concluding we reiterate a few caveats and recommendations with respect to resource usage feedback in internets. First, network administrators should avoid charging end users on an usage-basis without understanding users' demand elasticity, the impacts of the charging policy chosen, and the technical overhead of doing so. Secondly, stub ADs and end users should be urged to develop tools necessary to manage their communication budgets before usage sensitive charges to stub ADs are introduced widely. Moreover, effective TOS support mechanisms should be employed in the network to fully exploit the benefits of statistical resource sharing. Finally, in a network environment that supports multiple TOS, it may prove most beneficial to introduce usage feedback first for the most demanding applications (performance sensitive) only – it is likely to have a significant impact on network usage efficiency and at the same time will more likely impact users that can afford the expense.

Acknowledgments We are very grateful to Bob Braden, Vint Cerf, Alex McKenzie, Yakov Rekhter, Scott Shenker, and the anonymous referees for comments provided on an earlier draft.

References

[1] Amer, P., et. al. *Local Area Broadcast Network Measurement: Traffic Characterization*, **University of Delaware Tech report 86-12**, January 1986.

[2] Boiteux, M. *Peak Load Pricing*, **Journal of Business**, 1980, volume 33, p 157-179.

[3] D. Clark, *Policy Routing in Internet Protocols*, **RFC 1102, SRI Network Information Center**, May 1989.

[4] A. Demers et al, *Analysis and Simulation of a Fair Queueing Algorithm*, **Proceedings of ACM SIGCOMM**, September, 1989.

[5] D. Estrin and G. Tsudik, *Security Issues in Policy Routing*, **Proceedings of 1989 IEEE Symposium on Security and Privacy**, May 1989.

[6] D. Estrin, J. Mogul, G. Tsudik, *Visa Protocols for Controlling Inter-Organizational Datagram Flow*, **IEEE Journal on Selected Areas in Communications**, May 1989.

[7] D. Estrin, *Policy Requirements for Inter Administrative Domain Routing*, **RFC 1125, SRI Network Information Center**, November 1989.

[8] D. Feldmeier, *Empirical Analysis of a Token Ring Network*, **Massachusetts Institute of Technology Technical Report MIT-LCS-TM254**, January 1984

[9] *The Tragedy of the Commons*, **Science**, December 13, 1968, volume 162, p 1243-1248.

[10] S. Hares, D. Katz *Administrative Domains and Routing Domains, a Model for Routing in the Internet*, **RFC 1136, SRI Network Information Center**, December 1989.

[11] S. Heimlich, *Traffic Characterization of the NSFNET National Backbone*, **Proceedings, USENIX, January 1990**.

[12] M. Lepp and M. Steenstrup. *An Architecture for Inter-Domain Policy Routing* **DRAFT RFC**.

[13] ISO *OSI Routeing Framework*, **ISO/TF 9575**, 1989.

[14] D. Piscitello and M. Kramer, *Internetworking Using Switched Multi-Megabit Data Service in TCP/IP Environments*, **Computer Communication Review**, Vol. 20, No.3, July 1990.

[15] J. Postel, *Internet Protocol*, **RFC 791, SRI Network Information Center**, September 1981.

[16] J. Postel, *Internet Control Message Protocol*, **RFC 792, SRI Network Information Center**, September 1981.

[17] R. Wilson, *Efficient and Competitive Rationing*, **Econometrica**, January 1989, volume 57, no. 1, p 1-40.

A Hierarchical Domain Concept as a Main Part of an OSI Accounting Model [*]

Ernst N. Bötsch

Leibniz-Rechenzentrum

Barer Straße 21, D-8000 München 2, Germany

e-mail: boetsch@lrz.lrz-muenchen.dbp.de

Abstract

As a consequence of the recent growth of computer networks and their increasing complexity, a mechanism to structure a large number of systems is urgently needed. The concept should also be suitable to specify areas of responsibility.

This paper proposes a domain model for this purpose in the field of OSI accounting management. In addition, the model is used as the basis of a naming scheme and of rules for scopes of validity / responsibility.

The concept is mainly influenced by needs of network administrators.

Keywords: Domains, Accounting Management, distributed systems, managed objects.

1 Introduction

The large and still increasing number of computer systems and their growing connectivity necessitates domain models in several areas of (OSI) network management (e. g. security and accounting) and in distributed communication applications (e. g. ISO-Directory and MHS).

In this paper domains are used as a means of describing groups of objects which are subject to a single accounting management policy (a more formal definition will be given in section 3).

The objects to be managed primarily consist of chargeable resources in communication components, but also in computers (hosts, mainframes, etc.). The latter is desirable because an accounting model for networks should be flexible enough to integrate local accounting. Therefore, the term 'system' includes network components, hosts, workstations, PCs, etc.

---

[*]This paper is the result of work conducted within a research project at the Institut für Informatik of Technische Universität München, which is sponsored by Siemens AG, Munich.

The grouping mechanism will follow the structuring method common in most larger organizations since an accounting policy is determined by the organization running the systems.

In the following some advantages of domain models are given which motivate the urgent need for an adequate domain concept (at least in the area of accounting management):

- The concept of domains is a suitable mechanism to divide large numbers of systems into manageable groups. This is necessary because it is impossible to treat each component or user as an independent entity for management purposes.

  Domains then extend the containment hierarchy ([ISO 10165-1]) to the level of computer networks.

- Within groups of homogeneous (or similar) systems in one organization (or organizational unit) it is desirable to have objects which are valid in more than one system. A login e. g. (including user rights, quotas and accounts) may be valid in all workstations of a department. Of course, the problem can be solved by defining a separate login in each host. But this solution is inconvenient for the user (e. g. multiple passwords). It may also lead to inconsistencies of management data (rights, quotas, etc.) with respect to all physical objects (e. g. the *individual* logins in the separate hosts) forming one logical object (e. g. the *one* domain login which is valid in all hosts of the cluster).

  Therefore, it is useful to have objects at the hierarchical level of groups of systems. With help of the notion of domains, relations between separate systems can be established. In consequence, a mechanism to integrate such objects into an extended containment hierarchy (see above) is supported, which will ease naming and addressing. Additionally, a good domain model will offer mechanisms to keep copies of management data (e. g. local login information) consistent. These copies may improve availability and performance.

- Algorithms to distribute management information in a group of systems may be more easily specified if they are based on structures and mechanisms supplied by domains.

- An appropriate domain model can be used to represent interrelations between *and* within organizations in the management of heterogeneous OSI networks.

The purpose of this paper is to propose a hierarchical domain model which forms an important part in a flexible accounting scheme and offers the benefits mentioned above. It may also serve as a basis of discussion for OSI service and protocol development in the field of accounting and other management areas.

After this introductionary section requirements, concepts and objects of the underlying accounting model are presented in section 2. Section 3 introduces the domain model. In section 4 a naming scheme and a concept to define scopes of validity are proposed which take advantage of the hierarchical domain structure. The last main section discusses the relation of the model to other domain concepts within ISO-OSI and the integration into an abstract classification. Finally, a short summary is given.

2 Requirements and Concepts of the Accounting Model

This section introduces important requirements and constraints which an accounting model has to cope with. It also presents concepts and managed objects of the accounting scheme which are the basis of the domain model.

An introductory discussion concerning accounting is contained in [HegChy 86, HegChy 87]. This summary here is detailed in [HegBöt 88, Bötsch 90].

Among others, a flexible accounting model for heterogeneous (OSI) computer networks has to cope with the following requirements:

- The association of usage data to the originator of the resource consumption must be possible ('originator orientation').
- The model should be suited for multi-network and multi-organization environments. These requirements mean that the concept must be able to handle large numbers of systems and to represent interrelations between *and* within organizations (i. e. it must be possible to manage and treat entities like groups, laboratories, departments, etc. separately).
- An accounting scheme should be able to integrate the accounting of the local system and be flexible with respect to chargeable resources and services.

The underlying accounting model identifies the following *managed objects* in the area of OSI accounting management:

- The **user record** (in the following 'login') is the conceptional representation of a single user and consists of a distinguished name (in order to uniquely identify it), access rights, quotas, personal information, etc. Every authentification is based on the login.
- The **charge number** is used as the basis for charging the consumption of resources. Every user has the permission to use at least one charge number. It is a special identification for accounting purposes and it is introduced to facilitate a billing separately for every project in which the individual employee or user participates.

  A budget is associated with a charge number serving as a resource independent quota.
- The **accounting resource** (briefly 'resource') is the model's view of a chargeable communication (or computing) resource in real world. The resource consists of a unique name, the price of the resource (or a price table in case of more complex resources), a switch to activate/deactivate charging of the resource, etc.
- The **accounting manager** (briefly 'manager') is the *logical* entity which is responsible for all services and actions dealing with accounting. The manager may be implemented as a process, as a part of the operating system or as a combination of both. Every communication for accounting purposes between systems shall be performed by managers by means of standardized protocols.

Not only objects, but additionally a procedural term has to be provided: The **accounting policy** (briefly 'policy') consists of data types, mechanisms, rules for (accounting) currency units, parameter values, etc. which are dealing with accounting.

3 Hierarchical Domain Model

A new domain model for an OSI accounting scheme is proposed in the following.

Main accounting domains (briefly 'main domains') are informally defined as groups of managers with a common accounting policy. Normally the shape of a policy only depends on the organization owning or operating the considered systems since the accounting policy represents the overall policy of the organization. Additionally, the relevant company may specify areas of responsibility and security areas concerning accounting (usage data, user rights, quotas, etc. may be kept secret).

To cope with these requirements and to represent these *inter-organizational* structures, main domains are formally introduced as objects with the following properties:

- All managers communicating in a network are divided into several groups with the affiliation to a policy as structuring criterion. Furthermore, the groups must have the following properties in order to become main domains:
  o In each group there must exist *exactly one* policy.
  o The union of all groups must result in the complete set of all managers.
  o Single groups must not overlap.

  Normally all managers of a company will form one main domain. Only in special cases (if the requirements on the policy are too divergent), the managers of an organization will be split up into two (or seldomly more than two) main domains.

  In order to distinguish main domains representing (a part of) an organization, each main domain must have a unique name, which is called 'relative distinguished name' (RDN) (similar to the ISO-Directory, [ISO 9594-1]).

- The rule 'every system is managed by *exactly one* manager' also defines a partitioning on the set of all systems. Naturally, this particular manager also determines the main domain to which the system belongs.

  A separate main domain has to be introduced for systems which are operated by two or more organizations (exceptional case).

- The structuring concept (i. e. main domains) results in a new kind of tasks for managers. All managers are split up into two logically independent types in consequence of this new quality:
  o Within a system *all* accounting services and tasks specific to this particular system are performed by a 'system manager', which is called **slave manager** (briefly 'slave'). This slave is completely responsible for the system and no other manager may perform accounting actions in it.

    Normally the RDN of a slave should be the name of the host in which the slave resides.
  o A 'domain manager' called **master manager** (briefly 'master') will be responsible for *all* objects not belonging to a specific system (e. g. a login which shall be valid in every host of a main domain). The master also manages the topology of the main domain and supervises all slaves of it. In consequence, it needs extra privileges.

    Finally, every master will be identified by the special RDN ' ' (the empty string).

A slave represents a single system (an element of the main domain) and a master a complete main domain from the accounting's point of view. But the logical distinction between a slave and a master shall not prescribe a physical separation of these entities in a particular implementation.

In every host there is at least a slave and configuration dependent additionally one (or more) master(s) (multiple masters may exist in case of subdomains; see below).

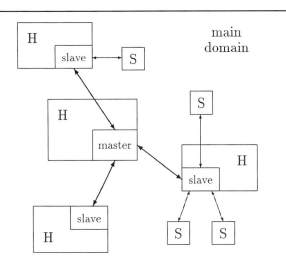

H: Host (with manager), S: System (without manager)

Figure 1: Relations between systems and managers

The possible relations between systems, managers and main domains are clarified by the example in figure 1. There, a host 'H' has sufficient (computing) capabilities and capacity to run a manager (normally an independent process). The systems 'S' contain (or are themselves) resources to be charged, but are not capable to have an own manager. Therefore, they are managed by a slave in a more powerful system 'H' and are merely subsystems or components of 'H' from the accounting's point of view. Printers, small computers (e. g. PCs) or dedicated computers (e. g. front ends with a special operating system) are examples of such a system 'S'.

Every communication for accounting purposes between different systems is performed by managers.

Group structures and RDNs of managers should be adopted from other applications or management areas if possible. This will help to minimize the number of names and structures. E. g. a slave may get its name from the host in which it resides and a main domain may be identical (with respect to the systems it contains) to a configuration domain (all computers of an organization) or parts of the MHS or ISO-Directory tree.

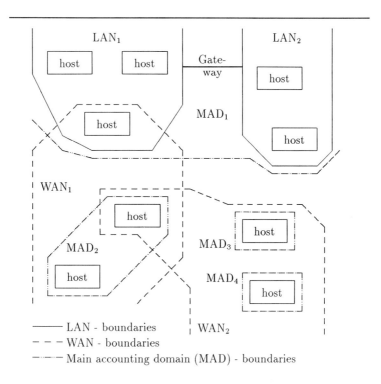

Figure 2: Relations between accounting and configuration domains

The example in figure 2 demonstrates the possible relations between main domains and other management domains (here configuration domains): Subsets, supersets or overlapping groups of hosts. As already mentioned, it is desirable to have identical structures or at most sub- or supersets in order to minimize the number of topologies. This goal is quite realistic since other structuring mechanisms (like the ones of MHS and ISO-Directory) are also mainly influenced by relations between and within organizations.

The principle of main domains introduces a means to group the (possibly) large number of connected systems into more easily managable units. But this might be not sufficient in large organizations with hundreds or even thousands of systems (communication components and computers ranging from PCs to mainframes or even supercomputers). Therefore, a mechanism for substructuring called **accounting subdomains** (briefly 'subdomains') with the following properties is proposed:

- All subdomains of a given main domain (or recursively another subdomain) D divide the managers of D into several groups, which must not overlap with each other (analogously to the definition of main domains). But unlike to main domains, the union of all subdomains of D need not result in the set of all managers of D. This is necessary because a system may belong to a main domain/subdomain, but to none of its subdomains.

A given domain will simply be called **accounting domain** (briefly 'domain') if it is not necessary to distinguish the type of the domain (main domain or subdomain).

All subdomains of D must be uniquely identifiable by a RDN within the scope of D and are called 'son-subdomains' of the 'parent-domain' D. A subdomain itself may be devided into subdomains and so on to any arbitrary depth.

Normally the definition of subdomains will follow the hierarchical substructuring of organizations into 'subsidiary company $\rightarrow$ branch office/establishment $\rightarrow$ department $\rightarrow$ subdivision', etc. The structuring criterion is then the affiliation of a manager to such an organizational unit. Therefore, the mechanism of subdomains is mainly used to represent *intra-organizational* structures.

Main domains represent the top level domains in the accounting domain hierarchies. Parameters of a policy which are free at a given hierarchy level of a domain D may be fixed in a subdomain of D and/or additional rules may be established.

- As in a main domain, all managers of a subdomain are divided into one master and slaves of any arbitrary number (even no slave if no system directly belongs to the subdomain).
- Every system managed by a slave of a subdomain also belongs to the subdomain.

4 Naming and Scopes of Validity / Responsibility

The proposed domain model shall permit arbitrarily nested domain hierarchies. Therefore, the following problems have to be solved if a manager m lies within a set $\mathcal{D}$ of more than one domain (in the main domain in any case and in at least one of its subdomains):

- To which of the domains of $\mathcal{D}$ is m assigned to, i. e. which concrete terms of the policy must m obey?
- Management of groups of systems is easier if there are objects belonging to the entire group (like logins in a Yellow Page Cluster of workstations). Which of such 'domain objects' belonging to one of the domains of $\mathcal{D}$ may m manipulate?

The domain hierarchy was designed to represent intra-organizational structures. Therefore, a fixed assignment of each manager to one of the domains in the hierarchy is sensible. Analogously, m is only permitted to manipulate domain objects of that particular domain. The relation 'is assigned to' can be derived from the structure of subdomains in a natural way if the following property holds: A manager m is assigned to a domain D iff m belongs to D, but to none of the subdomains of D. Systems 'are assigned' to subdomains analogously to the definition of domains.

The visualization of the parent/son relation between the subdomains of a main domain results in a domain-tree (with domains as nodes and edges depicting the relation). A second type of nodes representing managers (and the systems managed by them) is connected with edges showing the 'is assigned to' relation to 'domain nodes'. The root of the (bipartite) tree is the main domain itself.

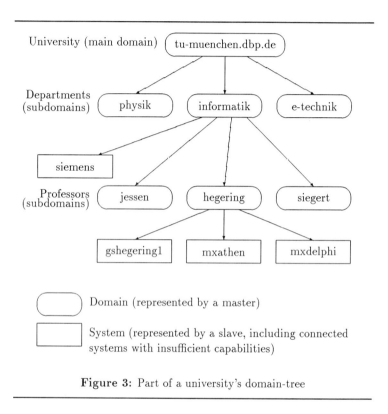

Figure 3: Part of a university's domain-tree

The example in figure 3 presents a part of a possible domain-tree of a university (here Technische Universität München; masters are omitted to improve clarity). As demonstrated, a domain (here the main domain) need not have systems which are assigned to it. In consequence, it must be possible to run a master in a host which is assigned to a subordinate subdomain.

The domain-tree can be used to obtain globally (or relatively with respect to a given node in the tree) unique names for domains and managers. This is performed by the usual mechanism of concatenating the locally unique RDNs of all domains (and the final manager if required) following a path in the domain-tree.

This naming scheme can also be used for addressing the objects since it generates globally unique identifiers.

'jessen,informatik,tu-muenchen.dbp.de[1]', 'siemens:informatik,tu-muenchen.dbp.de[2]' or 'mxathen:hegering,informatik' are examples of such names (cf. figure 3).

Figure 3 also demonstrates the possibility of adopting a topology (or parts of it) from other applications: The two upper levels of the tree in the example are borrowed from the MHS hierarchy and are extended by the level of professorship.

Domains have been defined as groups of managers. Therefore, it is sensible not to introduce a separate object type (i. e. a new managed object) to specify the domain structure. Instead, it is sufficient to hold the necessary structure information as part of the data which makes up the accounting managed object 'manager' of a master.

Naming and scope of responsibility have been determined for managers in the previous part of this section. A similar definition for other accounting objects (i. e. logins, charge numbers and resources) heavily depends on the particular object variant:

- *Objects specific for a particular system* ('system objects') are assigned to a system and the slave managing it. The objects' scope of validity is naturally restricted to that system. Only the particular slave is responsible for the object and may manipulate it directly.

  Each system object must be identifiable by a unique RDN within the slave's scope of responsibility. Global names are derived from the domain naming scheme in a usual way. E. g. the administrator login 'root' in the UNIX host 'mxdelphi' in figure 3 is 'root@mxdelphi:hegering,informatik[3]'.

- *Domain specific objects* (like a login or a charge number for a cluster of workstations) are uniquely assigned to a single domain in the domain hierarchy (and, of course, also to the master of the domain as its representative).

  Such 'domain objects' are *implicitly* at most valid in all systems *lying in* the domain (not only in systems which are *assigned to* the domain). This rule is conforming to the semantics of subdomains to represent intra-organizational structures. Additionally, an *explicit* filter which restricts the scope of validity may be connected with a domain object (e. g. a login may only be valid in Sun workstations) in order to provide more flexibility.

  The master of the domain is solely responsible for domain objects and may manipulate them.

  Like system objects (see before), a domain object needs a unique RDN within the naming scope of the master managing it. As usual, globally unique names like 'boetsch@hegering,informatik[4]' (a login) or 'network-research@informatik' (a charge number) are derived.

  By means of the proposed naming scheme, it is easy to decide if a system lies within the scope of validity of a domain object: This is true iff the domain-part of the (global)

---

[1] The domains' RDNs (least significant first) are separated by ',' in the string notion of the examples.
[2] The RDN of a manager in front of a name is separated by ':'.
[3] The object's RDN in front of a name is separated by '@'.
[4] Notice that the manager-part is omitted to indicate that a master with its fixed RDN ' ' is involved.

object name is a postfix of the domain-part of the (global) system/slave name.

In most cases domain objects are logins or charge numbers which are used for the same purpose as accounts in a Yellow Page Cluster. But resources may also facilitate management. A domain resource e. g. may be used to *consistently* specify the prize for computing time of all Vax workstations in a cluster.

As defined before a manager may only manipulate those objects which are assigned to it. The following extensions are useful in order to be more flexible and to obey the semantics of organizational structures:

- Every master which is superordinate to a manager m (i. e. m lies in the domain of the master) has the *implicit* permission to manipulate all objects of m. This also conforms to the hierarchical responsibility in organizations.
- *Explicitly* specified managers may have special rights (i. e. to collect usage data or to adjust budgets).
- Orders from another manager may (but need not necessarily) be handled as part of remote operating. Of course, it must be possible to authenticate the originator of orders as an administrator with sufficient rights.

But in all cases, a manager m can only *directly* access objects which are assigned to it. Every manipulation by another manager has to be done in an *indirect* way via orders to m. This rule facilitates management since concurrency control is not necessary at the conceptual level. Additionally, it helps to keep accounting information consistent.

The naming scheme proposed in this section is an extension to the rules of [ISO 10165-1] ('containment hierarchy') onto arbitrarily nested management domains. Principles of the ISO-Directory ([ISO 9594-1, ISO 9594-2]) and the MHS ([X.400, X.402]) are also taken into account.

5 Classification of the Proposed Domain Model

A possible classification of concepts for management domains is given in [SloMoff 89a, SloMoff 89b]. This section will show how the proposed domain model for the accounting management fits into this classification. There are also some remarks about domain concepts in other areas of OSI communication.

A domain is defined as a group of objects having a common property or relation, as in [SloMoff 89a]. However, the proposed structure is not primarily based on 'normal' objects (like charge numbers, logins, systems, etc.), but on the accounting managers in the domains. Grouping rules and responsibilities are defined to cope with the requirement of representing relations between and within organizations (cf. section 2). In consequence, domains are 'subset domains[5]' in terms of [SloMoff 89a] with an additional hierarchy between the subdomains of a main domain.

The mechanism of '[SloMoff 89a]-subdomains[6]' is not chosen since it prohibits global

---

[5] Subdomains are viewed as subsets in [SloMoff 89a]. There, a manager can also see all objects of all subdomains.

[6] [SloMoff 89a]-subdomains are viewed as single objects in the superordinate domain. A manager cannot see objects belonging to subdomains.

control. But if required, the behaviour of [SloMoff 89a]-subdomains can be obtained by a suitably adjusted policy.

Main domains and all subdomains of the same level in a domain-tree are disjoint. The responsibility rules impose a 'transitive (management) relation[7]' on the managers. Objects are 'indirectly managed[8]'. This gives a high degree of control to managers, at expense of little overhead.

Finally, a domain/system is managed by a *single* entity. As an advantage, mechanisms for concurrency control are not necessary at the conceptual level. But in a concrete implementation, a manager may be realized by multiple processes which, of course, have to be synchronized.

Domain models are also specified in some other fields of OSI-communication:

- MHS knows a rudimentary concept of administrative and private management domains (in this sequence), which allows a two-level structuring ([X.400, X.402]).

- A detailed and hierarchical concept is also proposed in the ISO-Directory framework ([ISO 9594-1, ISO 9594-2, ISO 4045]). But there, a subordinate manager may have absolute authority over his area.

- A similar (but not so detailed) domain model is specified by the security management papers ([ISO 3337, ISO 3614]).

- The need for domains is discussed in the paper dealing with OSI systems management ([ISO 10040]).

6 Summary and Current Work

A method to structure large numbers of computers cooperating in a network is urgently needed in the area of accounting management. This paper proposes a hierarchical domain model of arbitrary depth which is suitable for this purpose. The model is mainly influenced by requirements of network and accounting administrators.

A naming scheme and rules for scopes of validity / responsibility are based on the domain concept.

A prototype of a manager conforming to the domain model is currently implemented.

Acknowledgements

The author wishes to thank Prof. Dr. Heinz-Gerd Hegering — the principal investigator of the research project — for enabling the Ph.D. Thesis ([Bötsch 90]) which formed the basis of this paper, and all other members of the project for useful discussions and exchange of ideas.

---

[7] A manager has access to objects which are assigned to subordinate domains. Access rights are transitive.

[8] Only the manager m to which an object is assigned can access it directly. Manipulations by other managers have to be performed *indirectly* via orders to m.

References

[Bötsch 90] Ernst Bötsch, *Ein umfassendes Abrechnungsmodell für ein integriertes Netzmanagement*, Ph.D. Thesis, Technische Universität München, March 1990.

[HegBöt 88] Heinz-Gerd Hegering and Ernst Bötsch, "A Proposal for an Originator-oriented Accounting Scheme in OSI-like Multinetwork Environments", In J. Raviv, editor, *Computer Communication Technologies for the 90's*, Proceedings of the Ninth International Conference on Computer Communication, pages 294–298, Amsterdam, New York, Oxford, Tokyo, November 1988, ICCC, North-Holland.

[HegChy 86] H.-G. Hegering and P. Chylla, "Access Control and Accounting as Part of LAN Management", Proceedings of EFOCLAN, Amsterdam, June 1986.

[HegChy 87] H.-G. Hegering and P. Chylla, "Benutzeridentifikation und Abrechnungsdienste in einer verteilten Systemumgebung", In N. Gerner and O. Spaniol, editors, *Kommunikation in verteilten Systemen*, volume 130 of *Informatik-Fachberichte*, pages 102–114, Berlin – Heidelberg – New York, 1987, Springer.

[ISO 10040] ISO/IEC, *Information Processing Systems — Open Systems Interconnection — Systems Management Overview*, DP 10040 (JTC 1 / SC 21 N 3294), February 1989.

[ISO 10165-1] ISO/IEC, *Information Processing Systems — Open Systems Interconnection — Management Information Services — Structure of Management Information — Part 1: Management Information Model*, DP 10165-1, May 1989.

[ISO 3337] ISO/IEC, *Liaison Statement to SC 21 / WG 4 Architecture and SC 21 / WG 1 Security Architecture on the Subject of Security Management Domains and Security Policies*, JTC 1 / SC 21 N 3337, January 1989.

[ISO 3614] ISO/IEC, *Working Draft Access Control Framework*, JTC 1 / SC 21 N 3614, May 1989.

[ISO 4045] ISO/IEC, *Working Document on the Extended Directory Information Models*, JTC 1 / SC 21 N 4045, November 1989.

[ISO 9594-1] ISO/IEC, *Information Processing Systems — Open Systems Interconnection — The Directory — Part 1: Overview of Concepts, Models and Service*, DIS 9594-1, May 1988.

[ISO 9594-2] ISO/IEC, *Information Processing Systems — Open Systems Interconnection — The Directory — Part 2: Models*, DIS 9594-2, May 1988.

[SloMoff 89a] Morris S. Sloman and Jonathan D. Moffett, "Domain Management for Distributed Systems", In Branislav Meandzija and Jil Westcott, editors, *Integrated Network Management, I*, pages 505–516, IFIP TC 6/WG 6.6, North-Holland, May 1989.

[SloMoff 89b] Morris Sloman and Jonathan Moffett, "Managing Distributed Systems", Internal report of the Domino Collaborative project, September 1989.

[X.400] CCITT/ISO, *Message Handling System and Service Overview*, CCITT Recommendation X.400 / ISO DIS 10021-1, November 1988.

[X.402] CCITT/ISO, *Message Handling Systems: Overall Architecture*, CCITT Recommendation X.402 / ISO DIS 10021-2, November 1988.

DESIGN OF AN OPEN NETWORK BILLING APPLICATION

B. E. Ambrose[1] and D. O Mahony[2]

[1]BROADCOM, Kestrel House, Clanwilliam Place, Dublin 2, Ireland.

[2]Computer Science Department, Trinity College, Dublin 2, Ireland.

Abstract
This paper describes an architectural design for billing of open network communications services. Presented here is a general architecture for the monitoring of network usage and accumulation of billing information. The approach is based on OSI principles. A practical implementation using CMIS and FTAM has been undertaken to test the design.

1. INTRODUCTION

On single user computers, there is no need to carry out billing for use of resources. However many computers are multiuser, and therefore have some provision for billing of users. Witness the terminology that is associated with mainframes where users are said to have accounts on computers. Job accounting is also carried out, whereby the CPU usage and usage of other resources such as disks and input/output is noted.

When it comes to use of the network however, things are not as straight forward. Networks by definition involve more that one computer, and so some cooperation is needed between computers in order to carry out monitoring of network usage and billing. This is made more difficult by the fact that the computers communicating may be of different types, using a standardized protocol between each other in order to transmit information. The accounting models may be different on the two computers, so there must be agreement as well on how the management of the accounting is to be carried out.

Accounting management is just one part of network management. Network Management is a topic of much debate in the communications standards arena today. The focus for standards development has shifted from protocol standards for interconnection of networks to network management standards for managing multivendor multidomain multiservice networks.

The questions that are of concern to network managers are (i) what communications resources should be chargeable ? (ii) what are the performance implications for the transfer of billing information ? (iii) how does network interconnection affect billing? This paper attempts to answer some of these questions.

To date much work in developing data communications standards has been done by the ISO organization. ISO have also outlined a work program for development of network management standards to manage diverse computer networks. However progress in this area is slow, as models of the managed networks have to be developed and full agreement reached between the major actors on the contents of network management standards. In particular, the work program for ISO will not produce a full international standard for accounting management of computer networks until the second quarter of 1992 [4,15]. Other parts of the ISO network management standards will be available sooner. In this paper, the process of billing will be examined in detail and a general architecture for management applications will be proposed. An example of the use of this architecture for developing billing of communications services running on a TCP/IP network is then reported upon and some conclusions drawn regarding issues still to be resolved.

2. MODEL OF BILLING AND ACCOUNTING MANAGEMENT

Clearly it is not necessary to standardize all aspects of billing. The issuing of bills is at the discretion of the manager gathering the accounting management information. The only aspects of billing that need to be standardized are those that involve communications over the network. The following model gives an insight into those aspects of accounting management that involve communication over the network. Given a tariff policy, billing can be subdivided into 6 basic functions and (at least) 6 optional functions. These are:

BASIC FUNCTIONS

Function 1:	Setting Tariffs
Function 2:	Usage Monitoring
Function 3:	Gathering of billing information
Function 4:	Billing
Function 5:	Payment Control
Function 6:	Setting up/Removing Accounts

OPTIONAL FUNCTIONS

Function 7:	Management Decision Making Support (Summaries of Billing Statistics)
Function 8:	Setting Detailed Billing on or off
Function 9:	Barring of Users
Function 10:	Setting Accounting Limits or Quotas
Function 11:	Providing Advice of Charge to the user
Function 12:	Cost Accounting (Analysis of costs in order to set tariffs)

A detailed description of these activities would take a number of pages, and not add much value to the list. However two general observations can be made. Certain parts of billing such as function 4 which concerns the physical issuing of invoices and function 5 which tracks payments received against bills issued are not generally considered to be candidates for standardization. However J. Hall and M. Turnbull in [8] have proposed an accounting and resource management scheme for the Internet that carries out billing and payment control using a new Internet protocol.

Secondly note that usage monitoring (Function 2) does not imply billing. The accounting manager is free to keep track of network resources using the facilities of accounting management, without having to issue bills for use of these resources. For the accounting manager, the ability to examine usage patterns associated with network resources means that abuses of resources can be spotted. For example a hacker could have broken into a network and be using it as a gateway to other networks. Also unusual traffic could be spotted. For example bugs with mailing systems whereby routing loops are created, can be spotted with the aid of accounting management. Finally usage monitoring provides help with planning of the network.

3. SYSTEM OVERVIEW

Communications between computers can be based on the OSI Reference Model [2]. This divides network functions in seven layers and specifies the functionality of each layer. ISO have provided specifications of the services at most layers. The layers of most interest to this paper are the transport layer (which is to be billed) and the application layer (which carries out the billing).

The implementation of the billing application was carried out on a network that uses TCP/IP protocols. The concept of layers is used by the TCP/IP protocols but these layers do not conform exactly to the ISO definition of the layers. The Transmission Control Protocol (TCP) was specified by the U.S. DoD and provides end to end transport services for networking applications. To access the underlying network, the TCP uses the DoD Internet Protocol (IP). The IP is a datagram, or connectionless service which makes no assumptions regarding the reliability of the communications subnet and uses a simple checksum mechanism. This requires higher layer protocols (e.g. the TCP) to handle the end-to-end reliability.

The TCP/IP protocols are based on the Arpanet Reference Model (ARM) [12]. This uses an underlying client server paradigm for services. A server in general listens on a well known port, awaiting a connection. Clients are then free to connect to the server. After some negotiation at connection time, communications can take place between client and server until one or other releases the connection. The TCP/IP protocols do not imply a client/server model, but this model is frequently implemented using TCP/IP protocols.

The hardware used in the implementation was a network of four SUN workstations, running SUN OS 3.5. This provided a UNIX BSD 4.2 environment. The network was a Ethernet Local Area Network, running Ethernet at the lower three layers. Both local area and wide area communications were considered in the implementation.

In this context TCP/IP communications were provided by abstractions called sockets. Operating system calls were available to set up and release sockets across the network. Calls were available to servers to allow them to listen at a given port for network traffic. If a connection attempt was directed to a port, operating system calls were available to accept the connection and communicate over the connection.

For example, if a request is made by Network File Service software for a file, the request takes the form of establishment of a socket to facilitate communications between two processes. The socket is a mechanism for Inter Process Communications (IPC) under Berkeley Unix 4.2 BSD, which the Sun uses. When sockets are established between processes that are separated by a network, some sort of communications protocol must be used, for example TCP/IP. This in turn relies for its access to the network on the Ethernet driver that is present in the Sun. There is some interface software on the Sun to handle the Ethernet driver.

All in all, there are many places that a request for network services could be trapped. In practice, the most convenient point to trap the connection is as the socket level. There is a one to one correspondence between use of the network and use of sockets (as all processes under UNIX 4.2 BSD will establish sockets in order to communicate with each other) but at the same time information about who is using the service is still available. The alternatives have drawbacks. If it was decided to perform billing at the Ethernet Driver interface, one would be guaranteed to trap all use of the network, but one would not have sufficient information about who is using the network to generate usage records, without some large address translation processing overheads. Alternatively if one carries out billing higher up the protocol stack, perhaps at the applications level, one could not be guarantied to trap all usage of the network, as each application program would have to be modified in order to record usage.

4. BILLING ASPECTS OF PRESENT DAY NM SYSTEMS

The problems of network management have already been solved for a number of specific network types. In this section, the management of SNA, TCP/IP and X.25 networks is examined with specific reference to billing. This represents no more that a sample of the set of NM systems available today.

(i) IBM Netview

The IBM approach to network management is defined in the Network Management Architecture (NMA). Netview is one example of IBM's Network Management Architecture [5,6,7].

The IBM Netview product can be used to manage IBM Systems Network Architecture (SNA). SNA network environments have evolved from centralized single host, single vendor entities into a distributed control, multiprocessor, multivendor, processing environment. SNA's job is to manage end-to-end sessions between logical units (LUs) and to physically manage network path control and data link control resources.

In the Network Management Architecture, Performance and accounting management is defined as that part of NMA that quantifies, reports, and controls the utilization and charges associated with network components. Utilization monitoring keeps tabs on network resource server utilization. It generates unsolicited problem notifications if preset threshold values are exceeded. Accounting Management records the use of network resources in an effort to properly allocate costs.

Netview Release 2 was announced in June 1987. It allows some fault management, configuration management and performance management facilities. The three basic modules are a command facility (low level interface), a hardware monitor, and a session monitor. The session monitor is the most interesting from the billing point of view as it provides the facilities to monitor virtual routes (VRs), returning such information as domain name, Physical Unit (PU) type, sent and received (sequence number of the last Path Information Units (PIUs) sent and received) and time (time that Virtual Route status was received by session monitor).

The Netview Performance Monitor is built on top of the Session Monitor. It is an application that monitors, records, and graphically displays network performance and utilization. Data that is reported to the host includes:

- Session accounting;
- Immediate or deferred accounting collection especially for accounting for voice communications over token ring local area networks.
- Session accounting byte and PIU thresholds for unsolicited data;
- Whether backup Network Performance Monitoring sessions are defined;
- The number of half sessions over which session accounting will be provided;

A session awareness PIU is sent at the condition of LU to LU setup, and byte and PIU counts are updated whenever session traffic flows. If these counters exceed user specified thresholds, threshold counters are sent to the Network Performance Monitor as unsolicited data. In all cases, session awareness data and the last set of session counters are sent to the Network Performance Monitor at session termination period.

(ii) CMIS on top of TCP/IP

The U.S. Department of Defence TCP/IP protocol suite forms the basis for a worldwide network known as the Internet. An internet addressing scheme enables users on any host connected to the Internet to talk to users on any other host. In April 1988, the Internet Activities Board (IAB) designated two different network management protocols with the same status of "Draft Standard" and "Recommended". One of these protocols is Common Management Information Services and Protocol over TCP/IP (CMOT). This protocol has been elaborated in a memo of April 1989 [3]. The other network management protocol was SNMP.

The Common Management Information Service (CMIS) is a draft International Standard from ISO which provides a generic set of management services to network management applications. This set of services allows the applications query and set management information using an ISO Remote Operations protocol, as well as defining other useful operations. The management information is stored in a conceptual "management information base". A separate memo of May 1990 [13] defines the management information base for TCP/IP.

The guidelines for inclusion of objects in the first draft of the MIB were:

1) An object needed to be essential for either fault management or configuration management.
2) Only weak control objects were permitted (by weak, it is meant that tampering with them can do little damage). This criterion reflects the fact that the current network management protocols are not sufficiently secure to do more powerful control operations.
3) Evidence of current use and utility was required.
4) An attempt was made to limit the number of objects to about 100 to make it easier for vendors to fully instrument their software.
5) To avoid redundant variables, it was required that no object be included that can be derived from others in the MIB.
6) Implementation Specific objects (e.g. for BSD Unix) were excluded.
7) It was agreed to avoid heavily instrumenting critical sections of code. The general guideline was one counter per critical section per layer.

CMOT contains some recommendations on how to use CMIS to query and set certain protocol related information, for example, routing tables residing on a particular host. Unfortunately CMOT does not give much guidance on how to implement billing applications as (i) events and actions associated with managed objects are not defined (e.g. setting up or releasing of a TCP/IP connection may be an event) and (ii) usage logs listing use made of the network on a per user basis are not defined. In conclusion although CMOT allows manipulation of management information associated with TCP/IP networks, it is still lacking as far as definitions of management information associated with billing is concerned.

(iii) Billing Policy for X.25 public data networks

Billing arrangements for X.25 Public Data networks are covered for the most part by CCITT Recommendations, in particular the "D" series of Recommendations [1]. Recommendation D.10 suggests that tariffs should be split into components, one to reflect the cost of access and the other to reflect the cost of usage, based upon:

- User class;
- User Facilities;
- type of switching;

- volume of data and/or duration of a call;
- Distance;
- Time (peak/off peak periods);
- route;
- other functions;

The recommendation states that the gathering of information required for charging should normally be the responsibility of the calling Administration, however if the interconnection of dissimilar networks occurs at the called administration, then the latter should be responsible for providing the calling administration with the necessary data for charging and accounting. (This is an unusual situation, the equivalent of asking the customer to keep his own bill).

Recommendation D.11 applies these principles to X.25 packet switching networks and considers the following packets chargeable:

- data packet;
- interrupt packet;
- call request/call incoming packet;
- reset request/reset indication packet (for certain causes);

Recommendation D.12 suggests that volume charging is carried out on the basis of segments of 64 octets. The number of segments is a data unit should be rounded to the next highest whole number. Packets which are not data packets can be treated as a single segment.

5. ACCOUNTING DESIGN GUIDELINES

It is important in designing network management standards that are to be acceptable to all vendors that (i) the standards are complete and contain enough functionality to carry out what is required (ii) the standards are simple and do not entail the implementation of an overly complex system in order to conform (iii) the standards are open to evolution and the advent of new technology and new products.

The definition of a Management Information Base (MIB) is one step towards the adoption of management standards which are complete, simple and open to evolution. The MIB contains definitions of the objects to be managed and the attributes and events associated with these objects. The managed objects are not necessarily real objects. They represent instead a management view of real resources.

There are guidelines published by ISO which apply to the definition of objects to be found in the management information base [9,10]. These guidelines can be used to provide guidance about the managed objects to be defined in the billing domain.

In general, the provision of options in managed object definition is discouraged, on the grounds that interworking becomes more difficult as the number of options increases. Layer groups should strive to achieve consistency of statistics across layers. Candidate features of the (N) layer for which statistics may be recorded are:

- Local errors;
- Successful peer-to-peer exchanges;
- Peer-to-peer failures;
- Service Rejections;

Counters should be kept for protocol data units, rather than service data units. Layer groups should avoid unnecessary duplication or overlap of statistics.

ISO DP 10040 [11] "Systems Management Overview" states that both CMIS and FTAM can be used for the purposes of transporting management information. It would seem that FTAM is more suited to the transfer of bulk billing data than CMIS, but this question still has to be decided by the standards bodies.

In general usage monitoring will take place close to the node where usage is taking place. At some point in time all the billing data will have to be gathered from the relevant nodes in the network, in order to prepare a bill. The architecture of the billing system has a strong influence on the volume of data transported over the network to effect billing.

6. DESIGN

A case study was selected for implementation in order to test the use of ISO protocols. The case study had to (i) use CMIS and (ii) align with available documentation regarding ISO accounting management. A realistic scenario was chosen. This meant mapping as closely as possible to the network management activities carried out on the network that the implementation was using. This also meant that interviews with the network managers of the network would be required. The network available was a small network of SUN workstations. Billing of both Local and Wide Area Networks was considered in order to assess performance and openness.

The scenario chosen for the case study was billing in a University environment. The network manager is in charge of issuing bills to the various departments and users that use the network. Some interviews were carried out with network management personnel in Broadcom, which resulted in the set of requirements given in Table 1.

Table 1
Billing Requirements Example

Service Components

1) Accounts Reconciliation. Reconciliation of public network bills (e.g. X.25 bills) with TCP/IP network usage as monitored in the University network itself.
2) Issuing Bills. Billing of users or departments for use of network resources, particularly with respect to number of bytes transferred over the network.
3) Detailed Billing. Option of turning on or off detailed billing for diagnostic purposes on a particular host for a particular user.
4) Accounting Limits. Option of removing network services from a user or department if they pass over certain agreed maximum usage figures. Conversely it should be possible to restore network service to the user.

The resulting architecture is shown in figure 1. A central billing application program was provided and on each host a network usage accounting daemon (netacctd) was set up. The netacctd program monitored the usage of the network and from time to time reported to the central billing manager. CMIS and FTAM were used for communications between the netacctd program and the central billing manager.

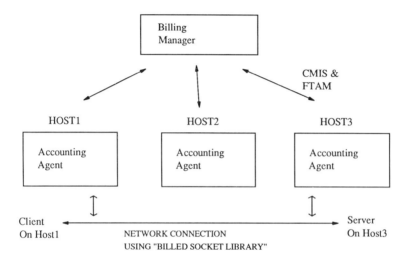

Figure 1 Implementation Architecture

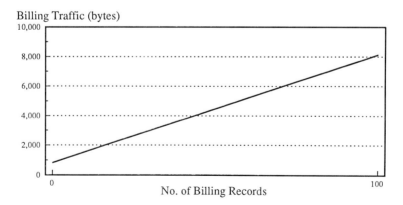

Figure 2 Billing Traffic as a function of Billing Records

CMIS was used to add/remove users from a list of users to which detailed billing or barring of service applied. FTAM was used for transfer of billing records. The format of the billing record was specified in ASN.1 as part of a FTAM document definition.

To monitor network usage, the netacctd program communicated with a billed socket library. The idea was that the billed socket library replaced the standard socket library that is provided under UNIX BSD 4.2 and allowed billing to take place. System V messages were used for Inter Process Communications between the modified billing routines and the netacctd program. It was found that various calls apart from those directly concerned with sockets had to be trapped. For example, the "exit" system call, which causes a program to terminate, has an indirect effect of shutting down the sockets associated with that program, and therefore had to be trapped.

Looking at the traffic and performance implications of billing for use of network resources, figure 2 shows a graph of how the network traffic associated with billing increases with the number of billing records gathered. Interestingly this information was obtained by billing the billing system itself. The amount of billing traffic is highly dependent on the coding of the billing information and the patterns of network usage. In this design, the netacctd program recorded:

<Date and Time of Communication>, <User Identification>, <Remote Host Identification>, <Number of bytes sent and received>, <Time Connected in seconds>

The minimum amount of information that must be stored depends very much on the tariffs. In this implementation, Requirement 1 dictated that accounts reconciliation must be possible between the local network accounts and the public network accounts. This meant that a complex tariff scheme (e.g. that used for public X.25) could be envisaged. Thus the netacctd program was provided with guidelines for when usage records associated with a given instance of communication could be amalgamated. In general, records for all local traffic for a given user to a given host were amalgamated. All remote traffic carried over the public network by a given user to a given host, during the same time of day (e.g. 9.00 - 5.00) was amalgamated. It was possible to turn off this amalgamation using the detailed billing option. Users were identified by a combination of userid, groupid and process id allowing groups of users to be billed as a unit, or single operating system processes to be monitored. From time to time, the billing information would be transferred to a central location.

Figure 3 shows a graph of the traffic on the network over a typical day and figure 4 shows a breakdown of this traffic. This information was obtained by using commands available on the SUN workstation to monitor traffic on the network.

Average figures of 12,000 packets per hour were obtained. This is an enormous amount of information and it would be unfeasible (and uninteresting) to keep details of each packet. The storage requirements for recording details of packets assuming a billing record size of 100 bytes and a 8 hour day would be 76.8 Mbytes/day.

Relating number of packets transferred over the Ethernet to number of (socket) connections can be difficult. The analysis of network traffic contained in Figure 4 gives some insight into the difficulties. The display of protocol related information gives the number of packets of each type of protocol that were detected over a test interval. Note that nd is the network disk protocol, tcp and udp are parts of the TCP/IP protocol suite and other represents system protocols. Only packets labelled as tcp and udp would be billed by the billing implementation. The other packets are "system" protocols and involve no user level processes. No calls to the socket library are involved in setting up these connections. It is difficult to estimate given these statistics the number of connections that would take place during a typical day (possibly of the order of 1% of the number of packets transferred). Instead an alternative approach can be taken based

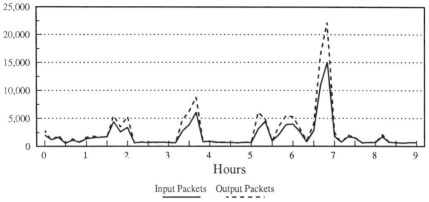

Figure 3 Ethernet Traffic

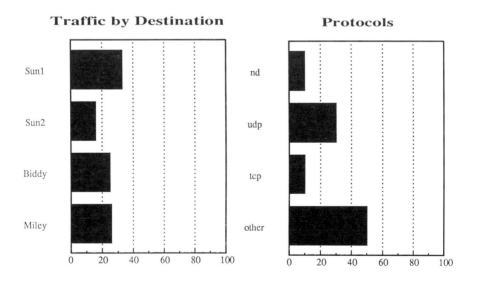

Figure 4 Analysis of Ethernet Traffic

on the information required to compute the bills.There is a tradeoff between degree of detail of bills and the amount of billing information that must be kept. In approximate form, this could be stated as:

Billing Information (kbytes) = (average billing information per account assuming a lumped bill) x (No. of Tariff Classes) x (No. of accounts) / (information storage efficiency factor)

where the Tariff Class takes into account the different tariffs that will be applied depending on time of day, destination, etc. It is important to note that accounts are billed rather than individual users. An account could correspond to a single user, a number of users or a single process. The efficiency factor takes into account that in gathering billing information, it is more convenient to carry out some processing of the information centrally rather than locally. For example, rather than distribute the details of all the accounts that exist, it is more convenient to keep this information in a central location, and locally just record the userid, groupid and processid involved in the communications instance. This decreases the efficiency of the implementation with regard to storage of information but means that the netacctd program is less complex and can complete its work faster.

In this implementation the average billing information per account assuming a lumped bill was found to be 8 bytes. The information storage efficiency was found to be 0.10. This means that 80 bytes were used to store a billing record that could theoretically be stored in 8 bytes.

7. COMPARISON OF MANAGED OBJECTS

This section compares managed objects defined in the implementation of the billing system with those suggested by recent ISO work on Accounting Management, as contained in a Working Document[15] that became available when the implementation was complete. This was issued November 1989 and became available July 1990. It becomes more definitive about managed objects for accounting than the December 1988 version. ISO suggest two classes of managed object for Accounting Management :

- The accounting meter control object class;
- The accounting meter data object class;

It is suggested that there is an accounting meter for every OSI resource which is to be monitored by accounting management. In addition the following information items are talked about but not rigorously defined:

- The accounting record (which is created by a meter data object and may be contained in an accounting log).
- The quota record (which places a limit on the use of OSI resources by individual users)
- The accounting log;

The ISO approach roughly corresponds with the approach taken in this project. The barring of users amounts to definition of the accounting management quota record. The detailed billing of users corresponds to the setting of the recording trigger attribute of the accounting meter control object class. The gathering of billing data using FTAM is equivalent to the retrieval of the accounting meter data object via a CMIS m-Get or via its own reporting trigger being activated.

The ISO accounting management approach of using CMIS to access the accounting meter data objects, rather than FTAM, makes for a more economical implementation in terms of having to implement one protocol rather than two. However all the added functionality of FTAM e.g. the ability to carry out remote manipulation of the virtual filestore and the ability to recover gracefully from file transmission errors is lost.

It should be stated that an ISO working document (WD) is at the first stage of the ISO standardization process. Major changes can and do occur between issues of working documents. This occurred in this case as there was no definition of the managed objects manipulated by CMIS in the Working Document used as the basis of this implementation whereas the later CMIS document had such a description. The next step in the ISO process of standardization, that of production of a draft proposal (DP), gives a far more stable document.

8. OPEN ISSUES

Some issues still have to be resolved. In particular the question of what happens if the host computer is providing a transit (intermediary) facility to other hosts that are setting up an internet connection. This transiting will not be seen and monitored by the "Billed Socket Library" as currently implemented, because there will be no calls to the billed socket library on the machine that acts as transit. Monitoring transit connections would entail changes to the UNIX kernel that implements the TCP/IP protocol.

Ideally it should be possible to define the transiting facility as a type of "service" offered between administrations, which could have tariffs, users and accounts associated with it, as with the end-user "use of network resources" service. However because it is an inter administration service, there would be additional security implications in the design of the protocols, i.e. only those administrations that are authorized (and therefore can be billed) should be allowed to use the service. Authentication information would therefore be needed in the protocols.

A second problem is that any applications that have a store and forwarding functionality (e.g. mail) need their own billing mechanisms as two separate network connections are being used to handle a single message. Thus these protocols have to be designed to incorporate transmission of accounting information.

A third problem is that in this implementation, only the requirements of the accounts manager were considered. The accounts manager has responsibility for billing and receiving payments. Other higher level managers may impose requirements on the system e.g. that it provide cost accounting information to help with planning of tariffs. Similarly other departments may have requirements, e.g. the legal department may require that the software contain some billing test functionality. These requirements have not been considered in this implementation, but should be considered when drafting standards.

9.0 CONCLUSIONS

This design met its goals of providing an open network billing application. CMIS and FTAM were used for communication of management information between the nodes on the network. This management information included billing information and control information about detailed billing and removal of service. An efficient implementation of a billing application on top of a TCP/IP network was produced.

There is a lot of work to be done by the standards bodies before a standard billing mechanism becomes available. The main areas to be developed are (i) extension of management information bases to describe actions and events. This is necessary because either the manager or agent must take responsibility for accumulation of usage information and the transfer of this information between the two must be carried out by means of actions or events. (ii) Definitions of managed objects associated with billing e.g. attributes of the billing agent and the means of metering the use of the network.

Final agreement may be difficult to reach because it depends on agreement of principles of billing policy across the entire spectrum of network managers. In fact there are probably as many billing policies as network managers. Also policies should be open to continuous evolution.

The final goal of the standards bodies should be to define managed objects for CMIS that are generic and satisfy the billing requirements (e.g. detailed billing of users etc) and to define a format of billing information that is equally generic.

10. REFERENCES

1) CCITT D Series Recommendations, Volume 2, Fascicle II.1, Blue Book 1984.
2) ISO, Information Processing Systems - Open Systems Interconnection - Basic Reference Model, IS 7498, October 1984.
3) Warrier, U., Besaw, L., The Common Management Information Services and Protocols over TCP/IP (CMOT), RFC 1095, Network Information Centre, DDN, April 1989.
4) NETMAN, EC RACE Project 1024, Report on Network Management Architectures, Annex 6 ISO Draft Standards Evaluation, Deliverable 4, December 1989
5) Routt, T.J., SNA network management: What makes IBM's Netview tick?, Data Communications, June 1988, pp. 203-227.
6) Kanyuh, D., An Integrated Network Management Product, IBM Systems Journal, Vol. 27, No. 1, 1988, pp. 45-59.
7) Fernandez, J., SNA and OSI: Which manages multivendor networks best?, Data Communications International, April 1989, pp. 85-94.
8) Hall, J., Turnbull, M., Accounting and Resource Management in an Internet Distributed Operating System, Integrated Network Management Conference, IFIP, May 1989.
9) ISO, Information Processing Systems - Open Systems Interconnection - Management Information Services - Structure of Management Information - Part 4: Guidelines for the Definition of Managed Objects for CCITT Applications, DP 4065, December 1989.
10) ISO, Information Processing Systems - Open Systems Interconnection - Procedures for the Operation of OSI Registration Authorities, DP 9834, December 1989.
11) ISO, Information Processing Systems - Open Systems Interconnection - Systems Management Overview, 2nd DP 10040, December 1989.
12) Cerf, V.G., Cain, E.A., The DoD Internet Architecture Model, Computer Networks and ISDN Systems 7(10), October 1983, pp. 307-318.
13) Rose, M.T., ed. Management Information Base for network management of TCP/IP-based internets: MIB- II, RFC 1158, Network Information Centre, DDN, May 1990.
14) Krall, G., SNMP Opens New Lines of Sight, Data Communications, March 21, 1990, pp. 45-50.
15) ISO, Information Processing Systems - Open Systems Interconnection - Accounting Management Working Document - Third Version, ISO/IEC JTC1/SC21/WG4 N875R, November 1989.

IX

MANAGEMENT INFORMATION

Distribution of Managed Object Fragments and Managed Object Replication: The Data Distribution View of Management Information

S. Mark Klerer and Roberta S. Cohen
AT&T Bell Laboratories
Crawfords Corner Road
Holmdel, New Jersey 07733

1. Introduction

This paper describes techniques for coping with object oriented management models in a distributed environment. We assume an environment in which CMIP[1] and the ISO management functions are used as the specific management protocols. Logical and physical resources (e.g., circuits, applications, modems, processors) are modelled as managed objects[2] in this environment, each represented by data that can be manipulated by local and remote application programs and communicated among open systems. Complex resources, such as networks, can be modelled as a collection of individual resources, as well as a managed object with a unique data representation. But concepts for object aggregation, relationships such as the IS-PART-OF relationship, have been slow to emerge for use within object modelling. To the extent that such concepts are available today, they are invariably applied at the level of the defined object. Such an approach, while valuable for conceptualizing complex resources, does not allow for units of aggregation below the object level, as, for example, when the data elements of a single object are maintained in a distributed fashion.

Network Management places unique requirements on managing systems, requirements which can often be met by optimally using data distribution and data reduction techniques. Maintaining data close to where it will be used, for example, is a useful technique of management, but one that often creates unusual data distribution needs when the same data are needed in multiple places. At times, the need for widely distributed data is counter balanced by the need for very small, efficient implementations of network resources-- implementations that can offer only minimal data maintenance. Both of these needs, data distribution and highly efficient, small implementations are addressed here through our proposal for object replica and object fragment handling capabilities.

Two types of users are defined for the network management information base (the MIB); the network management staff (NMS) and the database administration staff (DAS). The network management staff is the primary user of the MIB, and the MIB should, therefore, be designed primarily to meet this user's requirements.

The database administration staff (DAS) is responsible for the distribution of data and the maintenance of the database. The DAS is also responsible for assuring that the MIB meets the performance requirements as determined by the network management staff. To accomplish these tasks the DAS must be able to distribute and/or replicate information as dictated by performance and reliability requirements.

Ideally the design choices made by the DAS should be transparent to the network management staff. Specific replicas accessed and specific distribution of management information is of no interest to the network management staff.

The paper discusses a technique that allows distribution and replication of management information so as to preserve transparency to the network management staff while at the same time allowing the network management information base administration staff to gain access to individual object fragments and specific replicas of object information.

2. The Data Distribution View of Managed Objects

In many cases practical implementation considerations will dictate that managed objects be visible from more than one system. Such considerations result in requiring that systems be able to act as virtual locations of managed objects. This can be accomplished by supporting multiple replicas of a managed object and by providing capabilities that will allow a system to assemble object information if not contained within the system.

The use of replicas may be required for performance reasons. The speed with which retrievals or modifications can be made will be improved when multiple replicas are available. Similarly, reliability is enhanced by having multiple replicas of managed objects. However, maintaining multiple replicas requires additional care to assure that the views offered by different replicas are consistent and it also requires management capabilities that allow for the identification of consistency problems.

Replication of management information might be avoid if one allowed systems to create chained management requests and obtain managed object information on behalf of managers. This would allow maintenance of a centralized distributed database but would impose potentially serious performance penalties and may not be feasible for large networks.

2.1. Managed Object Replicas

2.1.1. Characteristics of Replicas

A replica is a copy of the managed object specification. For network management purposes it is desirable that the number of replicas that exist or the particular replica that has been accessed be transparent to the network management staff. However, the network management information base administrator (DAS) must have the ability to access a particular replica and manage it.

This gives a replica a dual nature. On the one hand, it serves as a window to the managed object; on the other, it acts as a data structure.

When used as a window to the managed object, a particular replica need not be identified. The managed object is accessed by specifying the managed object class and its instance. The network management authority may provide access to the managed object through any replica available. When accessed in this mode each replica provides a means of obtaining information about the resource, as represented by the managed object, or exercising control on the resource, as represented by the managed object. In this way, each replica acts as if it were the managed object.

It also should be possible to access a particular replica of the managed object in this mode. If a specific replica is specified, the DAS or NMS has the ability to verify that control can be exercised through that particular replica.

In order to allow manipulation of replicas as data structures, all managed object replicas are considered to be allomorphic to a new object class, called the *"Data Object Class"*. In this view replicas must be identified individually and can be managed as parts of a distributed database.

Individual replicas can be identified by specifying the managed object name and an associated replica suffix as part of the name. When accessed in this mode all operations pertain to the data structure and not the resource as represented by managed object. In this way, CMIS[3] services can be used to manage replicas. The following semantics are associated with operations on replicas:

    i.   In response to a create operation, a replica duplicating the managed object description will be created. No real resources required to support the managed object will be allocated. Also a "shadowing" mode and "shadow" source will have to be specified (see below) as part of the create request.

        The create request will also have to identify the destination of any control information that is sent to the replica. This destination can be another intermediate system or the system that contains the actual resource to be controlled.

    ii.   In response to a delete operation, the data structure and identifier associated with a particular replica is deleted. This operation has no impact on the underlying resource.

        When the last replica is deleted, the resource that was represented by the managed object still exists and provides service, however, all management knowledge of that resource has been lost; that is, from an open systems viewpoint the resource has become unmanageable.

iii. In response to get operation, a read of the current values as available in this particular replica is obtained.

iv. In response to a set operation, the values stored in the replica will be changed without impacting the managed object. This service may be used to force an update in a replica.

In essence, when accessed as a data object, the relationship between the replica and the resource is severed.

2.2. The MIB Directory

In order to allow management users to locate and access managed objects, a directory type of service must be available that supports determination of the location of all managed objects. This directory service may itself be a distributed application.

The MIB directory should make available at least one address for each managed object that has been instantiated. When the address of a specific managed object is requested, the MIB directory may return the address of a particular replica, without identifying which replica was chosen. When data structures are requested, the MIB directory should provide the capability for locating specific replicas, thus when the address of a specific replica is requested, this address should be returned.

The MIB directory can be optimized for the use of management. Such optimization can include the ability to determine the replica closest to the application inquiring about its location, the ability to avoid replicas that have become inconsistent and the ability to select a replica that can presently provide the best service. It should be noted that such optimization requires that the MIB directory have some knowledge of the state of each replica and that this functionality allows the MIB directory to take on the characteristics of a "*trader*" as defined in ODP[4].

Figure 1 shows how use may be made of the MIB directory to locate particular managed objects and replicas. The figure also shows how a front end system may be used to make the full complexity of the management data distribution transparent to the user. The front end system is serving a particular user and is designed so as to appear as the virtual location of all the managed objects that are of interest to that user. Management requests are sent to the front end system which automatically uses the MIB directory to access the managed object or replica, as required.

2.3. Updating of Replicas (Shadowing)

Two modes of updating replicas can be defined. One is event driven and the other relies on polling.

2.3.1. Event Driven Shadowing

Where replicas are to be kept tightly synchronized, an event report based shadowing service should be provided. This requires that event reports be sent from a replica that is already closely coupled to the resource, whenever an attribute in the object changes. The object emitting these notifications, thus, serves as master copy for that particular replica.

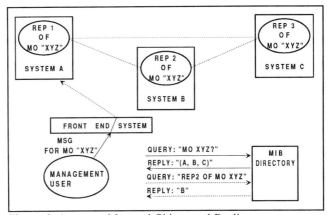

Figure 1. Access to Managed Objects and Replicas

In event driven shadowing it is the responsibility of the system receiving the event reports to assure that the MIB is updated on the basis of the information contained in the event report. It must be noted that this is more complicated and different than mere logging of event reports. To provide a reliable shadowing service, confirmed event reports should be used. If no confirmation is received, an event may be generated and sent to a DAS location to indicate that the database may have become inconsistent. The DAS may then take steps to assure that that particular replica is put in an appropriate state.

Since all replicas may provide access to the managed object, it is possible to have more than one place through which a change may be initiated. Therefore, to avoid looping of "shadow" messages it may be desirable that each replica only have one source from which it obtains shadow information and that shadow information not be sent to a system that was the source of the information on the basis of which the replica was updated.

Figure 2 shows an example of a change on managed object "XYZ" that is accessed via replica 2, resulting in event reports being sent to the systems containing replicas 1 and 3 and in the updating of these replicas.

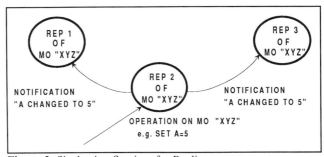

Figure 2. Shadowing Services for Replicas

2.3.2. Shadowing via Polling

An alternative method is for the system storing the replica to periodically poll a master copy for the latest values of the attributes. Notice that if the polling is done in response to an incoming request for information about this object, then this service becomes similar to "*chaining*". This kind of shadowing is more likely to be used where near real time (as distinct from actual real-time) alignment is acceptable, (e.g. collecting of billing information).

In shadowing via polling it is the responsibility of the system doing the polling to assure that the requested updates are received. The system being shadowed is unaware that it is being shadowed. If a response is not received, an event report must be sent to a DAS location indicating that the database may have become inconsistent.

Finally, a combination of polling and event driven shadowing may be used. Some attributes may be updated based on attribute change notifications, and others will only be updated if polled.

3. Object Fragment Distribution

In some cases, implementation consideration may dictate that less than a whole managed object instance (i.e. less than a whole replica) be stored in a single open system. In those cases where this is done, one or more than one open system may appear to be the virtual location of the managed object definition. An open system that appears to be a location of a managed object specification will be called a "*Managed Object Server* (MOS)." Any system that either stores or can assemble a whole replica can assume the role of a MOS.

A network management user may access any system that is a MOS in order to obtain information about or control the underlying resources for which the MOS has access to managed object specifications.

3.1. Object Fragments

In order to allow distribution of characteristics of a managed object the concept of an object fragment is introduced. Object fragments may themselves be viewed as objects, however, they differ from managed objects in that no specific object class is associated with a fragment. Instead a fragment type is defined as being merely a piece of an object class specification that has an identifier associated with it.

Object fragments can be used to accommodate the need to allow the storage of attributes, behavior and the exercise of control in the most appropriate system. This distribution, may for example, allow the definition of a fragment that contains only a behavior specification. Such a fragment would be useful when a simple device is to be controlled, in which the attributes must have a specific relationship to each other, but where the device itself cannot enforce this relationship. A fragment, stored in a more complex front end, could be responsible for enforcing these consistency constraints.

Figure 3 shows three cases where it may be useful to break a managed object into fragments and to distribute the fragments in different systems.

Case 1 represents the case of a modem that has been modeled as a managed object. The modem managed object has parameters that can be set and monitored and also requires that certain relationships be maintained between parameter values for different modes of operation. To support the implementation of simple modems, the modem managed object has been split into two fragments; one acting as sink for control information and the other fragment

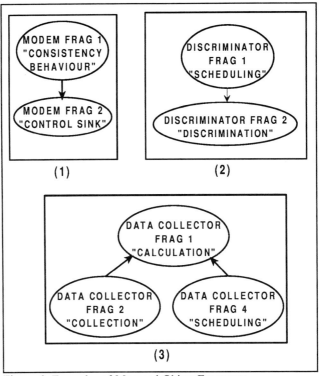

Figure 3. Examples of Managed Object Fragments

enforcing consistency constraints. Allowing the consistency constraint fragment to be located remote from the basic control fragment allows the design of modems that due not enforce any constraints and allows the behavior to be controlled remotely. This particular fragmentation of an object is of special interest as it results in a fragment that only contains a behavior specification. It should be noted that in order to enforce the constraints the behavior fragment must have access to the attribute values of the control fragment. This may imply that a single management operation, i.e. a set, may have to be decomposed into a get and set operation or converted to a filtered set or may require that a complete replica be kept at the system that maintains the behavior fragment.

Case 2 represents a discriminator managed object that has been split into two fragments. One fragment provides the capability to filter event reports based on specified criteria to be satisfied by the content of the potential event report; the other fragment provides the capability of scheduling the activity time of the discriminator. Again, this fragmentation allows the design of simple discriminators that do not perform scheduling and whose scheduling is determined by the remote fragment. Note that this implies that the remote fragment has access to the state attributes of the discriminator and can manipulate these to obtain the desired behavior.

Case 3 represents a statistical data processing managed object that allows for the collection and scheduled summarization of management information. This managed object has been split into three fragments. One fragment schedules the activity, the other fragment collects the appropriate parameter values and the third fragment performs the statistical computation.

3.1.1. Characteristics of Object Fragments

Object fragments provide the DAS with the capability of distributing managed object replicas in an arbitrary fashion. The key characteristic of a fragment is that the fragments constitute a decomposition of the managed object and that any customized behavior that has been associated with an attribute or other managed object characteristic is retained as part of a fragment.

Objects fragments can be standardized or proprietary (possibly defined dynamically).

Standardized fragments can be defined ab-initio, or may correspond to already existing building blocks. Existing building blocks that can be used as fragments are:

  i. Conditional packages in which the package object identifier used for the package would serve the purpose of fragment type identifier. The detailed behavior of the fragment type is, thus, determined by the package identifier and the managed object class in which the package is installed.

  ii. Inherited or allomorphic object classes in which the object class of the allomorphic object is used to identify the fragment type. The behavior of the fragment is determined by the allomorphic object class identifier and the managed object that includes the characteristics of the managed object class.

  Notice that this is not equivalent to an allomorphic invocation of the object. In an allomorphic invocation the object exhibits the behavior of the allomorphic class; when invoked as fragment of a managed object, the actual managed object behavior is exhibited. Also when a managed object is required to act as an allomorph, the managed object class field is used to request that allomorphic behavior be exhibited. When fragments are accessed the actual object class is used.

Proprietary fragments may defined by assigning unique identifiers to arbitrary decompositions of object classes. Schema management could allow dynamic decomposition of object classes, the only requirement being that a system acting as MOS must be able to determine where all fragments are located and which set of fragments constitute the managed object in its entirety.

3.2. Naming of Object Fragments

In order to allow the DAS to manage individual fragments, individual fragments must be identifiable. In general it is possible not to assign identifiers to individual fragments, but to access a system and request all information that it has about the object. This approach is possible as long as only one fragment of the managed object acts as an assembler of fragments. If more than one fragment has that ability, provision must be made for suppressing such assembly in order to avoid looping. This can be accomplished by assigning identifiers to individual fragments or requiring that all the attributes to be retrieved are explicitly identified. Where fragment naming is used, a fragment instance can be identified by appending the fragment type to the managed object instance name. This provides for a consistent naming plan across managed objects, replicas and fragments.

3.3. MOS Functions

In order to be able to function as a MOS an open system must be able to provide the following functionality:

i. determine which open systems contain object or fragment specifications for all objects for which the system is a MOS;

ii. distribute control information to relevant objects and fragments.

iii. assemble all relevant object fragments.

3.3.1. The Object Fragment Directory (OFD)

In order to enable the MOS to assemble all relevant object fragments the MOS must have an Object Fragment Directory (OFD). The OFD must contain the decomposition of each object into its actual fragments and the identification of the open system(s) that provides access to that fragment.

3.3.2. Assembly and Distribution of Information

For retrieval operations, the MOS must retrieve information about each object fragment it does not store by obtaining that information from a system that has that information. This is accomplished by using the OFD to determine which fragments are to be assembled and the systems to be accessed. Of course, for a particular management operation it may not be necessary to access all fragments.

In order to allow network managers to control the resource, the MOS must distribute the control information to the appropriate object fragments.

Figure 4 shows the relationship between the managed object, its fragments and the OFD.

3.4. Fragments and Objects

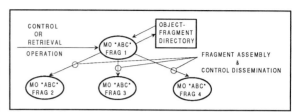

Figure 4. Fragment Assembly

It should be noted that fragments are objects and are manipulated as objects from the standpoint of the data administrator. However, from the viewpoint of management they are viewed as fragments of a larger object - i.e. the managed object. The reasons for adopting this approach is to make clear that there are some key differences between the requirements for the definition of managed objects and those for the definition of fragments.

First, and most important, managed objects can be instantiated by themselves. Fragments are meaningless except when instantiated as part of a managed object.

As a consequence of the above, the final behavior of a fragment type is determined by the object class of which the fragment is a part. In essence, this implies that when fragments are modeled as object classes, these classes themselves are not directly instantiable but require that the managed object as a whole be instantiated and that the appropriate behavior is assured. A managed object class may be viewed as a composition of fragment types with customized behavior added to make these fragments into a coherent whole. When the fragments are then again distributed, these fragments maintain their customized behavior.

In terms of ODP definitions, a managed object with fragments may be viewed as a set of distributed objects with specific interfaces and contracts defined among these objects to assure correct behavior.

Second, the reason for defining fragments as a concept separate from managed objects is to make it clear that these objects are subservient to the managed object view and to allow the definition of the generic relationship that exists between managed objects and fragments. The existence of fragments must be transparent to the management user. Therefore, the naming of fragments is hidden from the network management user; and for ease of maintaining the name space, a naming strategy derivative from the managed object name is employed.

4. Conclusion

This paper proposes an approach that allows separation of network management and database administration concerns in a distributed management environment. The resulting model allows the data distribution aspects of the MIB to remain invisible to the network manager user. The relationship between replicas and managed objects is shown in Figure 5 and the relationship between fragments and managed objects is shown in Figure 6.

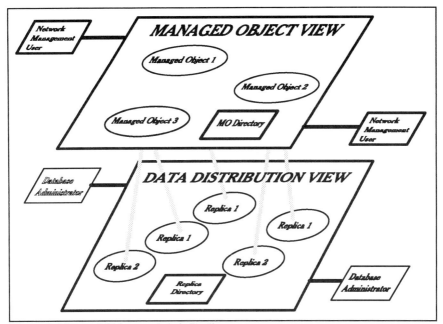

Figure 5. Managed Objects and their Replicas

Replicas must be as numerous as the need for near-instaneous information demands. A resource may be represented by managed object specifications that are found on a number of different open systems, and several schemes for shadowing the information contained in managed object replicas were discussed here. Satisfaction of the need for information nearness, however, does not meet the requirement within some management enterprises for very simple implementations. This need is met by object fragments shown in Figure 6.

Object fragments, like object replicas, are also associated with managed objects but represent only a portion of each managed object specification. Fragments are distributed on the basis of efficiency of implementation and local custom.

Needless to say, fragments may be used to represent replicated managed object specifications. Although complex from the standpoint of the database administrator, the principles of object fragmentation and object replication lend themselves to use in concert and together, provide a powerful arsenal of modeling tools.

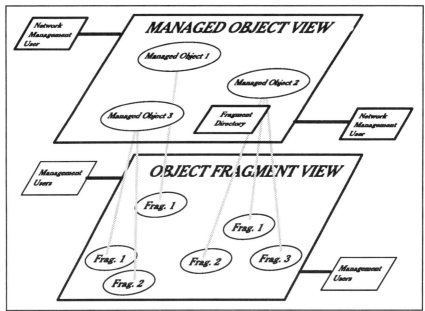

Figure 6. Managed Objects and Their Fragments

5. References

[1] ISO/IEC 9596 : 1990, *Information technology - Open Systems Interconnection - Management Information Service Definition - Common Management Information Protocol*

[2] ISO/IEC 10165-2, *Information technology - Open Systems Interconnection - Structure of Management Information - Part 2: Definition of Management Information*

[3] ISO/IEC 9595 : 1990, *Information technology - Open Systems Interconnection - Management Information Service Definition - Common Management Information Services*

[4] ISO/IEC JTC1/SC21 N4888, *Draft Basic Reference Model of Open Distributed Processing - Part 2: Descriptive Model*

IX

MANAGEMENT INFORMATION

A
Modeling Aspects

Design Concepts for a Global Network Management Database

Robert F. Valta
Leibniz-Rechenzentrum
Barer Str. 21, D-8000 München 2, Germany
Email: valta@lrz.lrz-muenchen.dbp.de

Abstract

One of the most important components within an integrated network management system is a conceptual network database that provides a repository for all management-related data. Design and implementation of such a database need careful considerations. First of all, requirements for such a database have to be investigated. These comprise the type of information to be stored in the database as well as the structure of the information. Considering the complexity of the database a model of a network must be developed that can be used as a starting point for the design of the database. The model proposed in this paper uses attributed layered graphs as a formal tool to describe a network. Our formal approach enables the modeling of the different layers of a protocol hierarchy. We demonstrate the use of such a database for the example of a configuration management application.

Keywords: Management Information Storage, Network Modeling, Attributed Graphs, Network Database, Management Domains

1 Introduction

During the last years, network management has become an important area in data communications. Manufacturers and user groups spent a great deal of effort to design management architectures suited to manage large heterogeneous networks. One of the most important goals is to overcome *component management systems* that allow management of components from a single vendor only. In a heterogeneous environment component management systems lead to a great variety of management tools that differ largely in functionality and usage. Besides, each component management system refers to its own network database, which includes configuration data, polling lists, network maps, component status etc.. Using these different systems and maintaining their network databases consistent requires a lot of manpower for the operational and administrative staff. Inconsistencies between the different network databases often cause communication problems, for example if configuration data were changed in an incorrect way, or increase time for fault correction if the network topology is not documented correctly. This variety of tools and databases impedes the handling of the huge amount of data that is generated or consumed by intelligent network equipment.

A solution to this problem is provided by integrated management systems where all management data are represented within a conceptual global network database. Instead of maintaining their individual network databases, all management tools access a single database (see for example DEC's EMA where a Management Information Repository is defined as a central component of the Director ([2])). By defining a vendor and protocol independent interface for accessing the database the network's heterogeneity can be hidden. Furthermore such a database is used to enable cooperation between tools from different functional areas ([9]).

A lot of work remains to be done to investigate the design of such a database systematically and in detail. The next section gives the requirements for such a design. In section 3 a layered attributed graph is proposed as a tool for modeling different layers of a communication network. Integration of the model into current standardization work is outlined. Section 4 gives an example how this model can be used by a configuration management application. The last section summmarizes implementation status and further research work.

2 Requirements for a Management Database

In the following chapter the requirements for a management database are developed. Most of the requirements arise from the view of a network administrator to such a database.

2.1 Relation to OSI MIB

The Management Information Base is defined within the OSI Management Framework as a conceptual data store for all OSI-relevant management data within a single open system ([6]). It has a tree-like structure and is based on a containment hierarchy for all objects within an open system.

Most configuration tools support a port-oriented configuration, where the configuration of a system's communication ports (e.g. its IP-ports, HDLC-ports) is done by one configuration task for each system. For an open system such a configuration task results in a management information tree.

To ensure correct communication and cooperation of systems that are configured in such a way an administrator must keep in mind all dependencies between the systems. These dependencies are strongly influenced by the position of the system within the whole network (e.g. the links it is connected to, its neighbours, whether it is a router or gateway, the routes it has to support).

As a consequence a network database must include object types which are not part of the containment hierarchy of a system but represent dependencies between different systems, e.g. communication links, subnets, distributed applications. Moreover, relationships between these objects and the resources within a MIB must be defined (e.g. a protocol-entity supports a link).

2.2 Application Programming Interface (API)

One of the main topics of network management research is the development of adequate interfaces for application programmers, administrators and operators. The interfaces that are defined as part of management standardization (CMISE- or SNMP-related services) are not well suited, because they are protocol-dependent and provide low functionality to a network administrator or an application programmer. Another approach is to derive an API from a protocol-independent information model. In [12], an SQL-based API is defined by representing the resources of a network as tables of a relational database. Some management architectures, for example HP Openview ([5]), provide a CMIS-like API that is based on the object-oriented OSI information model.

Current APIs only allow access to objects within one system. To create and maintain a network-wide information base more advanced services are necessary for adding new components, changing network topology, adding subnets, and so on. Such services may affect several resources in more than one system. It is important for a service user to understand the effects of a service call, which can be achieved by describing the semantics of that service.

2.3 Multi-Functional Usage

To prevent all network management tools from individually maintaining their private network databases, the central database should be accessible for tools from different functional areas. The central database must provide a kernel of management data that are of interest to all areas, e.g. topology data. For each functional area it must be possible to add specific data to this kernel. The following example for a packet-switching network gives an idea about the general and the area-specific information:
- Kernel data: switches (id,type), endsystems, links (protocol), connectivity.
- Configuration data: link parameters (speed, timers), system interfaces.
- Fault-relevant data: reliability of links and switches.

Maintenance and interpretation of area-specific data must be done by the area-specific tools, while kernel data should be maintained by a separate database management tool. Such a tool supports functions to
- check the consistency of network topology,
- provide a user-friendly interface,
- inform management tasks about changes in network topology,
- coordinate activities of tasks from different functional areas (e.g. a change of performance data could be of interest to configuration management for allocating optimal network routes).

2.4 Automation

The amount of data generated by network devices overwhelms the human network administrator. Therefore some automation of management activities is necessary. Besides employing techniques from the field of expert systems (e.g. rule-based systems for fault diagnosis) automation can be supported by representing a sufficient amount of knowledge about a network in the network management database. This knowledge comprises, for example, information about the relationships between distributed applications, underlying transport networks, and the physical network. Such relationships can be used as a diagnostic structure context for automized fault management or as a basis for performance tuning and capacity planning.

2.5 Support for Different Views of Abstraction

One of the fundamental structuring principles of today's networks is horizontal layering, providing different views of abstraction:
- geographical view (site of components, structured by countries, areas, cities, buildings, floors, rooms)
- physical view (components and media with their connectivity and physical characteristics)
- protocol view (protocol-specific attributes for LAN-segments, subnets, etc.)
- application view (client-server relationships)
- administrative view (structuring according to administrative responsibilities within the network, domains)

As these different views are not independent of each other, relationships between the views must be stored in the database. The idea of relating information from different layers and views was also stated by [8] especially in the field of configuration management.

3 Modeling of a Network

Important conclusions from the above requirements are:
- The design of a network database can not be done in an ad hoc fashion. First an adequate model for such a database must be carefully developed.
- It is essential to provide mechanisms that allow the description of networks for management purposes as a basis for automation and better understanding of management operations.

In the following, a model for describing a network is proposed. It is based on attributed graphs which provide a well-known and widely understood basis for our modeling. In addition, different graphs, each of them representing only a single layer of a network, are set into a formal layering relationship that leads to what is called a *Layered Attributed Graph (LAG)* as an appropriate formal basis for developing a network management database.

3.1 A Graph-based Modeling Tool

Graphs as an aid to model networks are used to
- develop routing algorithms,
- analyze communication processes,
- estimate the effects of equipment failure (isolation of network elements),
- optimize communication cost and reliability.

Using graphs is advantageous since graph theory provides a rich set of well-developed methods and algorithms that can all be applied to problems in the area of data communication and network management (connectivity, route optimization etc.).

One limitation arises from the fact that usage of graphs is always restricted to isolated aspects of network management, where only a single layer of a network is modeled. However, for our purpose we need to combine several graphs, where each graph models a single layer of a network. A first approach towards such a view was developed by [3], which seperates an information transport graph from an operations support graph.

A graph that represents a network must have the following characteristics:
- Nodes represent systems within our network. Systems include endsystems as well as systems that are used for communication purposes only (repeater, bridges, modems, gateways etc.).
- Edges represent relationships between systems. Different types of relationships are distinguished:
  - physical connectivity: an edge represents a communication media (a coax-cable, a fibre-optic cable etc.).
  - communication connectivity: an edge represents the ability of its adjacent nodes to communicate via a communication protocol (e.g. a CSMA/CD-edge, HDLC-edge, IP-edge).
  - cooperation connectivity: an edge represents the ability of its adjacent nodes to cooperate within a specific distributed application (e.g. directory-edge, file-server-edge, configuration-management-application-edge).
  - organizational connectivity (domains): an edge represents an organizational unit to which its adjacent nodes belong (e.g. a naming domain or a fault management domain).
- More than one relationship can exist between different systems. Therefore, a set of nodes can be connected through more than one edge.
- Relationships are not restricted to a pair of systems, but can exist between any number of nodes (hypergraph). In reality, the number of systems adjacent to one edge depends on the type of relationship an edge represents. Edges with no adjacent nodes are possible.
- Different edges with the same nodes adjacent to them are not independent of each other. The basic relationship between such edges is *layering*, which means that most edges assume the existence of one or more subordinate edges, e.g. an application edge relies on a transport edge, a transport edge on a network edge an so on. Thus meaningful protocol hierarchies can be modeled.

3.2 Definition of a Layered Attributed Graph

Now we are in a position to exactly define a *Layered Attributed Graph (LAG)*. Although the formal definitions might be a burdon on the reader, they are essential for implementing the model. In addition they provide means to determine the semantics of configuration services, which allows to automize configuration tasks. However, for a first reading the formal expressions can be omitted.

In a first step we define an attributed graph AG as an extension to what is commonly known as Hypergraph (see for example [1]).
An attributed graph AG is defined as a triple AG = (V,E,a) with
- V is a finite set of nodes
- E is a finite set of edges with $E \cap V = \emptyset$
- $a : V \times E \to \{True, False\}$ with

$$a(v,e) = \begin{cases} True & \text{if node v is incident with edge e} \\ False & \text{otherwise} \end{cases}$$

- A set of node attributes V_{att} and a set of edge attributes E_{att}. For each element $V \in V_{att}$ and $E \in E_{att}$ some function $proj_V$ and $proj_E$ respectively, supplies attribute values for individual nodes and edges. For each attribute, the allowed set of values is determined by a function *vset*.

The definition of an AG does not talk about dependencies between edges. The following definition of a Layered Attributed Graph (LAG) introduces the layering of edges and provides rules that take care that layering is done in a correct way.

A Layered Attributed Graph LAG = (V,E,a) is an AG with some additional properties:
- Two different types of nodes are distinguished: end systems V_E and transit systems V_T (e.g. gateways, routers) with $V = V_E \cup V_T$.
- A distinct edge attribute E_{prot} exists with set of values $E_p = vset(E_{prot})$ and projecting function *prot*: $E \to E_p$. E_p contains all the possible relationships between systems in the network to be modeled, e.g. E_p = { TCP, IP, LLC1, CSMA-CD, 10BASE5, 10BASE2}.
- For E_p there exists an irreflexive, asymmetric relation $R_{layer} \subset E_p \times E_p$. R_{layer} represents the permitted layering of relationships, e.g. a TCP/IP-protocol stack: $R_{layer} = \{$(TCP,IP),(IP,LLC1), (LLC1,CSMA-CD), (CSMA-CD,10BASE5), (CSMA-CD,10BASE2)$\}$.
- The transitive closure $T(R_{layer})$ is strictly ordered in E_p. Relative to this order, there is a non-empty set $E_{min} \subset E_p$ of minimal elements. The elements of E_{min} can represent different communication media, e.g. $E_{min} = \{$10BASE5, 10BASE2$\}$.
- For any edge with $prot(e) \in E_p \setminus E_{min}$ a sequence $(v_0, e_1, ..., e_n, v_n)$ is called a *route* $R^{(e)}$, if the following properties hold:
  - $a(v_{i-1}, e_i) \wedge a(v_i, e_i)$ for i = 1,...,n.
  - $e_i \neq e_j$, for $i \neq j, 1 \leq i, j \leq n$.
  - $v_i \neq v_j$, for $i \neq j, 0 \leq i, j \leq n$.
  - $(prot(e), prot(e_i)) \in R_{layer}$, for i = 1,...,n.

- $v_i \in V_T$, for i = 1,...,n-1.
- The following *Layering Rule* holds:
$\forall e \in E : prot(e) \in E_{min} \vee$
$\quad \forall v, w \in V : \quad a(v,e) \wedge a(w,e) \Rightarrow$
$\quad \quad \exists Route R^{(e)} = (v_0, e_1, ..., e_n, v_n) : v = v_0 \wedge w = v_n$

The *Layering Rule* states that for every non-minimal edge e with more than one node adjacent to it, all these nodes are connected through at least one route. The existence of such a route and all the nodes and edges which belong to the route is a prerequisite for edge e to exist. Thus layering of attribute values according to relation R_{layer} induces a layering of edges in the graph (see [4] for an example).

The layering of edges according to relation R_{layer} is based on the protocol layering as it is, for example, defined by standardization profiles (e.g. MAP/TOP) or communication architectures (e.g. TCP/IP-protocols, SNA, DECnet).

Overall, we have the following relationships between the elements of our graph:
- $R_{connected} \subset V \times E$:
  connected(v,e) means that node v can communicate or cooperate with all the other nodes connected to edge e.
- $R_{edge-map} \subset E \times E$:
  edge-map(e,f) means that edge e relies on edge f, e.g. an X.25PLP-edge on a HDLC/LAPB-edge.
- $R_{node-map} \subset E \times V$:
  node-map(e,v) means that edge e relies on node v, for example when e is an X.25PLP-edge, v could be an X.25-switch that must be configured in such a way that connectivity is provided for all the end systems connected to e.

As a next step, we can define the basic services that allow us to construct and manipulate a LAG representing a network. These services include the
- creation and removal of nodes and edges.
- creation and removal of the above relationships.
- information about existing objects and their relationships.

3.3 Extended Model

Our definitions provide only a basic framework for the design of a management database. Only two elements (nodes and edges) are defined so far, which is of course not enough to model a network in sufficient detail. One important extension is the inclusion of object-oriented techniques quite similar to those adopted by standardization organizations for modeling the resources of an open system.

The following steps are required to integrate object-oriented techniques to the LAG (see [11] for details):
- NODE and EDGE are defined as object classes and the relationships are defined as generic relationship-types, representing the possibility to create a relationship between two objects of certain object classes. Generic relationships are not defined as

object classes because such basic concepts as naming and state management are not applicable.
- Relationships as part of the LAG model can be mapped on OSI relationship attributes as defined in [7]. Creating or deleting a GAG-relationship would result in changing values of two objects' OSI relationship attributes.
- Refinement of the object-classes EDGE and NODE is done in an attribute-driven way. For EDGE, the main attribute is *prot* which describes the semantics of an edge. Refinement is done by defining new object classes for all possible values of *prot* (e.g. (CSMA/CD)-EDGE, (IP)-EDGE, (Directory)-EDGE, (Fiberoptics)-EDGE, ... are all subclasses of EDGE) with new specific attributes, for example protocol-timers, association-context.

  Similar refinement can be done for object class NODE with an attribute *realtype* that allows integration of vendor-specific properties (e.g. DEC/VMS-NODE, BSD/UNIX-NODE).
- Rules are defined as part of an object class definition to formalize dependencies between attributes.
- New object classes must be defined and integrated. This includes, for example, ENTITY and APPLICATION-PROCESS, which can be refined in the same way as edges (((CSMA/CD)-ENTITY, (DUA)-PROCESS, ...). Integration into the graph model is accomplished by new relationship-types, for example CONTAINS(NODE,ENTITY), CONNECTED(ENTITY,EDGE), MAP-ENTITY(ENTITY,ENTITY) with rules that describe how classes and relationship-types fit together.

  Example: Before a node n can be connected to a (prot)-edge e a (prot)-entity en must be created with the relations $contains(n, en)$ and $connected(en, e)$.
- For all relationship-types rules must be defined that must be considered when creating a relationship. The *Layering Rule* in the definition of the LAG provides only one basic rule for relationship-type CONNECTED(NODE,EDGE). Examples for rules are:
  - Line speed is an attribute of an edge representing a physical connection (e.g. a V.35-line). When attaching a system to such an edge, the current value of the line's speed attribute must be used to configure the system's line port: CONNECTED(V.35-EDGE,NODE) has the rule: `edge.speed = node.speed`
  - Rules could be defined for a relationship MAP-EDGE(EDGE,EDGE) to represent dependencies between layered edges. For example a communication timer of an upper edge could be adjusted to an optimal value depending on the type of edge which the upper edge relies on.
  - For all nodes connected to one edge (e.g. an IP-edge) the communication addresses of the nodes must be unique (attribute dependencies).
  - All physical links which a Token-Ring-edge relies on must form a ring (topological dependencies).

The original LAG together with its extensions forms a complex model. The usage of the LAG as a basis provides a means to structure and formalize all static information about the resources of a network.

4 Management Applications and Domains

In this chapter we will illustrate how distributed management applications can be described using the LAG. The edges that represent the application in our graph are quite similar to what is commonly known as a *management domain*. This allows us to formalize domain concepts, to store information about domains, and to manipulate them with the services defined in the previous chapter.

From the various definitions for the term *domain* (see [10] for a comprehensive discussion of domain concepts) the following one is most suitable in our context: A domain is "that set of objects that are directly or indirectly managed by a single management center". As there are different management functional areas, different management domains can be distinguished, e.g. a fault management domain or an accounting management domain. In every domain, we have at least one manager that has some responsibilities towards the other systems of that domain. To meet these obligations the manager provides area-specific services, e.g. within a fault management domain, where it is responsible for fault detection and isolation, it collects and analyzes error reports. If the tasks of a manager within a domain are distributed among several systems, subdomains and submanagers must be introduced.

There are two reasons to include domains in our model:
- The role a system plays within a domain influences the configuration of the system (e.g. communication resources for error reporting, configuration of report discriminators, application software for storing and filtering events). Since our database serves as a basis for system configuration, all information that is necessary to configure application processes should be stored in the database.
- Domains themselves are objects within a network and should be subject to manipulation by a network administrator via a well-defined interface.

The following example shows, how domains can be described in terms of the LAG.

One important function within the area of configuration management is the genereration and distribution of networking software. The whole process can be separated in several subtasks, which
- store information about the configuration of a network e.g. topological information, routes. This is done within a LAG-based configuration database.
- generate software for particular systems.
- store versions of generated software for systems.
- transport software to the system it was produced for.
- load software into a system's memory and activate it.

In our example, the whole task is divided in two subtasks:
- System A stores the database, generates software files and sends them to submanagers.
- A submanager S stores software files for agent systems and loads them into the memory of agents when necessary (e.g. for a system reset or after a configuration change).

The whole domain "SW-Administration" is structured into one subdomain "SW-Generation" and several subdomains "DownLineLoading". To describe this configuration within a LAG, three types of edges are necessary:
  SW-ADM-EDGE, SW-GEN-EDGE, SW-LOAD-EDGE
with the relationship-types
  MAP-EDGE(SW-ADM-EDGE,SW-GEN-EDGE) and
  MAP-EDGE(SW-ADM-EDGE,SW-LOAD-EDGE).
The edges that represent domains are mapped onto edges representing communication-relationships, e.g. on FILE-TRANSFER-EDGE and LOAD-PROTOCOL-EDGE.

Corresponding to a system's role within a domain, different types of management processes are installed:

- GEN-PROCESS that stores the network's configuration database and generates software.
- STORE-PROCESS that stores software files and initiates downline loading.
- LOAD-PROCESS that accepts downloaded software.

Besides using this information for the configuration of management processes, a process of type GEN-PROCESS can access the information to control the task of SW-administration. Assuming that a configuration change in system Z makes it necessary to load new software into the system, the following steps must be executed:

- Process *startp* (of type GEN-PROCESS) in node A accesses the database and determines to which subdomain system Z belongs (search for an edge of type "SW-LOAD" that Z is connected to).
- As an edge *load1* was found, A determines the responsible manager within domain *load1* (search for a node that contains a process of type STORE-PROCESS and that is connected to edge *load1*) and finds node S.
- A generates new software by reading configuration data about Z from the database and sends it to S.

Thus by accessing domain information in the database, the managing system A is independent of configuration changes. When new systems are added to a domain or when the domain structure is changed, no reconfiguration is required for the management software in system A.

5 Conclusion and Status of Work

The *Layered Attributed Graph* defined in this paper provides a flexible and powerful mechanism to model a network as a basis for the design of a management database. Besides modeling single systems, as it is done for example by OSI (MIB-tree), it enables the description of the entire network. Formal rules represent dependencies between resources of different systems and layers and provide a basis for the automation of management actions. The LAG is independent of any specific functional area and can be used by all management applications thus avoiding redundant data storage by the applications themselves. Additional information specific to functional areas can be added by refining existing object classes.

A prototype of the LAG has been implemented using PROLOG to implement the rules and a relational database system to store the information. With the help of a menu-driven interface, one can construct a network with consistent configuration information for the individual systems. This information can be used to generate correct configuration software for network components. The current prototype is still restricted to a small number of object classes and the equipment from one specific vendor only.

Further work includes:

- use of the LAG as a base for an automized diagnosis system:
  When communication errors are detected by users in the application layer, iterative testing of the underlying edge stack and transit systems could result in isolating the problem.
- improvements to the prototype:
  The modeling of a network described above is conceptual and nothing is stated about the implementation of the database. To verify the usefulness of the concepts and to gain more experience, an implementation for a larger heterogeneous computer network, the network of the Leibniz-Rechenzentrum, is planned. This poses the problem of how to use different database tools for adequately storing the management information (ORACLE SQL-Database, CAD-database for cable management, object-oriented database as part of HP Openview with a CMIS-like interface, QUIPU X.500-Dircetory to store organizational data, file systems for storage of large configuration data).
- investigation of dynamic concepts:
  The above design concepts were mainly developed for storing *static* network management information. However, they provide a good basis for the systematic investigation of the dynamic aspects of management information (network events, network versions). Currently research is performed to develop a uniform method for handling different time aspects of network management information (short-/medium-/long-term). One idea is to add a time dimension to the three-dimensional LAG, which would lead to a new four-dimensional structure.

6 Acknowledgement

This paper presents the results of research conducted within a research project at the Technische Universität München that is sponsored by Siemens AG, München. I would like to thank H.-G. Hegering, the principal investigator of the project, and all other members for valuable discussions and improvements.

References

[1] Claude Berge, *Graphs and Hypergraphs*, North-Holland, 1976.

[2] Digital Equipment Coporation, *Enterprise Management Architecture - General Description*, EK-DEMAR-GD-001, 1989.

[3] D. E. Diller and R. F. Merski, "An Architecture for Providing Universal Access to Network Equipment from Distributed Support Centers", In: *Proc. of the IEEE INFOCOM '87*, p. 245 – 250, San Francisco, Calif., March 1987, IEEE Computer Society.

[4] Heinz-Gerd Hegering and Robert Valta, "Describing an OSI Network Configuration - Problems and Possible Solutions", In: *Proc. of the ICCC '88*, p. 288 – 293, Tel Aviv, Israel, November 1988, North-Holland.

[5] Hewlett Packard, *HP Openview Network Management Server - Programmer's Guide*, 36989-90001, October 1990.

[6] ISO/IEC, *Information Processing Systems - Open Systems Interconnection - Structure of Management Information - Part 1: Management Information Model*, DP 10165-1, September 1990.

[7] ISO/IEC-CCITT, *Information Processing Systems - Open Systems Interconnection - Systems Management - Part 3: Relationship Management Function*, DP 10164-3 (X.732), May 1990.

[8] Yuko Murayama, Graham Knight and James Malcolm, "The Scope of Network Configuration Management", In: I. N. Dallas and E. B. Spratt (eds.), *Issues in LAN Management, Proc. of the IFIP TC6/W6.4A Workshop on LAN Management*, p. 131 – 146, July 1987, North-Holland.

[9] Dejan Sirovica and James A. Malcolm, "Recnik: A Virtual Database for the Coordination of Distributed Network Management Services", In: I. N. Dallas and E. B. Spratt (eds.), *Issues in LAN Management, Proc. of the IFIP TC6/W6.4A Workshop on LAN Management*, p. 161 – 172, July 1987, North-Holland.

[10] Morris S. Sloman and Jonathan D. Moffett, "Domain Management for Distributed Systems", In: *First International Symposium on Integrated Network Management*, p. 505 – 517, Boston, Massachusetts, May 1989, North-Holland.

[11] Robert Valta, *Entwicklung einer Methodik zur Beschreibung von offenen Rechnernetzen als Grundlage für integriertes betreiberorientiertes Netzmanagement*, Dissertation (Ph.D. Thesis), Technische Universität München, 1990, in German.

[12] Unnikrishnan S. Warrier and Carl A. Sunshine, "A Platform for Heterogeneous Interconnection Network Management", In: *First International Symposium on Integrated Network Management*, p. 13 – 24, Boston, Massachusetts, May 1989, North-Holland.

Composite Managed Objects

Andreas Dittrich

GMD FOKUS, Hardenbergplatz 2, D - 1000 Berlin 12, Fed. Republic of Germany
E-mail: dittrich@fokus.berlin.gmd.dbp.de

1. Introduction

Within the framework of the BERKOM-Project (BERliner KOMmunikationssystem) [1], where the German Bundespost and the State of Berlin are cooperating in a study to develop applications and end systems for a future broadband fibre-optic network (B-ISDN), the BERMAN sub-project is concerned with the management of distributed applications within B-ISDN. In addition to the conceptual work, BERMAN is realizing a *Basic Management Support System* (BMSS) [2], which permits the interchange of management information and the invocation of management operations on application components which exist locally or remotely.

The BMSS is based on OSI Management thus enabling cooperation with other management systems which support this standard. The OSI Management Information Model [3] describes the characteristics of a resource at a very abstract level, by means of a *managed object* (MO). However, there is no consideration within the model for the relationship between the logical managed object and the concrete realization of a resource. The concept of *composite managed objects* (CMO) attempts to fill this void in that managed objects are constructed in such a way that existing relationships between managed objects and resources can be mapped as relationships between the parts created during MO decomposition.

The basic ideas of the CMO concept are discussed in section 2 which assumes that the reader is familiar with the OSI Management Information Model. Section 3 shows the benefits of composite managed objects using trading as an example.

2. The concept of Composite Managed Objects

An important part of OSI Management is a dedicated information model whose basic building blocks are *managed objects* (MO). This concept has been refined within BERMAN to include the concept of *composite managed objects* (CMO), in order to meet the requirements of a real implementation. First the reasoning leading to the development of the CMO concept is described. Following is a discussion of the elements of the CMO concept, whereby knowledge of the OSI Management Information Model is assumed. Finally, the relationship between the parts are viewed within the constraints of a possible realization.

2.1. Relationships between MOs and the associated Resources

Because the OSI Management Information Model does not support the modelling of the possible relationships between managed objects and associated resources, it is possible to describe managed objects at an abstract level, so that a unified logical view of one resource is possible independent of the implementation aspects. This simplifies the definition of MO classes.

The implementer of an MO, on the other hand, is also interested in the relationship between an MO and its associated resources, because of the effects these might have upon the realization design. Because modern operating system architecture is based upon the concept of processes, it is assumed below that logical resources are represented by one or more processes, or that access to physical resources is only possible via a corresponding process. The primary difference between the view of the specifier of an MO class and the view of the implementer of an MO, is that the former views a resource only in the form of an MO, while the latter views real processes, and must consider which relationships exist between the MO and the associated processes.

This problem becomes evident when regarding standardisation activities. In order to support management in a truly distributed and heterogeneous environment, managed objects must be defined in an implementation-independent manner which is usually done in standards by defining abstract entities and assigning one or more MOs to them. However, within an implementation these entities can be realized by a set of processes, so that the information represented within the attributes of an MO can be distributed among different processes. The MO implementer requires guidelines in order to decide which attributes and notifications should be included within a process, and when an MO is split, how the fiction of an atomic MO, as required in the standards, can be maintained.

In addition to this *distribution* aspect, other relationships can be identified between MOs and associated processes. In some cases it is not feasible to couple the *lifetime* of an MO, or a specific element of an MO, to the lifetime of the application process upon which it is based. This requirement can be important, for example, when an error leads to the termination of a process. Within the framework of fault management, however, it is certainly necessary that after the occurrence of an error, information stored within an MO, which can provide an answer as to the cause, should be accessible.

In addition to the case where parts of an MO may exist beyond the lifetime of the process, it is also possible that parts of an MO should exist before the corresponding process is created. This means that the initiation of a *Create* Operation need not automatically include both the generation of an MO and the creation of an associated process, but rather that the process may be started later as needed, for example as the implicit result of the execution of a special *Action* operation upon an MO. Therefore, for reasons of optimization, a process cannot always be available, but due to the existence of the corresponding MO, it is possible to recognize that the process (and the functionality it provides) is in principle available. Also in this case, information concerning the resources may be provided to a management system.

The aspect of lifetime implies that parts of an MO, which are only loosely coupled to the existence of a process, need not actually be part of the process, but can be stored separately. On the other hand, there is the aspect of the *dynamic alteration* of attributes. Attributes with quickly changing values are not suitable for "separation" from the processes which cause the alteration, because implicitly a copy of the attribute is created which resides outside the process, resulting in an unacceptably complicated updating process.

2.2. Composite Managed Objects

Presented below is a description of how the above-mentioned relationships between Managed Objects and the associated processes are modelled by means of the *composite* MO (CMO) concept. In order to maintain the object-oriented approach for the mapping of the relationships between MOs and processes, which is pursued in OSI with the concept of managed objects, it is also necessary to model processes by means of objects.

Because processes may be considered to be the realization of resources, they are modelled as *resource objects* (RO). A process should be represented by precisely one RO. An m-to-n relationship may then exist between MOs and ROs: On the one hand, an MO may be associated with several ROs (=processes); on the other hand, a process can enable access to several (abstract) resources, so that several MOs may be associated with it. The latter would be the case, for example, if the functionality of the four lower layers of the OSI Reference Model were implemented by means of one process. Assigned in this case to the RO would be the MOs of the four layer entities, together with additional MOs, representing the connections on the various layers.

The basic idea behind the CMO concept is to divide what within the OSI Management Information Model is considered to be an atomic managed object, into distinct *parts*, so that the modelling of the relationships, discussed above, between MOs and resources can be derived from the *composite* relationships between these parts. A part, like an MO, is defined by attributes, notifications and action operations. The level of atomicity is shifted within the CMO concept from the entire MO to the characteristic elements of an MO. This means that each attribute, notification and action operation defined for an MO is assigned to exactly one part.

Two distinct types of parts exist: *Master Parts* and *Subaltern Parts*. Every CMO must contain a Master Part which provides access for the management application to the CMO. This means that all the attributes, operations and notifications defined for the MO, are available at the interface to the Master Part, so that the fiction of an atomic MO is preserved for the management application.

Every RO, which is associated with the original MO, may contain exactly one Subaltern Part of this MO. A Subaltern Part possesses the functionality required to ensure access to the attributes and to process actions it contains, and to signal the occurrence of internal events.

Internally, the Master Part knows about the assignment of each attribute to a single Subaltern Part, so that it can forward the corresponding read and write operations for these attributes. The Master Part also recognizes the Subaltern Part in which the required functionality for the execution of an *Action* Operation is to be found. During execution of a *Create* operation, the Master Part is created first; it can then initiate the generation of the Subaltern Parts. The *Delete* operations are also processed by the Master Part, which first ensures the deletion of the Subaltern Parts, before it terminates itself. The Master Part is informed by the Subaltern Parts about the occurrence of internal events, and it sends the generated notifications to the local management system, which then makes a decision concerning additional processing (for example, forwarding the notification to a special management application).

The distribution problem mentioned in section 2.1 can now be solved by separating an MO into parts associated with the existing resource objects and by adequately assigning attributes, notifications and actions to these parts. Adequately means considering that the required lifetime of attributes coincide with the existence of the RO that the part is contained in. In addition attributes which have a highly dynamic nature should be assigned to those ROs (or their parts) which supply the attribute's value.

2.3. Managed Object Repositories

With the current CMO concept, only 1-to-n relationships between MOs and ROs can be mapped. It is still impossible to model m-to-1 relationships between MOs and ROs. This should be supported by the implementation of *Managed Object Repositories* (MOR). By combining the CMO and MOR concepts, any m-to-n relationship between MOs and ROs can be mapped.

A MOR contains a set of MOs and administers the containment relationship between these MOs. In other words, the set of all MORs within a system is the realization of the MIB concept. A MOR also provides operations for manipulating the MOs stored within it. The functionality of these operations is similar to the elements of the *Common Management Information Service* (CMIS) [4]. This means there are operations:
- for the reading and altering of attributes,
- for the creation and deleting of MOs
- for initiating actions
- for sending notifications.

A MOR supports the selection of specified MOs by means of *scoping* and *filtering*, and ensures the mapping of its operations upon corresponding MO operations. It can control the access to MOs and ensure the synchronization of operations upon several MOs. Except for the handling of communications with a management application, a MOR provides the functionality of an agent of a managed open system as described in OSI Management.

Two types of MORs are distinguished: *Systems Managed Object Repositories* (SMOR) and *Application-integrated Managed Object Repositories* (AMOR). A SMOR is a special component of the management system and exists independently of the availability of any RO. It can contain those parts of an MO, especially the Master Part, which should be available before or after the existence of a RO.

In contrast, an AMOR is integrated into a process. Both MOR types are able to store not only parts of a CMO, but also whole MOs. Such MOs, when stored in the SMOR, are called *Systems MOs*. MOs, which are completely integrated within exactly one AMOR, are called *Application-integrated MOs*.

For reasons of simplification, within the BERMAN project, the assumption has been made that within one system, exactly one SMOR, but any number of AMORs can exist. This SMOR serves as the central access point to the MOs of a system. In order to avoid consistency problems no Subaltern Parts but only Master Parts can be stored in the SMOR (see Figure 1). If the MOs or parts of MOs, which are addressed by a management operation, are not available within the SMOR itself, then the SMOR forwards the operations to the corresponding AMOR. This requires that every MOR has been assigned a unique name within the scope of a single end system.

These restrictions affect which types of MOs are subordinate in respect to the containment relationship. The roots of the containment tree are formed by the Systems MOs in the SMOR. Subordinate to the Systems MOs can be other Systems MOs or Master Parts of CMOs. The assignment of an AMOR to the SMOR occurs either by means of at least one CMO, i.e. by the composite relationship between the Master Part in the SMOR and the Subaltern Part in the AMOR, or by the containment relationship between a Systems MO and an Application-integrated MO.

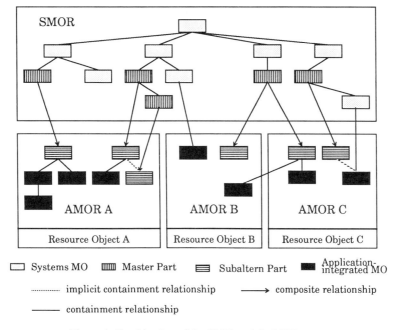

Figure 1: Combination of the CMO and the MOR concepts

3. Trading

The importance of trading has been widely recognized so that trading has become a topic of international standardization activities, e.g. within the framework of ODP [5]. Within BERKOM the sub-project BERCIM [6] is responsible for providing an infrastructure [7] which supports trading and which is based on Directory and management services [8]. Below, a possible scenario for trading, which is being developed by BERCIM and BERMAN in cooperation, is briefly described. It will then be shown how the CMO concept can be used within this context.

3.1. Scenario

The trading scenario is based upon the assumption that a set of *servers* offer their services via a coupling of local networks within a broadband network serving as a backbone. A *service user*, which may be a human or an application, then faces the choice of determining which server is most suitable to provide the desired service. Because this is a very complicated task, requiring extensive knowledge concerning the existence, location and characteristics of specific services, it is advantageous to provide the service user with support via a special application, called a *trader*, permitting the network administration to influence the use of available services by modifying the selection and operation criteria of the servers. This allows work to be more evenly distributed among the individual systems, and generally increases the performance of the entire system.

The selection of a *server* is based upon *service qualities*. Differentiated here are the *static* service qualities, whose value generally does not change as long as the service is available, and *dynamic* service qualities, which can fluctuate frequently over the lifetime of a server. For a printer service, for example, the static characteristics might be the resolution, the supported paper formats, the definition of graphics and colour capabilities. Dynamic service qualities might be the current status of the printer service (i.e. pending, processing, out of paper, etc.), and the number of jobs which still must be processed.

Static service qualities are suitable for storage in the Directory [9], while the dynamic service qualities are best provided by a management system. It is beyond the scope of this paper to discuss the data which must be stored in the Directory in order for the trader to fulfil its duties. For more information, please refer to [10]. Discussed below are only the steps which are related to management:

1. By issuing a Directory query the trader will receive a list of servers which meet the static service qualities. From each of these servers the trader requests information concerning its dynamic service qualities via the management system.

2. Based upon these answers, the trader can determine which of these servers can fulfil the specified dynamic service qualities. If several servers come into question, the trader makes a decision based upon an internal selection strategy. Then the trader, aided by the management system, concludes a contract with the selected server which for a specified period of time precludes a deterioration of the dynamic service qualities and reserves the necessary resources.

The trader provides the name of the selected server to the service user. The service user then establishes a connection with the server and transmits his job.

3.2. Management Aspects of trading

In order that dynamic service qualities of a server can be requested via the management system, the server must define a *Server* MO, in which the dynamic service qualities and other information are represented in the form of attributes. Additionally, it is assumed that for each job which the server is currently processing, or holding for processing, a *Job* MO exists, which represents a manageable active or pending job and which is a subordinate of the server MO.

A generic MO class *Server* could contain the following attributes:

- *serverId*: This is the naming attribute of the server MO. For this reason, only the operation *Get attribute value* is permitted.
- *operationalState*: This attribute is also read-only. It can possess the values *Enabled* (the service is available, but is not currently processing any jobs), *Active* (the service is currently processing one or more jobs, and has capacity available for additional jobs), *Busy* (the service can not process any additional jobs) and *Disabled* (due to faults, the service is not functioning).
- *administrativeState*: Permissible values for this attribute include *Unlocked* (the service may be used), *ShuttingDown* (the service is processing only already accepted jobs, and is not accepting any new jobs) and *Locked* (the service has completed all jobs and is not accepting any new jobs). This attribute may be read by management applications or manipulated using the *Replace attribute value* operation.
- *serviceRequestsPending*: This attribute specifies how many jobs the service has queued for processing. It is a read-only attribute.
- *serviceRequestOrder*: This *set-valued* attribute contains the names of the MOs representing jobs still to be processed by the service. The sequential order in which they are stored determines the order in which the jobs will be processed. The following operations may be executed upon these attributes: *get attribute value*, *replace attribute value*, *add member* and *remove member*.

Also defined for the generic server MO is the Action operation *AcceptJob*, with which the trader can query the server as to whether he will accept a specified job for processing.

A generic MO class *Job* shall, for reasons of simplicity, possess only the following three attributes, which are read-only for the management system:

- *jobId*: This is the naming attribute for the MO.
- *jobState*: This attribute shows whether the corresponding job is presently being processed (*Active*) or being held for processing (*Pending*).
- *serviceUser*: This attribute identifies the service user for this job.

Additionally, both object classes inherit attributes from *top*, the highest object class in respect to the inheritance hierarchy. Of these attributes, in this example, only the attribute *objectClass* is considered. In respect to the containment relationship, the server MOs should be subordinated to an object of the class *ManagedSystem*, directly below the root of the containment hierarchy.

Service-specific jobs have to be derived from the two generic object classes *ServiceProvider* and *Job*, in which additional dynamic service qualities may be defined. In the following example (see Figure 2), it is assumed that for a printing service, the object classes *PrintServer* and *PrintJob* are defined in this way, but only the above-mentioned attributes are considered in this discussion.

In order to determine which MOs should be realized as composite MOs, implementation aspects must be considered. In this case, a constraint is that a process which realizes the printing service shall only exist if there is actually a job to be processed. This means, however, that it must be possible to query the dynamic service qualities (status, number of queued jobs), without the existence of a printing process. As a result, MOs of the class *PrintServer* should be realized as CMOs,

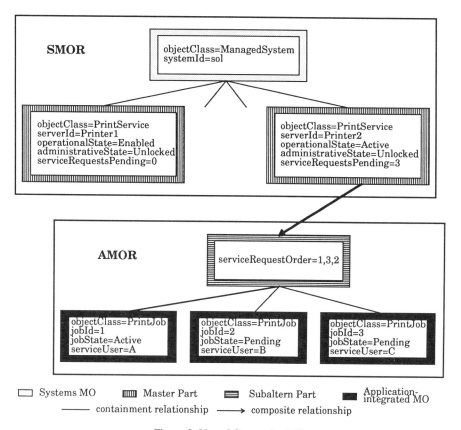

Figure 2: Use of Composite MOs

whereby the attributes *operationalState, administrativeState* and *serviceRequestsPending* must be stored in the Master Part, because this part of a CMO is independent of the existence of an associated application process.

In order that the CMO can be addressed at all by a management application, the attributes *objectClass* and the naming attribute *serverId* must be stored in the Master Part. Because frequent alteration cannot be expected for these attributes, the aspect of alteration rate should not prohibit storage in the Master Part.

The attribute *serviceRequestOrder* is, on the other hand, only of interest if jobs are queued and a management application wishes to manipulate the sequential order of the processing. Because it is assumed that a printing process exists if a printing job has been accepted, and that therefore an AMOR also exists within the printing process, this attribute can be stored in a Subaltern Part. This also implies that by default an empty list can be returned by the Master Part if this attribute shall be read while no printing process is available.

For MOs of the class *PrintJob*, there are no constraints in respect to implementation. For this reason, they can be realized completely within the AMOR of the printing process as application-integrated MOs. If it is desired that, in the event of an error leading to the disappearance of the printing service, another printing service takes over the job, it is recommended that these MOs should be realized as Systems MOs, which in respect to the containment relationship, should be subordinate to the Master Part of the *PrintServer* MO. In this way, the information concerning the job would be preserved, independent of the existence of a printing process.

If a printing service is installed, an entry must be made in the Directory, in order to make the server known, and in the system, where it provides its service, an MO must be created. Because during installation of a service, there cannot yet be any jobs, and therefore no printing processes need yet exist, only the Master Part of the CMO needs to be created.

Within the framework of the example presented here, a service user queries the trader for a printing service in order to print a PostScript file. Via the Directory, the trader learns that two servers can process PostScript documents, and that these are represented by the MOs {*systemId=sol; serverId=Printer1*} and {*systemId=sol; serverId=Printer2*} (see Figure 2).

By means of a query via the management system, the trader can ascertain the dynamic service qualities of the two servers. Because the printing job should be executed as quickly as possible, the server is sought with the fewest number of jobs for processing, whereby an upper limit is set for the value of the attribute *serviceRequestsPending*. For a more precise method of determining the workload of a printing service, additional information must be available within the MOs such as total size of all jobs.

Because it is not sensible to consider printing servers which are not operational, the servers are queried as to whether their *operationalState* possesses the value *Active* or *Enabled*, and whether the *administrativeState* is *Unlocked*. Using the scoping and filtering functionality of CMIS, the following query may be made:

M-Get (BaseManagedObject: objectClass=ManagedSystem; systemId=sol,
    Scope: *first level subordinates,*
    Filter: *objectClass=PrintService &*
                    administrativeState=Unlocked &
                    (operationalState=Active | operationalState=Enabled) &
                    serviceRequestsPending < =2,
    AttributeIdList: serviceRequestsPending)

As a result of this query, the trader receives via the management system (see Figure 2):

    ManagedObject: *objectClass=PrintService; serverId=Printer1,*
    AttributeList: *serviceRequestsPending=0*

Should more than one suitable server be found, the trader selects one based upon the values of the *serviceRequestsPendings* attributes. Thereafter the trader requests, by means of the *Action* operation *AcceptJob* defined in the MO, whether the values transmitted for the dynamic service qualities are still valid, and whether the server is willing to take on an additional job.

The *AcceptJob* operation is forwarded from the management system to the Master Part. This determines whether a printing process exists. If not, the Master Part initiates the generation of the printing process, whereby implicitly the generation of the corresponding AMOR is initiated, and this action is forwarded to the AMOR. The printing process then checks, based upon the job parameters provided (e.g. size of the job), whether it can take on the job, and provides its decision as a result of the *Action* operation. If it accepts the job, it reserves the required resources, which in the case of a printing service could be performed by placing an entry in the printer queue.

The trader provides the Directory name of the selected service to the client, so that the client can transmit its job to the server. If the server accepts the job, it will generate a corresponding *PrintJob* MO. This will be deleted when the job is completed. If there are no more jobs to be processed, the printing process is terminated, and only the Master Part is retained in the SMOR. The Master Part will only be deleted when the printing service is taken out of service completely.

4. Summary

The CMO concept is an approach aimed at creating the missing correspondence between the OSI Management Information Model and the requirements of a real implementation. Composite managed objects are generated by the decomposition of an MO into a Master Part and any number of Subaltern Parts. CMOs and "ordinary" managed objects can be combined into a Managed Object Repository which supports their handling and which defines an interface to a managed application.

The modelling of the relationships between the MOs and the represented resources is thus simplified since the following dependencies between MOs and resourcescan be considered:

- *Distribution* of the MO functionality over several resources,
- different *lifetime* for MOs and resources,
- different *alteration rate* of attributes.

It must be pointed out that the application of the CMO concept upon a defined MO cannot occur independently of implementation issues. On the other hand, the CMO concept provides guidelines for the development of new applications, such as how applications can be implemented to provide efficient access for the management of distributed applications.

The CMO concept is not fully developed in all of its details. In respect to a realization of the CMO concept, important topics within the BERMAN project are the definition of interfaces for a MOR and the two part types and how the interaction both among themselves and with the application process can be supported. However, the MO implementer remains responsible for the coupling between a part and the associated RO which must be solved individually, i.e. only the cooperation between parts of a *composite* MO can be supported by a management system.

5. References

[1] - Popescu-Zeletin, R. et. al., "A Global Architecture for Broadband Communication Systems: The BERKOM Approach", *Proceedings of the Workshop on Future Trends of Distributed Computing Systems in the 1990's*, pp. 366-373, IEEE, Hong Kong, September 1988.

[2] - "Guide to the Basic Management Support System (BMSS), Version 1.0", BERMAN Project Deliverable 3, Berlin, December 1990.

[3] - *Information Processing Systems - Open Systems Interconnection - Management Information Services - Structure of Management Information - Part 1: Management Information Model*, ISO/IEC DIS 10165-1, June 1990.

[4] - *Information Processing Systems - Open Systems Interconnection - Common Management Information Service Definition*, ISO 9595, January 1990.

[5] - *Working Document on Topic 4.3 - Function and Interface Definitions*, ISO/IEC JTC1/SC21/WG7 N4885, July 1990.

[6] - "Spezifikation für eine verteilte, fehlertolerante CIM-Struktur im Berliner Kommunikationssystem (BERKOM). Bericht zum fünften Meilenstein. Band 1: Übersicht", BERCIM Project Deliverable 5, Berlin, September 1990.

[7] - Tschammer, V., Wolitz, A., Hall, J., "Support for cooperation and coherence in an open services environment", *Proceedings of the 2nd Workshop on the Future Trends of Distributed Computing in the 1990's*, Cairo, Egypt, September 1990.

[8] - Hall, J., Tschichholz, M., "Management in a Heterogeneous Broadband Environment", *Proceedings of the IFIP TC6 / WG 6.6 2nd International Symposium on Integrated Network Management*, Washington, D.C., USA, April 1991.

[9] - *The Directory - Overview of Concepts, Models and Services*, CCITT Recommendation X.500, December 1988.

[10]- "Guide to the BERKOM Directory System, Version 1.0", BERMAN Project Deliverable 3, Berlin, December 1990.

A Model for Object Relationship Management

K. Klemba[a] and M. Kosarchyn[b]

[a]Information Networks Group, Hewlett-Packard, 19420 Homestead Rd., Cupertino, CA 95014

[b]Information Architecture Group, Hewlett-Packard, 19046 Pruneridge Ave., Cupertino, CA 95014

ABSTRACT
Object relationships represent associations and constraints among resources that are relevant to the management of the resources. We propose a model for managing object relationships that has the following key characteristics: supports aggregation of heterogeneous objects, models the relationship itself as an object, enforces separation of relationship semantics from relationship structure, and uses a directed graph as a common structural representation for all relationships.

KEYWORDS   Relationship, Management, Model, Graph, Object

1 Background

ISO's international standard for Open Systems Interconnection uses a model for management information that is object-oriented[1]. Any resource that is to be managed is modeled as a Managed Object, more precisely, as an instance of a class of Managed Object. Managed Object classes are formally defined using standard Templates. Consequently, any management architecture which is based on the OSI model must support this Managed Object paradigm.

While there are a number of interesting issues surrounding Managed Objects this paper focuses on the issue of managing relationships among Managed Objects. It proposes a model for managing relationships which is itself object-oriented and is consistent with the ISO management information model.

1.1 What are Relationships Among Managed Objects?

Relationships are aggregations of objects along with the associations that exist among the object members of the aggregation. Objects are aggregated on the basis of satisfying an identified relation, one with associated semantics. Among the properties that characterize a relationship are its set of object members, the semantic definition describing the rules and behavior, and a representation structure to indicate how the members are ordered with respect to each other.

1.2 Other Approaches For Managing Relationships

Before considering our approach it is helpful to recognize other approaches that have been pursued regarding relationship management.

1.2.1 A Unique Solution for Each Type of Relationship

One approach to managing relationships is to devise a unique solution for each type of relationship. Within the ISO standards for Systems Management two relationships have been dealt with uniquely. The *is-a* relationship is handled by the Derived From clause in the Managed Object Template. Using an entirely different mechanism, the *is-part-of* relationship is handled by a separate Name Binding Template.

Faced with the possibility of hundreds of different types of relationships it is not reasonable to develop unique methods for each one. Furthermore, techniques such as those prescribed by ISO to support the *is-a* and *is-part-of* relationships are not as flexible as they should be, due to their side effects on registration and naming.

1.2.2 Every Object Supports Relationship Management

In some relationship management schemes it is necessary for each object involved in a relationship to play an active role in supporting relationship management. In an object-oriented architecture this implies that each Managed Object must support relationship Operations, Attributes, and/or Events[4]. Such things as maintaining relationship state, updating relationship data, and relationship auditing must be included in each object specification for each type of relationship the object is going to support.

Often Managed Objects are arbitrarily bound into relationships (e.g., *is-cooled-by, is-maintained-by, is-installed-at, is-incompatible-with*). It is difficult to anticipate all the types of relationships an object may be bound into in its lifetime. Consequently, it is unlikely that a complete list can be defined at the time the object class is specified.

1.2.3 Database-Oriented Solutions

Another popular approach to managing relationships uses relational database technology to provide both the specification semantics needed to define relationships and the structure necessary for representation and maintenance. The popularity of this approach is also due in part to the wide audience of database experience.

The database-oriented solution moves away from the object-oriented international standards for management of open systems. Consequently, the exchange of relationship information would have to depend upon vendor uniformity in database choices and defacto standards. We feel that a more open solution to relationship management and relationship information exchange is achievable within the object-oriented paradigm set fourth in the ISO Systems Management standards.

2 Method of Approach

The objective of our approach is to choose an object-oriented methodology capable of providing a significant degree of uniformity in relationship management. Additionally, the methodology must be compatible with the ISO international standards for Managed Objects.

2.1 Separating Representation Structure from Semantics

We have identified two primary aspects of relationships to be considered in developing a methodology for relationship management. First, there is the semantics of (or criteria for) establishing the relationship. Secondly, there is the structure (or syntax) for representing a relationship.

Managing the semantics of a relationship is quite different from managing its representation structure. The semantic aspect can also be dynamic over time and can, especially if influenced by the end-user, be varied within a single representation. For example, if the end-user is allowed to graphically indicate that two things are connected it may not be possible to "formally" define the end-users criteria. The best one can do is to say, "If the end-user says two things are connected? Then they must be connected!" Observing network operation activities affirms that indeed end-user defined semantics are both desirable and a regular part of day-to-day operations in the real world.

Hypothesis 1: Semantic management of relationships must be allowed to be deferred to developers and in some cases to end-users in order for relationship management to be useful in the broadest context.

Semantic relationship management will be addressed further in Section 3, while we continue here focusing on the representation structure. What if we could find a uniform representation for all relationships that promoted maintenance and analysis independent of semantics? Such stability in the representation structure for relationships would provide an opportunity for developing a set of manipulation capabilities for relationships that operate independently of the semantics.

Hypothesis 2: Common syntactic representation of relationships is possible and desirable in promoting a set of relationship management functions.

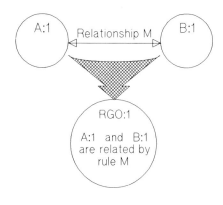

We have chosen a method for dealing with relationships among Managed Objects which uses a separate Managed Object to represent and manage the relationship structure. This technique defines a Managed Object class called <u>Relationship Graph Object (RGO)</u> which administers the structure of relationships between other Managed Objects. One premise of this method is that Managed Objects are optionally interested in the relationships into which they are bound. Another is that uniformity in relationship administration yields more leveragable relationship analysis. This approach is not to exclude other approaches but rather to provide a common open exchange paradigm for relationship management.

Hypothesis 3: The use of a separate class of objects for maintaining relationships provides a desirable level of flexibility and growth.

2.2 The Representation Structure

We have chosen to use a graph to represent a system of relationships. Drawing a relationship graph for Managed Objects implies that the vertices represent instances of Managed Objects and the arcs are drawn to represent the existence of a relationship in the direction indicated. Such a graph can be captured in the form of an adjacency matrix, a form which promotes manipulation and analysis of relationships.

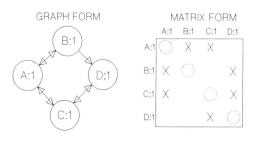

Much of the RGO description is based on the conceptual representation of a relationship graph as a matrix[3]. Consequently, allusions to matrices (diagonals, rows, and columns) should be interpreted in this context. Use of this conceptual view does not suggest that an RGO must be implemented as a matrix nor that the matrix representation must be persistent. On the other hand, it may very well be that a matrix is the best way to implement RGOs. Either way, this is for the developer to decide, however, RGO operations will support the matrix abstraction.

The ideas presented in this paper were influenced heavily by the desire to marry the relatively new object-oriented technology of Managed Objects with the more analytically developed and mature graph theory technology[2].

2.3 Common RGO Terms

At this point it might be useful to clarify a few terms used throughout the discussion of RGOs.

Party   The term party is used to refer to an object which is a member in an RGO. At any point in time a party in an RGO may or may not have a relationship with other parties in the RGO.

RelDir   RelDir is a parameter used to identify the directionality of a relationship arc between two parties in a relationship graph. There are three choices for RelDir: Fwd, Rev, and Bi.

Depth   Depth is a parameter that appears in the interface of the TestRels and FindRels operations and refers to path lengths where indirect relationships between parties in an RGO are of interest. By indirect relationships, we mean relationships that exist by the transitive property.

2.4 RGO Managed Object Class Template

```
Relationship_Graph        MANAGED OBJECT CLASS  (Partial)
    DERIVED FROM          top
    CHARACTERIZED BY
        ATTRIBUTES        InstanceID           READ
                          RelationshipType     READ
                          RGOUpdateID          READ
                          RGOPartyList         READ
        OPERATIONS
            CREATE        RGOCreate( RelationshipID,RelationshipDEF, OwnerID )
            DELETE        RGODelete( OwnerID )
            ACTIONS       AddParty( PartyList )
                          DelParty( PartyList )
                          AddRels( Party, PartyList, RelDirection )
                          DelRels( Party, PartyList, RelDirection )
                          TestRels( Party, PartyList, RelDirection, Depth, Qualifier )
                          FindRels( Party, RelDirection, Depth )
                          RetrieveGraph( PartyList )
        NOTIFICATIONS
                          PartyRelChange
                          PartyListChange
                          RGOUpdateIDChange
    REGISTERED AS
```

2.5 RGO Operations
AddParty(PartyList)
DelParty(PartyList)
These operations add or delete one or more parties to an RGO independent of relationship arcs. From the matrix perspective a row and column corresponding to this party is added to or deleted from the matrix.

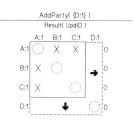

AddRels(Party, PartyList, RelDir)
DelRels(Party, PartyList, RelDir)
These operations add or delete directed relationships between **Party** and each party in **PartyList** to an RGO. If **RelDir** is *Fwd* rows are marked, *Rev* cols are marked, *Bi* both rows and cols are marked. If **Party** appears in **PartyList** the diagonal is marked.

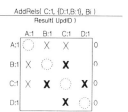

TestRels(Party, PartyList, RelDir, Depth, Qualifier)
This operation tests for the existence of a relationship between **Party** and each party in **PartyList**. Tests are conducted according to **RelDir** for as many arcs as indicated by **Depth**. A TRUE is returned if the **Qualifier** percentage of the parties in **PartyList** are found to have a relationship with **Party**.

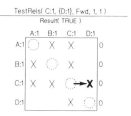

FindRels(Party, RelDir, Depth)
This operation finds all the parties related to **Party** according to **RelDir** for the number of arcs less than or equal to **Depth**. A PartyList and the RGOUpdateID attribute are returned. **RelDir** can also be *None*, to return parties NOT related.

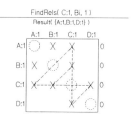

RetrieveGraph(Partylist)
This operation retrieves a relationship graph of the parties indicated in **PartyList** (an empty **PartyList** means all parties in the RGO). The following are returned: RGOUpdateID, RGOPartyList, and RGOMatrix (r1c1, r1c2, ... , i.e. matrix elements in row order).

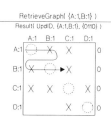

2.6 RGO classes

What we have described here is a basic RGO. It is intended that there be available a hierarchy of specializations of this basic RGO, derived according to the following criteria:

- every RGO class inherits the graph-oriented structure of the basic RGO

- specializations are distinguished from one another by their interfaces, i.e., by the set of graph-related algorithms that their operations implement[3,6,7]

- these groups of operations are formed according to their anticipated usefulness for providing the kind of capabilities needed in a particular problem domain

For example, various path-finding operations such as "all paths between party A and party B", "shortest (or other weighted) path between A and B", etc., may be grouped together to form the interface of an RGO that supports semantic relationships where path analysis is meaningful. We emphasize that all RGO classes inherit from the basic class so that the common structural representation (a matrix) be preserved for every relationship. This is a critical aspect of our model.

3 Supporting Relationship Semantics

As stated earlier, one of the major goals of our relationship object model is to provide an underlying structure that is capable of supporting many different kinds of relationships, and many different purposes for using those relationships. Hence, it is desirable to keep this underlying structural model as "generic" as possible, so that all semantics are imposed from outside of the common structure.

It is, however, important that support for semantic interpretation and manipulation of relationships is provided, or the user of the relationship object will be faced with a tedious and error-prone activity of constant translation between actions and events occurring in the structural graph-theory-oriented model and their meaning in the context of a particular application.

To provide a relationship modeling service that is more usable, provision for semantic relationship interfaces must be made. In our model, this is provided through another set of object classes, the <u>Semantic Relationship Objects (SRO's)</u>. SRO's represent the relationship of interest in a particular context. This means that when a relationship is discerned during the course of the design of a system, the designer thinks in terms of the SRO interface, rather than in terms of the RGO interface. Similarly, when an application is being developed and a relationship among system components and objects representing resources is found to represent information relevant to the application, an SRO is used to model that view of the relationship. For example, suppose that a dependence relationship has been identified in the course of designing a software installation module that can update not only a particular software component, but also all components that depend on the one being explicitly updated. Then the actions that need to be performed on the relationship object will be something like "List components that depend on", or "Add the following component and its dependents" and so on. The application designer, in other words, need not be concerned with the graph-oriented interfaces of the RGO's, but rather with objects that reflect the semantic relationship that makes sense in the particular application area.

We have thus identified two primary requirements for handling semantics of relationships:

- semantics need to be considered separately and implemented on top of the graph-oriented structural model, and

- the objects that developers and end users design and access for relationship information and manipulation must have interfaces that are meaningful in the context of the specific application and therefore be natural for them to use.

3.1 SRO's and RGO's work together
As mentioned earlier, the SRO is an object class that is defined when the design of a system or an application has identified a relationship that needs to be represented. A key characteristic of our model is that after the SRO interface has been specified, it is then implemented by using an "appropriate" RGO.

3.2 SRO Object Classes
The interface of an SRO is designed in very much the same manner as the interface of other objects that emerge from design activity. When a relationship is discerned, an object representing that relationship is denoted, and its responsibilities as a cooperating component of the system or application start to emerge. Services that the object is responsible for exporting for use by other objects are identified and translated into an interface for the object by specifying the attributes, actions and events that the object contains. The interface thus defined will clearly be in terms of the semantics of the relationship.

Having defined the interface of the SRO, one chooses an RGO whose operations can be used to implement those of the SRO. Recall that the RGO stores the state of the relationship and has the underlying structure common to all relationship objects in the system. The SRO is implemented by wrapping its interface around one or more RGOs. The (semantic) interface of the SRO is exposed to the system as the relationship object, yet the RGO, with its structure common to that of every other RGO in the system, lies underneath it in support. For example, if the relationship that needs to be supported in a system is the connectivity of network nodes, the interface of the semantic relationship object, the SRO, will include operations such as "list all nodes that can be reached from a node". To implement this interface, we can choose the path-oriented RGO cited earlier. The "list all

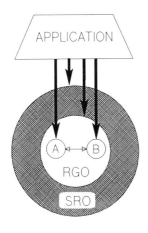

nodes that can be reached from a given node" operation can be implemented simply by asking for all parties that can be reached by a path from a vertex in the graph, which represents an object that is bound in the connectivity relationship. The SRO uses a (single) instance of the path-oriented RGO, and maps the "list all nodes that can be reached from a given node" operation to a PathBetween operation of the RGO that returns a list of objects that can be reached from the given object by following links in the graph to any depth (i.e., by any path).

Once the SRO implementation is complete, all objects in the system can access relationship information through the SRO interface, which is meaningful and natural, and not concern themselves with the graph structure that is supporting the relationship in a way common to all other relationships in the environment. As with the object members themselves, access to the RGO itself is not prevented by this "encapsulation".

4 Example

To clarify the relationship between RGO's and SRO's and how they work together, it is useful to consider some of the different types of semantic relationships that arise in system and application development, and the kinds of RGO's that support them. In particular, we consider what kind of graph-related operations support the extraction of semantic information of a particular kind from a relationship modeled as a graph.

Semantic relationships that are common include connectivity, equivalence with respect to some characteristic, such as administrative domains or workgroups, client-server associations, containment, dependence, among others. The following example illustrates a semantic relationships implemented using a RGO.

4.1 Domains as SRO's

An example of a relationship that arises in the management of distributed environments is the domain relationship, where heterogeneous objects representing resources are bound together on the basis of responsibility or authority [5]. The reason for domains and the different types of policies that may be applied to domains generate requirements for the structure of a model capable of representing the domain relationship and the operations that the relationship object must support. One view of the domain concept[5] defines a domain to be an aggregation of the resources only. This definition imposes the following requirements: the domain object must be capable of binding objects of different types, the domain object must not restrict access to the member objects themselves, domains should themselves be objects and therefore be candidates for being bound into other domain objects (forming subdomains), domains can overlap, and objects must contain information about domains to which they belong, as certain operation requests will require this knowledge and enumeration of all domain contents may not suffice to answer this question in a local environment.

The RGO model satisfies all of these requirements. We have already discussed the capability of binding heterogeneous objects into an RGO, the lack of restriction of access to object members themselves, the fact that RGO's are themselves Managed Objects and the model ignores any distinction between them and any other kind of Managed Object with respect to object membership. RGO's can overlap with respect to their object membership, and the RGO model does not prevent objects from tracking the relationships of which they are members, although it deliberately does not require it. A possible list of operations that a domain object needs to export [5] follows:

- Include object in a domain
- Remove object from a domain
- List objects in a domain
- Other operations to read/write domain attributes

If we take these to comprise the interface of a Domain SRO, we need to choose an RGO interface that contains the kind of operations that can be used to implement this domain-specific interface. The example basic RGO defined in section 2.4 will suffice for this interface. This being an equivalence relationship, we don't need to use the directionality of the graph nor concern ourselves with less-than-n-ary relationships, as all n objects bound together in an equivalence relationship together represent an n-ary relationship.

To implement the Domain SRO interface, we map the SRO operations to the RGO operations in the following way:

For all the operations, the domain is specified by application of the operation to the particular instance of the SRO that represents that particular domain.

- Include object in a domain => AddRels(Party,PartyList,Both), where Party represents the object to be added and PartyList consists of all other parties currently in the RGO.
- Remove object from a domain => DelParty(Party), where Party represents the object to be removed.
- List objects in a domain => GetAttribute(RGOPartyList).

Another view of a domain defines an administrative domain to be a representation not only of the resources that are grouped together but also of the administrative roles that are allowed to manage some or all of the resources in the domain. An administrative role in a domain can be, e.g., administrator, technical support, or end-user. This role is associated with resources, which are those resources that are managed by the role.

If we represent such a domain as a relationship object, then we have an aggregation of objects representing resources and objects representing roles, and additional information can be gleaned from the single relationship object. For example, the following operations can be added to the domain SRO class interface:

- List all resources that a role manages
- List all roles that can manage a resource
- List all resources that either of roles A or B can manage

In this case, the basic RGO will again suffice to represent the structure that the SRO needs, but we will make use of the directionality of the graph and the SRO operations will not all map directly onto RGO operations. In the graph that conceptually represents this domain, roles have edges that originate with them and terminate with resources, so that the directionality of the graph is used. Additionally, the design of the SRO operation "List all objects that either of roles A or B can manage" specifies an expression using the results of two operations made on the RGO:

let A_OBJECTS = list all objects that a role manages(A)
    B_OBJECTS = list all objects that a role manages(B)

return (A_OBJECTS intersected with B_OBJECTS)

The implementation of this expression can be contained in the SRO class rather than in the RGO class, where it closely mimics the algorithm described above. On the other hand, one might create a new RGO specialization that adds the operation

FindCommonRels(PartyList, RelDirection, Depth) which returns all the relationships (in RelDirection and up to Depth) that the parties in PartyList have in common.

If this choice is made, and a matrix is chosen for the implementation of the graph, the performance of the more complex operation in the interface of the domain SRO can be substantially enhanced by using fast matrix row bit-masking operations.

5 Conclusion

In this paper, we propose the use of a common representation and methodology for dealing with all types of Managed Object relationships. The proposed approach emphasizes separation of the structure of relationship representation from the semantics, and the use of OSI standards for defining Managed Objects. The value of separating representation structure from semantics is multi-fold. It allows for the possibility of standardization of the representation structure which in turn promotes cross relationship analysis. It also allows virtually unbounded freedom in relationship semantic definition.

The model that is used for relationship representation is a graph. We use the matrix as the conceptual representation of a graph to carry out graph operations. Further, the relationship graph is modeled as a relationship graph object that maintains relationship information, as opposed to requiring objects to track all relationships into which they are bound. Finally, we propose support for relationship semantics with SROs which encapsulate the graph-oriented operations of RGOs. This allows the application developer to deal with relationships in terms of the unique relationship semantics while retaining the benefit of the common underlying relationship representation.

6 References

[1] ISO, Information Processing Systems - Open Systems Interconnection - Management Information Services - Structure of Management Information, ISO/IEC DP 10165-1, 5/89.

[2] HP ING/NAL, HP OpenView: HP Network Management Architecture Overview and Models, Version 4.0, 12/89.

[3] Steward, D. V., Systems Analysis and Management: Structure, Strategy, and Design, Petrocelli (TAB), 1981.

[4] ISO, Information Processing Systems - Open Systems Interconnection - Systems Management - Part 3: Relationship Management Function, ISO/IEC DP 10164-3, 2/89.

[5] Sloman M. & Moffett J., Managing Distributed Systems, Domino Report A1/IC/89/1, September 1989.

[6] Deo, N., Graph Theory with Applications to Engineering and Computer Science, Prentice-Hall, 1974.

[7] Dornhoff, L. & Hohn, F., Abstract Modern Algebra, Macmillan, 1978.

[8] Kleene, S.C., Mathematical Logic, Wiley, 1967.

[9] ISO, Information Processing Systems - Open Systems Interconnection - Working Draft of the Configuration Management Overview, ISO/IEC JTC1/SC21 N3311.

IX

MANAGEMENT INFORMATION

B
Distinguished Experts Panel
(Same as VII-B)

IX
MANAGEMENT INFORMATION

C
Implementation Aspects

OSI Management Information Base Implementation

Subodh Bapat

Racal-Milgo, MS E-204, 1601 N. Harrison Parkway, Sunrise, FL 33323, USA

Abstract

The Management Information Base (MIB) is the conceptual repository of all management information stored in an OSI-based network management system. The definition of a MIB, which is the subject of many standardization efforts, describes a conceptual schema containing information about managed objects, their attributes, and relationships between them. A MIB definition does not describe how a MIB may be implemented.

This paper addresses and highlights issues involved in implementing the Management Information Base in an OSI-based network management system product. Aside from the object model framework itself, many other considerations are important in a MIB implementation. These include suitability of the MIB platform for the selected object model, the architecture for distributing MIB information, mechanisms for ensuring MIB data integrity, extensions of the standard MIB object model to allow value-added applications, maintenance issues including MIB schema changes, as well as MIB backup and recovery. This overview may provide some benefit to MIB designers of network management system products.

1. INTRODUCTION

The Management Information Base (MIB) is the conceptual repository of all management information stored in an OSI network management system. The MIB defines the set of all managed objects visible to a network management entity.

The MIB is in general an interface definition - it defines a conceptual schema which contains information about specific managed objects, which are instantiations of managed object classes. Further, the schema embodies relationships between these managed objects, specifies the operations which may be performed on them, and describes the notifications which they may emit. The definition of a MIB, which is the subject of many standardization efforts [1], [2], does not in any way specify how such a MIB may be implemented.

This paper presents the considerations involved in implementing the Management Information Base in an OSI network management system product. The term *Management Information Base* will be used interchangeably to mean both the abstract model of management information which is visible at the interface, as well as its implemented form which allows the persistent storage of this information.

2. THE OBJECT MODEL

For a MIB to be implemented, its object model must be well-defined and complete. The process of modeling involves the following steps:

1. Selection of a modeling paradigm
2. Identification of resources that need to be modeled as objects for management purposes
3. Identification of attributes of the managed objects
4. Identification of relationships between the managed objects
5. Description of object behavior, i.e. operations on managed objects and the notifications emitted by them.
6. Formalization of the above information in an unambiguously defined syntax.

In addition to the above, a diagrammatic representation of the object model, or subsets thereof, is often found in practice to be a useful aid to comprehension.

2.1 The Modeling Paradigm

An important modeling paradigm that is used in the definition of a MIB is *Object-Oriented (OO) Modeling*. The benefits of object-oriented modeling are many[3], chief among which is the imposition of a formal conceptual framework to guide the thought process of the system designer. In an OO model, MIB objects are organized in a class hierarchy, such that they exhibit an *Inheritance Relationship* from the superclass to the subclass.

Entity-Relationship (ER) Modeling is another modeling paradigm [4] which can be used in a MIB definition. The ER paradigm is highly suited for defining generalized relationships between MIB objects. Parts of the OO model can be subsumed within the ER model - the OO notion of inheritance can be modeled through the **is-a** relationship. For example, if the `lanMACBridge` object class is a subclass of the `equipment` object class (such that it inherits all its attributes in the OO model), it can be represented as a `lanMACBridge` **is-a** `equipment` relationship in the ER model. However, the OO notion of methods which operate upon objects is not modeled in the ER framework. On the other hand, the ER concept of a generic relationship is not explicitly available in an OO model; it is nevertheless meaningful and often necessary to introduce such relationships as a semantic constructs [5].

Both object-oriented modeling and entity-relationship modeling are well-understood concepts and have been dealt with at length in many excellent references in the literature [3], [4], and will not be described here. Current OSI-based standardization efforts concentrate largely on OO modeling, with supplementary information being supplied through ER diagrams.

From a productization standpoint, however, each has specific advantages and disadvantages in its ease of implementation in an actual MIB; this is discussed in a later section.

2.2 Object Identification

This phase consists of explicitly identifying the resources which need to be modeled as objects for management purposes. The choice of what constitutes an object is often obvious; at other times, it is not always clear. Generally, every resource that is used in communication, and which must be described in order that its behavior be monitored or influenced, must be modeled as an object. This may be a network communications resource, or may be a logical or physical resource not involved in communicating data, but is still required for management purposes (such as a location or a vendor object). In general, object decomposition is performed in a manner so as to permit as much independence between objects as possible.

2.3 Attribute Identification

This phase consists of explicitly identifying what aspects of the above resources need to be modeled as object attributes for management purposes. Even though a particular apsect may be part of the complete description of the resource (e.g. the color of the exterior), it may not be necessary to model it as an object attribute for management purposes if such an attribute is either not useful, not relevant or not meaningful in management activities.

2.4 Object Relationships

The OO notion of superclasses and subclasses naturally stems from the *Inheritance Relationship* between objects, in which the subclass inherits attributes and methods from its superclass, and may further refine and specialize the methods. Such inheritance relationships can be modeled and implemented in a MIB using a well-understood, rigorous framework.

In addition, networks require the modeling of *Containment Relationships*, in which some object instances may be physically or logically contained within other object instances. Containment relationships are different from inheritance relationships, because they do not imply inheritance of attributes from the containing object class to the contained object class. For example, an equipment object instance may be contained within a location object instance; this models containment, and does not imply that the equipment object class inherits attributes from the location object class. Further, the equipment object instance may itself contain a portCard object instance, which may contain a port object instance; in no case are attributes passed down as they would be in a superclass/subclass relationships. Containment relationships need to be explicitly identified for purposes of object addressing and naming. The OSI Directory [6] could be used to store names and addresses of managed object instances, though such a mapping between containment hierarchies and The Directory has not yet been defined or standardized.

Other types of relationships between objects are also important and need to be modeled as well. *Connectivity Relationships* between objects are important in modeling network topology and for determining routing. *Domain Relationships* must be modeled for clearly defining object boundaries and administrative responsibility. *Authority Relationships* are necessary in defining roles of management entities, for example, a management entity acting in a manager role participates in an *Authority Relationship* with a management entity acting in an agent role. Several other types of relationships exist and are useful for management purposes; these must all be modeled and reflected in a MIB implementation.

2.5 Actions

The specification of an object includes a description of what methods may act upon the object. In the MIB object model, the methods constitute operations, such as **CREATE**, **DELETE**, and **ACTION**, which may be performed on the managed object. A MIB implementation must have the capability of storing this information so that the validity of a requested type of **ACTION**, for example, may be checked against the targeted managed object class.

2.6 Formalization

All object modeling information above must be presented in a formal manner, either notational or pictorial. The formal presentation of a MIB's object model must clearly identify every informational element such that it unambiguously translates into whatever implementational form is selected.

The technique chosen to formalize OO object definitions and inheritance relationships is ASN.1 macro notation [7], which has successfully been applied in many standards efforts [8].

The technique used to formalize ER relationships has generally been pictorial [4], which assists in defining the schema to implement those relationships in a MIB.

3. IMPLEMENTATION OF THE MIB

3.1 Platforms for Hosting a MIB Implementation

A MIB implementation, which essentially is a store of persistent information required for management purposes, can be hosted on several types of platforms:

1. Object-oriented databases
2. Relational databases
3. Flat-file databases
4. Proprietary format databases
5. Firmware

6. Volatile memory
7. Combinations of the above

Generally, MIB information in a system product is distributed. A typical configuration may include a disk-based relational database on a central management station, static object attributes stored in a PROM contained within the managed resource, and dynamically changing information in RAM memory accessible to software processes. These processes could be system management applications, agents, or proxy agents, and could be resident in either the central management station, in intermediate management devices, or within the managed resource itself. Figure 1 depicts a typical such configuration.

3.2 Architectural Principles for Partitioning MIB Information

An important architectural issue which must be resolved during the system design of the product, is the partitioning of MIB information across the entire management solution. The following discussion assumes that a minimal configuration for a network management system product consists of a management station with some persistent storage capability (referred to as a central management station database), and that this management station is connected to network resources of interest, which are modeled as managed objects.

In many cases, network resources are capable of being downloaded or strapped for certain management attributes (e.g. configuration management parameters, fault management parameters). Further, they are capable of being queried as to current settings of those parameters. In such cases, it is not strictly necessary to store this information in the central management station, although it may be desirable to do so from a redundancy perspective.

Certain other network resources may be capable of storing MIB information, not just about themselves, but about other managed objects as well. For example, the internal database in a PBX or an intelligent multiplexer may already possess some information about its immediate topology, and about the communication capabilities of its nearest neighbors. If such a resource can be queried for this information, then this information need not be stored in a central database. In other instances, information may already be stored within a managed resource (e.g., a management entity may choose to store an **eventReportingSieve** object within a managed communications resource, so that sieving may be done locally). Such information, too, may or may not be stored in a central database.

On the other hand, there are pieces of information which are most optimally stored in a central database - this typically includes system-wide information, such as a complete directory of managed objects, and network-wide topology information. While this information could possibly be distributed, the access speeds required for such applications as graphical depiction, zooming and panning of network topology, often necessitate the storage of this information in the central database.

The following architectural principles, then, may be applied towards distributing MIB information in a product implementation:

1. The internal information storage capabilities of every managed object instance must be determined. If the network resource can store non-volatile information (e.g. either in firmware, via hardware straps, or on an internal disk, etc.) and is capable of being queried for that information, those attributes need not be stored in the central database on the management station.

2. If the managed object can store only volatile information which needs to be supplied to it (e.g. through its front panel or downloaded on boot-up) then copies of this information must be stored in a central database (as it may be lost in an unreliable situation.)

3. System-wide information, such as the Directory, network-wide connectivity, high-level security management functions for administrative authentication, historical data to generate performance and trending statistics, as well as usage information required for triggering accounting management applications, is most optimally stored in the central database.

4. In a hierarchy of management entities (e.g. proxy agents in software, mediation devices in hardware), where the intermediate entity also has the capability of storing persistent information, consideration should be given to storing MIB data in the intermediate entity if such data is not frequently required by management applications. This is so that the number of queries the intermediate entity needs to make of the central database in the course of its normal operation is minimized, thus reducing traffic on that link, and the transaction load on the central database.

5. Reliability and fault-tolerance requirements of the network management system product may require that redundant copies of MIB information be stored in both a managed object's internal database as well as the central management station. The tradeoff for this redundancy could be an increase in the requirements of database capacity and processing power on the central management station, affecting product cost.

In several productized network management systems, a combination of the above techniques is used, with some degree of redundant storage of MIB information. Redundancy is often desirable for the purposes of recovery and backup of the entire MIB. However, the MIB then essentially becomes a distributed database, and this leads to the concomitant problem of maintenance of consistency between replicated copies of information. It is then incumbent on the application entities to ensure atomic updates of each copy of MIB information. Further, either an access control mechanism or a deadlock detection and recovery mechanism must be in place, in the event that conflicting updates are simultaneously applied to different replicated copies of the same MIB information.

Determining the optimal partitioning of MIB information in an actual network management system product is best accomplished after analysis results obtained from actual experimentation. Such analysis should consider the tradeoffs between application response time, bandwidth cost, the nature of application queries, and the frequency of queries. Different heuristics will emerge depending on the primary use the network management system product is subjected to. In addition, the reliability and redundancy requirements on the product will also influence the partitioning of MIB information.

3.3 Management Station Databases

The choice of database type on the central management station is important. Some issues involved here are described:

3.3.1 Object-Oriented Databases

Object-oriented databases are a strong candidate platform type for storing the MIB object model. The advantages are as follows:

1. Object-oriented databases are naturally suited for storing MIB information, because the MIB itself is formally described as a set of abstract data types in an object-oriented class hierarchy.

2. In addition to the ability to model object description through storage of attributes, they also have the ability to model interface behavior, through the ability to store methods.

However, there are some drawbacks with this choice:

1. Although object-oriented database technology is expected to achieve stability soon, it may not be mature enough at this point to be considered in a product implementation.

2. The definition of an access language for querying and manipulating object-oriented databases is not a standard.

3. The MIB object class hierarchy, as defined in several standards documents [8], [9], when graphed in an inheritance tree, is rather broad and shallow. This means that the depth of superclass/subclass relationships between MIB object classes is of low order (and typically is about 2 or 3), whereas the number of high-level MIB object classes that derive from {top} is large. Therefore, although inheritance relationships are important, the number of such inheritance relationships in the current OSI-based MIB models may be low enough so as not to make an object-oriented database implementation an absolute necessity.

4. Performance characteristics of object-oriented databases are not well documented. Techniques such as *Class-Hierarchy Flattening* to improve performance have been used to improve performance, but are not particularly advantageous to MIBs because the MIB class hierarchy is itself shallow. No standard benchmarks are available for measuring performance of object-oriented databases embedded within a productized application system.

3.3.2 Relational Databases

Relational databases are a popular choice for central management station implementations of MIBs. The advantages of relational technology are many:

1. Relational database technology is mature, stable, and enjoys the support of several large companies.
2. Relational databases are usually accessed using a standard access language (SQL) [10].
3. Well-defined mechanisms exist to translate an Entity-Relationship model into normalized schema within a relational database.
4. Application activities, such as report generation and ad-hoc querying, are easily prototyped and implemented in a relational database.

On the other hand, relational databases suffer from the drawbacks that:

1. They are not well-suited to storing Object-Oriented models. Both inheritance relationships and containment relationships are essentially tree structures, which must be "flattened" into tabular relations for storage within a relational database. Although this is certainly possible, tree traversal through relational tables involves a sequence of several join operations, which are awkward to perform in SQL.
2. Many other types of relationships also require joins, which can cause performance degradation unless the ancillary access structures (such as indexes and clusters) are finely tuned for the type of queries made.

3.3.3 Other Database Types

Flat-file databases and other proprietary format databases can be tailored very specifically to a MIB object model or a subset thereof, and can be optimized for performance. However, the disadvantage is that network management applications are more complex to develop, since now they must be aware of the physical layout of the data under the proprietary format in order to manipulate it. This increases not just development time, affecting the productized system's time-to-market, but may also further complicate the maintenance effort.

3.4 Translation of MIB Object Model into Schema

A crucial step for arriving at a MIB implementation from an object model, is the translation of the object model into a set of data structures (collectively termed *schema*) which are understood by the platform on which the MIB is implemented.

An object-oriented database is well suited to storing both the attributes and methods that define a MIB object; however, a translation is necessary from the ASN.1 macro notation (e.g. **M-OBJECT-CLASS** macros defining a managed object, **ATTRIBUTE** macros for defining object attributes, **NAME-BINDING** macros to define addressing [8]), into the corresponding data definition language of the object-oriented database system. This will define the object schema which will contain instances of each managed object class.

For relational databases, such a translation results in SQL **CREATE** commands to generate the relational tables and views for storing MIB object classes. Depending on the level of normalization desired, some attribute information could be stored redundantly in order to avoid excessive joins while traversing the class hierarchy. A limitation of most relational database products is that they do not have the capability for storing methods (actions which may be performed on object instances, expressed as code fragments) associated with the object class.

3.5 Attribute Integrity Issues

An area of concern in a MIB product implementation is integrity of attribute data, which must be assured to minimize the potential for MIB corruption. Attribute integrity can be assured by many different mechanisms, some of which are described below.

3.5.1 Consistency of Replicated Attributes

Many MIB implementations contain replicated attribute data. In some situations, replication occurs within the scope of a single management domain - for example, managed object attributes may be stored both internally within the managed object as well as in the central management database. In this situation, consistency can be assured by having a well-defined concurrency control mechanism, which assures simultaneous and atomic updates of all copies of attributes. This can be achieved by appropriate advisory or mandatory locking mechanisms. Alternatively, if it is tolerable to allow minor windows of inconsistency, a periodic polling mechanism can be instituted where the subordinate instance of the attribute frequently queries the superordinate instance to determine whether they are synchronized; if an inconsistency is detected, the subordinate instance of the attribute will update itself.

In a situation where attributes are replicated in the central management databases of peer-level management entities sharing the same conceptual schema[11], [12], consistency can be effected by the use of CMIS primitives. When a managed object changes state, the management entity responsible for that object issues an **M-SET** primitive [13] to the peer management entity, thus causing it to update its internal copy of the attribute. This message may be confirmed or unconfirmed depending on the specific CMIS service used.

Sometimes it is necessary to create redundant copies of information within the same central management database. Although this violates principles of normalization, selective replication often helps run-time performance. In this instance, consistency must be assured by appropriately constructing the application code responsible for maintaining this information, or through mechanisms supported by the host database platform.

3.5.2 Dependency Integrity

Often, object instances have dependencies on each other, an example of which is an *Existence Dependency*. This concept formalizes the notion that a contained managed object instance participating in a containment relationship, is existence-dependent on the containing managed object instance (e.g a **t1AggregateCard** object instance cannot exist unless its containing **t1Multiplexer** object instance exists.) This implies that the contained managed object instance cannot be created, until the containing managed object instance has been created (e.g. with an **M-CREATE** CMIS primitive [14]).

In a MIB implementation, such dependencies can be enforced using *Referential Integrity* mechanisms. These mechanisms assure that references between objects (semantically understood to be relationships, e.g. a containment relationship in this example) are consistently maintained. This is usually accomplished with the aid of *triggers*, which provide the ability to manipulate data across a set of instances over the entire schema, in a cascaded manner. Thus, when a **t1Multiplexer** managed object instance is deleted following the receipt of an **M-DELETE** CMIS primitive, triggers can be used to ensure that all other managed object instances (e.g. **t1AggregateCard**) contained within that **t1Multiplexer** are automatically deleted.

More complicated dependencies can be enforced using *stored procedures* within the MIB implementation - for example, the ability to detect and prevent cyclical or reflexive containment relationships.

3.5.3 Domain Integrity

Some object attributes are required to possess only one of a finite, enumerable set of values. For example, T1 Multiplexer ports may only operate at speeds at the T1 rate or one of a restricted set of predefined subrates. A MIB implementation must therefore enforce that its representation of the **speed** attribute of all **t1AggregatePort** object instances can only possess one of the acceptable set of values. This can be accomplished using *Domain Integrity*, which instructs the host database to check that any attempts to set the attribute to an unacceptable value (outside its domain) are flagged as error conditions. This requires that the domain of acceptable values for the attribute be specified at schema definition time, and requires the translation of the ASN.1 macro notation for attributes **WITH ATTRIBUTE SYNTAX <syntax-name>** into the appropriate schema definition construct for attribute domain integrity.

4. EXTENSIONS TO THE STANDARD MIB

In addition to the managed object classes defined by various standards, a real MIB implementation may require other types of information to be stored in order to increase the power or usefulness of the product. The following are examples of such extensions.

4.1 Extended Security Management

A MIB may be required to include security information about access to various managed object instances, as well as mechanisms for performing authentication for operations on specific managed object instances. Also necessary may be security information for access control to the network management system itself, including direct control and manipulation of the MIB itself. This is especially necessary if the system is a multiple user system, and involves defining to the MIB the judicious grant of authorizations to query and manipulate stored information.

4.2 User Configuration

The storage of administrative information about users of the communications network in the MIB may add additional value to the system. For example, when several different mutually exclusive groups of users are using the same communications resources (as in an X.25 packet switched environment), *Closed User Groups* must be configured and maintained within the MIB. Users and user groups can themselves be managed objects which must be created, maintained and deleted in the MIB.

4.3 Configuration Histories and Profiles

An extended MIB may maintain past configurations of object instances, so that a history of configuration changes (manual or automatic) can be determined for selected object instances, objects can be restored to their previous configuration states from the current state.

It may also choose to store typical configuration profiles for managed object classes, whether or not current instances of that object class actually conform that profile. This would provide a Configuration Management application the ability to merely select one of a set of predefined profiles, while configuring a specific managed object instance. It would also allow automatic scheduling of configuration changes for managed objects, which may take place at predefined times in a programmed manner. For example, a managed object could conform to one configuration profile for peak hours of operation, and another configuration profile for offpeak operation. Storing the various possibilities for configuration profiles for the object in the MIB, makes it possible to schedule this reconfiguration in an automated manner.

4.4 Problem Management

A network management entity choosing to implement a complete Problem Management, Tracking and Resolution system, linked to Fault Management information, would require an augmented MIB. This would include the ability to translate Event Reports into trouble tickets or incident reports, assign job codes based on those reports, recognize and categorize types of problems, group related incidents together, institute automatic escalation of problem severity with time, track progress of problem resolution, interface with Accounting Management applications which provide credit for lost service, and close out, delete, and/or archive past problems for historical purposes.

4.5 Extended Performance Management

Some users of a network management system may require a complete Performance Management system, including tracking resource utilization, predicting potential overload situations, allocating and managing bandwidth, generating trending statistics, and funneling this information to appropriate network planning applications. The storage of such information within a MIB would facilitate these functions.

5. MAINTENANCE OF THE MIB

Maintenance of the MIB is an ongoing process in the lifecycle of a network management system product, and maintenance issues must be considered during the MIB implementation phase.

5.1 MIB Information Redistribution

As the network management system evolves, it may become necessary to redistribute MIB information within the system. This could become necessary for performance and architectural reasons as well as to take advantage of new technology.

If a new version of a managed object class is introduced, whose instances can now locally store attribute information previously located in the central management database, it may be desirable to redistribute the location of this information in order to reduce traffic between agent and manager processes.

Excessive traffic patterns or query loads on the central management database can cause performance degradation, and may require that MIB information be repartitioned to alleviate this problem. This could be achieved by devolving some subset of the MIB schema down to an intermediate management entity, such as a proxy agent, which would then be responsible for maintaining that information.

5.2 MIB Schema Modification

MIB schema modification is required in situations where the evolution of the system requires that new managed object classes be supported, or new capabilities and functions are added to existing managed object classes.

When a new managed object class is introduced, the schema must be augmented to support the new managed object. This requires updating appropriate Directory information to recognize the new object class as a supported type. The Directory must be architected in a manner that minimizes the impact on application software of additions and deletions of supported object types, and attempts to maintain as much backward compatibility as possible.

In a situation where new functionality is added to an existing managed object class, the schema could be impacted in various ways, which include new attributes to be added to the object's representation, or a change in the attribute's domain of values, etc. For example, if a communications resource is upgraded to operate at higher speeds, the domain of acceptable values for its **speed** attribute needs to be changed. Dynamic schema modification may also occur during the process of schema negotiation between peer management entities.

The impact of such a change is not restricted to the MIB schema alone, because the application processes concerned with configuring, validating and using this information will also need to be upgraded to reflect the new attributes or new range of allowable values for existing attributes.

6. PERFORMANCE

A MIB implementation must ensure that it meets adequate performance criteria, i.e. the response time required for a MIB-oriented transaction must be well within the system specifications. The performance criteria a MIB must satisfy vary with the nature of the managed objects, the size of the managed network, its distribution architecture and its connectivity with peer management systems.

6.1 MIB Transaction Load Mix

Most MIBs will experience different types of transactions depending on the network management primitive received. Figure 2 shows the types of MIB transactions resulting from CMIS requests. The following are examples of the type of MIB transaction triggered based on the CMIS primitive received:

1. **M-GET**: MIB information retrieval
2. **M-SET**: MIB attribute value update
3. **M-CREATE**: MIB object instance creation
4. **M-DELETE**: MIB object instance deletion
5. **M-EVENT-REPORT**: MIB event instance creation or attribute value update
6. **M-ACTION**: may result in a MIB attribute value update as a side-effect

Other CMIS primitives may also result in MIB transactions being generated as side-effects.

6.2 Performance Tuning

The response characteristics of MIB operations are very sensitive to the transaction mix. For example, a network system prone to frequent failures will generate a large number of event reports to its management system, requiring the MIB to perform a large number of event instance creations. On the other hand, a management system whose primary use is to generate reports, will be frequently subject to information retrieval queries from the MIB. A MIB tuned to deliver optimal performance under one type of transaction mix will in general not be optimally suited for another type of transaction mix. In tuning a relational database, for example, appropriate placement of indexes on tables can be very effective for a large number of retrieval operations; but these same indexes may hamper performance if the system is subject to frequent object instance creations and deletions.

The MIB performance must be sized for the maximal possible load on the system. During occasions of catastrophic network-wide failure, a burst of a very large number of alarms can be expected simultaneously from many managed objects; the corresponding load on the MIB will display the characteristics of high-demand OLTP (On-Line Transaction Processing) activity. The MIB as well as its access interface must be designed to handle such bursty, high-volume peak demands without loss of significant operational data.

7. BACKUP AND RECOVERY

Like the managed network, network management systems themselves are not immune to failure; this makes it critical to have in place a mechanism for backing up and recovering information in the MIB.

The backup mechanism consists of two distinct parts: (i) Schema Recovery, and (ii) Content Recovery. Schema recovery consists of retrieving from an implemented MIB merely the data structures which provide the templates within which information stored. Content recovery consists of recovering the actual information stored in the MIB, e.g. specific instances of managed objects. In a distributed MIB, content recovery involves querying every distributed element of the MIB - whether in the central management database, agent entities or within the managed object itself - in order to capture the complete picture of the network in a regeneratable form.

Schema recovery need not be performed unless the schema has changed. Schema changes could occur during system software upgrades, and are not expected to happen frequently. On occasions like incremental backups, only content recovery is sufficient. On the other hand, when it is desired to create another instance of the network management system for controlling a different set of managed objects, only schema recovery is sufficient, because the second system will presumably create a different set of managed objects within the same schema.

8. CONCLUSION

This paper has attempted to address and highlight some of the important issues in implementing a MIB in an OSI-based network management system product.

Aside from selecting an object model framework for the MIB, many other issues become important for a MIB implementation. These include suitability of the MIB platform for the selected object model, the architecture for distributing MIB information, mechanisms for ensuring MIB data integrity, extensions of the standard MIB object model to allow value-added applications, maintenance issues including MIB schema changes, as well as MIB backup and recovery. This overview may provide some benefit to MIB designers of network management system products.

REFERENCES

1. ISO 10165, "Information Processing Systems - Open Systems Interconnection: Management Information Services - Structure of Management Information"
2. OSI/NM Forum Release 1, "Object Specification Framework", Forum 003
3. "Object-Oriented Concepts, Databases and Applications", ed. Won Kim and Frederick Lochavsky, Addison-Wesley/ACM Press, 1989.
4. "Fundamentals of Database Systems", Ramez Elmasri and Shamkant Navathe, Benjamin/Cummings, 1989.
5. Proceedings of OOPSLA 87, "Relations as Semantic Constructs in an Object-Oriented Language", Rumbaugh, James, Pages 466-481, Oct 1987.
6. CCITT Draft X.501 / ISO DIS 9594, "The Directory."
7. ISO 8824, "Information Processing Systems - Open Systems Interconnection: Specification of Abstract Syntax Notation One (ASN.1)".
8. OSI/NM Forum Release 1, "Library of Managed Object Classes, Name Bindings and Attributes", Forum 006
9. Committee T1M1.5, "A Generic Network Model", T1M1.5/89-010R2
10. "An Introduction to Database Systems", C.J. Date, Addison-Wesley.
11. OSI/NM Forum Release 1, "Forum Architecture", Forum 004
12. OSI/NM Forum Release 1, "Shared Management Knowledge", Forum 009
13. OSI/NM Forum Release 1, "Protocol Specification", Forum 001
14. OSI/NM Forum Release 1, "Application Services", Forum 002

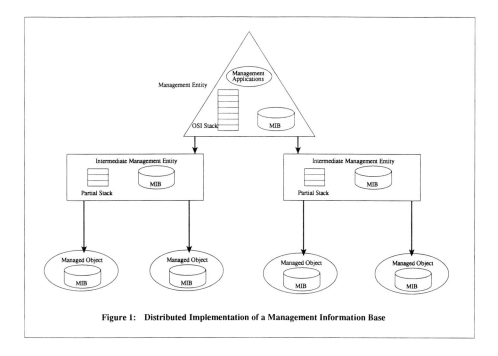

Figure 1: Distributed Implementation of a Management Information Base

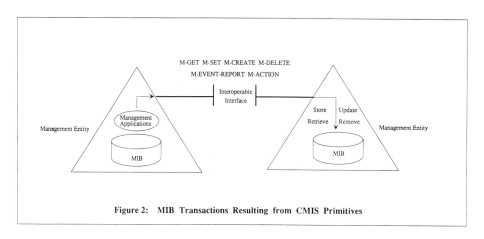

Figure 2: MIB Transactions Resulting from CMIS Primitives

The Concept of the Network Management Information Base in CNM, the TRANSDATA Network Management Scheme

Claudia Rauh

Siemens Nixdorf Informationssysteme AG
STO NC412
Otto-Hahn-Ring 6
8000 MÜNCHEN 83
Tel.: 089/636 49572
FAX: 089/636 46931

Abstract

The growing demands on network management resulting from the increasing trend towards networking led to the conception of a new network management system for data communication systems known as Communication Network Management (CNM). In creating this new concept, high priority was given to achieving a perspective encompassing the network as a whole, as opposed to a system-specific view. This aim led to the development of a uniform and consistent Network Management Information Base. This paper examines the thinking behind the global information base, and the interface which permits access to the data it contains.

Keywords

Communication Network Management, Network Management Information Base, global information base, object oriented, ISO network management standards, static and dynamic information, network perspective, system perspective

1 Introduction

Calls by network operators for availability of information are placing new demands on network management, while at the same time the automation of manufacturing processes requires the setting up of new networks. Increasingly, the various problems are dealt with by installing special computers from a variety of different vendors. Users of these networks expect them to provide high levels of efficiency, security and availability. It was for this reason that Siemens Nixdorf data communications technology experts have developed a new network management concept, called Communication Network Management (CNM)(see [1]). This new network management scheme was designed to provide support for TRANSDATA® (the Siemens Nixdorf network architecture) networking and allow integration of OSI systems from other vendors.

The concept is based on the following structural principles:

1. All information relevant to network management is held in a shared information base, in the form of objects with attributes and the relationships between these objects.

2. An object-oriented interface is provided for accessing the information base. This interface permits the definition of reactions to events taking place within the network, and thus supports the implementation of complex network management applications.

3. Compliance with ISO network management standards enables the integration of heterogeneous systems.

4. Standard applications are provided for the most important network management tasks, such as configuration management and performance management.

5. A graphical, object-oriented user interface is provided for the network administrator.

This paper is concerned predominantly with points 1 and 2, although point 3 is also considered. Enhancements in line with user requirements have been introduced, taking the OSI standards as a starting point.

2 The Structure of the NMIB

A network management interface should be implemented around object-oriented methods. There follows an explanation of the term "object-oriented" as used in this context.

Every management activity can be regarded as the performing of certain operations on a set of managed objects (see [2]). The set of these objects represents the network as seen by the network management facility, and the objects are thus distributed over the network.

Network management objects (NM objects) have a name by which they can be addressed, and are described by a set of attributes. In addition, certain relationships exist between NM objects. There are also rules applicable to objects, their attributes and relationships, which determine such factors as, for instance, the conditions under which attributes can take on particular values.

These network management objects as a whole constitute the Network Management Information Base (NMIB).

Operations performed on the NMIB concern either object information, or the objects themselves. Operations carried out on the object information include the modification of attributes and the defining of relationships, while operations involving the objects themselves include, for example, the activation and deactivation of those objects.

There are two kinds of management information, namely static information and dynamic information.

Static information is, by its very nature, long-lived. It describes on the one hand the network configuration as a whole, and on the other hand the individual components. Such information is held in background storage, the medium employed being a software product from Siemens Nixdorf known as the Entity Relationship Management System, or ERMS, based in turn on a relational database.

Dynamic information refers to those objects required by the network components during operation, or those which only come into being during operation (connections would be one example of this). Such object information typically corresponds to the real object, and is stored in main memory.

The database is thus not stored using a uniform medium, neither is it held on one single system. However, applications will be provided with uniform access facilities to information stored in different objects, via the program interface. For the standard objects, modelling uses the Entity Relationship Model. Applications will be offered the opportunity of expanding this model in line with their requirements.

In addition, life-cycle phases will be made available, along the lines of the version concept familiar from software management. A number of phases can be provided for the planning and installation of networks. The phases will differ in terms of the rules subject to checking, i.e. with regard to guaranteed consistency.

There will be one phase for networks currently in operation and one for archiving purposes.

An important aspect of the Network Management Information Base is that the information to be managed and maintained in a consistent state relates to an entire network and not just to a single system. This represents a shift from a system-oriented perspective to a network-oriented one, and thereby enables a network administrator to manage a large network efficiently, and to perform extensive consistency checks.

2.1 The Integration of Dynamic Information into the NMIB

Dynamic information is also included in the NMIB's static schema. This means that all dynamic object types are described, along with their attributes. The schema also describes how this dynamic object information is to be accessed.

The NMIB thus holds the static and dynamic object information in its entirety. In particular, the NMIB contains the global description of the network.

Each system in the network includes what is known as the Management Information Base (MIB), which contains all information on NM objects that is relevant to that system. The MIB holds information about objects which are required during operation of the communication layer and reflect the network from the point-of-view of the system, or which have purely local relevance.

A distinction is drawn between the logical MIB and the physical MIB. The former contains the information which is logically assigned to a system, while the latter consists of the information actually stored on the system in question.

2.2 Distribution of Object Information

A network typically contains different kinds of systems. In CNM three levels of systems are distinguished.
Firstly, Basic systems; these systems are only able to store dynamic information about themselves.
Secondly, Extended systems; these systems include an ERMS, where they may store the static information which describes the system itself and its environment.
Thirdly, the Network Management Center; the information for the whole network is stored in this system.

It is possible to keep several copies of the static object

information. These copies are known as replicas, and serve as an efficient means of providing object information on several systems.
Access to replicas is restricted to read operations. All other access requests are routed to the system holding the so-called master copy. Any modifications are distributed to the replicas by this system.

Objects also have an owner. This is the system on which the object was originally created. The master characteristic can be delegated by this system to another system. An object's owner system never changes though.

The diagram below illustrates the distribution of the NMIB.

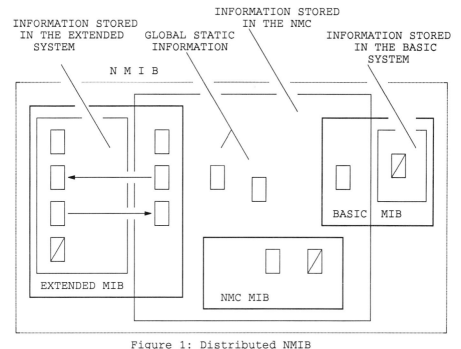

Figure 1: Distributed NMIB

☐   Static object information

▨   Dynamic object information

⟶   is a replica of...

NMC   Network Management Center
Extended system System with ERMS
Basic system   System without ERMS

The thick lines in Figure 1 are the boundaries of the logical MIB, which are not equivalent to an endsystem.
The thin lines are the limits of the physical MIB, and these limits correspond to the real storage of the information in one endsystem.

3 Data Model for the Definition of Networks

The Entity Relationship Model, which is directly supported by the ERMS, is used for modelling the data in the NMIB. This model is often used in other projects as a basis for modelling an MIB (see [3]). The ER model permits the description of objects and of the relationships between them. The following should be stipulated in the modelling: the object types, the attributes of an object type, and the possible relationships between object types.

To allow full use to be made of the rule mechanisms provided by the ER model, a high degree of specialization in the object types is recommended. One could thus define either an object type "system" with the attribute "operatingSystem", or a separate object type for each operating system ("sinixSystem", "bs2000System", etc.). The second of the two alternatives has the advantage of permitting the options for installing applications on systems to be specified directly via the possible relationships, without needing to make checks according to supplementary rules along the lines of: "if operating system = SINIX®*), then static MIB = ERMS". In addition, it is possible to specify more precisely which attributes are of importance.

Using the pure ER model, it is easy to end up with a multitude of object types, which no longer represent the true situation with any accuracy due to the impossibility of expressing similarities between different types of objects.

Many statements apply equally to all systems, and it is desirable to formulate these with the expression "system", without needing to enumerate all the systems. This is of vital importance regarding possible expansions (introduction of a new operating system).

What lends itself here is the class concept associated with object-oriented methods, which is widely used in the field of expert systems and also supported in [4]. The term "class concept" here denotes the ability to assign object types to a class. Similarly, classes can be assigned to other classes. The attributes of a given class are inherited by its subordinate object types or classes.

The central aspect of the modelling is the logical communication structure, around which is grouped the hardware.

The starting point for the data model are the object classes "system" and "subnet".

"Systems" are connected to "subnets" via "subnetEndPoints". Addresses are attributes of "subnetEndPoints" (which correspond to SNPAs in the ISO world).

There are various types of systems and subnets, and the connectivity options between them are governed by different rules, and expressed through relationships.

Examples of "systems" include: OSI systems, SINIX®*) systems, terminals. Subnets permit systems to communicate with each other.

Subnetworks such as X.25 networks, X.21 dial-up networks, ISDN, dial-up telephone networks, LANs, and dedicated lines provide a foundation and specify a profile, where this is not negotiated for the specific connection.

This example will be used to illustrate same of the concepts presented in section 2 of this paper.

The object type "isoLAN" is one of the object types in the "subnet" class. In the same way "isoLanEndPoint" is derived from the "subnetEndPoint" object class and "sinixSystem" from the "system" object class.

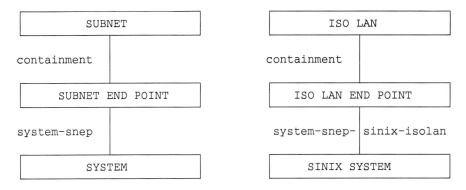

Figure 2 ⟶specialisation⟶ Figure 3

Figure 2 shows how the object classes mentioned above are related and Figure 3 shows the same situation as Figure 2 with specialized object types.

"SubnetEndPoints" are contained in "subnets" and, "isoLanEndPoints" are contained in "isoLans". Between "subnetEndPoint" and "system" there exists a relationship, called system-snep, from which a special relationship, called system-snep-sinix-isolan, is derived which connects "isoLanEndPoints" to "sinixSystems".

Managed objects belonging to the classes "isoLan" and "isoLanEndPoint" are stored in the database of the NMC because this is global information. Managed objects belonging to the class "sinixSystem" may be stored in the NMC or in the Extended system they describe, with the possibility of having a replica in the other system, see Figure 1.

An example of an object class where the managed objects would always be stored in the system where they are contained is "transportConnection". This class describes a transport connection with the attributes it needs while it is in use, for example: "calledTsapAddress".

Insofar as it is not defined by the "subnet", communication between systems is described by means of object types such as communication relationships.
The attributes of communication relationships are protocols and profiles. An example of such an object type is "endSystemRelIso" with the attribute "transportProtocolClass". The systems that are related to the same object of the "endSystemRelIso" class may communicate using the protocol class defined in the "transportProtocolClass" attribute.

More precise description of systems is supported by object classes local to specific systems (system components), for example line buffers. There are also network-specific object classes, for example LANs connected via bridges which form a subnetwork comprising sub-LANs and bridges.

4 The Interface to Network Management Information Services

As already described, network management in CNM consists of a set of operations performed on network management objects. For the network management applications, these operations form a service interface known as the Interface to the Network Management Information Service (INMIS). INMIS was conceived with the following underlying objectives:

- The type and nature of the services should be oriented towards or adapted to OSI management. The internal interfaces should already be prepared for OSI management, in that there should be OSI services and, as far as possible, OSI objects at the program interface. This approach ensures that CNM can also be integrated into an open systems environment. In the longer term, it will also be possible for TRANSDATA® networks to be administered within such open systems internets.

- The INMIS interface is a uniform basic interface for future network management applications.

- The interface should be so designed as to minimize the number of "adjusting screws" for the user, so that, for example, appropriate default values and settings should be determined and adopted automatically.

- Consistency, syntax and values or value ranges should all be checked in the same fashion and as extensively as possible. Such checks should guarantee global consistency across the entire network. The network management interface as a whole will henceforth be known as the Interface to the Network Management Service, or INMS.

Various groups of service can be differentiated within the INMS.

One particular range of services is defined by the ISO as Common Management Information Services (CMIS) (see [5-6]). In addition, there are higher-level services which ISO designates Systems Management Functions (SMF) (see [7-13]). Internally, these latter build on the CMIS services. There is also a further set of services at a higher level, which go beyond the standardized range, these being needed for additional management requirements within CNM. These are known as TRANSDATA® Systems Management Functions.

It is possible to summarize the interface as follows:

Services for session control:
These services are used to open and close network management sessions. They further enable requests to be grouped together into transactions. Only by using transactions is it possible to perform modifications whose individual stages result in a state of inconsistency within the database, but which, when completed, restore it to a consistent state. Rule checking is suspended during transactions.

Services for schema management:
These services enable the definition and deletion of object classes, together with their attributes and relationships, as well as the obtaining of information on existing classes. In order to be able to set up an NMIB, it is first necessary to create a schema which allows the information to be transferred into the information base in a standardized form. With the aid of these services, the user can extend the schema, according to his or her requirements, while at the same time retaining its consistency.

Services for object management:
These services make it possible to carry out certain operational functions on objects, namely the creation, deletion and renaming of objects, and searching for objects which fulfill certain conditions.

Services for attribute management:
These services are used to display and modify attribute values, as well as to check that values are permissible, or that mandatory values have been specified.

Services for relationship management:
These services make it possible to define, delete, modify, display, and check relationships.

Services for the distribution of object information:
These services make it possible to set up or cancel object copies (replicas), and to control the authorization to modify these replicas. This improves the availability of the information.

Services for rule management:
These services permit the introduction and removal of rules, as well as the output of information on them. Rules support the user when storing information in the information base. They facilitate the automatic generation of certain entries, thereby taking the first step towards the introduction of expert systems for network management purposes. Such services are TRANSDATA® SMF services.

Services for the reading and generation of printable files:
These services provide input/output facilities for editable files in the NMIB.

5 Architecture of the Network Management Services

The Network Management Services comprise a Network Management Kernel, a Network Management Event Forwarding Element (NMEFE), a Network Management Information Service Element (NMISE), and a Network Management Communication Service Element (NMCSE).

The Network Management Kernel manages and logs the requests which it distributes to the agent systems. In addition, it performs an authorization check and monitors network management applications. The Network Management Event Forwarding Element receives notifications, converts them into events, and forwards them to the applications which requested them. The Network Management Communication Service Element passes non-local requests to other computers in the network. The Network Management Information Service Element is both managing process and agent process for the Network Management Kernel. The NMISE is a collection of different servers whose function is to map the INMIS access operations onto the various object representations and implementations. Access to static and dynamic object information is via separate servers. The execution of management operations on the objects themselves can be initiated via this interface. ERMS access is provided as a standard feature. This module permits access to the static part of the information. The access module also handles those calls that involve the schema. It is part of the NMISE. The NMIS is the Interface to NMEFE, NMISE and NMCSE.

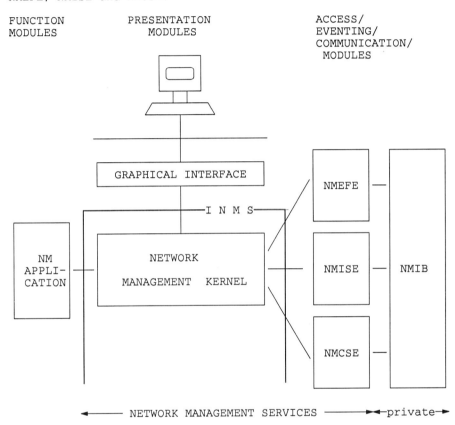

Legend for the architecture diagram above:
- INMS   Interface to Network Management Service
- NMCSE   Network Management Communication Service Element
- NMEFE   Network Management Event Forwarding Element
- NMIB   Network Management Information Base
- NMISE   Network Management Information Service Element

6 Summary

This paper has dealt with the implementation of a Network Management Information Base. This information base uses the object-oriented approach espoused by OSI, along with the Systems Management services. This suggested a design which takes account of new implementation techniques. A network perspective has been introduced in addition to the system perspective. Furthermore, services for management of the NMIB schema have been provided. The possibility of formulating rules permits the information base to perform a consistency check each time an access operation is carried out. This provides the foundation for new network management applications with a high degree of security.

7 Footnote

*) = SINIX is the UNIX System derivative of Siemens Nixdorf Informationssysteme AG
UNIX is a registered trademark of UNIX System Laboratories, Inc.

8 Abbreviations

ER model	Entity Relationship Model
ERMS	Entity Relationship Management System
INMS	Interface to Network Management Service
INMIS	Interface to Network Management Information Service
MIB	Management Information Base (ISO terminology)
NM	Network Management
NMC	Network Management Center
NMCSE	Network Management Communication Service Element
NMEFE	Network Management Event Forwarding Element
NMIB	Network Management Information Base
NMIS	Network Management Information Service
NMISE	Network Management Information Service Element
SMF	Systems Management Functions
SNPA	SubNetwork Point of Attachment

9 References

1. Dr. Karl Beschoner: Architecture of Communication Network Management (CNM), incorporating OSI Network Management. Paper presented at SAVE conference, spring '90
2. ISO IS 7498-4 Management Framework
3. Integrated Network Management, Boston 1989
4. ISO DIS 10165-1 Management Information Model
5. ISO IS 9595 Common Management Information Service
6. ISO IS 9596 Common Management Information Protocol
7. ISO DP 10040 Systems Management Overview
8. ISO DP 10164-1 Object Management Function
9. ISO DP 10164-2 State Management Function
10. ISO DP 10164-3 Relationship Management Function
11. ISO DP 10164-4 Alarm Reporting Function
12. ISO DP 10164-5 Event Report Management Function
13. ISO DP 10164-6 Log Control Function

MINT: an OSI Management Information Support Tool

T.Nakakawaji, K.Katsuyama, N.Miyauchi and T.Mizuno

Information Systems and Electronics Development Laboratory,
Mitsubishi Electric Corp. 5-1-1 Ofuna, Kamakura 247 JAPAN
E-mail: nakawaji@isl.melco.co.jp

Abstract

OSI management comprises two aspects: (1) management information aspect (the information that exists in Management Information Base or MIB), and (2) management protocol aspect (the management protocols that are needed to access and manipulate management information). OSI management standards define management information in the form of template. We have developed a tool called MINT (Management INformation support Tool) which generates access library software for the management information from the network management information defined by the template notation. Software for network management applications can be developed efficiently by using MINT. In this paper, we present the design criteria and functionalities of MINT.

1. INTRODUCTION

ISO and CCITT have defined a Basic Reference Model for Open Systems Interconnection (OSI) to allow cooperation among heterogeneous systems. Conformance to OSI requirements and effective interconnection have been achieved in several organizations which promote OSI. Due to the possibility of interconnection of many OSI-based architecture, the needs for effective, standardized, management procedures is growing steadily.

Network Management standardization covers two complementary aspects: (1) the nature and structure of the necessary information for the purpose of Network Management [1] [2], and (2) the appropriate Network Management services and protocols for accessing this information [3] [4]. The necessary basic services and protocols have already been standardized as an Application Service Element. On the other hand, the necessary management information heavily depends on the structure and intended usage of the managed network or the kind of the communication system. Therefore, it is difficult to provide precise standard definitions for all layers and the standardization process is still in progress. To define many kinds of management information uniformly, a precise standardized notation, *template*, has been developed [5]. From an implementation standpoint, software for both the managing as well as the managed systems must be built efficiently, according to the management information actually available within the management domain.

We have developed a tool called MINT (Management Information support Tool) that generates management software from the definition of the management information by the template notation.

2. MANAGEMENT INFORMATION AND TEMPLATE

Information required to support network management is defined for each managed object, and stored in the Management Information Base (MIB). Management information is also exchanged between the managed and managing systems. The network management model is shown in Figure 1.

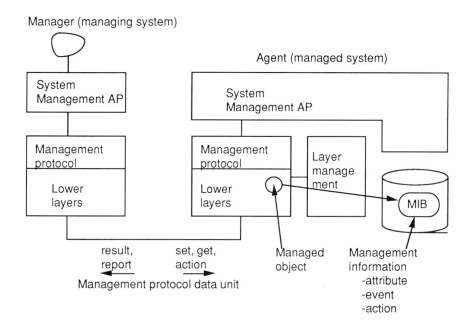

Figure 1. Network management model.

For a managed object, three types of management information exist: (1) attributes contained in the managed object, (2) events emitted from the managed object, and (3) actions to the managed object. For instance, a transport connection is an object, while the maximum and current numbers of retransmitted protocol data units are attributes. Overflowing the maximum retransmissions number is an event, and aborting the connection is an action. To define these management information formally, the template notation has been developed. There are 9 kinds of template: object class template, attribute template, behaviour template, notification template, name biding template, specific error template, group attribute template and conditional package template. An object class template defines management information for a managed object and it points the other templates. Figure 2 shows an example of templates.

```
Object Class Template

ExampleObjectClass    MANAGED OBJECT CLASS
    DERIVED FROM       ISO/IEC 10165-2:top
    CHARACTERIZED BY:
        BEHAVIOUR DEFINITIONS  CommunicationErrorBehaviour;
        ATTRIBUTES             QOS-Error-Cause GET;
        OPERATIONS   CREATE with-automatic-instance-naming;
                     DELETE deletes-contained-objects;
        NOTIFICATIONS          CommunicationError;
    REGISTERED AS {ObjectClass 1};
```

```
Attribute Template

QOS-Error-Cause              ATTRIBUTE
    WITH ATTRIBUTE SYNTAX   AttributeModule.QOSErrorCause;
    MATCHES FOR             Equality;
    BEHAVIOUR               QOSErrorBehaviour;
    REGISTERED AS {AttributeID 2};
```

```
Behaviour Template

CommunicationErrorBehaviour    BEHAVIOUR
    DEFINED AS   The CommunicationError notification is
                 generated by .....
```

```
Notification Template

CommunicationError              NOTIFICATION
    BEHAVIOUR               CommunicationErrorBehaviour;
    WITH DATA SYNTAX        EventModule.ErrorInfo;
    WITH RESULT SYNTAX      EventModule.ErrorResult;
    REGISTERED AS {NotificationID 3};
```

Figure 2. An example of templates.

Management information must be defined for many different objects, depending on the type and scale of the network. Objects representation may also vary according to the management application purposes (e.g., configuration management, performance management, accounting). Therefore, the standards so far have focused on the definition of a common, high-level, abstract, notation allowing one to define various objects and attributes as well as various behaviours, actions, etc. Generic objects, such as entities and connections have been defined in the standards [6]. However, specific objects (e.g., objects specific to a given layer) are still to be defined and integrated in the relevant protocol standard documents.

Since management information includes information related to operational as well as other aspects, collecting it is not a simple task. It is thus necessary to define which information is necessary according to the network structure and operation as well as to the purpose of network management. While software for protocol operation can be re-used in many different configurations, that of network management information is often to be implemented on a case by case basis. Therefore, it is requested to develop the software efficiently.

3. STRUCTURE OF SMAP SOFTWARE

We have attempted to provide structure of SMAP(System Management Application Process) software promoting effectiveness and efficiency during the development process. We, therefore, proposed the following design criteria for SMAP software:

(1) Dependency on the target system, management function, and managed object should be minimized.

(2) It should be possible to generate part of SMAP software directly from the relevant information specified in the template.
Thus we must construct the SMAP by considering management independent information as well as information relevant to a specific management purpose.

(3) SMAP software should provide high performance during the various management operations:
- when collecting relevant management information
- when receiving the request to retrieve management information from the manager
- when reporting events to the manager

(4) SMAP software should be designed according to an object-oriented approach to maintain flexibility and extensibility in the model of OSI network management.

Considering above design criteria, we designed SMAP software as follows:

(1) Independency between dependent part and independent parts on management information is achieved by designing the SMAP software as a set of three modules as shown on Table 1.

(2)The MIB access module is generated by the tool (MINT).

Table 1
SMAP modules

Module name	Processing	dependency to the management information
management protocol processing	handle PDUs which are exchanged between manager and agent	independent
managed object tree management	manage the managed object tree (containment tree)	dependent
MIB access	store management information and provide access interface to it	dependent

(3) In order to achieve high performance for MIB access, the MIB access module is implemented as a library, which is linked with the layer management module and management protocol module respectively. The software structure of SMAP is shown in Figure 3.

(4) A managed object has been directly mapped to an object in an object-oriented language. An object-oriented interface to access and manipulate the managed objects has been defined; e.g., we defined some procedures (or methods) such as "create object", "delete object", and "pass message to object".

We used an object-oriented language, superC [7], which we developed. Characteristics of superC are as follows:
  (a) It was defined as an extension of C, and implemented as a preprocessor for C.
  (b) It supports inheritance.
  (c) It offers two kinds of message binding: dynamic binding done at runtime, and static binding done during compile-time.

4. MINT

The MIB access and management modules for the managed object tree are dependent on management information. The MIB access module consists of a set of definitions for each managed object. MINT can generate the module automatically from the template definition. Using MINT, the layer management and management protocol module can deal with the managed objects' features without needing to know the internal structure of the managed object. MINT uses yacc/lex to analyze the syntax of the template definition since the templates definition is still being standardized, and thus could change. The relationship between the SMAP and MINT is shown in Figure 4.

In the following section, we describe the relationships between the input and output of MINT.

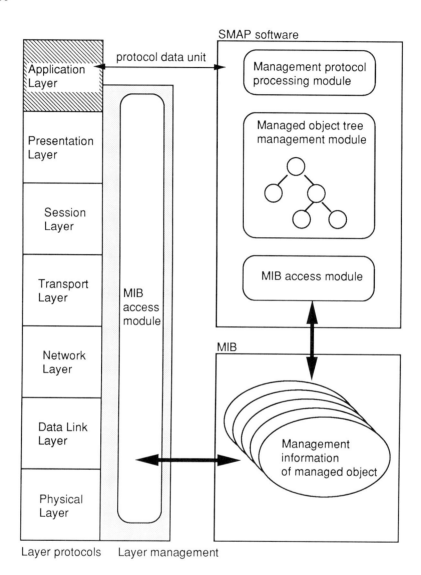

Figure 3. Software structure of SMAP.

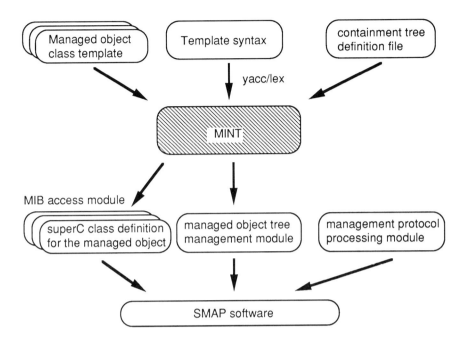

Figure 4. Relationships between MINT and SMAP.

4.1 Class definitions for the MIB access module
4.1.1 Managed object name
MINT generates a class definition, in superC, for each managed object. There is a direct correspondence between the template of a given managed object and the superC class definition. The name of a superC class definition generated by MINT is the class label of the corresponding managed object.

4.1.2 Superclass label
A superclass label of an object class template indicates the name of its superclass. MINT generates it as a superclass name in the superC class definition. In the template notation, it is possible to specify a label denoting multiple superclasses for a given object class template. However, superC does not support multiple inheritance; so we have to specify only one superclass. This point is for further study.

"Allomorphism" allows one to treat an object of a given class as if it were an instance of a superclass. For instance, an instance of a Transport Connection Class 0 can be seen, and treated, as an instance of some more general Transport Connection. However, superC does not support that feature. We have solved that problem by adding the value of "object class" of the superclass to the "object class" attribute values.

4.1.3 Management information(instance variable)

Instance variables in the class definition of a given managed object also constitute management information. Attribute type and value, notification, and action have been defined as instance variables.

The attribute template of a given class is retrieved by using the attribute label present in the object class template. Data related to this attribute are also generated as instance variables of the class definition written in superC.

Since the attribute syntax is defined in ASN.1(Abstract Syntax Notation One) [8] [9], we use APRICOT [10]. APRICOT is a tool which processes ASN.1 syntax and provides a development and test environment for OSI application layer protocol software.

For notifications, events selected according to the discriminator are reported to the manager. To realize this mechanism, the information for notification are also generated as instance variables.

An action can be implemented as a message accepted by an object, but an action request by the manager and the action itself must be reflected within the layer protocol processing. Considering these points, the management protocol module sets the action requested by the manager in the instance variable, and the layer protocol module reads it and operates accordingly. The action type and its argument are generated as instance variables by MINT according to the behaviour template.

4.1.4 Creation and deletion of managed objects

Both creation and deletion of a managed object are realized as methods of the object. A method for creation works as follows:
- it first gives an initial value to each attribute.
- it then copies the attribute values from the referenced object if it is specified in the creation label of the managed object class.
- it creates an instance of the superclass and sets initial values.
- finally, it sets the notification information to indicate the object creation.

A method for deletion works as follows:
- it first deletes the instance of the superclass.
- it then sets the notification information to indicate the object deletion.
- it finally deletes the object itself.

4.1.5 Retrieving management information

Retrieving management information is realized as a method of a class definition. In this method, an instance variable which is related to the indicated attribute is retrieved, and the appropriate "result" is returned. When a filter is specified, the filter is analyzed, and the appropriate "results" matching the filter are returned.

4.1.6 Setting management information and notification

Setting management information is realized as a method of the appropriate object. In this method, an instance variable is set to the appropriate value. In the case of setting an attribute value, the property list of the attribute as defined in the class template is retrieved to check whether the attribute can be written or not.

Notifications may occur as the result of setting an attribute value, such as a threshold value being reached. In this case, the relationship between the attribute and the notification is described in the behaviour template. The behaviour template is written in natural language. Therefore such relationship is not machine processable. We added a simple syntax to the behaviour template so that we could describe the relationship between the attribute and notification in a machine processable form.

When an attribute value is set in the layer management module, conditions defined in the behaviour template are checked, and if a notification is required, it is set in the notification information. In the management protocol processing module, the notification from the layer management module is received; the instance is accessed, filtered with the discriminator, and it sends the notification to the manager. As such the notification processing resulting from the setting of an attribute is realized by the method to set management information.

4.2 Managed object tree management module

The managed object tree management module manages the containment relations between managed objects. This module has the following features:
- it manages the containment relations between the managed objects which are created.
- it automatically names the managed object.
- it analyzes the "scope" parameter of the management protocol service primitive.

This module is generated from the file containing the definition of the managed object containment tree and from the name binding templates.

4.3 Class definition generated by MINT

Relationships between the template and the class definition are shown in Figure 5. Specific error templates, group attribute templates, and conditional package templates can not be supported in MINT since their use is ambiguous and therefore not suitable for machine processing.

5. CONCLUSION

We have developed a management information support tool, MINT. We can develop some network management software effectively according to the definition of management information by using it. The generated programs are written in an object-oriented language, superC, so that the object-oriented model used in the OSI management documents is effectively implemented in a straightforward manner.

6. ACKNOWLEDGMENT

We would like to thank Prof. Pierre Mondain-Monval for reading and commenting on this paper.

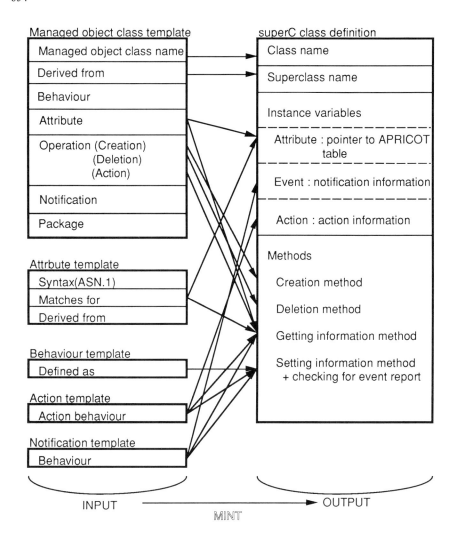

Figure 5. Relationships between the template and the superC class definition.

7. REFERENCES

1. ISO DP 10165-1: Structure of Management Information Part 1: Management Information Model (1989).

2. J.Tucker: A Common Approach to Managed Objects, IFIP Integrated Network Management, I, pp.159-165 (1989).

3. ISO 9595: Common Management Information Service (1990).

4. ISO 9596: Common Management Information Protocol (1990).

5. ISO DP 10165-4: Structure of Management Information Part 4: Guidelines for the Definition of Managed Objects (1989).

6. ISO DP 10165-2: Structure of Management Information Part 2: Definition of Management Support Objects (1989).

7. T.Mizuno, et al: COTTAGE: Systematic Method for the Development of Communication Software, IFIP WG6.1 Protocol Specification, Testing and Verification, VIII, pp.269-280 (1988).

8. ISO 8824: Abstract Syntax Notation One (ASN.1) (1988).

9. ISO 8825: Abstract Syntax Notation One (ASN.1) - Basic Encoding Rules (1988).

10. T.Nakakawaji, et al.:Development and Evaluation of APRICOT(Tools for Abstract Syntax Notation One), Proceedings of the Second International Symposium on Interoperable Information Systems ISIIS '88, pp55-62 (1988).

AUTHOR INDEX

E. Adams	171	T. Ikuenobe	413	D. Panno	425
L. Aguilar	565	A. Johnston	403	G. Pavlou	259
B. Ambrose	747	D. Johnson	301	A. Pras	109
M. Anwaruddin	679	K. Kappel	3	G. Raeder	135
J. Balfour	625	K. Katsuyama	845	L. Rahali	469
S. Bapat	817	S. Kheradpir	371	C. Rauh	833
G. Berkowitz	403	W. Kiesel	245	J. Read	625
G. Bochmann	77	Y. Kiriha	157	M. Rose	9
E. Bötsch	735	K. Klemba	801	M. Saito	189
P. Brun	625	S.M. Klerer	763	S. Sanghi	285
P. Brusil	3	G. Knight	259	R. Sasaki	189
W. Buga	343	H. Kobayashi	189	W. Schödl	493
E. Carter	213	M. Kosarchyn	801	J. Schröder	493
V. Cerf	609	A. Kouyzer	147	P. Sen	465, 481
A. Chandna	285	A. La Corte	425	S. Sengupta	285, 639
S. Chum	413	L. LaBarre	227	P. Senior	359
J. Cohen	185	H. Latin	541	J. Sipos	531
R. Cohen	763	L. Lau	531	M. Sloman	595
D. Cohrs	119	A. Lazar	653	P. Smyth	505
J. Davin	703	L. Lecomte	77	M. St. Jacques	531
K. Deiretsbacher	245	B. Lemercier	201	J. Statman	505
M. Densmore	313	P. Linington	553	D. Stevens	531
J. Dia	213	A. Lombardo	425	W. Stinson	371
A. Dittrich	789	K. Lutz	397	D. Stokesberry	73
E. Duato	201	D. Mahony	747	C. Strutt	577
A. Dupuy	639	P. Manson	29	G. Sundstrom	371
R. Durst	327	N. Matsumoto	45	M. Suzuki	189
J. Embry	29	S. Mazumdar	653	M. Sylor	57
M. Erlinger	271	K. McCloghrie	339, 703	A. Tanaka	45
D. Estrin	719	G. McElvany	403	P. Toft	625
M. Feridun	691	B. Meandzija	3	S. Towers	519
M. Frontini	519	P. Mellor	625	Y. Trodullies	359
P. Fuhrer	403	K. Meyer	301	M. Tschichholz	613
N. Fujii	383	D. Milham	29	R. Valta	777
D. Gaiti	469	B. Miller	119	A. van den Boogaart	147
J. Galvin	703	N. Miyauchi	845	S. Walton	259
G. Goldzmidt	95	K. Mizuguchi	189	G. Wetzel	285
R. Goodman	505, 541	T. Mizuno	845	O. Wolfson	639
S. Goyal	439	J. Moffett	595	T. Yamamura	383
B. Gray	403	P. Mondain-Monval	77	T. Yasushi	383
J. Griffin	519	K. Morino	45	S. Yemini	95
J. Hall	613	R. Nagai	189	Y. Yemini	95, 639
S. Harris	359	S. Nakai	157	L. Zhang	719
S. Hasegawa	157	T. Nakakawaji	845	A. Zolfaghari	413
P. Hong	481	D. O'Sullivan	359	D. Zuckerman	397
P. Hyland	625	G. Oliver	505		
Y. Ihara	157	S. Palazzo	425		